PROCESS ENERGY CONSERVATION

Edited by
Richard Greene
and
the Staff of Chemical Engineering

McGraw-Hill Publications Co., New York, N.Y.

Copyright © 1982 by Chemical Engineering McGraw-Hill Pub. Co.
1221 Avenue of the Americas, New York, New York 10020

Printed in the United States of America

Library of Congress Cataloging in Publication Data

Main entry under title:

Process energy conservation.

Includes index.
1. Chemical plants—Energy conservation.
I. Chemical engineering.
TJ163.5.C54P76   660.2'8   82-1234
ISBN 0-07-010697-5 (case)   AACR2
ISBN 0-07-606819-6 (paper)

Contents

Introduction

This book is a guide that will help you to save energy in your process plant. Here are scores of articles that will show you methods to cut your electricity bills, improve boiler performance, design new plants to be energy-efficient, and help you in many other tasks.

The articles here appeared in *Chemical Engineering* from 1973 through 1981. It was in 1973 that energy began to become expensive—and energy conservation started to become a major part of the chemical engineer's job. Before then, of course, energy costs were not considered to be so important in plant design and operation. In fact, some of us can still remember performing energy-cost calculations after all other design considerations had been handled!

Included in this book are some more-recent news stories, as well as the usual engineering "how-to" articles. We feel that these stories indicate the ways people in the chemical process industries are thinking to save energy. Such direction should be helpful to you in the years ahead.

All the articles that appear here are in their original form. No attempt has been made to update the numbers or methods used in them. We trust you will be aware of any changes that have taken place since the time of original publication, and adjust for them when applying the ideas contained in this book.

The volume is divided into three parts. The progression is from the general to the specific. There is practical information for use by engineers in any process plant, as well as specific tips for many applications:

Section I: Overall Strategies and Ideas for Saving Energy—This section contains general tips for energy conservation. Information is provided on recent energy-related news, fuels, how to set up an energy-conservation program, and how to estimate and optimize energy needs.

Section II: The Energy-Efficient Plant—Here are methods to design and run plants and systems economically. Tips are given for new and existing plants. Also included are articles on how to accomplish distillation and save energy, and how to cut energy in steam systems. In general, this section covers the larger and more general aspects of process-plant energy conservation.

Section III: Equipment and Materials—This part is a guide to specific materials and types of equipment. The sections include refractories, insulation, boilers and furnaces, heat exchangers, motors and pumps.

These three parts contain a wealth of information and experience that will save you money. The methods have been proven by experience and were reported by leading experts. As energy costs continue to soar, this book will help you to keep your expenses down.

Section I
OVERALL STRATEGIES AND IDEAS FOR SAVING ENERGY

Energy-related meeting focuses on new routes

A coal-gasification process that needs no desulfurization equipment, and a flue-gas desulfurization technique that requires no gas pretreatment, were highlighted.

☐ Among the rich selection of energy-related papers presented at the 16th Intersociety Energy Conversion Engineering Conference held last month (Aug. 9-14) in Atlanta, Ga., two that described new processes, and another that told of advances in wood-gasification technology, had special interest for an audience filled with engineers knowledgeable in energy conservation.

Japan's Sumitomo Metal Industries, Ltd., unveiled a coal-gasification route that boasts high yields of a raw gas that does not have to undergo desulfurization. Georgia Institute of Technology (Atlanta) gave details* of a low-cost flue-gas-desulfurization (FGD) process based on the electrochemical concentration cell concept. And researchers at the University of Missouri-Rolla described a fluid-bed wood-gasification reactor that may be used to produce low-Btu gas for a municipal power plant.

LOW-SULFUR GAS—The Sumitomo process (*Chem. Eng.*, Aug. 24, p. 19) appears to be very similar to another molten iron coal-gasification route (*Chem. Eng.*, Apr. 6, p. 10), recently announced by Humboldt-Wedag AG (Cologne, West Germany).

Both processes employ a molten-iron bath for gasification. This reportedly makes the chemical reactions easier to control (by assuming characteristics more similar to a liquid/liquid reaction than to a solid/gas reaction, as in most gasification routes).

In the Sumitomo technique, pulverized coal, along with oxygen and steam, is blown through a water-cooled lance held above the bath, which has a temperature of 1,500-1,600°C. The hot metal cracks the coal, releasing hydrogen; the coal's ash and carbon dissolve in the iron.

Dissolved carbon reacts with the oxygen feed to form CO, and with the steam to generate additional hydrogen via the water-gas reaction.

The product gas is said to have a heating value of about 2,600 kcal/Nm³. The raw gas is composed of more than 90% CO and H_2 (hydrogen ranges from 29-33%, and CO from 60-64%).

The process has a gasification yield of over 98%, according to pilot-plant results. And the route generates a raw gas that contains only 10 to 300 ppm of total sulfur. S. Okamura, manager of Sumitomo's research and development department, points out that the molten iron has the ability to dissolve the sulfur present in the coal, forming FeS. This is removed in the slag. Because the desulfurization capability is higher than 90%, Sumitomo claims that no desulfurization facilities are needed for cleanup.

The company has operated a 60-ton/d batch deslagging pilot unit since the spring of 1980 at its Kashima steel works in Japan. In July, the firm started up a continuous pilot unit with an average capacity of 60 tons/d and a maximum of 90. A $74-million demonstration plant will begin operating in 1983-4; this facility will have a capacity of 500 dry metric tons/d of coal and will use three gasifiers, each containing 125 tons of molten iron (two of the gasifiers will be operational; the other will be on standby).

Okamura states that, with coal costing about $65/ton, the product gas would cost $7.50 to $8.50/million Btu.

ELECTRIC ANSWER—For the past two or three years, Georgia Institute of Technology has been using a small test-cell to investigate the electrochemical removal of sulfur dioxide from stack gases. The idea, according to authors Dan Townley and Jack Winnick of the Institute's Chemical Engineering Dept., can be applied to SO_2 emissions from such sources as power plants, ore smelters, sulfuric acid plants, and Claus units.

The Institute's cell is of the electrochemical-concentration type, in which a reactive species is present at both electrodes, but at different concentrations, so that the reaction at one electrode is the reverse of that at the other. The cell itself is a high-temperature, molten-salt version that uses the eutectic of lithium, potassium and sodium sulfates as the electrolyte.

At atmospheric pressure, the cell can operate in two different modes. In the driven one, a carrier gas (air) is fed to the anode to carry away the reaction product (SO_3); this requires consumption of electricity. In the reducing-gas mode, hydrogen or another gas reacts with the sulfate present to produce H_2S and water. Since energy to run the cell is obtained from hydrogen oxidation, this mode consumes no electricity.

According to Winnick, the system doesn't generate sludge, doesn't require flue-gas pretreatment, and doesn't take a pressure-drop penalty. All this makes it cheaper than conventional FGD routes. In a power plant application, says Winnick, wet-scrubbing FGD units would have operating costs ranging from 1.5 to 2.0 mills/kWh, vs. about 0.5 mills/kWh for the electrochemical route operated in the driven mode. This does not include a credit for sulfuric acid produced (concentrated SO_3 is the anode product).

The authors conclude that for a 500-MW power plant that burns 3.5%-sulfur coal, the current needed for 90% SO_2 removal is 13×10^6 amp. At a cell potential of 0.8 V, this amounts to approximately 2% of the plant's power, which compares favorably with conventional FGD processes requiring up to 6% of a plant's power (e.g., lime scrubbing).

Winnick has patented the process, and the Institute now seeks funds to run pilot-plant tests.

FLUIDIZED WOOD—At the meeting, researchers from the University of Missouri-Rolla outlined the operation of a semicommercial-size fluid-bed

* The Institute's paper was not read at the conference, but is included in the proceedings.

Originally published September 21, 1981

wood gasifier. This pilot unit, operated with support from the U.S. Department of Energy, has a capacity of 2,000 lb/h of wood and produces a low-Btu gas (about 160 Btu/ft.^{3}).

In the system, a 40-in.-dia. fluid-bed reactor is fed wood, as sawdust (see figure). Preheated air is used to fluidize the sand bed. The air also provides oxygen needed for the gasification reactions. Product gas contains about 7% hydrogen and 5% methane.

The Missouri-Rolla unit has also been tried on a high-Btu mode, in which the product gas is recycled to the reactor, and a catalyst—potassium carbonate or wood ash—is used to increase methane output. However, says Virgil J. Flanigan, a professor of mechanical engineering, they have had "little success with catalysts."

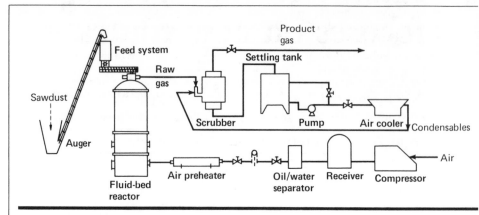

Wood-gasifying scheme features a fluid-bed reactor

A feasibility study for a 100-ton/d unit, using the low-Btu technology, has been completed for the city of Detroit Lakes, Minn., which would use the product gas in a powerplant. If the plan is accepted, the gasifier could go onstream in two years. Fixed cost for the facility would be $900,000.

Reginald I. Berry

Energy topics pervade AIChE meeting

Processes and hardware aimed at saving energy were featured. Among the routes covered in sessions were methods for coal gasification, and production of ethanol, methane and methanol.

☐ Energy and most everything related to it are still very much in the news, to judge from the agenda of the 72nd Annual Meeting of the American Institute of Chemical Engineers (AIChE), held in San Francisco, Calif., Nov. 25-29, 1979.

The event, which attracted over 3,800 registrants (a record), devoted two sessions to the topic—one dealing with energy conservation, the other with alternative-energy processes.

TIPS FOR SAVING—A capital-cost reduction of 5 to 20% was claimed for a special evaporator design described by Richard C. Bennett, division manager of Swenson Div. of Whiting Corp. The unit replaces the conventional cylindrical vapor head with a special elbow that connects the heater with the condenser.

Swenson's elbow separator/evaporator uses centrifugal and gravitational forces to separate vapor and liquid components of an evaporated fluid,

eliminating the need for a bulky vapor head. Bennett said that the equipment provides vapors as clean (liquid-free) as those leaving a conventional vapor head, and is capable of operating satisfactorily over a wider range of evaporation rates than do conventional units.

Two other papers—by D. A. Austin of Resources Conservation Co. (Seattle, Wash.) and A. H. Beesley of Aqua-Chem, Inc. (Milwaukee, Wis.)—dealt with vapor compression evaporation as a conservation method. In these systems, mechanical work is substituted for the heating steam required in single- and multiple-effect evaporators, thereby providing lower energy consumption and operating costs.

In distillation, R. M. Stephenson and R. F. Anderson of the University of Connecticut (Storrs) gave examples of minimum and actual energy requirements for common industrial

separations, and made recommendations as to where modifications of conventional distillation systems could lead to improved energy efficiency.

Thermal energy storage was examined from a how-to standpoint. Among the options were the use of paraffins, phase-change materials and winter-chilled cold water stored in aquifers.

PROCESS-ORIENTED PAPERS—On the alternative-energy-process theme, several symposia were held that reviewed the status of various technologies, including process developments and economics for:

■ Production of ethanol from wood, wheat straw, corn stover, and high-sugar-content plants by fermentation, and from certain aquatic plants by acid hydrolysis.

■ Production of methane from mass cultures of certain algae, from bioconversion of animal manures and agricultural crop residues, and from anaerobic digestion of municipal solid waste.

■ Production of various chemicals, including methanol, oils, chars and ammonia, from the thermal conversion (pyrolysis) of various biomass feedstocks.

In addition, the technical and economic ramifications of underground coal gasification merited an entire session. D. R. Stephens, of the University of California's Lawrence Livermore Laboratory (Livermore, Calif.), outlined the encouraging results of field tests to date, which indicate that the technology offers a "relatively low-cost, environmentally sound

were given on diverse synfuels technologies, including high- and low-Btu coal gasification, coal liquefaction, and production of gasoline from coal by molten zinc-chloride hydrocracking. As described by C. R. Greene of Shell Development Co. (Houston), the route, in a single step, produces 4.5 bbl of gasoline per ton of coal processed. Gasoline is removed as a vapor and the spent zinc chloride is regenerated by vaporization in a fluidized-bed combustor. Currently operated in a 1-ton/d demonstration unit, the route is jointly funded by the U.S. Dept. of Energy, Continental Oil Co. and Shell Oil Co.

A symposium entitled "International Synfuels Technology" highlighted coal-conversion developments in five countries—South Africa, Australia, the United Kingdom, West Germany and The Netherlands.

Japanese thrift spawns energy-saving processes

With no natural fuel resources to speak of, Japan has had to pare energy consumption considerably since 1973. Savings realized in this decade are expected to come from process development work aimed at raising yields.

Daicel Chemical's new acetic acid plant will obtain methanol from CO

☐ It's been seven years since Japan, like other industrialized nations, was hit by the oil crisis, which caused the worst postwar slump experienced by that nation's chemical and petrochemical industries. All along, Japanese engineers have been doing wonders in cutting energy and raw materials. And now they are starting the 1980s with a more ambitious program aimed at further reductions.

During phase one of the conservation effort (1974-78), companies imposed relatively low-cost measures—

what Nobuto Ohta, general manager of research and development for Mitsubishi Petrochemical Co., calls "turning off the lights"—that nevertheless yielded substantial results (see box). But the savings planned for the present decade will be keyed to process-oriented research and development. In effect, Japanese firms will be trying to cut raw-material consumption further by improving process yields.

MORE FOR LESS—The chemical industry will have to work harder to obtain (smaller) savings in coming

years. For example, while Sumitomo Chemical Co. achieved a 15% energy cut during 1977-79, its next three-year target has been pegged at 12% (including a 5% cut this year). Similarly, Mitsubishi Chemical Industries Ltd. says it expects to save only an annual 2% during 1979-81. And Asahi Chemical, which reported a 40% saving since 1973, is trying for a 3-5%/yr conservation over the next few years.

Some of the gains made during the early slump years will indeed be difficult to match. Take the case of Teijin Ltd., which has reported a 38% boost in the production of polyester, nylon and acetate for the last six years, with no increase in energy consumption. Also during that period, Ajinomoto Co. has notched a 15% raw-materials saving (sugar-cane molasses, acetic acid), plus cuts in the consumption of steam (48%), electricity (15%) and water (50%)—all in the manufacture of monosodium glutamate.

The conservation efforts have enabled oil refiners to curb production somewhat. According to Japan's Petroleum Assn., oil refining in fiscal 1979 (ended in March) dropped 2.4%—to 21,392 million L—from the 1978 figure. Fuel oil sales underwent a bigger decline—3.4%, to 21,466 million L. Meanwhile, product inventories rose 23.6%, to 15,360 million L, by March 1980.

WORKING ON PROCESSES—In phase two, Mitsubishi Petrochemical will aim for a 5% cost cut, equivalent to $24 million/yr (in 1978 dollars) during 1979-82. The firm, while not increasing its R&D staff of about 100, will try to slash raw-material costs by improving yields of processes for: ethylene, benzene, styrene monomer, ethylene oxide, acrylic acid, higher alcohols, low- and high-density poly-

Originally published June 30, 1980

Editor-in-chief Calvin S. Cronan interviews Asahi Chemical's Maomi Seko

ethylene, and polypropylene. Mitsubishi has already spent a considerable amount on making its styrene monomer process more efficient; the new route saves steam consumption by producing the monomer at lower than normal pressures.

A number of Japanese companies, including Mitsubishi Petrochemical and Nippon Petrochemicals Co., are working on gas-phase, fluid-bed, low-pressure and -density polyethylene routes along the lines of Union Carbide's Unipol process. As for polypropylene, Sumitomo Chemical Co. has developed a method that consumes no solvent or water and needs no de-ashing (*Chem. Eng.*, May 7, 1979, p. 42). This route will probably make its commercial debut in 1982, as part of a complex to be built in Singapore.

The brand-new polypropylene technology will not be the only improvement in evidence at Singapore. Says Takeshi Hijikata, president of Sumitomo Chemical, "We will make the Singapore complex, particularly the ethylene plant, the very best in terms of [reduced] feedstock and energy consumption."

(Ethylene units completed in Japan before the 1973 oil crisis consume some 10,000 kcal/kg of product ethylene. But industry sources say that Showa Denko K.K. and Ukishima Petrochemicals Co. cut those requirements to 6,000 kcal/kg at 300,000- and 400,000-m.t./yr plants they, respectively, built in 1977 and 1978. "At present, they are the best here," admits Hijikata.)

HELP FOR LICENSING—Of course, energy and raw-material conservation are not the only motives behind the process improvement drive—the Japanese also want to maintain their licensing competitiveness abroad. Says Susumu Takao, director of production and engineering for Nippon Zeon: "We cannot let our guard down. Others will quickly make similar improvements, and we are engaged in neck-and-neck competition."

Takao has a point, because his firm's butadiene extraction route is a close competitor of a process owned by West Germany's BASF. The economics of both are thought to be similar, though the Japanese method requires more steam, less electricity.

Now, Takao reports that steam requirements have been reduced from 3.2 m.t./ton of product to about 2 m.t. for new plants. In existing facilities, Nippon Zeon has managed to reduce steam consumption to 2.4 m.t./ton of product, mainly by heat recovery. Earlier this year, the Tokyo-based company concluded a licensing agreement with Mexico's Pemex—the thirtieth such contract for the firm's butadiene technique.

Lack of competition, according to Takao, has been responsible for the slow improvement (in terms of energy consumption) of Nippon Zeon's isoprene extraction process. To be sure, steam requirement has been nudged down from 12 m.t. to 10 m.t./ton of product. But this doesn't satisfy Takao, who quips, "I urge engineers to assume there is a competitor in the isoprene field, too."

CHLOR-ALKALI FRACAS—Takao's urging is one that Maomi Seko, executive vice-president of Asahi Chemical Industry Co., can easily dispense with. His firm and Asahi Glass Co. are experts in membrane processing for chlor-alkali production—a highly competitive field in which both companies have separately developed proprietary technology.

The Asahi processes will likely benefit from the Japanese government's

Post-crisis steps—modest spending yields savings

According to a survey by the Japan Chemical Industry Assn., 52 leading chemical firms spent about $230 million on conservation during 1974-78. The payoff was a drop in energy consumption from 206,544 billion kcal (equivalent to 20,863 million kiloliters of fuel oil) in 1973 to 189,363 billion kcal (equivalent to 19,128 million kiloliters of fuel oil) in 1978. The 1978 figures were reached despite a 5.8% rise in energy consumption for pollution abatement systems—mainly desulfurization and denitrification.

Responding to the JCIA questionnaire, the 52 companies, which operate 228 chemical plants, said there is room for a 9% saving during 1980-85. This would amount to a cut of 19,900 billion kcal, equivalent to approximately 2 million kiloliters of fuel oil. A capital investment of $544 million will be needed to reach the six-year goal.

Although a precise comparison is impossible, JCIA says that the chemical-industry savings rate of 17.2% during 1973-78 looks good on an international level, and compares favorably with other Japanese industries. In the same time-span, cement makers realized an estimated 18.5% energy conservation, and petroleum refiners saved an estimated 11.2%, while the steel industry achieved a reduction of only 9.0% (equivalent to 5.6 million kiloliters of fuel oil).

order to chlor-alkali producers to switch the remaining mercury-cell capacity (1.7 million m.t./yr, out of a total of 4.4 million m.t./yr) to non-mercury routes by December 1984. The Ministry of International Trade and Industry has requested both firms to expand their capacities by 1982, so that their plants can serve as model large-scale facilities.

Meanwhile, last year, in a London symposium, Seko compared the power consumption of Asahi Chemical's membrane process with that of other routes (per metric ton of caustic soda). His figures: 2,703 kWh for the membrane technique, 3,100 kWh for mercury cells, and 2,500 kWh for asbestos diaphragm cells. In an interview, Seko predicted that his company's process would eventually lower electricity consumption to 2,000 kWh. And the executive noted that the technique already produces more than 10,000 m.t./yr of caustic soda per cell, which surpasses the two other methods.

In addition to improvements in chlor-alkali processing, Asahi Chemical has bettered catalyst performance. The firm has recently licensed its high-density polyethylene technique to Exxon Chemical Co., which intends to build a 400-million-lb/yr plant at Mt. Belvieu, Tex. It will feature a highly active, modified Ziegler-type catalyst.

"New catalysts," notes Seko, "will play important roles in energy and raw-material conservation. Development of catalysts is one of the wisest ways because you do not have to change the plant's chemistry."

Following this approach, Mitsubishi Chemical has developed a more selective catalyst for its oxo gas process, said to achieve a 30% energy saving and more than a 10% cut in raw-material consumption. The revamped process is being offered for licensing. The firm also has improved its terephthalic acid route by emphasizing prevention of solvent (acetic acid) oxidation. As a result, acid consumption has been reduced by 50%.

ALUMINUM AND PVC—Sumitomo Aluminium Smelting Co., which was formed to take over Sumitomo Chemical's aluminum business, and Shin-Etsu Chemical Co. are two other companies that have used process improvements to gain licensing business.

The former's modified Soderberg technique consumes 14,000 kWh/m.t. of aluminum ingot, while its pre-

Cell room of Asahi Chemical's chlor-alkali plant at Nobeoka

baked-anode process uses 13,500 kWh. This energy efficiency compares favorably with a world average of about 17,000-18,000 kWh/m.t. of ingot, and has helped—in combination with good environmental-protection features and labor economies—in securing $32 million in aggregate licensing income for the firm in recent years.

On the strength of the size (130 m³) and design of its polymerization reactor, Shin-Etsu has become the leading polyvinyl chloride producer in Japan. Through its subsidiary Shintech Inc., the firm is expanding its business in the U.S. Licensing elsewhere includes a 144,000-m.t./yr plant for Shell Chimie, S.A.; two 200,000-m.t./yr facilities for China, to be completed in Shengli and Nanking by 1982; and a 50,000-m.t./yr plant for Mexico's Primex S.A., also for completion in 1982.

MULTIFACETED TECHNIQUE—Attempts to curb naphtha consumption are getting a boost from Texaco's partial oxidation process, which Ube Industries Ltd. licensed years ago. Suitably modified, the Texaco-Ube process makes syngas from vacuum residue, and Tetsuro Yoshimura,

manager of the plant engineering department's technical group, sees great possibilities for producing ammonia, methanol, and carbon monoxide.

Ube, which has used the route to make ammonia for years, recently clinched a contract to export three 1,000-m.t./d vacuum-residue gasification plants to China. The Texaco-Ube process also will fit in well with plans by Daicel Chemical Industries Ltd. to start up a 150,000-m.t./yr acetic acid facility this month at Aboshi, west of Osaka. Raw-material methanol will initially be purchased from outside sources, but a CO plant built by Ube will eventually be the supplier.

Yoshimura hopes that the gasification process will find even wider application—e.g., as a source of hydrogen in refineries. "Refinery gas and naphtha, now the two main sources, can be better employed in making products with higher value-added," he says. Ube already uses the knowhow in the small-scale production of oxalic acid and derivatives. And Daicel is planning to use CO feed to make 1,4-butanediol.

Shota Ushio
World News (Tokyo)

CPI firms map strategy for energy-saving plans

Such measures as steam and heat recovery, cogeneration of electricity, and use of coal and other materials for fuel and feedstocks figure prominently in many firms' energy-conservation manuals.

☐ What can the chemical process industries (CPI) do to conserve energy? The question can, of course, be answered in a thousand different ways. To pinpoint where the CPI are focusing their energy-saving efforts, CHEMICAL ENGINEERING interviewed a group of experts in the matter, and complemented the information with data from the Second Pacific Chemical Engineering Congress (held Aug. 28-31 in Denver, Colo.), which devoted a morning and afternoon session to such topics as how chemical process and equipment design can be modified to reduce energy consumption.

Most interviewees agreed that the CPI are concentrating on the following three approaches:

■ The development of new technology—i.e., of energy-efficient processes and operating techniques.

■ Increased integration of processes—e.g., more emphasis on electricity cogeneration, recovery and reuse of steam and heat now wasted, and interprocess transfer of energy.

■ A drop in consumption of crude oil and natural gas, to be achieved by a combination of energy-conservation procedures and an increase in the use of coal and other materials for fuel and feedstocks.

While pursuing this three-pronged plan of attack, many CPI firms are discovering that many of the old ground rules from the era of cheap, abundant natural gas and crude oil no longer apply. For instance, because it has always been traditionally desirable to hold down the capital cost of projects, energy-efficient process design has been, until recent years, too expensive to be worthwhile.

Now, however, it is more likely that a heavy capital outlay to ensure high efficiency is more than paid back in fuel-cost savings over the life of a project. And this is not the sole reward of more-efficient processes: in many instances they can further lower manufacturing costs by increasing product purity and yields.

HEAT THRIFTINESS—Although heat recovery plays a big part in efficient process design, there have been virtually no significant modifications made in recent years in the design of heat-exchange equipment. Such noted manufacturers as Pfaudler Co. (Rochester, N.Y.), Thermxchanger, Inc. (Oakland, Calif. and Wiegmann & Rose International Corp. (Richmond, Calif.) agree that they haven't been making changes in basic equipment design, nor are they contemplating any.

The manager of the heat-transfer section of a large U.K. engineering design firm explains why. "Energy conservation in the CPI," he says, is much more a function of process design than of heat-transfer equipment design. By the time a flowsheet has been drawn up and finalized, there is relatively little gain to be made in equipment design. Don't forget, this technology is fairly mature."

This sentiment is echoed by B. D. Coffin, manager of power-plants engineering for H. K. Ferguson Co. (Cleveland, Ohio), who adds, "We are applying the same equipment today as we did before, but with more diligence, to capture waste heat that was previously lost. There is more consideration given to insulation, equipment sizing, pumps and motors . . . We are striving for optimum efficiency through proper consideration and operation at design levels."

What these experts are saying is that heat recovery today is based on the use of available equipment—units that may have been too expensive when fuel costs were low. Notes Milton J. Buffington, regional manager of Bumstead-Woolford Co. (Los Alamitos, Calif.), an engineering and construction firm specializing in energy and power: "A good deal of energy conservation results from old design techniques applied under a new economic banner."

Buffington cites the example of one

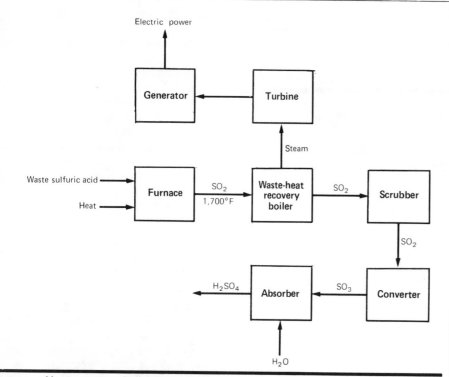

Heat recovery fuels nearly all electricity generating needs at a Stauffer H_2SO_4 plant

large U.S. oil refinery, which is installing a waste-heat unit on its CO boiler. About 900,000 lb/h of exhaust gas leaves the boiler at an average temperature of 600°F. The installation of a waste-heat boiler and an economizer at the back end of the unit produces saturated steam at about 175 lb/in^2 and approximately 300°F.

What makes these additions attractive is the current cost of fuel oil (about $2.40/million Btu), when compared with the price tag of a few years ago ($1.50/million Btu). The refiner that put in the heat-recovery equipment is saving about 45 million Btu/h, which amounts to slightly more than $100/h. At an installed cost of $1.25 million to $1.6 million, these units can be amortized in little over a year, at today's energy costs. (Of course, the pay-back time would shorten if fuel oil became more expensive.)

PINCH THOSE PENNIES—One firm that has benefited from selective equipment- and process-modifications is Union Carbide Corp. (New York City). M. A. Williams, energy-conservation manager of the firm's Chemicals & Plastics Div., says that, since 1972, Carbide has managed to save the equivalent of nearly 12.8 million bbl of crude oil—a saving of $64

million, assuming a price of $5/bbl—through an investment of $20 million. This year, adds Williams, the company expects to save about 2.5 million bbl (worth about $30 million, assuming an oil price of $12/bbl) by means of changes that will cost about $10 million.

Typical of the alterations that Carbide has made this past year are the installation of new trays in a distillation column of a plant in Brownsville, Tex., and the rerouting of overhead vapors from an ammonia column at a Taft, La., plant.

The novel trays have variable and directional slotting, and box-type promoters, which improve distillation efficiency. Savings in steam, due to less reflux and heat input to the column, exceed $700,000/yr—the equivalent of 53,000 bbl/yr of crude oil.

By installing some piping at a cost of $60,000, Carbide has been able to reroute overhead vapors from the ammonia column directly to an existing recycle compressor, rather than through a condenser and vaporizer. The resulting savings in steam and water will run to some $375,000/yr, or more than 32,000 bbl/yr of oil.

Lyman Gilbert, director of control operations at Environmental Data

Corp. (Monrovia, Calif.), points to the importance of proper instrumentation when trying to conserve energy. Says Gilbert: "New instrumentation techniques allow automatic, continuous flame analyses, so it is possible to optimize air/fuel chemistry to get peak fuel-burning efficiency. *In situ* measurement of flame chemistry of each burner results in an evenly balanced furnace. In one system for the multi-burner boiler of a pulp and paper mill, the saving exceeded $2 million/yr, and the complete installation was made for only about $500,000."

A TIGHT FIT—"The recovery of large quantities of low-level heat rejected in processing probably represents our single greatest design challenge," said Robert I. Taylor, of Exxon Research and Engineering Co. (Florham Park, N.J.), at the Denver meeting. "This loss is typically about three to five times what we lose from furnace stacks."

Another speaker at the Congress—Harold Huckins, vice-president of Halcon Research and Development Corp. (New York City)—noted that various studies show that more than 75% of the heat transferred in industrial processing units that Halcon designs is at 350°F or less.

Recovery of this heat is very much a part of new installations, according to Nosh Mistry, manager of energy systems for Betz Environmental Engineers Inc. (Plymouth Meeting, Pa.). He says that energy utilization is very efficient in new plants, since high efficiency is built into them during the design phase. But older units, designed when energy was cheap, pose a problem: the lack of available space impedes the installation of often-cumbersome, low-temperature heat-exchange equipment.

Consider, for instance, the experience of the Western Kraft Paper Group (Oxnard, Calif.). The company had a boiler, built in 1938, which was nevertheless in good condition. But the flue-gas exit temperature was 500°F, resulting in a significant loss of heat. Western Kraft wanted to install a heat-recovery device, but the plant layout left almost no room for the necessary equipment.

Despite the space limitations, management finally decided to install an economizer in an outlet duct to reduce exit-gas temperature to 300°F. The work took two days—and it was laborious. As plant engineer Earl Shook puts it, "We had to squeeze it in with a shoehorn."

Was it worth the effort? The total installed cost was $60,000. But heat savings are 2,035,500 Btu/h—equivalent to $43,978/yr, with fuel oil at 33¢/gal.

FOREIGN EFFORTS—European and Japanese observers feel that traditionally higher energy costs in those regions have led to the use of process designs and equipment that are at least as energy-efficient as those employed in the U.S., if not more so.

Several British firms, for instance, have installed waste-heat boilers to replace fired heaters on various units. According to an oil-refinery spokesman, use of the recovered heat results in energy savings equivalent to one-third of the total fuel oil previously consumed.

In Japan, where the energy problem is even more pressing, the search for ways to save takes on an aura of urgency. A source at Showa Denko K.K. (Tokyo) says that "a few percent of energy saving may be possible by turning off lights, etc., but cuts of 10% or more require process modifications and improvements."

The Japanese firm has certainly gone far beyond turning off lights. It reports energy savings of 20% in metal production at its 40,000-metric-ton/yr, high-carbon ferrochrome plant in Chichibu, near Tokyo. The plant features a special prereduction step for chrome-ore roasting. This is said to halve electricity consumption, compared with a conventional electrolytic furnace fired with heavy fuel oil.

Showa Denko reports another 20% cut in energy consumption, at its 300,000-metric-ton/yr ethylene plant at Oita, Kyushu Island. According to the firm, this is done mainly by recycling steam.

PROCESS INTEGRATION—Energy-efficiency improvements need not be confined to a specific piece of equipment or part of a process. As Halcon's Huckins points out: "The process designer and the company's management must look beyond the battery limits of an individual processing unit to maximize steam utilization and conservation. We believe there are situations today where it is economically attractive to add equipment that can reduce the overall purchased energy-consumption per unit of product. Where there is energy to spare, a plant can add units to efficiently consume the surplus energy."

"Obviously," adds Huckins, "in developing such complexes, consideration should be given to generating some or all of the power requirements, along with the steam-heating needs."

This advice has been followed by Stauffer Chemical Co. (Westport, Conn.), whose plant in Carson, Calif., has earned an Industrial Concern Award for energy conservation from the Southern California Gas Co. (Los Angeles).

Stauffer uses acid sludge—a waste product collected from local refineries—and sulfur to produce sulfuric acid. In early 1976, the firm connected waste-heat boilers to two acid reactors in which sulfuric acid is generated. The boilers use heat from those reactors to make steam, which drives a turbine generator to produce electricity. This scheme generates about 2.5 MW—nearly all of the plant's power needs. And the company has reduced its annual electric bills by $500,000.

The waste-heat boilers also allow Stauffer to reduce the amount of water needed to cool the acid-reactor exhaust before it is released to the atmosphere. This is an added bonus in the drought-stricken West.

Some companies are even turning a profit from the sale of excess steam. Big Three Industries, Inc. (Houston, Tex.), for example, sells more than 2 million lb/h of medium-pressure steam exhausted from turbines in its air-separation plant to industrial users in the Bayport industrial district; the steam goes to process heating.

TWO IN ONE—A fuel switch and a special cogeneration agreement are combined in a $15-million power-generating facility, to be built at Celanese Corp.'s (New York City) plant in Pampa, Tex. The firm will first spend more than $50 million to replace its natural-gas-fired boilers with coal-fired ones. Then a nearby utility, Southwestern Public Service Co., will install a 30-MW steam-turbine generator. High-pressure steam from the new boilers will run the turbine, producing all of the plant's power requirements, and effluent steam, at lower pressure, will supply the plant's process heating.

This way, Celanese will obtain both its electricity and process steam from one installation. And the company says that the plan will save up to 50% of the normal energy requirements for generating electricity.

FUEL ALTERNATIVES—Many firms are, of course, planning to reduce their use of oil and natural gas as fuels. William van der Hoeven, operations manager for energy systems at Union Carbide's Chemicals & Plastics Div., says that one of the company's major energy-related projects is fuel switching aimed at reducing dependence on natural-gas supplies, improving Carbide's fuel-cost position, and conserving gas for higher-value petrochemical-feedstock use.

"The long-term objective," says van der Hoeven, "is to shift as much of our gas-fired boiler capacity as is practical to more-abundant, less-expensive coal. In the interim, however, we are switching from gas to fuel oil in order to provide the time needed for planning major capital investments to use coal in all new facilities."

From now through 1983, Union Carbide plans to spend $140 million to wean its facilities from natural-gas consumption. The company has earmarked an additional $75 million for major energy-conservation projects aimed at further reducing its total fuel requirements.

The entire program is designed to cut gas needs from about 60% of the total fuel demands to about 10% in 1985. Use of byproduct fuels, residues, steam and electricity would account, by that time, for 40% of Carbide's total energy requirements.

A good example of how Carbide plans to achieve this last goal is the new boiler installed at the firm's Brownsville, Tex., site. The unit uses aqueous chemical wastes as primary fuel to generate 600-lb/in^2 steam for process use. Combined with the burning of other chemical wastes in conventional boilers, operation of the new equipment enables the Brownsville plant to use 97% of its wastes as fuel, cutting natural-gas needs by 3 million ft^3/d. This represents a fuel-cost saving of approximately $600,000/yr.

The Brownsville plant's new boiler blends chemical wastes—containing a heating value of 5,000 Btu/lb—with fuel oil in an 80:20 ratio. The final mix has a heating value of about 8,000 Btu/lb.

Philip M. Kohn

Converting gas boilers to oil and coal

A transition from gas-fired to oil- or coal-fired boilers requires extensive engineering to obtain satisfactory performance. Many factors must be considered, and new equipment is needed for the change.

Arlen W. Bell and *Bernard P. Breen*, KVB Engineering, Inc.

☐ Today, there is public pressure toward operating more boilers on coal—which is an abundant resource—rather than on oil or natural gas, both of which are scarce. Regulatory agencies will therefore ask plant managers and engineers to convert from gas to oil, or preferably to coal-burning boilers.

Natural gas and light distillate oils are in increasingly short supply, as proved by extensive shutdowns for lack of fuel, particularly in the winter of 1973. Even No. 6 fuel oil is considered a premium fuel due to increases in electric-utility demand.

Because of the complexities involved in each conversion, only elements to be evaluated and possible alternatives are presented here. Some relative cost estimates are also presented, but final decisions will undoubtedly be made for other reasons than the lowest steam-generating cost.

Considerations involved in boiler-fuel conversion are fuel availability, fuel purchase cost, environmental requirements, and socio-economic pressures.

For the boiler owner, the most important subject is cost. There may be overall increased product costs due to fuel-price increases, or production losses due to lack of fuel. Fuel costs to competitors may also prove to be significant.

Environmental requirements for low-sulfur oils and coals have further tightened the short- and long-term supply and have shifted geographical production plans. Environmental considerations have actually been responsible for the original shift to high natural-gas consumption, which is now being recognized as a waste of this valuable resource.

Socio-economic pressure on a local, regional or national level can also cause a change in fuel usage. There are strong justifications for maintaining local fuel sources in operation, or shifting to either more-economic or lower-sulfur fuels from a great distance (i.e., substitution of low-sulfur western coal for some of the higher-sulfur eastern coals.).

Technical problem areas

Many engineering problems must be solved in conversion of a gas- or oil-fired boiler to coal, but such

Originally published April 26, 1976

problems can be solved by several alternate methods. Such problems can be grouped into three major classifications: coal handling and ash removal, boiler and related equipment, and emission control.

Coal handling and ash removal—Coal presents some unique requirements in contrast to gas and oil. Coal-storage areas for 30–90 days' supply are normally required, and mechanical conveying equipment is needed for transport from the storage area to the hopper, which is usually sized for a day's supply. Unlike oil, coal combustion generates large volumes of ash that must be temporarily stored in hoppers and shipped out for removal on a daily or weekly basis.

Boiler requirements—Each of the three fuels—gas, oil and coal—imposes unique requirements on the design of the boiler: the burner(s); fuel-supply system; accessory support-equipment such as soot blowers, forced-draft and induced-draft fans; and control systems. Particularly important is boiler-furnace sizing, as well as convective-tube spacing. A boiler designed originally only for gas firing may be almost impossible to convert economically to coal firing.

Emission-control requirements—Emission-control equipment is required to some degree on all boilers. For coal firing, the most troublesome area is the removal of ash and carbon carryover from the flue gas. SO_x control is usually done by limiting the sulfur content of the fuel. In the future, more-elaborate flue-gas cleaning devices, such as wet scrubbers, may be needed. NO_x control is ordinarily performed by burner design and minimum excess O_2 operation. CO can be controlled by proper air distribution to the burner(s), or coal grate.

Comparative efficiency—In general, conversion from gas or oil to coal will involve some loss in efficiency. For example, on boilers without combustion-air preheating, steam-generating efficiencies based on gross heating value are in the range of 81% for oil, 78% for gas, and 76% for stoker-fired coal. Efficiencies should be established by tests on operating boilers in normal service. Claimed efficiencies based on heat-transfer area, or tests on new boilers from factories, can be considerably higher than long-term, actual operating efficiency.

Fuel costs are really a larger economic factor than

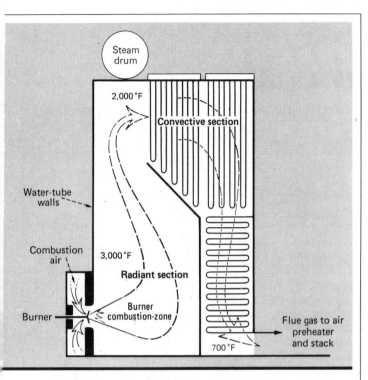

Sectional side-view of a vertical boiler   Fig. 1

(labels in figure:) Steam drum; 2,000°F; Convective section; Water-tube walls; Combustion air; 3,000°F; Radiant section; Burner; Burner combustion-zone; Flue gas to air preheater and stack; 700°F

just thermal efficiency alone. For instance, coal costs range from $0.50–$1.00/million (MM) Btu; oil from $1.50–$2.00/MM Btu; and gas from $0.34/MM Btu, for regulated interstate gas, to $2.00/MM Btu for unregulated intrastate gas.

Fuel supply and handling

Even though there may be sound economic and conservation reasons for boiler conversions, the logistics of fuel delivery and inplant handling can be severely misunderstood in the initial evaluation.

Outside plant boundaries—Inasmuch as gas-fired boilers receive their fuel through pipelines, there is very little interference with local developments. Oil and coal are more commonly delivered by tank truck, rail or barge. Therefore, switching from gas to oil, and particularly coal, may precipitate local traffic problems objectionable to nearby residents, or overextend existing rail or barge facilities. With the current financial collapse of several railroads, many local and spur rail lines will not be able to manage the load. And costs of a fuel-supply plan based on rail can drastically increase if switching to trucks is required.

Inside plant boundaries—Provisions will have to be made for oil tankage or coal-storage area within the plant's boundaries. The number of days' supply to be maintained on site as ready inventory has to be decided by management, as a hedge against strikes, severe weather, transportation-system breakdowns, or rapid changes in fuel availability. A 30–90-day supply of oil or coal is about normal.

Oil tanks usually require protective dikes to contain spills, as well as separation from buildings and equip-

ment. Coal-storage areas should either be well paved or well drained, and should be located near the boiler to minimize conveyor length. Drainage from coal piles may require a holding and settling basin as well, to avoid water pollution. Dust from coal storage or reclaiming operations may also create in-plant and neighborhood problems.

If gravity does not suffice, a heater and pump (plus a backup pump) will be needed to move oil from storage to a tank adjacent to the boiler.

Belts and bucket conveyors are ordinarily used to move coal into a hopper adjacent to the boiler. Some installations use small tractors to move coal into the feed chute and to keep the coal pile in shape. Coal is ordered in the needed size, but classification and regrinding equipment are occasionally used.

How a boiler works

The boiler (Fig. 1) consists of a furnace (radiant section) and tube banks (convective section). For gas, oil and pulverized-coal firing, burners are usually inserted in the front face. For stoker-fired coal, the combustion takes place on a grate across the radiant-section floor. An integral part of boiler design is provision for soot blowers in the convective section. Provision must also be made for removal of the ash that falls to the boiler floor or ash hopper. Ash in the flue gas is removed by precipitators or cyclones.

Furnace—Designed to take advantage of the high radiant-heat flux near the burner, or coal stoker, the furnace is basically a watertube-lined box, with either a tube-lined or refractory floor. Normally, gas requires the smallest furnace, due to its lower emissivity, which results in a relatively even heat-flux to the tube walls. Oil, having a higher emissivity, requires moving the walls of the boiler further from the burner, to even out the peak-heat flux.

Coal firing presents a different problem, namely, that the flue-gas temperature at the entrance to the convection section must be at least 100°F below the ash-softening point. Such a reduction in temperature can result in about a 15% increase in radiant surface area. Coal-fired furnaces are therefore larger than oil or gas units. Comparative furnace sizes for gas, oil and coal are, respectively: 1, 1.05 and 1.10 for width; and 1, 1.2 and 1.5 for length.

To meet the coal-firing requirements on a unit originally designed for gas-firing, would require either an addition to the radiant-heat-transfer surfaces, or a reduced load to the boiler. The reason for this is the additional burnout time needed for coal particles, as well as the needed heat absorption in the radiant (waterwall) section, so that flyash temperatures will be below their softening point, prior to entering the convective section. Such a requirement can be quantified in terms of furnace volume versus fuel type. For gas, the furnace should be sized approximately 60,000 Btu/(ft³)(h), while for overfed stokers the volume should be around 35,000 Btu/(ft³)(h). For coal, in particular, the furnace volume is strongly dependent upon coal type and ash properties.

Convective-section design—The convective boiler section is designed to extract the maximum amount of heat from the partially cooled (2,000°F) flue gas. The convective section should be as compact as possible to obtain the most efficient use of the construction metal and the plant space. Design factors are flue-gas velocities between tubes, tube spacing, and the use of fins. Gas, oil and coal need different designs for optimum operation.

For gas, flue-gas velocities of 120 ft/s, and extensive use of tube fins would be representative, whereas for oil, fuel-gas velocities of 100 ft/s and fewer fins (due to fouling) would be common. For coal, flue-gas velocities should not exceed 60 ft/s, due to the highly erosive characteristics of coal ash; and the tube fins should be spaced much further apart than for gas.

These design restraints indicate that gas-to-oil conversions, or alternate day-to-day use of these fuels on a boiler, is usually satisfactory. Oil or gas to coal conversion, however, requires either extensive modifications to the convective section, or a drastically reduced load (up to 50%) on the boiler.

Soot blowers—Due to the particulate matter carried in the flue gas through the convective section of the boiler, soot blowers are generally required for oil-fired boilers and are almost universally needed for coal boilers. An alternative for oil-fired boilers is the use of frequent water washes of the tubes, with the boiler out of service.

Tube deposits from fuel ash have several adverse effects, including reduced flue-gas to tube-heat transfer coefficients; reduced space between tubes (which results in high horsepower requirements for both forced-draft or induced-draft fans); and increased corrosion of tubes by selective chemical attack from ash constituents.

Soot blowers are either retractable from high-temperature zones (greater than 1,900°F); or rotary (non-retracting, with a 15-ft-limit length for a 30-ft-wide boiler) for temperature zones less than 1,900°F. Both types have high-pressure lances that use either air or steam to blast tube deposits off. Operator attendance is reduced if blowers are automated to operate individually on a preselected schedule. The boiler normally remains in operation during soot blowing. When using high-pressure water jets for water washing, the boiler must be taken out of service.

Because gas-fired boilers do not require soot blowers, conversion to oil or coal can present serious problems in clearances between tubes for soot blowers, and in external clearances (equal to one-half the boiler width on each side) for the placement of the blower retraction mechanism. For a gas-to-oil conversion, soot blowers could be omitted with some oils, such as light distillates. For coal firing, soot blowers would be required.

Burner requirements

The burner is intended to evenly mix the fuel and combustion air, to be capable of load changes, to provide stable ignition or "flame holding," and to require minimal operator attention. The burner design plays a large role in overall efficiency, safety and reliability of the boiler. Gas, oil and coal impose somewhat unique requirements for the burner.

Gas—Industrial gas burners are usually of the ring type, with flame-holding being controlled by adjustments of air registers for swirl, and a ceramic quarl or

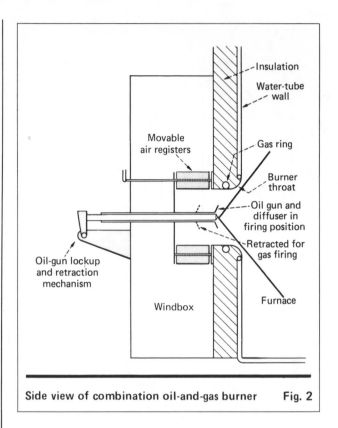

Side view of combination oil-and-gas burner   Fig. 2

burner "throat." A gas-fired igniter, flame detector(s), and a flame-failure shutdown system generally complete the burner package. Operation is relatively simple, and almost maintenance-free, which is one of the reasons for the popularity of gas firing.

Oil—Oil burners are geometrically similar to gas burners, with the oil gun being centered in the burner throat, and flame-holding generally established in the wake of a diffuser. Oil tips can be of different forms: straight mechanical (once through); constant differential (spill type); and steam atomizing (external and internal mixing). The type of atomizer selected depends upon boiler duty-cycle, fuel characteristics, availability of gas or diesel fuel for lightoff, and plant preference. A combination oil-and-gas burner is shown in Fig. 2.

Oil guns require more attention than gas ring-burners because, after shutdown, a steam purge is needed to prevent oil from freezing or coking in the tip and in the supply lines. The furnace should be visually inspected several times during each shift to verify flame quality and the absence of coking or of clinkers in the vicinity of the burner. Periodically, oil guns must be pulled for cleaning, and worn oil-tip components replaced.

Pulverized coal—Coal is delivered to coal burners (Fig. 3) from coal mills, in a finely pulverized form and suspended in a portion of the combustion air. The burner generally consists of a ceramic quarl, air registers, flame-shaping vanes, and a coal-supply tube centered on the burner throat. Pulverized-coal firing is ordinarily not used on boilers that produce less than about 200,000 lb/h of steam.

Operators must be concerned about flame quality,

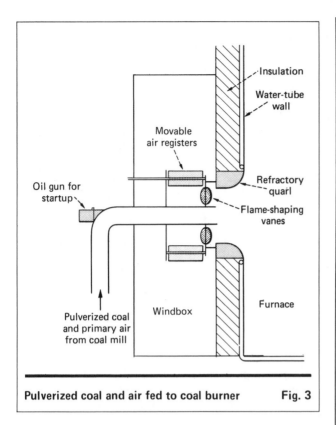

Pulverized coal and air fed to coal burner   **Fig. 3**

and the supply of fuel to and from the coal mill. The fuel-supply lines can erode rapidly from hard coal and contaminants, and often require annual patching or replacement. Lightoff is a more difficult and involved procedure than for gas or oil, often requiring the use of auxiliary oil or gas burners to preheat the furnace.

Stoker-fired coal—For coal firing in smaller industrial boilers, economics dictate one of the many variations of stoker designs, which require uniform distribution of the coal on the grate, and subsequent removal of the ash. Stokers can be broadly classified as underfeeding or overfeeding. The underfeeding stokers (Fig. 4) push coal up from a trough—in the center of a doubly inclined grate—so that the ash falls off the sides into ash pits. Frequently, the bed is mechanically agitated to maintain uniform coal flow. Typical capacity of an underfeeding stoker is 25,000 lb/h of steam flow. This limits its use to small plants; for large plants, multiple stokers must be used.

The overfeeding stokers are either spreader stokers (Fig. 5), or crossfed stokers, in which the fuel is mechanically thrown onto the grate, where burning takes place. Stokers are further classified according to their grate design: chain grate, traveling grate, or water-cooled vibrating grate. Heat-release rates are expressed in Btu/(h)(ft²), and vary with the coal used. For example, a spreader stoker with stationary or dumping grates is rated at 450,000 Btu/(h)(ft²); a stoker with vibrating, oscillating or reciprocating grate at 400,000–600,000; and one with a traveling grate at 750,000. The heat rate for crossfed stokers using chain or traveling grates is 450,000 Btu/(h)(ft²), and for those with water-cooled, vibrating grates it is 400,000.

Spreader stokers are the most versatile, being available in sizes from 5,000–400,000 lb/h of steam, and capable of burning virtually all coals, from lignite to high-rank bituminous. As the coal is mechanically thrown into the furnace, large lumps fall onto the grate in a thin layer and burn, while fines are usually burned in suspension.

The disadvantage of spreader stokers is that there is an appreciable carryover of combustible material in the flyash. However, carbon can be partially segregated from the flyash and can then be either reinjected into the furnace above the fuel bed or deposited on the fuel grate. This can raise boiler efficiency up to 3%.

Mass-burning stokers distribute coal onto a moving or vibrating grate through a hopper; ash falls into the ash pit at the far end. Almost any solid fuel except highly coking coals smaller than 1¼ in. can be burned, and sizes exist in the 6,000–200,000 lb/h of steam range. Typically, furnaces for mass-burning stokers include a long, rear furnace arch to direct the air-rich rear-end combustion gases forward. This enhances mixing with the highly fuel-rich gases from the front end.

Combustion air for both the spreader and mass stokers is introduced through the coal bed and also through overfire air-ports above the grate. This overfire air is injected at high velocity to obtain good mixing of combustion air with volatiles distilled off the coal.

Installation of a stoker in an existing oil- or gas-fired boiler entails, as a minimum, removal of the existing boiler front to accommodate the coal-feed system; mounting of the mechanical grate and drive system on the boiler floor; installation of an air-supply system under the grate, and overfire air above the grate; and installation of an ash pit. The placement of the ash pit for a spreader stoker makes it more advantageous for retrofitting purposes.

Fuel systems

Fuel systems increase in complexity from gas to oil to coal. For natural gas, a typical gas-main pressure of 50 psig is normal, but inasmuch as the burner requires around 10 psig, a stepdown regulator is needed. The boiler-control system opens or closes the control valve as load (or steam pressure) increases or decreases. With the exception of occasionally draining the condensate trap, the gas-fuel system is nearly maintenance-free.

For oil fuel, the system can become more complex, particularly for residual fuel. As a minimum, a filter set, booster pump, and control valve are required. For residual oil, insulated and steam-traced lines may be needed, as well as an oil header, a steam-purge system, and a recirculation system. If high-turndown, spill-type oil guns (wide-range mechanical atomizing) are applied, a return system is required, so another pump is ordinarily used. Oil systems generally require more maintenance than gas systems—because of filters, pumps, valves, and line cleaning—but a planned preventive-maintenance program makes an oil system nearly as flexible as one for gas.

Coal is more difficult to handle than gas or oil, because of the number of mechanical devices needed. Coal and associated contaminants are very abrasive, and all the handling equipment requires scheduled maintenance and planned component-replacement. A coal-handling system varies in complexity, depending upon whether the boiler is stoker-fed or whether it utilizes pulverized coal. Stokers require coal feeds within certain size limits provided by the supplier. For larger installations, a classifier and crusher for oversize lumps may be installed to improve fuel economy and minimize stoker jams.

For pulverized-coal firing, the fuel-supply system becomes mechanically much more complex. The coal is supplied to the pulverizer from the coal hopper by a variable-rate feeder. The coal is then finely pulverized and pneumatically conveyed by a portion of the combustion air to the burner.

Coal mills require heated air, which can either be supplied by the boiler forced-draft fan or through a separate primary air fan. If an exhauster is used to draw air through the pulverizer, it must resist the abrasion of coal particles. The heated air performs three tasks: it dries the coal, provides inertial energy for classification after the coal is initially ground, and transports finely ground coal to the burners.

Coal mills are sized on expected coal properties, i.e., heating value, grindability, moisture content, and desired ground-coal size. Departures from design conditions result in either oversized equipment or reduced capacity. Engineering judgment dictates some excess capacity. Typical pulverized-coal fineness is 70% through 200 mesh, with less than 2% larger than 50 mesh.

Coal mills can be classified in three types: ball or rod mills, roller/race pulverizers, and hammer mills. For abrasive coals, the ball or rod mill offers distinct advantages in lower overall cost. The coal is broken up by the impact from the tumbling balls, as well as from a grinding action within the mass of coal and balls. Speed is in the 2–25-rpm range.

Roller and race pulverizers are classified as either low (up to 75 rpm) or medium speed (75–225 rpm) units. The feed to these machines, which crush coal lumps between two rolling surfaces, is fed into the mill and circulated into the crushing zone by centrifugal force. When the coal is broken up, the finer particles are picked up by the air, carried to a classifier, and then conveyed to the burner. Oversized coal particles are recirculated back to the rolls. Races and rolls require replacement on an average of every two years, even though some units need annual replacement, and others last five years.

Hammer mills are high-speed machines (direct drive) that use hinged hammers to smash coal particles directly. Crushing is also performed between the hammers and fixed side-pieces. For small installations, these mills have the advantage of low capital cost per ton/h of output. They are also quiet, need only low space and direct drive, and are easy to maintain (hammer replacement). Because of the high speed, high-maintenance costs would occur with abrasive coals.

Auxiliary equipment

Requirements for auxiliary equipment are substantially different for gas and oil firing than they are for coal. Major items for coal burning include the furnace

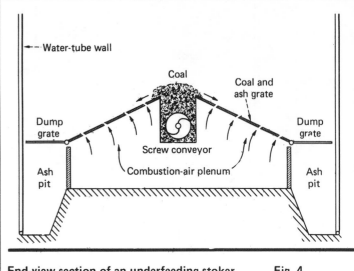

End-view section of an underfeeding stoker   Fig. 4

combustion-air-supply system, the air preheater, and the slag- and ash-handling methods.

Air fans—Gas- and oil-fired units are usually designed for pressurized firing. Combustion air is introduced into the furnace by a forced-draft fan, which supplies the total pressure required to overcome duct losses, burner-to-furnace pressure drop, convective-section pressure drop, and air-preheater pressure drop. As a result, the furnace operates under a positive pressure of 10–20 in H_2O.

For stoker firing, a pressurized furnace is impractical because it is difficult to maintain a sealed system. The coal-feed system and ash hopper present formidable sealing problems, which can be solved by using an induced-draft fan, following the air preheater. The furnace is then operated under a slight negative pressure of less than 0.5 in H_2O, with the I.D. of the fan carrying the pressure drop through the furnace, the convective section, and the air preheater. This additional fan would be mandatory on any gas- or oil-to-coal conversion. Also, a differential-pressure controller, to couple the forced-draft and the induced-draft fan operation, is required.

Air preheater—Although optional for stoker firing, air preheating is a must for pulverized-coal operation. Moisture content and mill air-flow determine the temperature necessary for pulverized-coal firing. If an existing preheater cannot supply the required temperature, a supplementary, direct-fired air heater may be necessary. Preheat temperature with stoker firing is normally limited to 350°F to avoid excessive maintenance of stoker parts. Air preheaters can improve overall efficiency of a boiler by approximately 2% per 100°F air preheat.

Air preheaters for gas firing use closely spaced, metal-mesh baffles (regenerative type), or closely spaced tubes (tube-and-shell type), constructed usually of mild steel. For oil firing, spacing is increased slightly, low-alloy material may be used, and provisions are ordinarily made for water-washing the oil-ash and sulfur

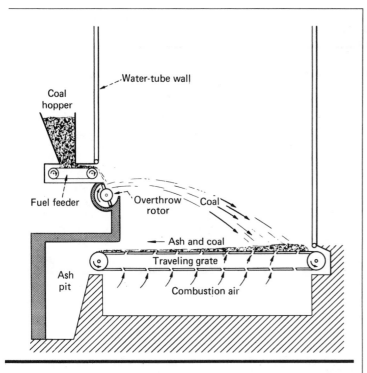

Side-view of traveling-grate spreader stoker   Fig. 5

compounds off the surface. Because erosion can be a major problem with coal firing, reduced flue-gas velocities are required, as well as the use of low-alloy steel for longer life of the elements.

Emission-control considerations

Poorly maintained and operated boilers emitting large amounts of CO and carbon are wasteful of fuel, and require more frequent maintenance due to fouling and corrosion. Once a specific fuel is chosen, emissions must be controlled by equipment selection and operating techniques.

Recent changes in laws—and interpretation of laws—have severely narrowed the extent to which an industry can deteriorate the air. As a result, it is necessary to consider all the emissions from a boiler as constituting a potential hazard. Such emissions include carbon monoxide, carbon, flyash, sulfur dioxide, sulfur trioxide, nitric oxide, nitrogen dioxide—and even water from combustion processes and cooling towers, if objectionable artificial clouds or fog are formed.

SO_2 emissions are directly related to the sulfur content of the fuel. Natural gas contains insignificant amounts of sulfur compounds; oil can range from 0.1–3%; and most residual oil sold in the U.S. is at least partially desulfurized, so that 0.5%-sulfur oil can be obtained at some premium price. Washed coal has some reduction in pyritic sulfur but, in general, desulfurized coal is nonexistent. Coal from other sources must be used if local coals cannot meet sulfur requirements.

Nitric oxide emissions occur from the combustion of atmospheric nitrogen in the high-temperature combustion zone, and with the partial fixation of organic ni-

trogen within the oil or coal. Although natural gas does not contain organic nitrogen, oil may contain up to 0.5%, and coals are in the range of 1% or above. Also, oils that have been partially desulfurized by either hydrodesulfurization, deep cracking, or chemical means, are reduced in nitrogen content. As a rough estimate, NO_x emissions on an industrial boiler would average 100 ppm for gas, 150 ppm for light distillate, 250 ppm for residual oil, and 500 ppm for coal. Pulverized-coal firing has higher NO_x emissions than stoker-fired coal.

Gas, which minimizes objectionable emissions, usually is limited to nitric oxides (NO and NO_2). And because process boilers have relatively low NO_x concentrations and do not represent major concentrated sources, they have fared well from the legislative standpoint. Gas-fired boilers have generally been well accepted by residential neighbors.

Oil-fired boilers present increased emission problems. During lightoff, load changes, and poor maintenance, very visible carbon plumes occur. Visible dark plumes are what the public associates with pollution, even though other pollutants are more hazardous.

With oil firing, SO_2 emissions depend upon the sulfur content of the fuel. Sulfur oxides are odoriferous, and SO_3—which constitutes 2–5% of the sulfur oxides—combines with moisture to form sulfuric acid mist. Of particular concern is the combination of this mist with the very hygroscopic oil ash. Such an agglomeration accumulates on the outlet flue-gas duct-surfaces and on the air preheater; soot-blowing then forces these deposits out the stack. Sulfuric-acid-laden particles tend to settle in the vicinity of the plant, destroying most plants and causing severe corrosion.

Coal firing presents an even more serious emission-control problem. With spreader-stoker firing, flue-gas ash could be as high as 10% of the weight of the coal fired. SO_2, SO_3 and NO_x must also be considered.

Flue-gas cleanup

Coal ash can be separated from the flue gas by cyclone collectors, electrostatic precipitators, fabric filters, or wet scrubbers.

With cyclone collectors—the simplest and most reliable equipment during operation—the flue gas is given a whirling motion, so that the particles are thrown onto the walls by centrifugal force. Cyclones are characterized by a relatively high pressure drop (about 5 in H_2O) and are effective on particles 3 microns and larger. They are also widely used on stoker-fired boilers, because these units generally have flyash of 10 microns or larger.

Electrostatic precipitators are used for finer particles—1 micron or larger—or where very high collection efficiencies (up to 99%) are needed. These devices first charge the particles by passing them across high-voltage wires, and then collect the particles on oppositely-charged plates. The plates are then periodically vibrated or rapped, so that the flyash will fall into a collection hopper.

Precipitators must be carefully engineered to a specific application, with design variables being flowrates (gas and flyash), particle-size range, gas temperature, and sulfur content of the fuel. The sulfur content is im-

Ash analysis and ash-fusion temperatures for representative U.S. coals

Properties	Low-volatile bituminous Pocahontas no. 3, W. Virginia	High-volatile bituminous No. 9, Ohio	High-volatile bituminous Pittsburgh, W. Virginia	High-volatile bituminous No. 6 Illinois	Utah	Subbituminous Wyoming	Lignite Texas
Ash, dry basis, %	12.3	14.10	10.87	17.36	6.6	6.6	12.8
Sulfur, dry basis, %	0.7	3.30	3.53	4.17	0.5	1.0	1.1
Ash analysis, weight %							
SiO_2	60.0	47.27	37.64	47.52	48.0	24.0	41.8
Al_2O_3	30.0	22.96	20.11	17.87	11.5	20.0	13.6
TiO_2	1.6	1.00	0.81	0.78	0.6	0.7	1.5
Fe_2O_3	4.0	22.81	29.28	20.13	7.0	11.0	6.6
CaO	0.6	1.30	4.25	5.75	25.0	26.0	17.6
MgO	0.6	0.85	1.25	1.02	4.0	4.0	2.5
Na_2O	0.5	0.28	0.80	0.36	1.2	0.2	0.6
K_2O	1.5	1.97	1.60	1.77	0.2	0.5	0.1
Total	98.8	98.44	95.74	95.20	97.5	86.4	84.3
Initial deformation temperature at which first rounding of apex cone occurs*							
Reducing, °F	2,900 +	2,030	2,030	2,000	2,060	1,990	2,130
Oxidizing, °F	2,900 +	2,420	2,265	2,300	2,120	2,190	2,070
Softening temperature at which cone has fused to spherical lump where height = width at base*							
Reducing, °F	—	2,450	2,175	2,160	—	2,180	2,130
Oxidizing, °F	—	2,605	2,385	2,430	—	2,220	2,190
Hemispherical temperature at which cone has fused to hemispherical lump where height = 1/2 width of base*							
Reducing, °F	—	2,480	2,225	2,180	2,140	2,250	2,150
Oxidizing, °F	—	2,620	2,450	2,450	2,220	2,240	2,210
Fluid temperature at which fused mass is nearly flat layer with maximum height of 1/16 in*							
Reducing, °F	—	2,620	2,370	2,320	2,250	2,290	2,240
Oxidizing, °F	—	2,670	2,540	2,610	2,460	2,300	2,290

* According to standard of American Soc. for Testing and Materials (ASTM) D1857.

portant, because the SO_3 that is emitted forms sulfuric acid, which condenses on the particles. This acid lowers the resistivity of the particles sufficiently to allow them to be electrostatically charged.

Also very popular for smaller installations are fabric filters. These are generally arranged as a series of cylindrical bags, on the outside of which flyash collects; this is periodically removed by pulsing or shaking the bags, one at a time. Removal efficiency can be very high on submicron particles, once a thin layer of ash has built up on the bags.

Wet scrubbers are also effective in particle removal (95–98%), but they have high-pressure drops (up to 20 in H_2O), and create more of a disposal problem, because the ash becomes a saturated sludge. In the future—if ground-limestone scrubbing systems become more widely used for SO_2 removal—the combined SO_2 and ash removal functions should increase the use of wet scrubbers.

Stack requirements, burner adjustments

Stack-height requirements may be modified by switching boilers from gas to oil or from gas to coal. The most common reason for requiring taller stacks is high SO_2 ground-level concentrations near the plant. This problem can be evaluated by existing techniques (for calculating plume rise and dispersion), modified by local atmospheric conditions. An outgrowth of installing taller stacks is greater boiler visibility, which perhaps make necessary more-stringent operating techniques than legally required, because of excess smoke during lightoff and rapid-load conditions.

Carbon carryover, carbon monoxide, nitric oxides (NO and NO_2)—and to a lesser extent, SO_3—emissions can be affected or controlled by burner adjustments. Gas, oil and coal present different opportunities for emission reduction.

With gas, sufficient air must be mixed with the fuel to permit complete burnup to eliminate CO. Too much air results in reduced efficiency, and high nitric oxide emission. Overly intense mixing also increases nitric oxide formation. Typically, burner adjustment, or tuning, involves adjusting overall air-to-fuel flow and changing the degree of air swirl (mixing energy).

For oil, the major emissions to be controlled are smoke and coke. Because more adjustments can be made to an oil burner than to a gas burner—including overall air-to-fuel flow, air swirl, position of the oil gun and diffuser with respect to the burner throat—changes can be made in the burner tip geometry (fuel-spray angle), and the fuel atomization technique used (straight mechanical, steam atomization, or air-blast atomization). Much current package-boiler research has been performed to seek minimum excess air levels, while avoiding coke. Such minimum levels result in reduced NO_x emissions and in the amount of SO_2 converted to SO_3.

Pulverized-coal burners may have almost as many adjustments similar to those of oil burners. Often, in addition to air-swirl registers, flame-shaping vanes are installed. Also, both the size-distribution of pulverized coal and the amount of carrier or transport-air can be altered. Again, best operation and lower emissions are obtained when adjustments are made to minimize excess O_2.

For stoker-fired coal boilers, adjustments are generally made to obtain higher carbon burnout of the fuel. The effects of overall air flow, overfire air flow, and ash recycle to the stoker have not been examined in detail to show how NO_x and SO_x emissions are changed. At the present time, it can be assumed that gaseous emissions cannot be substantially modified.

Fuel storage and handling

Oil can impose requirements on storage-tank vapor recovery systems, varying with fuel volatility and ambient or storage temperatures. Crude-oil storage, in particular, poses difficulties because of the high volume of volatiles and H_2S outgassing.

Coal presents a dust-control problem, which can generally be controlled to a reasonable level by proper design of handling equipment, such as enclosing conveyors and good housekeeping (for example, maintaining high volume-to-surface coal piles). In extreme cases, all storage might have to be in silos. Ash storage and removal present analogous problems.

Gas-to-coal conversion methods

The starting point of a gas-to-coal conversion is to obtain analyses of the candidate coals. It is impossible to design a boiler that can handle all coals. As a minimum, data should be obtained on the amounts of volatile matter, fixed carbon, sulfur, ash, ash-fusibility and softening temperature, heating value, caking or coking, grindability, and sizing available. Such characteristics determine furnace-resizing needs, tube spacing, soot-blowing needs, type of combustor (pulverized or stoker), emission-control hardware, and added site requirements.

Coal-ash characteristics are extremely important to avoid serious convective-section fouling, because the flue-gas temperature leaving the radiant section must be lower than the ash-softening temperature. The table presents a summary of fusibility temperatures and ash analyses for several representative coals.

Costs vary so widely, particularly on custom-retrofit work, that only a rough order of the cost of conversions can be obtained. After detailed drawings are prepared, estimates of ±20% can probably be made. Some rules of thumb for new, factory-assembled equipment are costs of $3/lb of steam generation for oil- and gas-fired boil-

ers. Field erection or modifications can easily cost twice as much as similar work performed in a shop.

Fuel savings on coal can be significant. For coal at $25/ton, with a heating value of 12,000 Btu/lb ($1.04/10^6$ Btu), as compared to oil at $10/bbl ($1.65/10^6$ Btu), or intrastate gas at $2.00/10^6$ Btu, fuel-cost savings of $140–230/h can result for a boiler of 200,000 lb/h steam output (gross firing rate of 240×10^6 Btu/h).

Three cases of gas-to-coal conversion or oil-to-coal conversion will now be discussed. These are in increasing order of difficulty: gas-fired boilers that originally operated as coal-fired units; older, larger boilers that have strong conversion potential; modern, shop-assembled package units designed optimally for gas or oil.

Reconverting back to coal firing

There are many older boilers, which originally operated as coal-fired units—utilizing multiple, underfeeding stokers, mass stokers, or spreader stokers. With the advent of cheap, plentiful natural gas, these were converted from coal to gas. Stokers were removed, ash pits were blocked or filled in, and gas burners installed. The use of natural gas practically eliminated soot blowing, as well as stack-gas cleanup.

The main items for conversion back to coal burning include: (1) rehabilitation or replacement of stoker; (2) verification that the system works as originally intended (this may require reinstallation of ductwork); (3) rehabilitation of ash-handling facility; (4) rehabilitation or replacement of soot blowers; (5) installation of equipment for flue-gas cleanup (particulates) to meet current emission requirements; and (6) consideration of the characteristics of the coal currently available versus the coal originally burned.

Older units with conversion potential

There is another class of boilers, which—although considered primitive by today's designs—should be able to be converted. These are relatively large-volume boilers designed for oil and gas firing, and induced or balanced draft, and probably incorporating soot blowers. For some coals, the boilers' convective-tube spacing would be adequate. Coal characteristics must be very carefully evaluated. Inasmuch as these boilers' furnaces are large enough, spreader-stoker equipment could be installed, although at some possible sacrifice in load. For units of 150,000-lb steam capacity or greater, pulverized-coal firing should be considered.

Major items required for conversion include: (1) installation of spreader stoker; (2) revision of ductwork to provide air through grates, and overfire-air through side-ports; (3) construction of ash pit, ash-removal system, and ash-recycle system; (4) installation of flue-gas cleanup equipment; (5) possible installation of added soot blowers; and (6) addition of air-supply capacity because of the additional ductwork and consequent increase of pressure drop, due to particulate removal.

Modern, shop-assembled package boilers

Package boilers are forced-draft, air-supply systems with high-heat-release furnaces, high velocities through the convective pass, all tubewall construction, combi-

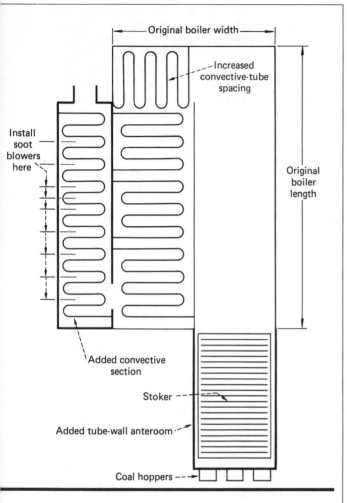

Package boiler converted from gas to coal firing   Fig. 6

tion of an entirely new coal-fired boiler, incorporating all the requirements for the intended coal.

Two techniques hold promise for gas-to-coal conversion: installation of pulverized-coal burner(s) in the position of the existing burners; and the addition of a refractory chamber, which should be large enough to accommodate a stoker.

For pulverized-coal firing, the following is required: (1) installation of coal mills; (2) installation of a pulverized-coal burner in the existing burner throat; (3) recontouring the floor of the furnace to facilitate partial bottom-ash removal; (4) installation of a stack-gas cleanup system; (5) installation of an induced-draft fan; (6) removal of a portion of tubes from the convective pass to reduce flue-gas velocity; (7) adding an economizer, or air-heater section, to replace the lost heating surface from the previous step and to provide sufficient preheated-air temperature; and (8) installation of soot blowers. A load reduction (up to about 25%) can be expected with this type of conversion.

For stoker firing, these would be necessary (Fig. 6): (1) installation of a combustion chamber to accommodate the stoker at the boiler front, including addition of sufficient heat-transfer surface; (2) installation of stoker and ash-removal systems; (3) removal of tubes from the convective section; (4) addition of tubes to the economizer section; (5) recontouring furnace floor; (6) installation of a special, soot-blowing system to clean tubes and floor surface; (7) installation of a stack-cleanup system; and (8) installation of an induced-draft fan. If all of these steps are properly performed, the rating (load) of the unit can be maintained.

From the foregoing discussion, it is apparent that extensive engineering is involved in changing from gas to oil or coal burning, particularly for the first few conversions of each boiler type. Prior to a final decision on conversion, similar, already-converted, boilers should be examined, and operating problems discussed with plant personnel.

nation oil or gas firing, and control systems designed for minimum operator attendance. Units intended for No. 6 oil normally have soot blowers, and horizontal convective passes that are adjacent to the furnace. The design is intended to obtain the most steam-generating capacity at the lowest cost, which results in a relatively lightweight, transportable boiler, with minimum onsite work required.

Prior to gas-to-coal conversion of a shop-assembled unit, serious consideration should be given to construc-

References

1. Babcock, G. H., and Wilcox, S., "Steam—Its Generation and Use," The Babcock & Wilcox Co., New York (1972).
2. Power from Coal, *Power*, Feb. 1974.
3. Leonard, R. R., "Influence of Fuels on Boiler Design," Riley Stoker Corp., Worcester, Mass., June 15, 1971.
4. "Boiler Rating Criteria for Nonresidential Boilers," Federal Construction Council, National Academy of Sciences, Washington, D.C. (1962).

The authors

Arlen W. Bell is associated with KVB Engineering, Inc., 17332 Irvine Blvd., Tustin, CA 92680. Before, he worked in aerospace, on projects such as aircraft and missile analysis, nuclear-weapon effects, structural dynamics, and research-facility design. He also has been program manager of nitric oxide reduction programs for various companies, and papers on this work have been presented at various professional-society meetings. He has also been program manager for burner development projects, and chemical desulfurization methods for oil and coal. He is a licensed mechanical, civil and structural engineer in California, and holds B.S. and M.S. degrees in engineering from the California Institute of Technology.

Bernard P. Breen is Vice-President of KVB Engineering, Inc., 17332 Irvine Blvd., Tustin, CA 92680. He cofounded this company with J. R. Kliegel, and has promoted KVB's growth in pollution reduction from utility boilers, gas turbines, and industrial combustors. This work has led to research and engineering in burner design, flame stability, and modified boiler operation. He has presented papers at meetings of the American Power Conference, Combustion Institute, and the Air Pollution Control Assn. He has a Ph.D. degree in chemical engineering from Iowa State University and is a licensed chemical engineer in California.

Wood: An ancient fuel provides energy for modern times

Wood contains no sulfur, can be processed like coal, is renewable, and is becoming increasingly economical. Small wonder that industry is more frequently turning to it as a fuel.

☐ After millennia of use, wood, it seems, is being rediscovered. Industries and utilities that never before even *considered* wood energy are beginning to plan major wood-fueled installations; meanwhile, long-time users of wood as a supplemental fuel, such as pulp and paper makers, are relying on it more than ever.

Of course, the interest in wood energy—both for home owners and for business—is rooted in other fossil fuels' rapidly rising prices and sometimes uncertain availability. Depending on plant siting, wood is cheaper in many cases than coal, oil or gas, and at least as abundant.

And the fuel's merits go past price. Like coal, wood can be burned directly, gasified to produce a low-to-medium fuel gas, or liquefied to make a fuel oil. Unlike coal and oil, wood contains virtually no pollution-producing sulfur.

There are, however, some limits to wood use. One is the absence of some kind of collection system to bring chips or residues to potential consumers. Also, the farther a wood-based plant is sited from a good source, the quicker the economics deteriorate because of the transportation costs involved.

RECENT DEVELOPMENTS—Despite these drawbacks, wood burning now provides the equivalent of 140,000 bbl/d of oil in the U.S., and the U.S. Dept. of Energy (DOE) predicts that this could double by the year 2000.

One of the largest wood-fueled installations to date was announced just last month by California Power and Light Corp. (Fresno). Some 700 tons/d of pelletized fuel will be burned in a new $70-million electric generating plant to be built in Madera, Calif. According to CP&L, the

40-50-MW facility will be the world's largest wood-powered electricity plant when it starts up in 1981. Construction is scheduled to begin in July, with Bechtel Power Corp. (San Francisco) expected to receive the engineering and construction contract.

Meanwhile, late last year, Dow Corning Corp. (Midland, Mich.) announced plans for a $30-million cogeneration plant that will burn 180,000 dry tons/yr of wood. The plant, which will start up in 1982, will be located in Midland and is set to provide 22.4 MW of electricity and 275,000 lb/h of 1,250-psi steam for the company's silicone-products facility there. Construction is due to start this month.

Another large installation also was announced last year. In Vermont, the Burlington Electric Dept. unveiled plans to build an $80-million, 50-MW electric generating plant that will feed on 500,000 tons/yr of green wood. When this starts up in 1983, it will be in the same league as the CP&L installation.

And pulp and paper companies are now taking even greater advantage of wood's fuel value. By 1982, for example, Chesapeake Corp. (West Point, Va.) will install a new 350,000-lb/h wood-waste-fueled boiler at its pulp and paper mill there. The unit will cut the site's oil consumption from 460,000 bbl/yr to 115,000 bbl; during normal operations, no oil will be used for steam- or power-generation.

EASING THE WAY—As utilities and industries switch to wood burning, a number of firms are attempting to smooth the transition.

To make it easier to use waste wood in existing coal-fired boilers, for instance, Bio-Solar Research and De-

velopment Corp. (Eugene, Ore.) has a patented process called Woodex that converts the material into 9,100-Btu/lb pellets. By itself, and through licenses, the company has put nine pelletizing plants onstream—eight in the past two years. Another ten plants are now under construction. Woodex, in fact, will fuel the new California Power and Light facility.

In the Woodex process, waste is pulverized, then dried in a rotary drum. The dried material is extruded at 30,000 psi, driving off more moisture and producing pellets about 3/8 in. long by 1/8 in. dia.; water content averages 10-14%, vs. up to 60% in the original wood.

Another pelletizing process is said to upgrade the heating value of wood to a level (13,000-14,250 Btu/lb) that makes it economical to ship the fuel across country—a previously unheard-of claim. Invented by Edward Koppelman of Encino, Calif., and developed in pilot-plant work at SRI International (Menlo Park, Calif.), the process reportedly will soon be commercialized; in fact, Koppelman says he'll be announcing a plantsite this month.

Meanwhile, to allow oil-fired boilers to be more readily switched to wood, Forest Fuels Manufacturing (Antrim, N.H.) introduced a wood gasifier last year that is designed to plug into existing furnaces.

INTEREST IN THE FORESTS—Much of this wood-burning activity stems from the skyrocketing prices of other fuels, and the ready availability of wood in most parts of the U.S.

As Table I shows, the annual Btu value of total unused wood residues is a sizable percentage of the usage of other fossil fuels. In the southern states alone, more than 1,600 trillion Btu/yr of wood residues are available—about 60% of the total fossil fuel used in that region in 1974.

"There are about 500 million dry tons/yr* of material that is wasted," says John I. Zerbe, program manager, energy research and development, for the U.S. Dept. of Agriculture's Forest Service. "It decays in the forest to be recycled back into the soil. About half of this material is recoverable at a reasonable price—about $40/ton." At approximately 17 million Btu/dry ton

*This is a more recent estimate than the one appearing in Table I.

Comparison of industrial fossil-fuel use and quantity of unused wood residues							Table I
	1974 Fossil-fuel use, 10^{12} Btu/yr					Total unused wood residues	
Region	Coal	Residual fuel oil	Distillate fuel oil	Natural gas	Total fossil fuel	10^6 DTE*	10^{12} Btu/yr†
Northeast	461.4	430.2	29.2	241.1	1,162	21.4	364.5
North Central	740.3	201.4	62.5	696.4	1,701	19.9	337.5
Southeast	124.8	240.8	17.6	108.2	491	43.6	740.9
South Central	184.5	168.1	14.5	1,795.3	2,162	51.0	866.4
Pacific Northwest	16.5	39.0	1.0	113.0	170	47.0	799.5
Pacific Southwest	2.7	58.8	3.3	250.2	315	15.0	254.4
So. Rocky Mountain	16.5	17.1	1.3	55.2	90	10.8	184.0
No. Rocky Mountain	66.4	26.0	6.6	99.8	199	3.7	63.0
U.S. Totals	1,613.0	1,181.4	135.9	3,359.2	6,290	212.4	3,610

*DTE=Dry tons equivalent.
†Conversion factor=8,500 Btu/dry lb, or 17 x 10^6 Btu/DTE.

Source: Battelle Columbus Laboratories

The economics of wood as a fuel				Table II
Price comparison (mid-1979 basis)				
Fuel	Estimated heat value	Boiler efficiency	Current price	Cost per million Btus, allowing for efficiency
No. 2 Fuel oil	139,000 Btu/gal	82.5%	$0.67/gal*	$5.84
Natural gas	1,000 Btu/ft³	82.5%	$2.36/$10^6$ Btu	$2.86
Wood 50% moisture (Wet basis)	4,300 Btu/lb	66.7%	$12/ton	$2.09
Wood 13% moisture (Wet basis)	7,490 Btu/lb	78.0%	$15/ton	$1.28

*This has gone up to about $1/gal as of April (*CE* estimate).
Source: Georgia Institute of Technology

of wood, that price converts to roughly $2.35/$10^6$ Btu—comparable with natural gas (see Table II) and cheaper than oil (currently approaching $5/$10^6$ Btu).

It's this bright price picture that is motivating the drive to burn wood. Dow Corning, for instance, expects to cut costs by 30% compared to $2.40/$10^6$ Btu for natural gas. The economics are so attractive, in fact, that the firm finds it will be worthwhile to pay to have wood harvested from privately owned forests, and to buy sawmill residues, and scrap material from commercial clearing and building-demolition operations.

Aside from economics and good supplies, wood has another advantage: it burns fairly cleanly. Because sulfur is not a normal component of wood, sulfur dioxide is not generated. And,

"compared to fossil-fuel-fired units of the same capacity, wood-fired boilers emit less oxides of nitrogen," points out John Milliken, of the U.S. Environmental Protection Agency's Industrial Environmental Research Laboratory in Research Triangle Park, N.C.

NOT FOR EVERYONE—Despite all those pluses, however, wood has enough drawbacks to limit its industrial potential.

"More industrial users have not pursued this fuel because there is no wood chip brokerage," states Elton H. Hall, senior research scientist in the Energy and Environmental Systems Assessment Section of Battelle Columbus Laboratories (Columbus, Ohio). Generally there is no infrastructure for getting wood chips or residues to potential users, he explains.

And transportation is another prob-

lem. "The energy costs for a 20-ton truck would almost preclude going over 20 miles," says Joseph G. Massey, assistant professor at Texas A&M University's Dept. of Forest Science. Indeed, transportation costs can run 5¢/ton-mile or more. And, since wood contains a great deal of water, the transportation costs get even worse on a cost-per-million-Btu basis.

Finally, wood's high moisture content also adversely affects combustion efficiency, with much of the heat produced going to drive off water: An 8,200-Btu/lb oven-dry wood typically rings in at just 4,000 Btu/lb on a wet basis. (As mentioned before, pelletizing processes are emerging to solve problems of high water content.)

OTHER WAYS—Although today's commercial applications simply involve direct burning of wood, gasification, liquefaction and pyrolysis may make inroads in the future.

Because existing oil- or gas-fired boilers and furnaces are not easily converted to a solid fuel, it would be more economical to gasify or liquefy wood than to use it in solid form, many experts believe.

Last summer, Mitre Corp. (McLean, Va.), under contract to DOE, completed an evaluation of wood gasification's potential. At that time, Mitre felt that low-Btu gas produced from wood would be cost-competitive in 1985. Now, says Mitre's Abu Talib, "if the current trend in oil price escalation is considered, the low-Btu gas . . . is cost-competitive today."

Wood gasification is not new—some of the technology, in fact, is decades old. But a bevy of newer, more-

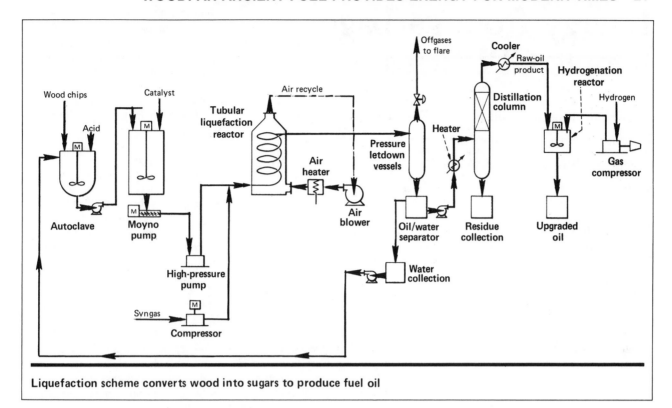

Liquefaction scheme converts wood into sugars to produce fuel oil

efficient processes are being developed that rely on catalysts and/or higher pressures.

Since 1977, for instance, DOE has funded the development of a catalytic gasification route at Battelle Northwest Laboratories. At 550°C and atmospheric pressure, steam is reacted with wood in the presence of a nickel catalyst to produce a gas containing 20-35% methane. This would then be methanated to make a substitute natural gas. The process will be tested with a fluid-bed reactor this year.

Wright Malta Corp. (Ballston Spa, N.Y.), also funded by DOE, plans to build a 4-ton/d unit this year to demonstrate its high-pressure, catalytic process that produces a 450-Btu/ft³ gas. Here, wood is fed to a stationary kiln that uses an auger to move the material. The reaction takes place at 330 psi and 1,100-1,200°F. According to the firm, 95% of the wood is converted to gas that contains 1% CO, 20% H_2, 25% CO_2, 18% CH_4 and 8% ethane and higher hydrocarbons.

Meanwhile, in Hudson Bay, Saskatchewan, a $250,000 wood-gasification demonstration plant has been operating since January 1979. Jointly funded by the province and the Canadian federal government, the gasifier has a feedrate of 290 kg/h of bonedry chipped wood and generates 4 million Btu/h of 150-Btu/ft³ gas. At the moment, planning calls for the testing of a unit that will drive a 65-kW generator.

PRODUCING OIL—Wood can also be liquefied to produce a Bunker C type of oil that can be used directly in furnaces.

DOE now spends about $25 million/yr on its wood programs, with much of the money going to a woodliquefaction project in Albany, Ore. Last August, its first barrel of oil was produced by a process developed by the University of California's Lawrence Berkeley Laboratory.

In the process, an aqueous slurry containing 25% wood is acidified and reacted at 180°F and 150 psi to form an emulsion of sugars (hexoses and pentoses). The stream then is reduced with a synthesis gas (a 50-50 CO, H_2 mixture) at 3,000 psi and 700°F.

From 100 lb of wood, 30-35 lb of 15,700-Btu/lb oil are obtained. The product, roughly half phenolics, has a specific gravity of about 1.1 and resembles No. 6 fuel oil. The researchers estimate that oil can be produced at $6/10⁶ Btu.

In Canada, researchers at the University of Toronto have hydrogenated wood in the presence of a Raneynickel catalyst (*Chem. Eng.*, Feb. 11, p. 55). Shredded wood is slurried in water and reacted with hydrogen at 1,500 psi and 340°C. One kilogram of wood yields approximately 400 g of oil having a viscosity of 5,000-8,000 mPa and a heating value of 35-38 MJ/kg.

Pyrolysis of wood is also catching industry's eye. Enerco Inc. (Langhorne, Pa.), for one, has sold two pyrolysis units—one to the Tennessee Valley Authority, the other to Forest Energy Co. (Beech Glen, Pa.). Both units have been designed to turn 3 tons/h of wood fiber into 40-50 gal of oil, 1 ton of charcoal and 7-8 million Btu of gas.

Other pyrolysis processes are coming on the market, too. Some, like the Thermex reactor of Alberta Industrial Development Ltd. (Edmonton, Alta.), and the Tech Air process (now being licensed by American Can Co., Greenwich, Conn.), introduce air into the reactor to partially oxidize the wood.

Reginald I. Berry

Energy from garbage tempts CPI firms

Municipalities will build many solid-waste-burning units in coming years, and chemical companies (barring one exception thus far) will probably buy energy products from such plants rather than invest in them.

☐ When Hooker Chemical Co's energy-from-waste plant at Niagara Falls, N.Y., becomes fully operational next month, it will supply enough steam and electricity to cover all the fuel needs and 20% of the power required by an adjacent Hooker-owned chlor-alkali unit. The 2,200-ton/d waste-fed plant will be—says the firm—the largest privately owned resource-recovery* unit in the U.S., and, according to a recent study, one of the first of a future wave of U.S. waste-fed facilities.

Indeed, a survey completed in July by the consulting firm Resource Technology Corp. (Cupertino, Calif.) for the American Iron and Steel Institute (Washington, D.C.) says that 179

* The term includes recovery of ferrous metals, aluminum, glass and energy.

units of this kind (mostly owned by municipalities) are under construction or in the planning stage. Right now there are only 17 energy-from-waste installations in operation; another 6 are "operable, but temporarily shut down," says Ronald D. Kinsey, president of Resource Technology.

It is the capital needed for such plants, plus some bad past experiences (see box) that will probably keep most chemical process industry (CPI) firms from investing directly in this new wave of units. In fact, with the exception of Hooker, which plans an even bigger energy-from-waste plant at Taft, La., CPI involvement so far has been limited to the purchase of steam, electricity, methane-from-landfill or refuse-derived fuel from the plant owners. Such is the case of B. F.

Goodrich, Arco, Shell Oil and others.

Other companies—e.g., Du Pont—having concluded that resource recovery holds little potential as an energy source, will neither invest in the plants nor buy their energy output.

SOME BUYERS—Donald A. Burge, Goodrich's manager of energy resources, believes that the volume of trash generated in the U.S., although seemingly large, is relatively quite small. He is urging Goodrich plants to enter into purchase agreements with municipalities because "I don't want others to beat us to this limited resource. If you think oil and gas are scarce, energy from municipal waste is much more so, even though people don't recognize it yet."

Since last fall, the company has been buying half of the steam produced from a garbage-fired plant in downtown Akron, designed to consume 1,000 tons/d of solid waste. Goodrich's share of the output is low in the winter months, when the city buildings need more heat, and high in the summer.

The firm is currently negotiating with an undisclosed municipality for a similar arrangement, but this will be a smaller, waste-burning, steam-generating facility. Earlier negotiations with the City of Louisville, Ky., for a 1,500-ton/d plant failed after the city decided against shutting down an operating incinerator. Goodrich then went on to expand its existing coal-fired boiler capacity there.

In New Jersey, Merck & Co. is discussing with officials from Union County and the City of Linden the construction of a $100-million resource-recovery plant on 21 acres of Merck property. If all goes well, the facility, designed to process 600,000 tons/yr of garbage, will go onstream sometime in 1984. Most of the superheated steam at 750°F and 850 psig will go to replace Merck's fuel-oil consumption of 400,000 bbl/yr.

Two plants belonging to Monsanto Co.—a plastics facility in Springfield, Mass., and the John F. Queeny chemicals plant in St. Louis, Mo.—also are exploring resource-recovery projects. At Springfield, a $50-$55 million installation would start up by 1984, producing about 160,000 lb/h of steam. Monsanto is considering using part of the output for cogeneration. As for the St. Louis-based plan, it is still in a very early stage.

Photo: William H. Rorer, Inc.

Hydraulic ram at left discharges ash from the burning of garbage

Originally published September 22, 1980

Why the CPI don't build energy-from-waste facilities

Despite the increasing number of resource-recovery facilities either under construction or planned, the U.S. lags behind other industrialized nations. According to DOE, for example, over 250 such plants are operating in Western Europe and Japan.

A general explanation for this gap, says DOE, is that until now, garbage disposal as landfill has been relatively inexpensive in certain areas, so there has been no incentive to look for alternatives. In addition, CPI firms have stayed away from direct involvement in garbage-fed installations for two reasons: (1) operating difficulties of some early plants built during the 1970s, and (2) the high cost of building such facilities.

Among the firms that have had to back out of projects are Occidental Petroleum Corp. (Los Angeles)—whose waste-to-oil pyrolysis unit at El Cajon, Calif., has been mothballed since 1978—and Monsanto Enviro-Chem Systems Inc. (St. Louis), which withdrew in 1977 from a project with the City of Baltimore, involving the use of Monsanto's Landgard process. Having made a number of modifications (especially to the air-pollution-control system), the city is now operating the plant.

As for plant costs, Hooker says that its Niagara Falls facility is twice as expensive as a comparable oil- or coal-fired power plant, mainly because of the additional equipment needed to handle, sort out and store the wastes. (Hooker's rationale for building hinges on the longterm availability of a local source of fuel.)

At least one of the reasons for the U.S. lag will soon disappear, as landfilling costs, swollen by the expense of meeting the requirements of the Resource Conservation and Recovery Act (RCRA), rise dramatically in coming years, according to Wade St. Clair—a spokesman for the Washington-based National Center for Resource Recovery, Inc. (a nonprofit research company in the solid-waste disposal field).

This push to build resource-recovery plants will get the positive backing of grants from both the U.S. Enviromental Protection Agency (for planning) and DOE (for feasibility studies and cooperative agreements).

At any rate, the energy that can be recovered from waste would not make any appreciable change in the U.S. energy picture. According to Hooker, if all the refuse produced in the U.S. in any given year were converted into fuel, it would represent only a 15-d oil supply, or 7% of the nation's total coal consumption.

REFINERY INVOLVEMENT—A 2,000-ton/d energy-from-waste facility will—by mid-1984—be supplying steam to an Arco Petroleum Products (a division of the Los Angeles-based Atlantic Richfield Co.) refinery in Carson, Calif. The plant, likely to cost more than $100 million, will be run by a resource-recovery company, Watson Energy Systems (Los Angeles), and use a patented system to prepare a refuse-derived fuel for immediate burning in four rotary kilns. The steam output (300,000 lb/h at 750°F and 685 psig) will be sent via a 9,000-ft pipeline to the refinery.

Another Watson Energy project is among a handful of those recovering gas from the anaerobic decomposition of organic matter in sanitary landfills.

Onstream since November 1978, it has been supplying 1.25 million ft³/d of gas (52-53% methane), obtained from a landfill in the Wilmington area of Los Angeles, to a Shell Oil Co. refinery. Shell mixes this with natural gas and refinery offgases for use in process heating.

USING ITS OWN GARBAGE—Rather than buy steam from municipalities, the pharmaceutical firm William H. Rorer, Inc. (Ft. Washington, Pa.) has been converting a portion of its own refuse into 90-psig steam for both process and heating purposes. Such plant wastes as cardboard, corrugated carton and wastepaper are burned in a pyrolytic, two-stage incinerator in an oxygen-lean atmosphere. The net heat output is 5.4 million Btu/h for a

10-to-12-h day, depending on production levels.

The $275,000 system is said to save over $100,000/yr by cutting oil and gas consumption by 25-28%, and lowering refuse-hauling costs. The return on investment, say Rorer officials, is in excess of 25%/yr.

FEDERAL HELP—Many of the 200 or so resource-recovery projects that are in some stage of development have been the beneficiaries of federal grants. And lately, the money has been flowing more freely from DOE coffers as the agency, responding to the Carter Administration's goal to stimulate synfuels production, has awarded about $200 million since July for feasibility studies and cooperative agreements (Chem. Eng., July 28, p. 39). An additional $300 million will be parceled out in coming months.

Among the recipients of the DOE largesse is Azusa Reclamation Co. (Azusa, Calif.), which wants to boost fourfold its current landfill-gas recovery, and has requested about $1.9 million to build a facility to use the gas for electricity production.

For the last two years, Reichhold Chemicals, Inc.'s resins plant in Azusa has been the sole customer for the landfill gas (about 10^{11} Btu/yr). Says Robert L. Beardsley, executive vice-president of operations for Reichhold Energy Corp., "We contracted for the gas because we thought natural-gas service would be curtailed, and that didn't happen. But we're now satisfied with the results, and we get the methane for 95% of the going rate for natural gas."

Rubber processor Dayco Corp. (Dayton, Ohio) says 6 of its 29 plants are either studying or designing projects to produce waste-derived steam. The goal of one plant at Walterboro, S.C., which has received DOE feasibility-study funding, is to stabilize energy costs that have tripled since 1975. "Overall, what we're looking for is energy-from-waste systems that establish a predictable fuel cost for the next 15 years," says Jerry Wood, corporate energy manager.

At Dayco's wholly-owned subsidiary, Colonial Rubber Works (Dyersburg, Tenn.), a joint venture with the city, now in the shakedown stage, will process 100 tons/d of municipal and industrial garbage to generate 20,000 lb/h of steam for plant use.

David J. Deutsch

Organizing an energy conservation program

Energy conservation has paid off in this large process-industry company. Here is how the conservation program was organized.

M. A. Williams, Union Carbide Corp.

☐ The success of an energy conservation program will depend principally on: (1) a complete investigation of existing operating departments for energy usage, primarily by adopting the "energy audit" concept (explained later), (2) incorporating energy savings in the design of new facilities, and (3) developing new processes that are less energy-intensive.

Plant organization

There are undoubtedly many ways to organize a plant energy conservation program. Our Chemicals and Plastics Div., which comprises fourteen plants, has adopted the concept of assigning one man in each plant to be responsible for promoting energy conservation ideas, monitoring progress, and in general, stimulating consciousness of energy usages and costs and of opportunities to save energy. The plant line-organization has the responsibility for developing, initiating, funding, and implementing energy conservation projects. Where large capital projects are concerned, then the other functions from engineering, research and development, and the commercial side are also intimately involved.

The engineer assigned as plant coordinator devotes an amount of his overall time to the program commensurate with the size of the plant and the maturity of the program; he reports to one of the assistant plant managers. We feel that this assistant plant manager plays a key role in the success of the plant program because of his position. He has the "clout" to get things done working with and through the other assistant plant managers and in his own direct area of responsibility.

It is also very important that the plant coordinator have sufficient time available to devote to the plant program. In addition, he should be allowed (and encouraged) to attend and participate in outside energy-conservation conferences. There sometimes is a tendency to load this person with miscellaneous plant projects. In most cases, the effectiveness of the energy conservation program will be proportional to the effort and time the plant coordinator is allowed to expend on it. In addition, the division's energy conservation coordinator assists in the plant program by providing guidance, promotional material, and ideas, and by making semiannual visits to each plant. Normally, the coordinator has a committee to help him.

The plant committee is composed of plant people representing engineering, maintenance, the shift organization, energy systems, production, etc. (see the organization chart). The committee meets periodically to review the status of the program and to discuss means of improving it.

The committee is also dedicated to stimulating, encouraging, and motivating plant personnel to save energy through publicity, periodic personal talks with the individual department heads, energy information presented to all production supervisors and operators, monitoring of the monthly energy-use product reports for all departments, etc. The committee also assists in funding examinations of potential energy-conservation projects, and gives general help in establishing a program in each department. All plant coordinators meet at least annually to discuss their programs and exchange ideas.

As in most companies, we measure our energy usage in millions of Btu—we convert our pounds of steam, million cubic feet of gas, kilowatt-hours of electric power, gallons of oil, tons of coal, all to the common base of millions of Btu, using published conversion factors or factors furnished by the supplier.

In order to be certain there is an active program in each operating department, one supervisor (in most cases, an engineer) is made directly responsible for the energy conservation program in his department. This has been found to be most effective, since the department head has many other duties and responsibilities that make demands on his time.

This department representative is the primary contact between the department and the plant energy conservation committee. He is also the focal point and contact for operating and maintenance personnel in the department on matters related to energy conservation. He encourages and furthers the department program by assisting in the development of ideas, making energy balances around the unit to pinpoint areas for intensive study, seeing that proper priority is assigned to energy conservation projects, reporting energy usages and

Originally published October 11, 1976

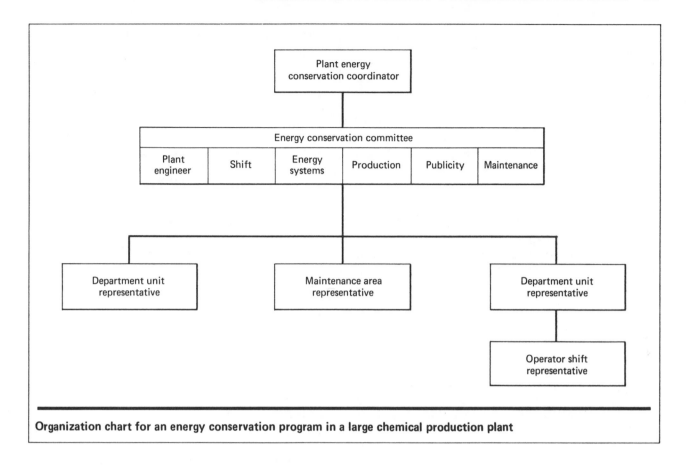

Organization chart for an energy conservation program in a large chemical production plant

costs, and publicizing the energy conservation program in general and also as it relates to his particular department.

Again, it is very important that the department representative be given sufficient time to devote to the department program. The department results will usually be related to the time he expends on energy conservation work. To assist the department representative in obtaining the full involvement of the shift organization, one operator on each shift is appointed as the energy conservation contact for the particular shift in the department. His job is somewhat analogous to the department representative but on a more limited scale. The important point is to reach all levels of personnel and obtain their involvement, only then will interest and commitment follow.

Energy-savings goals

All operating departments establish yearly dollar and million Btu energy-savings goals. From the sum of these goals coupled with the projected energy usage of the plant as a whole, the plant goal is set. To have the maintenance department actively participate in the program, besides having a representative on the plant committee, each maintenance area has an engineer assigned the responsibility for energy conservation in that area of the plant. These maintenance representatives form a small subcommittee that develops and implements programs designed specifically for their personnel.

Examples of their programs include insulation inspection and repair, steam-leak tagging and repair, oil reclamation, optimum consumption of lubricants through feedrate control and use of better-suited materials, and so forth.

Reporting

Each operating department reports quarterly on the progress of its energy conservation program. The report includes a list of completed energy conservation projects with dollar and million Btu energy savings and project cost plus a list of projects being implemented. The total of all the operating department reports in a plant then comprises the plant quarterly report which is sent to the divisional coordinator as well as to all other plants. A report by the divisional coordinator summarizing the plant reports is sent to higher management.

The plants are ranked in the consolidated report according to their savings in relation to their energy usage. Within a plant, the various operating departments report to plant management, also on a monthly basis, their energy usage per pound of product. This information is plotted and monitored for unfavorable excursions.

Besides the quarterly report, each plant is now being requested to submit monthly both its energy usage expressed as Btu per pound of saleable product produced and its total energy cost per sales value of the products produced.

Plant programs

The plant program is primarily oriented around optimizing the energy usages of the existing processes.

This is accomplished in many ways, ranging from simple, inexpensive changes brought about by discussions over a coffee table to elaborate, detailed engineering studies and costly modifications resulting from what we term "energy audits."

An "energy audit" is a study and evaluation of the energy consumption of a defined system by a team of competent and experienced engineers with the purpose of optimizing the efficiency of energy usage. A system can range in size from a single distillation column (with condensers and reboilers) to all the equipment of an entire plant.

Motivating the organization

However, the question always arises: How do you motivate the organization to generate and implement energy conservation projects? The success of the program depends on motivation.

Some of the things that have been helpful in developing and maintaining interest and enthusiasm are:

1. Management's unequivocal commitment to energy conservation from the top down.

2. Publicity for energy conservation. Use of internal publications, presentations on energy usages, costs, and forecasts to all supervisors, pamphlets mailed to the home and distributed at entrance gates, signs, talks to civic clubs, monthly newsletters devoted to energy conservation, contests for energy conservation ideas and symbols, etc.

Sitting down with the department head and his energy conservation representative on a periodic schedule to discuss both his program and your thoughts is one of the most successsful techniques to keep alive an active ongoing program.

We have quarterly luncheons also for all the unit representatives to discuss new approaches to energy conservation and hear about their programs. Visits by the plant coordinator to various operating control rooms to talk personally about conservation with the operators and maintenance personnel has proven worthwhile.

We feel that it is also important that energy conservation outside the confines of the plant be stressed. Employees, for example, consume a lot of energy in their daily lives, and the attitude you help create in the plant can spill over and have a beneficial effect on the community attitude toward energy use and savings. It is estimated that the residential and private auto segment consumes 37% of the U.S. energy budget compared to industry's 41%.

3. Tough and measurable goals for energy conservation should be established by individuals, by departments, and by plants and action plans developed to meet them. A monthly or quarterly reporting system can be used to monitor the progress toward the established goals.

4. An "energy usage per pound of product" factor expressed as Btu/lb for every operating department should be calculated each month and plotted against the plan. Such a plot will reflect the energy conservation efforts of the department and therefore act as a stimulus. Also, the plot can act as a monitor to show unfavorable excursions and encourage immediate in-

vestigation. It will also pinpoint the energy-intensive products made by a company.

5. Foster reasonable competitiveness between departments and plants by publishing results and ranking their efforts according to some established standard (example, Btu savings shown as a percentage of planned usage).

6. Provide ideas and information on various segments of energy-consuming facilities in order to assist in stimulating the development of projects. As as example, our central engineering department has produced 18 Energy Conservation Opportunity video tapes, each of approximately 20 minutes duration, covering such subjects as heat transfer, operation of distillation equipment, fired heaters, steam tracing, instrumentation, etc. Another example is a checklist of the various areas to consider for energy conservation ideas and projects.

7. Have operating departments make heat balances around their systems so energy-intensive segments can be pinpointed for priority study.

8. Use realistic future energy costs to justify and provide incentives for energy conservation projects. In our division, on major capital projects, we use the projected cost of energy over five years after estimated start-up of the unit. On small projects, we use the projected cost of energy two years hence.

Results

You measure how your program is progressing by the results realized. Your program is designed to achieve energy savings; consequently the amount you save, shown either by the reduction in your total energy costs and/or by the decrease in total energy usage compared to the goals you have established, is the yardstick by which to gauge the effectiveness of your program. Union Carbide has been gratified by the results of its energy conservation program in the Chemicals and Plastics Division.

This Division reported savings of energy equivalent to 2.5 million bbl of oil in 1972, 2.6 million bbl in 1973, and 2.8 million bbl in 1974. To give you some frame of reference, the U.S. is currently consuming 17.0 million bbl/day of petroleum products.

The Chemicals and Plastics Division's energy conservation index, which is a measure of our actual energy consumption in the year involved compared to what we would have used based on 1972 consumption, and which we report semiannually to the Manufacturing Chemists Assn. for submission along with the other chemical companies to the Federal Energy Administration, showed a drop to 0.975 in 1973 and to 0.925 in 1974.

The author

Milton A. Williams is Energy Conservation Coordinator for Chemicals and Plastics, Union Carbide Corp., Texas City, Tex. He is responsible for the energy conservation efforts of Chemicals and Plastics' manufacturing, engineering, research and development, and distribution functions, and has also served the company as Gulf Coast fuels coordinator, production manager of the olefins businesss area, production manager of the refined aromatics business. area and production representative of the normal paraffins businesss area. He holds a B.S. degree in mechanical engineering from the University of Minnesota, is president of the Gulf Coast Energy Conservation Society and is an AIChE member.

Energy conservation programs require accurate records

The production of multiple products in multiple plants of a single company poses a difficult and complex job of energy reporting and conservation. Here is a practical approach for a computer-based information system for processing such data and producing reports.

James E. Troyan, Olin Corporation

☐ Plant and corporate managements are acutely concerned with escalating costs and impending long-term shortages of energy. Hence, they encourage energy-saving projects and request frequent reports to indicate what progress is being made.

Managements are not alone in desiring such information. Mandatory reporting of energy usage and savings has become an integral feature of the U.S. Dept. of Energy's (DOE) Energy Conservation Program, which was initiated by FEA (Federal Energy Administration) under the Energy Policy and Conservation Act [1]. The DOE seeks continuing assurance that energy-saving programs in industry are approaching established targets. In the case of the chemical industry, the 1980 gross target is a 17% reduction as compared with base-year (1972) performance [2].

For small businesses, increased attention to energy reporting has involved little extra effort. The desired information on usage of fuels, power and other energy forms has been readily extractable from standard cost-accounting records. However, in large multiproduct, multiplant companies, meaningful energy accounting and reporting can be a more difficult and complex proposition.

To develop the necessary details on energy consumption and conservation progress desired by the large company and currently required by the DOE industrial program, computerized reporting is the obvious answer. Having decided upon the mechanized data-processing approach, an opportunity then exists to develop a versatile reporting system to meet sophisticated needs.

One approach to computerized reporting of energy-conservation efforts, based upon our experience in Olin's Chemicals Group, will be described. Our system covers 17 plant locations, with over 150 product or overhead departments, and represents hundreds of production items. The reports from this system provide energy-consumption information on a utility and product (or department) basis, and indicate the percentage improvement of utility usage for a current period over that for a reference or base period.

Requirements of a computer program

How then do we establish a computerized energy-reporting system? We must first define the specific information or answers that are desired, the frequency with which they are to be reported, and the source data that must be supplied to the computer to provide those results. Discussion of these basic needs with EDP (electronic data processing) specialists will disclose that several interdependent programs of instruction will be needed. Such programs would provide for the:

■ Specification of a standard format to be used in tabulating all input or entry information to be processed by the computer. Such data might be keypunched for entry into the computer memory (tapes or disks). Other machine-readable approaches that are available for data entry could also be used.

Each plant would be given identification coding unique to its location and covering individual product departments. All plants would use the standard format in listing output of the manufacturing, utility and overhead departments, the consumption of utilities and energy materials by these departments, and related reference information. The data layout would provide for identification of physical units of measurement, type of department (operating, utility or energy material), period involved, and computer transaction required (i.e., data to be added, replaced or deleted).

■ Conversion of input data into desired answers by using several generalized mathematical equations. Such equations become the core of the overall computer program. Their use in a repetitive sequence of calculations, carried out for every department or product being monitored, is called an algorithm. Specific expressions involved will be described later in this article.

■ Establishment of a reference file of conversion factors and of historical data for previous periods of departmental and plant performance. This must be provided in support of the necessary calculations and the production of desired reports.

■ Determination of appropriate computer instructions (logic) regarding the processing of data, including

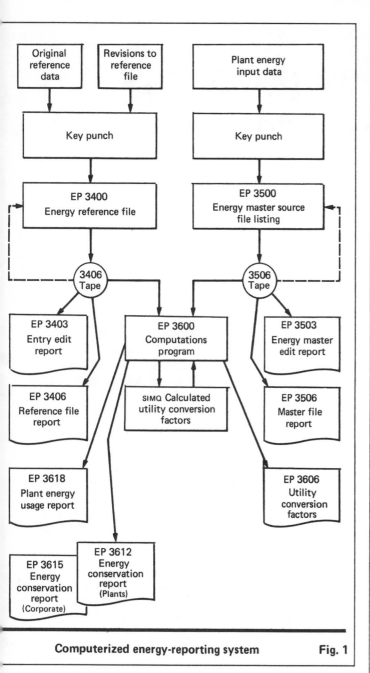

Computerized energy-reporting system   **Fig. 1**

calculations and tabulation of answers in computer printout form. Such reports would normally be prepared monthly, with year-to-date and 12-month cumulative total results as valuable options. The latter type is required in preparing information for the semiannual, voluntary energy-conservation report of the chemical industry to the DOE.

Programming the computer

In the proposed data-processing scheme, the foregoing essentials are established in three primary programs. These are:

Energy Program—Reference File   (EP 3400)
Energy Program—Master Source File   (EP 3500)
Energy Program—Computations   (EP 3600)

The required data format is followed in all three programs. Reference data and accumulating historical

information will be filed in EP 3400 and EP 3500, respectively, for future retrieval. Program EP 3600 will be used to carry out calculations and produce desired reports. A subroutine for calculating variable utility-conversion factors is provided in support of the EP 3600 operation.

There are four primary equations that will be used to describe the energy calculations. Eq. (1) expresses the total energy input consumed by specific operating, utility or overhead departments, and represents the summation of all utilities or energy materials multiplied by their fixed or variable conversion factors. Energy content of feedstocks is ignored, as specified in the DOE reporting procedure [3], where other exceptions are also explained.

$$E_p = \sum_{u=1}^{u=N} Q_u F_u + \sum_{m=1}^{m=N} Q_m F_m \tag{1}$$

where E = energy quantity, 10^6 Btu; Q = quantity of a product, utility or material in appropriate units; F = conversion factor of particular energy form, 10^6 Btu/unit; and the subscripts p = particular product or department, u = generalized utility form (steam, electricity, water, etc.), and m = energy material (oil, gas, coal, etc.).

For example, the energy used in producing a quantity of propylene oxide, $E_p(PO)$, would be:

$$E_p(PO) = Q_s F_s + Q_e F_e + Q_w F_w$$

where the subscripts s = steam, e = electricity and w = water.

Likewise, steam energy might be expressed as:

$$E_s = Q_o F_o + Q_e F_e + Q_{tw} F_{tw} + Q_{s_1} F_{s_1} + Q_a F_a$$

where the subscripts o = fuel oil, tw = treated water, s_1 = steam, and a = air.

Other utilities would be similarly expressed.

Eq. (2) represents variable utility-conversion factors. These are calculated as the sum of all energy inputs to a particular utility, divided by the total quantity of that utility. Examples of such factors include steam, generated electricity, and treated water.

$$F_{u_1} = \frac{1}{Q_{u_1}} \left(\sum_{u=1}^{u=N} Q_u F_u + \sum_{m=1}^{m=N} Q_m F_m \right) \qquad (2)$$

where u_1 denotes specific utility type, and m denotes energy materials.

For example, the treated-water (tw) conversion factor would be:

$$F_{tw} = (Q_s F_s + Q_e F_e + Q_w F_w)/Q_{tw}$$

Distributed steam (where s_1 is steam to make steam) would be:

$$F_s = E_s/Q_s$$
$$= (Q_o F_o + Q_{tw} F_{tw} + Q_e F_e + Q_{s_1} F_{s_1} + Q_a F_a)/Q_s$$

For cooling water, the factor might be:

$$F_w = Q_e F_e / Q_w$$

Energy savings are measured by comparing current-period energy consumption for a particular product to a comparison-base value, E_{pc}. E_{pc} is calculated from the unit energy demand established in the reference period (1972), expressed in Btu/lb, multiplied by the current production rate. This is shown as Eq. (3):

$$E_{pc} = \frac{E_p(Base)}{Q_p(Base)} \times Q_p(Current) \qquad (3)$$

The energy conservation factor or index, $\eta_{E.C.}$, is the fraction or percentage improvement in energy usage achieved during the current period over what would have been required had base-period efficiencies prevailed. Eq. (4) expresses this relationship:

$$\eta_{E.C.} = [E_{pc} - E_p(Current)]/E_{pc} \qquad (4)$$

More commonly, $\eta_{E.C.}$ is expressed as a percentage, i.e., $\eta_{E.C.}$ is multiplied by 100.

The EP 3600 computer program provides for the

Input data for plant energy to computer reporting system   Table I

PGM EP3503   ENERGY MASTER SOURCE FILE UPDATE REPORT   PAGE 16
RUN DATE 03/13/78 TIME 181725

PLANT 062

					— —PRODUCTION— —		— —ENERGY USAGE— —			
ACT	PLT	MO	YR	CSTCNTR	UM	QUANTITY	CC/MTRL	UM	QUANTITY	ERROR MESSAGE
A	062	02	78	CC070	T	2,512.00	CC100	MK	26.00	
A	062	02	78	CC070	T	2,512.00	CC120	MP	889.00	
A	062	02	78	CC070	T	2,512.00	CC130	MG	870.00	

PGM EP 3506   ENERGY MASTER SOURCE FILE LISTING   PAGE 151
RUN DATE 04/13/79 TIME 193925

PLANT 062

				— — —PRODUCTION— — —		— —ENERGY USAGE— —		
PLT	CSTCNTR	UM	MO YR	QUANTITY	CC/MTRL	UM	QUANTITY		
062	CC070	T	0278	2,512.00	CC100	MK	26.00		
		T	0278	2,512.00	CC120	MP	889.00		
		T	0278	2,512.00	CC130	MG	870.00		

(Similar data from previous months tabulated below.)

Report summarizes materials and utilities for plant   Table II

PGM EP3406   ENERGY REFERENCE FILE LISTING   PAGE 2
RUN DATE 02/14/78 TIME 184230   PLANT 052

		RCD–TYPE			OSHA/EPA CREDIT	— — — BASE PERIOD — — —		MMBTU/UNIT
PLT	CC/MTRL	USER/PGM	DESCRIPTION	U/M	(MMBTU)	PRODUCTION (MMLBS)	ENERGY (MMBTU)	FACTOR
052		OD 10			92,500			
052	CC010	OD 20		T		386.6	3,220,128	
052	CC030	OD 20		T		182.8	129,647	
052	CC060	OD 20		T		201.5	7,063	
052	CC070	OD 20		T		2.9	6,687	
052	CC106	OD 20		T		168.6	12,591	
052	CC107	OD 20		T		2.9	2,678	
052	CC116	OD 20		MP		42.3	7,691	
052	CC130	OD 20		MP		45.4	663,322	
052	CC140	OD 20		MP		.0	0	
052	CC210	OD 20		T		.0	0	
052	CC212	OD 20		T		.0	0	
052	CC220	OD 20		MP		4.4	32,798	
052	CC230	OD 20		MP		3.4	33,890	
052	CC260	OD 20		MP		2.6	27,147	
052	CC270	OD 20		MP		2.9	6,618	
052	CC970	OD 20		T		313.3	87,344	
052	CC640	UD 40	ELECTRICITY	MK				
052	CC650	UD 40	TREATED WATER	MG				
052	CC660	UD 40	STEAM	MP				
052	0002B	RM 50	HYDROGEN	MC				.3000
052	1002	RM 50	ELECTRICITY	MK				10.0000
052	1029	RM 50	NATURAL GAS	MC				1.0000
052	1271	RM 50	FUEL OIL NO 6	MG				145.0000
052	1272	RM 50	FUEL OIL NO 2	MG				136.0000

Abreviations used in computer printouts

CC	Cost center
CC/MTRL	Cost center/material
UM	Unit of material
OD	Operating dept.
UD	Utility dept.
RM	Raw material
MC = 10^3 ft^3	
MG = 10^3 gal	
MK = 10^3 kWh	
MP = 10^3 lb	
MM = 10^6	
T = ton	

Conversion factors for utilities consumption at plant　　　　　　　　　　　　　　　Table III

```
PGM EP3606                          ENERGY UTILITY CONVERSION FACTOR REPORT                    PAGE 9
RUN DATE 03/16/78 TIME 003046          PROCESSING PERIOD 02/78 THRU 02/78
                                                PLANT 062
```

		— — — — PRODUCTION — — — —				— — — — — ENERGY CONSUMPTION — — — — —			
PLT	CSTCNTR	DESCRIPTION	DATE	QUANTITY	UM	CC/MTRL	DESCRIPTION	QUANTITY	UM	MM BTU/UNIT
062	CC100	ELECTRICITY	02 78	2,928.00	MK	CC120	STEAM	6,679.00	MP	.0000
						1002	ELECTRICITY	2,627.00	MK	10.0000
062	CC120	STEAM	02 78	18,172.00	MP	003A	HYDROGEN	79,662.00	MC	.3000
						1029	NATURAL GAS	17.00	MC	1.0280
062	CC130	RAW WATER	02 78	43,232.00	MG	CC100	ELECTRICITY	292.00	MK	.0000

```
PGM EP 3606                         ENERGY UTILITY CONVERSION FACTOR REPORT                   PAGE 10
RUN DATE 03/16/78 TIME 003046          PROCESSING PERIOD 02/78 THRU 02/78
                                                PLANT 062
```

PLT	CSTCNTR	UM	MMBTU/UNIT	MESSAGE
062	CC100	MK	11.9741 ⎫	
062	CC120	MP	1.3161 ⎬	(Calculated variable conversion factors)
062	CC130	MG	.0809 ⎭	

solution of the foregoing equations. To summarize, the program converts input energies into a total product (or department) energy; and determines the percentage improvement of that total energy over the base-period energy.

Let us consider the diagram of Fig. 1, tracing the sequence of operations that the computer program will perform.

Reference file (program EP 3400)

Conversion factors—Fixed energy-conversion factors, usually representing the higher heating values of fuels or other energy-containing materials, are one of several groups of data stored in computer reference file EP 3400.

The typical listing of energy-material heating values (conversion factors) that follows was taken from reporting instructions published in the *Federal Register* [3]. The actual heating value of any energy source as purchased and consumed at the plant facility should be used for specific cases.

Energy material	Heating value, 10^6 Btu/unit
Natural gas	1.030/M ft^3
Propane	91.6/M gal
Butane	103.0/M gal
Fuel oil, No. 2	138.7/M gal
Fuel oil, No. 6	149.7/M gal
Coal, bituminous	23.6/t
Electricity, purchased	10.0/M kWh

where M = 1,000

For electric power, 10,000 Btu/kWh is stipulated as the conversion factor rather than 3,412 Btu/kWh. The larger value represents roughly the amount of energy for producing 1 kWh of electricity by a utility. When U.S. Government reporting began, the higher conversion factor was considered preferable by FEA. It continues to be followed in reporting by the petroleum and chemical industries.

Some of the fixed-value conversion factors may vary from location to location, depending upon the specific fuels involved—especially coal. If the heating value fluctuates frequently, it may be preferable for the plant location to report input data in millions of Btu. The conversion factor for such data in the system would then be 1.00.

Comparison-base energy—The actual energy in Btu used by each department and the corresponding production levels during the base year (1972) are also programmed into the EP 3400 reference file. The data are converted into an energy efficiency factor (Btu/lb). In turn, this is used in calculating comparison-base energy values for current reporting periods. The basic input data for 1972 will have to be processed through the EP 3600 program initially to obtain desired energy-reference information.

As new departments or products are added to a plant, energy consumption data for a reasonably stable initial operating period after startup would be similarly added to the reference file. Data on terminated operations should eventually be deleted from the file.

Nonproductive energy—Energy involved in nonproductive operations that are required to meet regulatory requirements may be deducted from plant energy totals if such operations were installed after 1972. This adjustment is in accordance with the DOE reporting procedure. Credit values (in millions Btu/period) attributed to OSHA/EPA are entered into reference file EP 3400 for subsequent use in correcting plant energy-conservation factors. As nonproductive operations are extended, the adjustment values in the reference file must be revised accordingly.

Historical data on energy usage and performance from previous periods constitute the largest reference file (EP 3500). It is maintained for developing year-to-date and 12-month running average information.

Computer in action: energy calculations

Let us assume that the systems-engineering advisers have created the necessary computer system, as de-

Typical plant results for energy consumption Table IV

		QUANTITY		CONV		DEPT PRODUCTION	UNIT CONS
DEPARTMENT--	ENERGY USAGE	USED	UM	FACTOR	(MMBTU)	(MMLBS)	(BTU/LB)
	COST CENTER / MATERIAL						
CC017	CC020 WASTE WATER TREAT	50273.00	MG	.0782	3931.34		
CC017	CC038 TREATED WATER	16407.00	MG	.4385	7194.46		
CC017	CC039 STEAM	105792.00	MP	1.5278	161629.01		
CC017	CC042 ELECTRICITY	1131.00	MK	10.2916	11639.79		
CC017	CC043 RAW WATER	32836.00	MG	.0120	394.03		
CC017	CC044 AIR	21836.00	MC	.0620	1353.83		
CC017	001E RESIDUE GAS	9971.00−	MM	1.0000	9971.00− 181563.31	15.61	11631
CC018	CC042 ELECTRICITY	180.00	MK	10.2916	1852.48		
CC018	CC044 AIR	2264.00	MC	.0620	140.36 1695.40		
CC018	CC046 C. C. W.	135632.00	MG	.0125	5306.89		
CC018	1029 NATURAL GAS	5270.00	MC	1.0070	8995.14	5.22	1721

Column (1) (2) (3) =(2) × (1) (4) (5) =(3)÷(4)

PGM EP 3618 RUN DATE 03/15/78 TIME 082749 — ENERGY COMPUTATIONS REPORT PROCESSING PERIOD 01/78 THRU 02/78 PLANT 075 — PAGE 23

scribed to this point. Trial runs have indicated that proper performance was being obtained with data previously worked up manually.

Let us now follow the process route to be taken by the input-energy data supplied from individual plants. Fig. 1 will facilitate our following the sequence of steps that are specified by the several computer programs.

Data entry

Production and utility quantities and energy usages are keypunched (or entered by other means) into the Master Source File EP 3500. Care is required at this point to assure uniform and precise handling of the data in accordance with the format written into the model.

Inconsistencies, errors or other deviations are automatically flagged by an error message. These are printed by the computer in an edit report (EP 3503) at the entry line that deviates from standard form. The EP 3503 printout is used to proof the accuracy of all input data, prior to carrying out the final computations in program EP 3600. Data entered into the record file are finally printed as Report 3506.

Examples of entries in the EP 3503 and 3506 files are shown in Table I, covering a single product department (listed as cost center, CC 070) for the month of February 1978 at plant location (coded) 062. Current input for February has been added in EP 3506 to data filed for prior months.

Any revisions to constants or other reference data (included in the EP 3400 file) are keypunched into the system at the same time that new energy-usage data are entered into EP 3500. After proofing of such input (in report EP 3403), the EP 3406 printout provides an updated source report. An example of EP 3406 is shown in Table II.

Having now established that current period information has been suitably entered into the computer and that references are updated, we are ready to carry out the calculations programmed in EP 3600.

Calculations for energy usage

Current-period input data on production and utility quantities must be converted into common terms (i.e., millions of lb, millions of Btu) to allow comparison with reference-base information. To accomplish this, the new utility-usage information must first be processed to obtain current variable utility-conversion factors. The computer model does this by using a simultaneous equation matrix or subroutine (SIMQ) that solves simultaneously equations whose generalized form is that of Eq. (2). Up to 10 unknowns can be handled in the SIMQ program used by Olin. When this operation is completed, conversion factors are tabulated in printout Report EP3606. See Table III for sample data tabulation and solution results.

Simpler plants using only those energy forms with fixed conversion factors such as purchased power and fuel gas have no need to calculate variable conversion factors.

The computer next determines total energy contributions from all energy materials and utilities chargeable to each of the production or operating departments. Fixed or variable conversion factors are used to convert various energy quantities to the common energy unit of measurement. Results are printed out as Report EP 3618. An example of typical plant results is shown as Table IV. Note that total energy input to each department is divided by its production quantity to obtain unit energy consumption. Table IV also shows the year-to-date (as of Feb. 1978) results.

In its final manipulation of the data, the computer algorithm obtains the conservation factor for each product center. The steps involved are as indicated by the column notations in Table V. Results are developed by using base-year data from file EP 3406, and current production and energy usage, as recorded in EP 3618. The difference between current energy usage and calculated comparison-base number is energy saved, and is represented as the conservation factor. The calculation is repeated for the total plant as the bottom line in

Conservation factors for each product center of a plant   Table V

PGM EP3612   ENERGY CONSERVATION REPORT   PAGE 7
RUN DATE 03/15/78 TIME 082425   PROCESSING PERIOD 01/78 THRU 02/78
                                                              PLANT 062

COST CENTER	BASE PERIOD 1972 TOTAL PRODUCTION (MMLBS)	BASE PERIOD 1972 TOTAL ENERGY (MMBTU)	BASE RATE UNIT CNSP (BTU/LB)	CURRENT PERIOD TOTAL PRODUCTION (MMLBS)	CURRENT PERIOD TOTAL ENERGY (MMBTU)	CURRENT UNIT CNSP (BTU/LB)	COMPARISON BASE EQUIV (MMBTU)	CONSERV FACTOR (%)	OSHA/EPA ENERGY (MMBTU)	REVISED FACTOR (%)
------OPERATING										
030	342.3	153003	446	52.44	21466.8	409	23391.5	8.23		
040	426.4	3502652	8214	66.41	476236.4	7171	545497.4	12.70		
050	198.2	139107	701	30.95	28896.7	933	21702.8	33.15—		
060	219.5	11221	51	32.38	3027.6	93	1651.5	83.33—		
070	40.3	6081	150	9.87	2408.0	244	1480.7	82.49—		
075	39.9	39181	2235	23.62	16555.4	701	52798.5	68.64		
------INDIRECT										
200 MAINTENANCE	12.0	6245	520	2.00	1256.9	628	1040.0	20.86—		
350 WHSE YARD & OUTSIDE	12.0	8380	698	2.00	1505.5	753	1396.0	7.85—		
360 ADMINISTRATION	12.0	10795	899	2.00	1788.6	894	1798.0	.52		
TOTAL PLANT		3928154			554217.3		650941.8	14.86	1116	15.03
Column	*(1)*	*(2)*	*(3)*	*(4)*	*(5)*	*(5)÷(4)*	*(6) = (3) × (4)*	*(7) = $\frac{(6)-(5)}{(6)} \times 100$*		

Table V (EP 3612), deducting from overall current energy input the OSHA/EPA adjustment.

Totals from individual plant locations are assembled in a final summary report EP 3615, which supplies average performance for the company as a whole, as shown in Table VI.

Problem areas with computer systems

Computer reporting can pose a few problems despite its advantage for the rapid processing of voluminous data. The first is what might be termed the rigidity of the program system. It calls for precise recording and entry of data in accordance with a standard format. Otherwise, data may not be accepted or erroneous results may be obtained. Our experience confirms that, without continued diligence, occasional report aberrations and consequent computer reruns are inevitable.

Second, the computer is not able to distinguish between valid and erroneous data. This is in contrast to manual reporting of energy usage, where the person doing the calculations is able to review the input numbers for consistency. Unless inconsistencies or faulty information are detected by the energy manager during a review of the data listed in the EP 3503 report, final results may be inaccurate.

A third problem relates to the computer program's inability to adjust equitably for variations in production volume or in product mix of a multiproduct department. With regard to volume, the computer model is designed to make a simple adjustment for the difference in volume when calculating the comparison-base energy for each department. The total energies for the current and base periods are thereby placed on an equivalent-volume basis for determining the conservation factor.

This would provide an equitable comparison except for one fact. The total energy of most products is composed of fixed and variable components so that average energy content in Btu/lb varies with the production level. Therefore, without any real energy-conservation action, an increase in production volume could lower the unit energy value (i.e., improve efficiency). Conversely, reduced volume typically yields a lower energy-conservation factor, other things being equal.

However, beware of exceptions—for example, the process where a secondary reaction causing a lower yield is exothermic, with recoverable heat credited to the variable energy of the department. As yield increases, less byproduct heat is available to credit, so the energy-conservation factor may drop.

Variations in the product mix can also influence conservation results, even when no real impact on energy efficiency has occurred. For instance, a department makes a series of polymers having individual energy values ranging from 500 to 5,000 Btu/lb, and averaging 2,000 Btu/lb. Over an extended period, changing market demands may alter this product mix, requiring more of the higher-energy polymers. If average energy content then increases to 2,500 Btu/lb, the conservation factor will drop by (2,500–2,000) 100/2,000 = 25%. To properly compensate, the base-energy information in file EP 3406 must be revised to reflect the altered mix of products.

Computerized reporting has on occasion shown inconsistencies in energy usage that could not be readily explained. In one case, investigation traced the problem to inaccurate 1972 base data, arising from lack of proper metering when the conservation programs were initiated. Here again, revisions to original base data have restored system integrity.

Other problems affecting the accuracy of energy results are covered in reporting instructions of DOE (Form FEA, U524-P-O) [3]. Reference there is made to certain operational changes whose energy impact may unfairly penalize an energy-conservation program.

Such changes are defined as identifiable variations in energy usage attributable to (1) factors beyond control of a company, such as those required by statutes, rules or regulations; (2) policies designed to require use of more-abundant energy (fuel) sources; or (3) natural

Final summary report shows totals for all plants   Table VI

	ENERGY USED (MMBTU)			CONSERV FACTOR (%)	OSHA/EPA ENERGY (MMBTU)	REVISED FACTOR (%)
PLT NAME	BASE PERIOD	CURRENT PERIOD	BASE EQUIVALENT			
031	473400.0	259918.2	197082.8	31.88–	1964	30.89–
052	4237604.0	1956530.5	1943211.6	.69–	53958	2.09
054	16353956.0	10214995.5	12159551.8	15.99	273000	18.24
057	12961.0	7632.7	15876.4	51.92	0	51.92
058	4686495.0	5400706.7	5514609.4	2.07	24378	2.51
061	8537575.0	4619645.9	5397780.0	14.42	30409	14.98
062	3928154.0	1959184.0	2272247.7	13.78	3907	13.95
068	39184.0	23189.9	52691.3	55.99	0	55.99
072	3007660.0	1082075.0	939445.1	15.18–	154875	1.30
075	9415036.0	4894160.0	4502797.9	8.69–	92143	6.65–
078	27394.0	9291.4	17362.4	46.49	0	46.49
085	309588.0	160616.9	243302.5	33.98	0	33.98
TOTAL	56809411.0	33897134.8	36190635.5	6.34	941044	8.94

PGM EP3615 RUN DATE 08/18/78 TIME 081939 — ENERGY – PLANT SUMMARY REPORT PROCESSING PERIOD 00/78 THRU 00/78 — PAGE 1

effects including changes in climate and raw-material quality.

Examples include the conversion to coal from fuel gas, use of lower-grade ore, the OSHA/EPA energy credit, variation in capacity utilization, and changes in the product mix. Also excludable are the abnormal energy demands associated with startup of new operations, until these have stabilized at commercial quality.

Application of the computer report

In addition to providing the basic answers on energy usage and conservation to satisfy minimal company and DOE requirements, the computer-reporting system supplies other data useful to the plant energy manager. For example, trends in monthly data for individual plant areas or products will confirm whether specific conservation projects are meeting objectives. These trends will indicate where operations require closer attention. Calculated variable utility-conversion factors will measure the efficiencies of generating steam or electricity.

By applying another computer subroutine, it is possible to extract and consolidate for further evaluation or reporting the quantities of specific energy materials used by individual plants. Data are taken from source file EP 3506. By combining such data with available pricing information, average costs of energy at various locations can be determined.

When the same product is manufactured at two or more company locations, comparison of their energy values may reveal a need for corrective action at one plant. Investigation may disclose a way for saving energy; or that utilities are being improperly allocated.

A more searching evaluation of energy consumption by individual product departments is possible by plotting monthly energy usage against production volume. The majority of such correlations are represented by straight lines, which can be expressed as:

$$y = mx + b$$

where y represents energy and x represents volume.

Projecting this line to zero production gives a positive intercept, b, on the Y-axis. This intercept is the fixed energy or base load of the operating unit. The slope of the line $m = (y - b)/x$ is a measure of variable energy (Btu/lb). In endothermic reactions, the slope is positive; in exothermic processes, it is negative.

Such analysis of a particular product in one plant, or as produced in different locations, may indicate potential approaches to improving energy efficiency. For instance, better insulation could reduce fixed energy, while optimized reflux ratios in distillation or increased waste-heat recovery could lower variable energy demand. Editor: Steven Danatos

Acknowledgement

The author acknowledges with appreciation the contribution of Olin's Management Systems Development Dept. in implementing the program on which this article is based. Special thanks are extended to B. A. Cunningham and N. Erickson for their assistance.

References

1. Energy Policy and Conservation Act, Public Law 94163, Section 371376.
2. *Federal Register*, p. 29642, June 9, 1977.
3. *Federal Register*, p. 32837, June 28, 1977.

The author

James E. Troyan is manager of energy and conservation at Olin Corporation, 120 Long Ridge Road, Stamford, CT 06904. Previously, he held various managerial positions associated with manufacturing, engineering and purchasing. His work experience includes 15 years in plant management with emphasis on start-up of new operations. He has a B.S. in chemical engineering from Case Institute of Technology, is a member of AIChE, and is a licensed professional engineer in Texas and Connecticut.

Presenting—the energy audit

Evaluating plant operations to see where energy outlays can be cut is now a business. Engineering firms and suppliers of equipment and materials are offering this service, but many companies prefer to do it themselves.

☐ Are your energy costs rising faster and higher than your flue gases? Do an energy audit. Or call in experts to do it for you. In any case, you will be evaluating plant operations and making changes—from putting insulation on a tank to extensively modifying the process—that will achieve specific goals in saving energy. For many firms who now offer this service, it has become the hottest new business in town.

Who does energy audits? A growing number of companies, as the partial listing in the table shows. Some are new firms in the energy consulting field. Others are well-known engineering firms or equipment and materials suppliers (many suppliers provide free audits to those who buy their wares) that have diversified into energy auditing. But regardless of how they got into the business, they all know they have a good thing going.

"Our energy consulting business has grown about 20 to 25% a year in the last four years," says A. W. Hogeland, vice-president and manager of energy engineering for Roy F. Weston Inc. (West Chester, Pa.)—one of the largest companies in the field. Profimatics, Inc. (Woodland Hills, Calif.), which has undertaken about 40 "energy surveys" since 1973, says these now account for a third of its revenues. No wonder some consulting firms have recently changed their names to reflect a new emphasis on energy consulting, says an oil company executive.

HOW THEY OPERATE—Energy audit firms vary in size and scope, ranging from one-person outfits to large engineering groups such as Weston—whose engineering staff of 20 can be boosted to 50 if necessary (the company has about 450 employees).

Company capabilities may be broad or narrow. For example, some specialize in pulp and paper mills because of their geographical location, while oth-

Originally published December 31, 1979

Partial roster of companies offering energy audit services

AES Inc. (Atlanta, Ga.)

Applied Resources Inc. (Medford, Mass.)

Assn. of Energy Engineers (Atlanta, Ga.)

Michael Baker Jr. of New York Inc. (New York, N.Y.)

Black and Veatch Consulting Engineers (Kansas City, Mo.)

Blount International Ltd. (Montgomery, Ala.)

Bovay Engineers, Inc. (Houston, Tex.)

Reese Brentzel, Consulting Engineers (Houston, Tex.)

Breen and Associates Inc. (Arlington, Tex.)

Brown and Caldwell, Consulting Engineers (Walnut Creek, Calif.)

Brown & Root (Houston, Tex.)

Burns and Roe Industrial Services Corp. (Paramus, N.J.)

Richard Bywaters and Associates (Dallas, Tex.)

The Cadre Corp. (Atlanta, Ga.)

Carnahan-Thompson-Delano Inc. (Oklahoma City, Okla.)

Cataudella Associates Inc. (Providence, R.I.)

Centec Corp. (Reston, Va.)

ChemDesign, Inc. (Houston, Tex.)

American Consulting Engineers' Council of Nebraska (Omaha)

Cost Reduction Inc. (Pittsburgh, Pa.)

Robert S. Curl & Associates (Columbus, Ohio)

Du Pont Co.'s Applied Technology Div. (Wilmington, Del.)

Ebasco Services Inc. (New York, N.Y.)

Energy Accounting Systems Inc. (Macon, Ga.)

Energy Appraisal Associates Inc. (Barrington, Ill.)

Energy Associates (North Quincy, Mass.)

Energy Consulting Services (Oak Forest, Ill.)

Energy Engineering Associates (Austin, Tex.)

Energy Resources & Planning (Chicago, Ill.)

Fish Engineering and Construction (Houston, Tex.)

Ford, Bacon & Davis Inc. (New York, N.Y.)

Fuel & Energy Consultants Inc. (New York, N.Y.)

J.F. Gaskill Co. (Pelham, Ala.)

Hoad Engineers Inc. (Ypsilanti, Mich.)

Hudson-McDermott (Houston, Tex.)

Hughes & Susemichel Inc. (Louisville, Ky.)

Raphael Katzen Associates (Cincinnati, Ohio)

King-Wilkinson (Houston, Tex.)

KPFF Consulting Engineers (Seattle, Wash.)

Kruse Associates (Jersey City, N.J.)

Lafayette Engineers (Lafayette, Calif.)

Litwin Corp. (Houston, Tex.)

Love, Friberg & Associates Inc. (Fort Worth, Tex.)

Natkin Energy Management (Englewood, Colo.)

Ottaviano Technical Services Inc. (Melville, N.Y.)

The Pace Company (Houston, Tex.)

Paoluccio Consulting Engineers (Modesto, Calif.)

Planergy Inc. (Austin, Tex.)

Profimatics, Inc. (Woodland Hills, Calif.)

Pullman-Kellogg (Houston, Tex.)

Recon Systems Inc. (Somerville, N.J.)

Robson & Woese Inc. (Syracuse, N.Y.)

Ross & Barazzini Inc. (St. Louis, Mo.)

C.A. Rubio & Co. (Houston, Tex.)

Schipke Engineers Inc. (Minneapolis, Minn.)

Simons Eastern Co. (Atlanta, Ga.)

S.I.P. Engineering and Construction (Houston, Tex.)

J.E. Sirrine Co. (Greenville, S.C.)

Walter F. Spiegel Inc. (Jenkintown, Pa.)

Stanley Consultants Inc. (Muscatine, Iowa)

Waterland, Viar & Associates, Inc. (Wilmington, Del.)

D.M. Weatherly Co. (Atlanta, Ga.)

Roy F. Weston Inc. (West Chester, Pa.)

Zinder Companies, Inc. (Washington, D.C.)

ers concentrate on a particular part of the plant or process. ChemDesign Inc. (Houston, Tex.), for instance, focuses on oil refineries, with special emphasis on improved heat recovery in crude and vacuum units.

How useful is the service provided by energy auditors? Although exact figures on cost savings are not available, interviews with consultants and industry spokesmen indicate that the firms are helping the chemical process industries (CPI) save about $1 billion/yr through projects completed over the past four or five years. Weston alone claims to have saved its clients at least $80 million through audits of some 60 process plants over the last four years.

A general rule of thumb is that an energy audit will probably prove beneficial for any plant built five or more years ago, when the engineering economics related to energy were quite different from what they are today. A case in point is the recent installation of insulation in several low-temperature tanks at the Paramount refinery of Douglas Oil Co. of California. Chief engineer Joe E. Pearson notes that the insulation would not have been cost-effective several years ago.

DOING IT YOURSELF—Often enough, when an audit seems the best thing to do, many CPI companies prefer to tackle it themselves instead of bringing in a consultant.

"There are mixed feelings about this," says Ray Doerr, manager of energy conservation for Monsanto Co. (St. Louis), which does most of its own energy auditing, but has used independent auditors on occasion. "Some feel our own technical people are more qualified because they know the process better, but others think it's good to bring in an outsider who doesn't have preconceived ideas."

Doerr adds that the independent audits of Monsanto plants have resulted in some worthwhile recommendations.

Milton A. Williams, president of Energy Engineering Associates (Austin, Tex.), believes that an auditor should not be brought in until a company has first done the obvious "housekeeping jobs." Williams, formerly energy conservation manager with the Chemicals and Plastics Div. of Union Carbide Corp. (New York City), also warns that while the recommendations from an audit may promise substantial savings, their implementation usually requires a large capital investment.

Representatives of operating companies and consultants generally seem to agree that energy auditors do not usually offer any ideas that a plant's own engineers could not come up with. However, one of the main considerations is that plant engineers cannot take the time from their regular duties to spend weeks or perhaps months on an energy audit. Also, the consultant is a specialist with varied experience, and his perspective is different from that of the plant's own engineers.

"Consultants come in with one objective, so they are able to get things done a lot quicker," observes Palmer C. Fusilier, manager of Mobil Oil Corp.'s Torrance (Calif.) refinery. "Also, they look at the whole system, whereas most of our people look at just the equipment they are working with."

Roger O. Pelham, vice-president of Profimatics, agrees: "We are not really bringing in anything that is technically new, but we bring in experience and an organized and fresh approach, plus the talent and the tools to do the job."

Profimatics' tools include special computer programs. The company also sells software that a plant may use to do its own analyses. ChemDesign also has its own computer programs, including a proprietary heat-transfer optimization system (HOTS).

HOW TO PICK AN AUDITOR — Selection of an appropriate auditor may be the most important step, since a good choice will likely minimize subsequent problems. Those who have employed auditors say they have sought advice from friends in other companies, trade and professional associations, and other consultants whose opinions they value. Advice may also be obtained from local offices of the U. S. Dept. of Energy and state energy offices.

Another source of information is the Assn. of Energy Engineers (Atlanta, Ga.), which was formed in September 1977 with the goal of developing energy engineering as a profession. Albert Thumann, executive director, says the association now has some 2,500 members, of whom about 38% work in industry, 20% are consultants, and 20% are with engineering and construction firms, and utilities.

Industry spokesmen stress that consultants seeking to do a plant audit should be asked for references on jobs they have done in similar plants. If the review is favorable, preliminary work may then start.

"The first thing we do is arrange a one-day visit for a plant walk-through and discussion with the client to see where he stands on energy costs," says Weston's Hogeland. "Then we sort out what appears cost-effective and establish a work schedule and cost of the audit. We feel it is important for a client to know how much he will spend."

THE COST INVOLVED — An audit may take from as little as a few weeks of one person's time to months of work by an engineering team, depending on the size of the plant and the scope of the work. Weston's biggest job—the analysis of a number of units in a large petrochemical complex—took about 10 months and cost around $340,000. However, typical costs range from a minimum of about $25,000 to around $150,000.

Even so, the bill for an audit is relatively small change compared to the cost of implementing its energy-saving recommendations. One refinery manager says his company has so far allocated roughly $500,000 for capital expenditures as a result of an audit that cost $50,000.

But the bottom line is the saving achieved. Energy consultants report savings that average 15-25% of their clients' energy costs. Payback times average two years or less, although many companies are now making investments that have a longer payback time—in anticipation of higher energy costs. ChemDesign says most of its clients are now interested in projects with a four- or five-year payback.

"The areas of biggest savings vary," says Profimatics' Pelham, "but we usually tackle heat exchangers first. Then we go heavily after the steam system and try to look at the total steam, more or less like a process. Next we look at the furnaces and check firing efficiency." Of course, an auditor's approach to a project may vary, depending on his preference or specialties, the nature of the job, or the client's needs.

Familiarity with plant processes is obviously an asset for the energy consultant, and some use their knowledge to suggest process changes that could improve plant efficiency. "We don't take for granted that something runs the way it ought to run," says Pelham. "We have been very successful in shutting down distillation columns, and in partially bypassing units."

ChemDesign says it can do process-design work for most refinery operations. The company uses its computer software to do fractionation studies and to split streams into different flows for greater efficiency. And Bovay Engineers, Inc. (Houston) recently changed a process to eliminate the need for hot water in rinse operations.

THE ANTI-AUDIT FACTION — Some companies feel that their own engineers are more qualified to do analyses than outsiders, or in any case, prefer that energy auditors not interfere with process operations. And there are some dissatisfied customers. A spokesman for a chemical company that had an audit done at a large complex complains that "we already knew about a lot of the things they came up with." Also, an audit of an oil refinery produced recommendations that were "so conventional, like putting economizers on heaters, when we were looking for something innovative," says an oil company spokesman. That particular audit came free from a supplier, but, adds the executive, it required "reams of data" that had to be gathered by plant personnel.

Others say they have benefited from audits by equipment and materials suppliers and point out that opinions may be obtained from more than one supplier.

One critic of a particular plant audit admits that while it produced few innovations, it provided a useful second opinion on "things we already knew about. In some cases this helps because it gives more substance to what you were thinking."

Gerald Parkinson

Forecasting energy requirements

It is possible to predict energy needs in a chemical plant with a simple equation based on the relationship between energy consumption and production rate.

William P. Jacobs, Jr., Monsanto Textiles Co.

☐ The basic equation is modified to allow for seasonal effects, energy conservation efforts and any variation in energy sources. Coefficients for the equation are calculated with the aid of a nonlinear regression analysis of historical data. Predictions may then be made for future periods, with a fair degree of accuracy.

Traditional methods

The rapidly escalating cost of purchased energy makes accurate forecasting of energy requirements a more and more urgent matter for the chemical process industries (CPI). However, a large number of variables affect the product energy rate (Btu/lb of product) for a given plant, and engineers have traditionally had difficulty in producing accurate (or, in many cases, credible) forecasts.

The plant energy director, therefore, may be caught in a difficult predicament. Conservative forecasts (high energy predictions) will be greeted with skepticism by plant management. On the other hand, overly optimistic forecasts (low energy predictions) may lead to subsequent disturbing budget variances.

The traditional method of energy rate forecasting involves first calculating the current rate, and then calculating the net effect of energy conservation projects to be brought online during the forecast period. This second figure is subtracted from the first to arrive at a forecast value. But, accuracy suffers when changes in production rates are incorporated into the forecast, and the method breaks down when forecasts for specific periods (e.g., September of the next calendar year) are required.

However, it is possible to forecast energy requirements with a fair degree of accuracy. The method that I will describe in the following paragraphs has been used successfully at our plant and, with minor modifications, should be adaptable to others. It takes the form of an energy rate equation. The inputs required for a prediction are the anticipated production rate, and the time period to be forecast. The coefficients for the equation require only historical energy data from the plant, and a nonlinear regression program for data correlation.

The energy consumption equation

The energy forecasting model itself fits together contributions from various sources to the overall energy-consumption rate. These are:
1. The relationship between production and energy rate.
2. Seasonal effects.
3. Energy conservation efforts.
4. Energy sources (electricity, fossil fuel, coal, natural gas, and so on).

Each of these factors is discussed in detail below.

The relationship between production and energy rate—In almost all CPI applications, the plant energy rate will be heavily influenced by the production rate. The model that I have found to be most useful is a simple linear function:

$$E_p = A_1 + A_2 P \qquad (1)$$

where:
E_p = Energy requirement of production, Btu
A_1 = Constant, fixed load, Btu
A_2 = Proportionality constant
P = Production, lb

Then, $E_p' = E_p/P$, where E_p' = energy rate, Btu/lb.

For a plant that produces more than one product, the function would appear in the form:

$$E_p = A_1 + A_2 P_1 + A_3 P_2 + \cdots A_{n+1} P_n \qquad (2)$$

where n = number of product classifications.

Check out the validity of this model if you have comparable figures on energy consumption versus production.

Seasonal effects—In plants where a substantial percentage of the total energy consumption goes to heating and air conditioning, a cyclical pattern will be apparent in the historical data. In our plant, the pattern has been accentuated by the fact that electricity is used to power cooling systems, whereas fossil fuel provides steam for heating. Electrical consumption peaks in the summertime, and fossil-fuel usage peaks during the winter months. Some type of sine function is required to model this pattern.

Originally published March 9, 1981

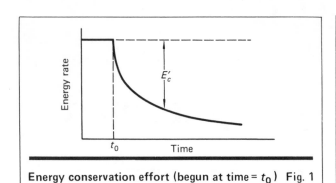

Energy conservation effort (begun at time = t_0) Fig. 1

One particularly good seasonal model [1] is:

$$E_s = A_3 \cos (2\pi t/12) + A_4 \sin (2\pi t/12) \qquad (3)$$

where:

E_s = seasonal energy requirement, Btu
A_3, A_4 = constants estimated by the regression
                program
$\tan^{-1} [(A_4)/(A_3)]$ = phase shift
$(A_3)^2 + (A_4)^2$ = amplitude
Period = 12, for a 12-mo cycle
t = the time period for which the forecast is being
    made, entered as a number (e.g., 108 =
    December 1980, 109 = January 1981, etc.).

Then, $E_s' = E_s/P$, where E_s' = seasonal contribution to energy rate, Btu/lb.

Energy conservation efforts—Most companies support an active program aimed at energy conservation. Although these programs vary widely in form and content, the aim of most of them is to (a) generate an atmosphere of energy awareness that will encourage conservation at all organizational levels, and (b) identify and implement energy conservation projects.

What effect has the conservation effort on the energy rate? While the form of the model may vary to fit an individual situation, there should be a decline in the energy rate with time. The experience at our plant is probably typical. Our formal energy-conservation program began after the 1973 energy crisis. Big, obvious energy-conservation projects were executed first. After these were implemented, smaller projects, or those with less-favorable economic returns, were implemented. As the energy rate was brought down, each successive increment of reduction became more difficult to achieve. This suggested a declining exponential function, as shown in Fig. 1.

The actual form of the function used in the model is one of the type:

$$E_c = A_5(1 - e^{-A_6 t}) \qquad (4)$$

where:

E_c = the energy contribution of the conservation effort, Btu
A_5, A_6 = proportionality constants
t = the time period for which the forecast is being made

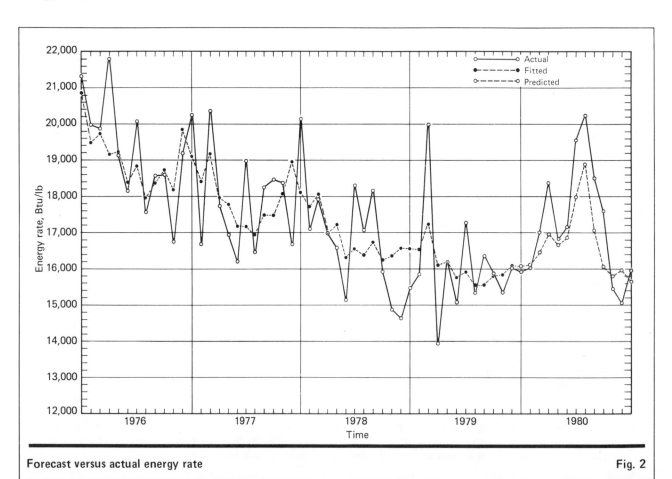

Forecast versus actual energy rate

Fig. 2

Then $E_c' = E_c/P$, where E_c' = energy conservation contribution to the energy rate, Btu/lb.

Energy sources—The energy mix concerns plants that use different sources of energy supply. In our plant, three sources are used—purchased electrical power, natural gas, and residual fuel oil.

The solution to the energy mix problem can be simple and effective. In our situation, I combined natural gas and residual fuel oil under the classification "fossil fuel." While there is a combustion efficiency difference, these two fuels are essentially alternates for each other in our operation. Where the availability of either is in question, a forecast of the actual mix of gas and oil is extremely unreliable. I forecast electrical energy consumption separately since electricity and fossil fuel are not alternates for our plant.

To summarize, formulate a combined model for alternate energy sources whenever the actual mix of these sources cannot be accurately predicted, and formulate separate models for those energy sources that have no alternates.

Fitting the pieces together—The energy consumption model can now be formulated by summing the separate parts:

$$E = A_1 + A_2 P + A_3 \cos(2\pi t/12) + A_4 \sin(2\pi t/12) + A_5(1 - e^{-A_6 t}) \quad (5)$$

Again, the energy rate may be derived by dividing energy, E, by production, P. Thus, $E' = E/P$.

In the case where energy mix must be addressed, there will be two (or more) equations. For our plant, for instance:

$$E_{elec} = A_1 + A_2 P + A_3 \cos(2\pi t/12) + A_4 \sin(2\pi t/12) + A_5(1 - e^{-A_6 t}) \quad (6)$$

$$E_{fossil} = B_1 + B_2 P + B_3 \cos(2\pi t/12) + B_4 \sin(2\pi t/12) + B_5(1 - e^{-B_6 t}) \quad (7)$$

and
$$E = E_{elec} + E_{fossil} \quad (8)$$

Correlating the data

Let us assume that you have now formulated a model similar to the one shown in Eq. (5). Where do you go to get all those coefficients? A nonlinear regression program will fit the model to your historical data and generate the coefficients for you.

The data should be representative of normal operations. If, for example, a wall were torn out for an expansion, causing a large air-conditioning loss during a particular period, the data for that period would simply be thrown out. Forecasting for periods in which production is predicted to be far outside the range represented within the historical data is to be done cautiously—if at all.

Most regression programs will output statistics, enabling you to determine how well your model fits your data. In the case of the equations developed for our plant, squared correlation coefficients were 0.68 and 0.78 for the fossil-fuel and electricity cases, respectively. These numbers indicate that the equations explain 68% and 78% of the variation in the data. This fit, while not exceptional, is good, and it does allow forecasting with a greater degree of confidence than that obtainable with less-sophisticated methods.

Using the model

Using the energy forecasting model is simple. The predicted production for the period in question is entered into the equation, and the period is entered as a number corresponding to its calendar time. The equation and its coefficients may easily be entered into a program for a calculator or a computer. Properly constructed regression models of this type may be rated as follows:

1. Good to very good for short term (1–3 mo) and medium term (3 mo–2 yr) forecasts.

2. Good for long-term (2 yr and over) forecasts [2].

One last word of caution—the energy rate forecast is only as good as the production forecast. One should, therefore, obtain the most accurate production forecast possible.

A model of this sort is not meant to be constructed and forgotten. As new data are collected, these should be fed into the correlation program for generation of new coefficients. The model is thus continuously kept up to date.

Fig. 2 shows energy data for our plant.

The solid line shows actual values of energy rate over the period 1976–1979.

The dashed line with solid points is the energy consumption equation (8) where real production rates have been used and the coefficients were generated with a nonlinear regression program, using the actual energy rate over the period 1972–1979.

These same coefficients were used to produce the dashed line with open points for 1980, using real production figures. This shows the predictive value of the model—the coefficients obtained from a previous time still allow the equation to approximate the actual figures for this later period.

Acknowledgement

In data correlation and model formulation, the assistance of A. W. Dickinson, Monsanto Science Fellow, is gratefully acknowledged.

References

1. McClave, J. T., and Benson, G., "Statistics for Business and Economics," Dellen Pub. Co., San Francisco, 1978.
2. Chambers, Mullick, and Smith, "How to Choose the Right Forecasting Technique," *Harvard Business Review*, 49, No. 4 (July-Aug. 1971).

The author

William P. Jacobs, Jr., is Process Modeling Supervisor for Monsanto Textiles Co. in Greenwood, S.C. His previous assignments with Monsanto included process control, quality control, energy conservation, and pollution control. Jacobs received a B.S.Ch.E. from Clemson University in 1970. He holds M.E.Ch.E. and M.B.A. degrees from the University of South Carolina. He is a Registered Professional Engineer in South Carolina and a member of AIChE.

Estimate energy consumption

Energy's spiraling cost has spurred interest in determining its consumption in processing. A correlation derived via regression analysis offers a means of comparing process energy efficiencies.

Jude T. Sommerfeld, Georgia Institute of Technology, and **Richard H. White**, *Cities Service Co.*

☐ Energy costs associated with the manufacture of industrial chemicals have risen sharply in recent years, and have become the subject of close scrutiny. Energy-conservation programs have been instituted by most companies, and in many cases energy savings of 15-20% or more have been realized.

Published data on energy consumed in chemicals manufacturing does exist, but only for the more important chemical commodities, and many of these data do not agree well. Essentially no information has been published on energy consumed in the manufacture of many important chemical products, such as specialty chemicals.

Possible energy-consumption correlations

It would be useful to be able to estimate at least roughly, from a minimum of basic information, the energy consumed in the manufacture of a new chemical product (or of an established product via a new process). This would be particularly valuable in the earlier stages of new-process development and evaluation.

Again, little of such information has been published. Several such correlations have been presented by Bridgwater [1]; parameters in these correlations include the plant's capital cost and the number of its functional units.

Originally published November 19, 1979

from heat of reaction

With the latter parameter, these correlations are somewhat similar to those presented by Allen and Page [2] for making pre-design estimates of plant capital costs.

Bridgwater suggests that the inclusion of other, more relevant parameters, such as a thermodynamic function, would improve his correlations. In this connection, a correlation was proposed some years ago to relate the selling price of industrial chemical commodities to their free energies of formation [3].

More recently, the concept of a recycle index of a chemical commodity was described [4]. This index is related to the recycle potential of a product, and takes into consideration the total material and energy consumed in its manufacture.

A likely candidate for correlation with specific energy consumption in chemical manufacturing is the thermodynamic heat of reaction for the primary reaction in a process. Logically, processes involving endothermic reactions will consume more energy than those based upon exothermic reactions. Of course, a large number of non-reactive, energy-consuming steps in the process (pumping, heat exchange, separations, compression, etc.) could significantly distort such a correlation.

Another point in favor of using heat of reaction in such a correlation is the ready availability of the required data. Heat-of-formation data (as opposed to free energies and entropies of formation) have been tabulated for a large number of chemical compounds, or else can be estimated from the bond structure and composition of the molecule. Thus, if a reasonable correlation with specific energy consumption could be found, it would be readily applicable to a large number of existing and proposed chemical processes.

Energy-consumption data sources

Because of the dynamic changes that have recently been occurring in energy usage in chemical processing, it was necessary for correlation purposes to select published data on energy consumption that have appeared over a rather short time span. Such data were thus extracted from: "Faith, Keyes and Clark's Industrial Chemicals" [5], "Practical Techniques for Saving Energy in the Chemical, Petroleum and Metals Industries" [6], and *Hydrocarbon Processing*'s "1975 Petrochemical Handbook Issue" [7]. Each of these reports energy consumption data for 20 to 30 chemical processes.

The energy consumption data presented in these works are generally in terms of different energy sources (e.g., electricity, natural gas, fuel oil, coal, steam, etc.). It has, therefore, been necessary to convert these data into a common energy unit of consumption, such as Btu/lb of production. The energy conversion factors of Sittig [6] were chosen for this purpose; these are listed in Table I. Note that in Table I the theoretical value of 3,412 Btu per kWh of electrical energy was used in order to avoid any double accounting.

The computation of a specific energy consumption is illustrated below for the manufacture of hydrogen fluoride by the acidification of calcium fluoride with sulfuric acid. On the basis of 1 metric ton (2,205 lb) of production of hydrogen fluoride, the values of consumption of various types of energy and their common equivalents are reported as [5]:

Type of energy	Amount	Equivalent, 1,000 Btu
Steam, 150 psig	1,750 kg	4,610
Electricity	220 kWh	750
Refrigeration	3.6×10^5 kcal	1,430
Fuel	2.3×10^6 kcal	9,130

Thus, the total amount of energy required to produce one metric ton of hydrogen fluoride via this process is equal to 15,920,000 Btu. The specific energy consumption (in Btu/lb of product) for this process is then equal to: 15,920,000 Btu/2,205 lb = 7,220 Btu/lb.

Similar calculations were performed for other chemical products used in the development of the correlation.

The energy value of raw materials is not directly included in the determination of specific energy consumption. It is, instead, incorporated into the correlation via the heat of reaction.

The heat of reaction was determined, via classical thermodynamics, from the heats of formation of the compounds participating in the reaction. Standard heats of formation at 298K and corresponding to the state of the material (solid, liquid, gas) at this temperature and at atmospheric pressure were selected for this purpose. The rationale behind this assumption was that probably most raw materials are received and products are delivered at ambient temperatures and low-to-moderate pressures. Also, heat effects associated with phase change, sensible

(text continues on p. 45)

Energy conversion factors		Table I
Type of energy or material	**Unit of measure**	**Btu equivalent**
Fuel oil	Barrel	6,287,000
Coal:		
Anthracite	Short ton	25,400,000
Bituminous	Short ton	26,200,000
Natural gas	1,000 ft^3	1,032,000
Coke	Short ton	26,000,000
Electrical energy	kWh	3,412
Steam, low-pressure	Pound	1,150
Steam, high-pressure	Pound	1,500

Data points used in the construction of the correlation

Product	Reactants	Standard heat of reaction (DHR) at 298K, Btu/lb	Specific energy consumption (SEC), Btu/lb			Deviation, Btu/lb
			Reference	Reported	Estimated	
Acetaldehyde	Ethylene and oxygen	-2,135	[5]	2,239	-936	3,175
			[6]	1,500	-936	2,436
			[7]	1,457	-936	2,393
Adhesives and sealants	Mixing and blending operations only	0	[6]	1,300	5,088	-3,588
Aluminum sulfate	Alumina and sulfuric acid	-280	[6]	2,420	4,168	-1,748
Ammonium nitrate	Ammonia and nitric acid	-783	[5]	2,197	2,613	-416
Bisphenol-A	Phenol and acetone	-145	[6]	7,810	4,606	3,204
			[7]	7,140	4,606	2,534
sec-Butyl alcohol	n-Butene and water	-329	[6]	5,250	4,011	1,239
			[7]	5,338	4,011	1,327
Chlorine (and caustic soda)	Brine	2,710	[5]	18,260	16,022	2,238
			[6]	17,000	16,022	978
Chlorine	Hydrogen chloride and oxygen	-614	[7]	100	3,121	-3,021
Cyclohexane	Benzene and hydrogen	-1,049	[6]	0	1,843	-1,843
			[7]	1,188	1,843	-655
Dimethyl terephthalate	Terephthalic acid and methanol	26	[6]	6,250	5,175	1,075
			[7]	5,128	5,175	-47
Ethanol	Ethylene and water	-412	[6]	6,290	3,747	2,543
			[7]	5,346	3,747	1,599
Ethyl acetate	Acetaldehyde	-714	[7]	1,802	2,819	-1,017
Ethylbenzene	Ethylene and benzene	-461	[5]	51	3,594	-3,543
Ethylene oxide	Ethylene and oxygen	-1,009	[5]	3,018	1,956	1,062
			[6]	-660	1,956	-2,616
Formaldehyde	Methanol and oxygen	-2,337	[6]	-400	-1,387	987
			[6]	-1,730	-1,387	-343
			[7]	-1,895	-1,387	-508
			[7]	-1,828	-1,387	-441
Hydrogen fluoride	Calcium fluoride and sulfuric acid	602	[5]	7,220	7,199	21
			[6]	6,680	7,199	-519
Hydrogen peroxide	Water	4,813	[5]	29,570	27,044	2,526
Isopropanol	Propylene and water	-376	[6]	8,190	3,861	4,329
			[7]	4,417	3,861	556
Maleic anhydride	Benzene and oxygen	-8,240	[6]	-7,635	-5,557	-2,078
			[7]	-5,859	-5,557	-302
Methyl acrylate	Acrylic acid and methanol	-489	[7]	6,390	3,506	2,884
Nitric acid	Ammonia and oxygen	-3,046	[5]	-554	-2,811	2,257
Oxygen	Air	0	[5]	3,293	5,088	-1,795
			[6]	1,460	5,088	-3,628
n-Paraffins	Kerosene	0	[7]	2,878	5,088	-2,210
Polyethylene (high-density)	Ethylene	-1,457	[6]	3,970	730	3,240
			[7]	1,893	730	1,163
			[7]	2,016	730	1,286
			[7]	402	730	-328
Polyethylene (low-density)	Ethylene	-1,457	[7]	2,367	730	1,637
			[7]	3,550	730	2,820
Polypropylene	Propylene	-877	[6]	2,940	2,337	603
			[7]	4,272	2,337	1,935
Polystyrene	Styrene	-289	[6]	2,250	4,139	-1,889
Polyvinyl chloride	Vinyl chloride	-660	[6]	6,020	2,982	3,038
			[7]	1,355	2,982	-1,627
			[7]	706	2,982	-2,276
Potassium chloride	Sylvinite	0	[5]	1,538	5,088	-3,550
Potassium hydroxide (and chlorine)	Potassium chloride and water	1,713	[5]	11,290	11,572	-282
Printing inks	Mixing and blending operations only	0	[6]	1,500	5,088	-3,588
Sodium carbonate	Sodium chloride and calcium carbonate	49	[6]	6,500	5,253	1,247

Table II

Product	Reactants	Standard heat of reaction (DHR) at 298K, Btu/lb	Specific energy consumption (SEC), Btu/lb			Deviation, Btu/lb
			Reference	Reported	Estimated	
Sodium silicate	Sodium carbonate and silica	275	[5]	3,076	6,030	−2,954
			[6]	5,000	6,030	−1,030
Sodium sulfate	Natural brines	7	[5]	2,560	5,111	−2,551
Styrene	Ethylene and benzene	11	[6]	7,020	5,125	1,895
			[7]	4,842	5,125	−283
Sulfuric acid	Sulfur, oxygen and water	2,305	[5]	9	−1,317	1,326
			[6]	−980	−1,317	337
Trichlorofluoro-methane	Carbon tetrachloride and hydrogen fluoride	98	[7]	1,526	5,419	−3,893
			[7]			
Urea	Ammonia and carbon dioxide	−954	[7]	1,337	2,114	−777
				1,171	2,114	−943
p-Xylene	Mixed xylenes	0	[6]	1,670	5,088	−3,418

Data points *not* included in the construction of the correlation

Table III

Product	Reactants	Standard heat of reaction (DHR) at 298K, Btu/lb	Specific energy consumption (SEC), Btu/lb			Deviation, Btu/lb
			Reference	Reported	Estimated	
Acrylic acid	Propylene and oxygen	−4,121	[6]	2,900	−4,488	7,388
			[6]	4,300	−4,488	8,788
			[7]	3,726	−4,488	8,214
Dimethyl terephthalate	p-Xylene, methanol and oxygen	−2,995	[6]	4,170	−2,717	6,887
			[7]	4,011	−2,717	6,728
			[7]	11,410	−2,717	14,127
Hydrogen	Coke and steam	9,620	[5]	94,900	60,557	34,343
Hydrogen	Propane and steam	7,988	[5]	136,500	47,881	88,619
Isoprene	Acetone, acetylene and hydrogen	−1,359	[7]	6,789	989	5,800
Lactic acid	Lactonitrile, sulfuric acid and water	−433	[7]	11,965	3,681	8,284
Phenol (and acetone)	Cumene and oxygen	−1,700	[6]	5,860	106	5,754
Phosphoric acid	Calcium phosphate, silica, coke and oxygen	−3,753	[5]	6,965	−3,979	10,944
Phosphorus	Calcium phosphate, silica and coke	9,978	[5]	22,130	63,516	−41,386
			[6]	23,790	63,516	−39,726
Polyacrylonitrile	Acrylonitrile	−624	[7]	29,870	3,091	26,779
Polyethylene (low-density)	Ethylene	−1,457	[6]	7,550	730	6,820
Polyisoprene	Isoprene	−460	[6]	17,048	3,597	13,451
Sodium (and chlorine)	Sodium chloride	7,691	[5]	25,590	45,718	−20,128
Sodium carbonate	Sodium chloride and calcium carbonate	49	[5]	18,900	5,253	13,647
Sodium chlorate	Sodium chloride	3,661	[5]	8,668	20,732	−12,064
Terephthalic acid	p-Xylene, oxygen and acetaldehyde	−4,354	[6]	1,540	−4,776	6,316
			[7]	2,709	−4,776	7,485
Tetrahydrofuran	Maleic anhydride and hyrdrogen	−1,895	[7]	9,592	−373	9,965

Correlation-based estimates of specific energy consumption

Product	Reactants	Standard heat of reaction (DHR) at 298K, Btu/lb	Estimated specific energy consumption (SEC), Btu/lb
Acetic acid	Acetaldehyde and oxygen	−2,276	−1,253
Acetic anhydride	Acetaldehyde and oxygen	−2,432	−1,592
Acetone	Isopropanol	518	6,894
Acetylene	Calcium carbide and water	−995	1,997
Acrylonitrile	Propylene, ammonia and oxygen	−5,526	−5,808
Adipic acid	Cyclohexanone, cyclohexanol and oxygen	−2,420	−1,566
Allyl chloride	Propylene and chlorine	−637	3,052
Ammonia	Nitrogen and hydrogen	−1,167	1,513
Ammonium chloride	Sodium chloride and ammonium sulfate	−56	4,901
Ammonium phosphate	Phosphoric acid and ammonia	−754	2,700
Aniline	Phenol and ammonia	−201	4,424
Aniline	Nitrobenzene and hydrogen	−2,571	−1,884
Aniline	Chlorobenzene and ammonia	−119	4,692
Benzoic acid	Toluene and oxygen	−2,406	−1,536
Benzyl chloride	Toluene and chlorine	−593	3,186
n-Butyraldehyde	Propylene, carbon monoxide and hydrogen	−887	2,308
Calcium carbide	Calcium oxide and coke	3,102	17,909
Calcium cyanamide	Calcium carbide and nitrogen	−1,550	488
Calcium hydroxide	Calcium oxide and water	−378	3,855
Calcium oxide	Calcium carbonate	1,364	10,132
Carbon disulfide	Methane and sulfur	700	7,560
Carbon monoxide (and hydrogen)	Coke and water	2,693	15,942
Carbon monoxide	Methane and oxygen	−548	3,324
Chlorobenzene	Benzene and chlorine	−499	3,476
Choline chloride	Trimethylamine, hydrochloric acid and ethylene oxide	−902	2,265
Cumene	Benzene and propylene	−396	3,798
Cyclohexanone (and cyclohexanol)	Cyclohexane and oxygen	−2,604	−1,952
Diphenylamine	Aniline and hydrogen chloride	−428	3,698
Dodecene	Propylene	−631	3,070
Epichlorohydrin	Allyl chloride, hypochlorous acid and caustic soda	−1,187	1,457
Ethyl acetate	Ethanol and acetic acid	−19	5,024
Ethyl chloride	Ethylene and hydrogen chloride	−478	3,541
Ethyl ether	Ethanol	−66	4,868
Ethylene	Ethane	2,101	13,245
Ethylene carbonate	Ethylene oxide and carbon dioxide	−666	2,964
Ethylenediamine	Ethylene dichloride and ammonia	−2,202	−1,087
Ethylene dichloride	Ethylene and chlorine	−945	2,140
Ethylene glycol	Ethylene oxide and water	−818	2,510
Ethylenimine	Monoethanolamine, sulfuric acid and caustic soda	−313	4,062
2-Ethylhexanol	n-Butyraldehyde and hydrogen	−797	2,572
Ethyl mercaptan	Ethanol and hydrogen sulfide	−432	3,685
Formic acid	Carbon monoxide and water	−265	4,216
Glycerol	Epichlorohydrin, caustic soda and water	−800	2,564
Hydrazine	Ammonia and sodium hypochlorite	−2,740	−2,227
Hydroxylamine	Nitric acid and hydrogen	−6,146	−6,076
Hydroxylamine sulfate	Hydroxylamine and sulfuric acid	−412	3,748
Melamine	Urea	1,627	11,211
Methanol	Carbon monoxide and hydrogen	−1,720	57

Table IV

Product	Reactants	Standard heat of reaction (DHR) at 298K, Btu/lb	Estimated specific energy consumption (SEC), Btu/lb
Methylamine	Methanol and ammonia	-333	3,998
Methyl chloride	Methane and chlorine	-478	3,541
Methyl chloride	Methanol and hydrogen chloride	-352	3,938
Methyl methacrylate	Methanol, hydrogen cyanide, sulfuric acid and acetone	-985	2,025
Methylethyl ketone	sec-Butyl alcohol	414	6,520
Methylisobutyl ketone	Acetone and hydrogen	-766	2,664
Monochloroacetic acid	Acetic acid and chlorine	-552	3,312
Monoethanolamine	Ethylene oxide and ammonia	-1,249	1,287
Nitrobenzene	Benzine and nitric acid	-509	3,445
Nonene	Propylene	-562	3,281
Perchloroethylene (and trichloroethylene)	Ethylene dichloride and chlorine	-1,063	1,804
Phosgene	Carbon monoxide and chlorine	-480	3,535
Phthalic anhydride	o-Xylene and oxygen	-3,758	-3,986
Polybutadiene	1,3-Butadiene	-580	3,226
Propylene glycol	Propylene oxide and water	-524	3,398
Propylene oxide (and t-butyl alcohol	Propylene, i-Butane and oxygen	-4,375	-4,800
Salicylic acid	Phenol, caustic soda, carbon dioxide and sulfuric acid	-731	2,769
Sodium bicarbonate	Sodium carbonate, carbon dioxide and water	-218	4,368
Sodium hypochlorite	Caustic soda and chlorine	-599	3,168
Trichloroethane	Ethylene dichloride and chlorine	-342	3,970
Vinyl acetate	Ethylene, acetic acid and oxygen	-1,019	1,928
Vinyl chloride	Ethylene dichloride	742	7,716
Vinylidene chloride	Trichloroethane	272	6,019

heating and compression are generally negligible in comparison with most heats of formation.

Standard heats of formation for the more common inorganic and organic chemicals can be found in many chemistry and chemical engineering handbooks and thermodynamics textbooks. For organic compounds, however, one of the most thorough compilations of heats of formation (for approximately 5,000 compounds) has been assembled by Stull, Westrum and Sinke [8]; this compilation was used extensively in this work. Heats of polymerization reactions were obtained from Rodriguez's "Principles of Polymer Systems" [9].

Heats of formation and reaction are generally tabulated in metric molar units. Because the specific energy consumptions were determined in units of Btu per lb of product, it seemed reasonable to represent the heats of reaction in the same units. Consequently, it was necessary to convert the units of the tabulated values into the desired units, using molecular weights and stoichiometric coefficients.

The conversion of the heat-of-reaction units is illustrated below, again using the production of hydrogen fluoride from calcium fluoride as the example. The reaction in this process proceeds according to the following equation.

$$CaF_{2(s)} + H_2SO_{4(l)} \rightarrow 2HF_{(g)} + CaSO_{4(s)} \qquad (A)$$

The heats of formation for the participants in this reaction are tabulated below:

Compound	Standard heat of formation at 298K, kcal/g-mol
$CaF_{2(s)}$	-290.3
$H_2SO_{4(l)}$	-193.91
$HF_{(g)}$	-64.2
$CaSO_{4(s)}$	-342.42

The heat of this reaction is thus equal to: $2(-64.2) - 342.42 - (-290.3 - 193.91) = 13.39$ kcal/g-mol, or 24,100 Btu/lb-mol. The molecular weight of hydrogen fluoride is equal to 20.01. Thus, the thermodynamic heat of this reaction is: [24,100 Btu/lb-mol]/[2(20.01 lb/lb-mol)] = 602 Btu/lb product.

Again, similar conversion calculations were performed on the heats of reaction for the other chemical products used in the development of the correlation.

Development of the correlation

A total of 89 data points were extracted from the three sources previously cited. For a number of chemical products, there was some multiplicity of reported energy consumption values; that is, for a given chemical product, there could be one or more such values (often considerably different) reported in each of the three sources.

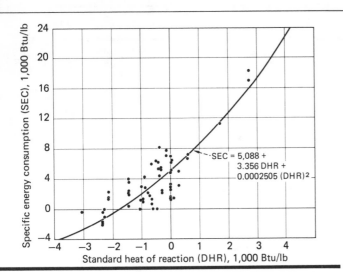

Raw data on specific energy consumption correlated with standard thermodynamic heat of reaction at 298 K

These data points were first analyzed by plotting specific energy consumption (SEC) vs. thermodynamic heat of reaction (DHR) (both in Btu/lb). It appeared that a concave-upward curve might reasonably fit these data points. Hence, it was decided to correlate these data via a simple second-order polynomial expression of the form:

$$SEC = A_0 + A_1 \times DHR + A_2 \times (DHR)^2 \qquad (1)$$

Standard least-squares regression analysis methods were used in this correlation procedure.

The final correlation that was developed was based upon 66 (or 75%) of the original 89 data points. The values of the constants A_0, A_1 and A_2 in this correlation are, respectively, 5,088, 3.356 and 0.0002505. Both SEC and DHR in Eq. (1) are in Btu/lb. The 66 data points, along with the estimated values of specific energy consumption and deviations for each data point, are listed in Table II. These data points and the correlation equation itself are presented in the figure.

The standard deviation for the 66 data points is equal to 2,141 Btu/lb. The maximum deviation is equal to 4,329 Btu/lb, which corresponds roughly to two standard deviations. Although this deviation is greater than one would like, it certainly is reasonable in view of the inherent scattering of the raw data. (Observe the differences in the reported energy consumption values for the same chemical product manufactured via the same process, such as ethylene oxide, formaldehyde, isopropanol, oxygen, high- and low-density polyethylene, polyvinyl chloride, sodium silicate and styrene in Table II.)

Although not evident from the figure, it can also be shown that this second-order polynomial correlation equation exhibits a minimum specific energy consumption (equal to −6,152 Btu/lb) when the heat of reaction is equal to −6,699 Btu/lb; that is, this correlation would then begin to predict larger values of specific energy consumption for more-exothermic chemical reactions. This behavior results from the simple fact that any second-order polynomial, such as in Eq. (1), with positive coefficients will always exhibit a minimum heat of reaction, namely at DHR equal to $-A_1/2A_2$.

In this particular case, the minimum is most likely spurious, and results from the scarcity of data points in the vicinity of the minimum. Indeed, the only data points used in this correlation for which DHR is more negative (and which, therefore, lie to the left of the minimum) are those for maleic anhydride (the most exothermic process listed in Table II). The inclusion of more data points (if they were available) in this region would probably have shifted the minimum farther to the left, perhaps to a large negative value of DHR corresponding to extremely exothermic and unrealistic chemical reactions.

Why data points were excluded

The 23 (of the original 89) data points that were eventually excluded in the development of this correlation are summarized in Table III. It can be seen from this table that each of these data points exhibits an error of at least two standard deviations (2 × 2,141 = 4,282 Btu/lb) from the correlation. This behavior served as the primary basis for excluding these points.

Some of the factors that might lead to this behavior are the following.

It is apparent from Table III that many of the processes listed actually consist of a sequence of two or more chemical reactions (e.g., dimethyl terephthalate, phenol, phosphoric acid, tetrahydrofuran). Also, in most of these cases, the large deviation from the correlation is positive; that is, the reported specific energy consumption far exceeds the estimated value. Thus, better agreement could perhaps be obtained if the specific energy consumption for that process or part of a process associated with a single reaction in the sequence could be correlated with the heat of that reaction. Unfortunately, energy consumption figures are generally not reported in such fashion for complex multiple-reaction processes.

The foregoing point can be illustrated in greater detail, using the manufacture of phenol (and co-product acetone) from cumene and oxygen as an example. The overall chemical reaction for this process is:

$$C_6H_5C_3H_7 + O_2 \longrightarrow C_6H_5OH + CH_3COCH_3 \qquad (B)$$

For Reaction (B), the standard thermodynamic heat of reaction at 298K is equal to −1,700 Btu/lb of phenol produced. From Table III, the estimated specific energy consumption for this overall process is only 106 Btu/lb, versus a reported value of 5,860 Btu/lb [6].

In fact, however, this process consists of two reactions in sequence. In the first of these, cumene hydroperoxide is formed from the peroxidation of cumene with oxygen:

$$C_6H_5C_3H_7 + O_2 \longrightarrow C_6H_5(CH_3)_2COOH \qquad (C)$$

This reaction is followed by the decomposition of cumene hydroperoxide into phenol and acetone products:

$$C_6H_5(CH_3)_2COOH \longrightarrow C_6H_5OH + CH_3COCH_3 \quad (D)$$

The heat of Reaction (C) is equal to −303 Btu/lb of cumene hydroperoxide produced. The correlation-based estimate of the specific energy consumption for this first reaction step in the process is then 4,094 Btu/lb of cumene hydroperoxide.

Similarly, the heat of Reaction (D) is equal to −1,210 Btu/lb of phenol produced. In this case, the correlation predicts a specific energy consumption value of 1,394

Btu/lb of phenol. It will be assumed that Reactions (C) and (D) proceed in quantitative fashion. Thus, the number of pounds of cumene hydroperoxide that need to be made in order to produce one pound of phenol is merely equal to the ratio of their molecular weights, namely, $152.19/94.11 = 1.617$. The total energy consumption required for the manufacture of one pound of phenol from cumene and oxygen, using the correlation-based estimates for the two steps of the process, is then:

$$SEC = (1.617)(4,094) + 1,394 = 8,015 \text{ Btu/lb of phenol}$$

This is admittedly higher than the reported value of 5,860 Btu/lb for the overall process. Nonetheless, in comparison with the value of 106 Btu/lb obtained from the correlation with consideration of only the overall reaction, this value of 8,015 is certainly much closer to the reported value. Similar results would be expected from decomposition of the other multiple-reaction processes in Table III into their individual reaction steps.

Estimate energy consumption

This correlation, Eq. (1), was used to estimate the specific energy consumption (SEC) associated with manufacture of 71 important chemical commodities for which reported data were not readily available. These 71 materials, along with the reactants that denote their mode of manufacture and standard heats of reaction, are listed in Table IV. Readers with specific energy consumption data on some of these materials may wish to compare their data with these correlation-based estimates. An attempt was made to minimize the inclusion of complex multiple-reaction processes in Table IV.

An interesting observation may be made from the various data in Tables II, III and IV: Most of the basic or starting materials for the chemical industry are manufactured via significantly endothermic processes with associated high energy consumptions (reported or estimated). Examples here would include calcium carbide, calcium oxide, chlorine, ethylene, hydrogen, phosphorus and styrene. Thereafter, the vast majority of derived downstream chemical products are manufactured via exothermic processes.

Possible improvements

There is no question that any reliable procedure for estimating energy consumption in the manufacture of chemical products should be based upon thermodynamic quantities, such as the heat of reaction. The major defect in the present correlation, as reflected in its relatively high standard deviation, is the fact that it takes no account of process complexity. Thus, for example, for either a process with no chemical reaction or a process with a reaction but little or no reaction heat (DHR $\cong$ 0), this correlation results in a specific energy consumption (SEC) value of 5,088 Btu/lb of product. This value is probably reasonable for complex separation processes, such as in a petroleum refinery, air separation plant or minerals processing facility. On the other hand, this value is probably high for a process consisting of just several steps and no reactor, such as simple mixing and blending operations.

Considerable improvement of this correlation could

then be made by somehow incorporating process complexity, such as Bridgewater has done but without thermodynamic considerations [1]. If process complexity were to be characterized by the number of major processing units, such a correlation would then have similar characteristics to the method of Allen and Page for pre-design estimates of capital costs of chemical plants [2]. Thus, such an improved correlation might have a form similar to:

$$SEC = (SEC)_0 (N/N_0)^k \qquad (2)$$

In Eq. (2), $(SEC)_0$ would represent the specific energy consumption corresponding to a "standard" processing operation with a total of N_0 major processing units (e.g., 10 to 20), and would be evaluated from an expression of the form of Eq. (1); N would be equal to the actual number of major units, and k would be a constant. In this connection, a tabulation of the number of major units for most or all of the processes listed in Tables II, III and IV would be very useful.

References

1. Bridgwater, A. V., How to Control Costs, *The Chemical Engineer*, Nov. 1973, p. 538.
2. Allen, D. H., and Page, R. C., Revised Technique for Predesign Cost Estimating, *Chem. Eng.*, March 3, 1975, p. 142.
3. Sommerfeld, J. T., and Lenk, C. T., Thermodynamics Helps You Predict Selling Price, *Chem. Eng.*, May 4, 1970, p. 136.
4. Mathur, P. K., and Russell, T. W. F., Chemical Recycle Can Save Energy, *Hydrocarbon Proc.*, July 1978, p. 89.
5. Lowenheim, F. A., and Moran, M. K., "Faith, Keyes and Clark's Industrial Chemicals," 4th Ed., Wiley, New York, 1975.
6. Sittig, M., "Practical Techniques for Saving Energy in the Chemical, Petroleum and Metals Industries," Noyes Data Corp., Park Ridge, N.J., 1977.
7. "1975 Petrochemical Handbook Issue," *Hydrocarbon Proc.*, Nov. 1975, pp. 97-224.
8. Stull, D. R., Westrum, E. F., and Sinke, G. C., "The Chemical Thermodynamics of Organic Compounds," Wiley, New York, 1969.
9. Rodriguez, F., "Principles of Polymer Systems," McGraw-Hill, New York, 1970.

The authors

Jude T. Sommerfeld, professor of chemical engineering at Georgia Tech (Atlanta, GA 30332), teaches process control, distillation, and reactor and process design, and has research interest in energy conservation. An industry consultant, he has had eight years of engineering and management experience with Monsanto Co. and BASF-Wyandotte Corp. A member of A.I.Ch.E., A.C.S., I.S.A. and N.S.P.E., he is a registered engineer in the State of Georgia. He received a B.Ch.E. degree from the University of Detroit, and M.S.E. and Ph.D. degrees in chemical engineering from the University of Michigan.

Richard H. White is a chemical engineer at Cities Service Co.'s Lake Charles Refinery (P.O. Box 1562, Lake Charles, LA 70602), where he works in the process automation department. He received a bachelor's degree in chemical engineering from Georgia Tech.

Energy

Analyses for the energy efficiency of a plant, a unit operation or a piece of process equipment are readily performed by using a single concept, "exergy." Here is the theoretical background for this concept and practical energy-optimization examples both for low-temperature separations in ethylene plants and for turboexpanders and compressors.

Victor Kaiser, Technip

☐ Common to all development in the chemical process industries (CPI) is the best utilization of the energy input to the process. To deal with such a requirement, the engineer has to select the methods that will yield a systematic approach to the problem.

Directly related to energy conservation are all of the following:
■ Yields of main products.
■ Recovery of main products.
■ Specific energy consumption.
■ Investment.
■ Operability.
■ Maintenance.

Resolving these six factors is our objective when treating an energy optimization problem. A new process or machine is a true improvement if we can show benefits in all of the above factors or, at least, in the sum of them.

The weakest link is the investment. We generally integrate this factor by conversion to monetary values. However, it is not certain that these values are representative of their energy and raw-material contents such as would be the case in a stable economy. Studies are conducted to evaluate such contents [1].

To be stressed in this article are operability and maintenance, which are equivalent to plant availability. These are of prime interest when a complete process study is performed. An improvement can easily prove illusory if these factors are not carefully evaluated.

Tools for the study

Exergy is the best single concept on which to base our energy-efficiency analysis, and is defined as *the maximum technical work that can be derived from a fluid or a system*. This concept has the advantage of being quite universal because it can be applied to chemical processes, combustion processes, biochemical or photochemical processes, and so on, as well as to physical systems. The exergy function is defined as:

$$E = (H - H_o) - T_o(S - S_o) \qquad (1)$$

It is important to notice the presence of a reference state, which is one advantage of the exergy function. It is evident that our conclusion about a specific process must depend on its environment—specifically on the cooling-medium temperature, T_o.

All CPI processes are open to the surroundings from which they receive feeds and energy and to which they discharge products, energy and effluents. Hence, the quality of these surroundings is of utmost importance to the process with respect to technical performance.

As long as we deal with physical transformations that

Originally published February 23, 1981

optimization

leave the molecular structure unchanged, we need only three parameters to define the surroundings:

1. Reference temperature, T_o. We assume this to be constant, no matter how much heat is exchanged with the surroundings. No heat can be exchanged with the surroundings at another temperature.

2. Reference pressure, P_o. This pressure is also assumed to be constant and independent of the amount of material exchanged with the surroundings.

3. Reference gravity, g_o. For processes where gravity forces are important, the acceleration due to gravity has to be set as a reference. For CPI plants, however, this is generally of no importance.

The three parameters are sufficient as long as we do not intend to study the dilution of chemical substances in the surroundings or extraction of these substances from natural resources. For example, in an air-separation plant, we will have to set a reference air composition. This leads to the necessity of defining all reference compositions in the surroundings of substances we deal with. The reference state has a large impact on the process evaluation.

We might fear that some fundamental difficulties would prevent us from finding unbiased reference states, leading to a lack of generality and an excessive sensitivity of the results toward the reference parameters. Fortunately, this is not true. Looking at the exergy function, Eq. (1), we notice that the reference values H_o and S_o will always cancel if we study exergy differences rather than absolute exergy values—the differences being calculated for steady-state systems. For such "flow systems" (i.e., open systems), this means that the material balance of each chemical species has to be maintained at all times without formation or loss or accumulation. This is true for industrial processes as long as chemical reactions are excluded. Any datum value can be selected for H_o and S_o for each pure compound separately, and the choice will have no influence on the exergy differences calculated.

If chemical reactions are involved, it is necessary to work with absolute entropies. Also, the datum, S_o, for each pure compound has to be selected in a manner so as to reflect the industrial reality. For example, in a combustion process where carbon dioxide is one of the products, the datum could be based on the ambient temperature and partial pressure typically encountered in combustion processes. This choice is not important if we compare such processes among themselves. In this situation, we can use the standard values for H_o and S_o (ideal state at standard pressure and temperature), and obtain precise conclusions about the value of each process relative to the others.

If the dilution of carbon dioxide in the surrounding air would be the process of interest, then another datum point should be selected, such as the condition of the carbon dioxide in nature.

From the preceding discussion, we can comprehend that the exergy concept always leads us to very carefully consider the technical reality related to the process.

Exergy analysis cannot be fully applied in the absence of computer programs that calculate the thermodynamic properties of mixtures and pure compounds over wide pressure and temperature ranges. For specific unit operations, the exergy fluxes can often be calculated from the defining parameters of the unit operation, with a minimum of calculations of physical properties of the material flows to and from the system. A few examples extensively used in the analysis of cryogenic separations will be given in this article.

The exergy balance

No exergy-conservation law applies for real systems. The exergy balance for a steady-state open system can be written as:

$$\Sigma E_{\alpha i} + \Sigma W_j = \Sigma E_{\omega k} + L \tag{2}$$

Eq. (2) expresses that: The sum of all incoming exergy due to material streams plus the sum of all work exchanged with the surroundings (+ sign for incoming work, − sign for outgoing work) is equal to the sum of all outgoing exergy due to material leaving the system, plus the lost work or exergy gap, L. L is also called the irreversibility, and represents the irretrievably lost work. It is no physically existing form of energy, but rather a measure of the amount of potential work destroyed by the system due to irreversible processes.

There are a few important rules that apply to the signs of the various functions. These rules are all deduced from the definition of exergy:

$$\text{If } P \geqslant P_o, \text{ then } E \geqslant 0 \tag{3}$$

Lost work, L, is always positive, so:

$$L > 0 \tag{4}$$

The actual exchanged work can be positive, negative or zero:

$$W_j \gtrless 0 \tag{5}$$

By using Eq. (4), we can introduce a concept (generally called effectiveness or reversibility) that measures how far we are from an ideally reversible system.

For each process or unit operation, the total supplied work and/or exergy must be calculated. Then, this total work can be compared to the exergy effectively received

Factors relevant to energy consumption		Table I
(Basis: 200,000-metric-ton/yr ethylene plant, naphtha cracking)		
Yield of olefins and aromatics	65-68	% by weight
Recovery of main products	99-99.5	%
Ethylene yield	30-35	% by weight
Total specific-energy input	7-8	kWh/kg of ethylene
Total compressor power	0.8-1	kWh/kg of ethylene
Heat of reaction, overall	1.5	kWh/kg of ethylene
Investment/kg of annual capacity	2-2.5	FF/kg of ethylene
Operability, major turnovers	2-3	yr
Availability, long term	90-95	%
Maintenance	3	%/yr of investment

by the processed stream and/or the work produced. This will become clear in the examples to be given. In general terms, it is possible to define effectiveness.

Looking at Eq. (2), we can define the effectiveness for an overall system if we focus the analysis on the technical work exchanged:

For $\Sigma(E_{\alpha i} - E_{\omega k}) > 0$,

$$\eta = -\Sigma W_j/\Sigma(E_{\alpha i} - E_{\omega k}) = 1/[1 - (L/\Sigma W_j)] \quad (6a)$$

For $\Sigma(E_{\alpha i} - E_{\omega k}) < 0$,

$$\eta = -\Sigma(E_{\alpha i} - E_{\omega k})/\Sigma W_j = 1 - (L/\Sigma W_j) \quad (6b)$$

Eq. (6a) and (6b) state that the effectiveness is the useful work or the exergy produced, divided by the work or exergy supplied to the system.

For example, if the purpose of the process is to upgrade the exergy of stream n (to the exclusion of any other stream), then:

$$E_{\alpha n} - E_{\omega n} < 0$$

Eq. (6b) applies (provided that $i, k \neq n$). Hence, the effectiveness, η, becomes:

$$\eta = -(E_{\alpha n} - E_{\omega n})/[\Sigma W_j + \Sigma(E_{\alpha i} - E_{\omega k})] \quad (7a)$$

$$\eta = 1 - L/[\Sigma W_j + \Sigma(E_{\alpha i} - E_{\omega k})] \quad (7b)$$

Eq. (6) and (7) are quite similar in mathematical form, but Eq. (6) applies to an entire system, whereas Eq. (7) puts stream n in particular focus.

Effectiveness, η, is always less than unity, but can be zero or negative. This simply means that the lost work can be larger than or equal to the quantity of useful work produced.

Let us now look at certain unit operations to find generally applicable relationships for calculating both the lost work and the effectiveness without calculating the exergy fluxes, E, of the various material streams.

Heat exchange

Here, the following relationships are important:

$$\Delta E_H = \Delta Q - T_o \int_{out}^{in} dQ/T = \Delta Q[1 - (T_o/T_{mH})] \quad (8)$$

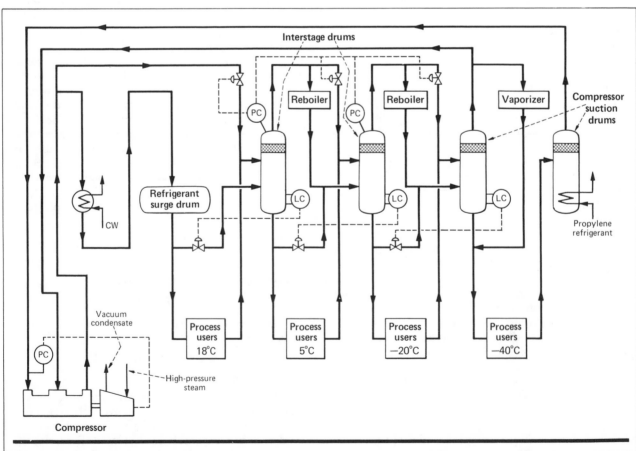

Typical propylene-refrigerant cycle   Fig. 1

For the hot side of the heat exchanger, differences are calculated as inlet minus outlet conditions, so:

$$\Delta Q > 0 \qquad (9)$$

If $T_{mH} < T_o$, then $\Delta E_H < 0$, $\Delta E_C > 0$, and the cold-side relationship is:

$$\Delta E_C = -\Delta Q[1 - (T_o/T_{mC})] \qquad (10)$$

Combining these relationships, we obtain the overall equation for the lost work as:

$$L = \Delta E_H + \Delta E_C = \Delta Q T_o \Delta T_m/(T_{mC})(T_{mH}) > 0 \qquad (11)$$

Based on Eq. (7b), we determine the effectiveness:

$$\eta = 1 - (L/\Delta E_C) = -\Delta E_H/\Delta E_C \qquad (12)$$

Closed-loop refrigeration cycle

For a closed-loop refrigeration cycle, effectiveness can be defined as the ratio between all outgoing exergy fluxes through the refrigerant users and the compressor power. By analogy to Eq. (6b), this can be written as:

$$\eta = -\Sigma\Delta E_{Ri}/W \qquad (13a)$$
$$L = \Sigma\Delta E_{Ri} + W \qquad (13b)$$

where:

$$\Delta E_{Ri} = +\Delta Q_i[1 - (T_o/T_{Ri})], \text{ and } \Delta Q_i > 0 \qquad (14)$$

ΔE_{Ri} is the exergy flux of the i-*th* cold producer at the refrigeration temperature, T_{Ri}. If the refrigerant is not

Nomenclature

E	Exergy flow, W
H	Enthalpy flux, W
L	Lost work, W
M	Mass flow, kg/s
P	Pressure, Pa
Q	Heat flux, W
S	Entropy flux, W/°K
T	Temperature, °K
V	Specific volume, m³/kg
W	Power, technical, W
δ	Recovery factor
ε, η	Effectiveness

Subscripts
C	Cold side of exchanger
H	Hot side of exchanger
i, j, k	Sequence numbers
m	Integral average
o	Reference state
R	Refrigeration
s	Isentropic
α	Inlet conditions
ω	Outlet conditions

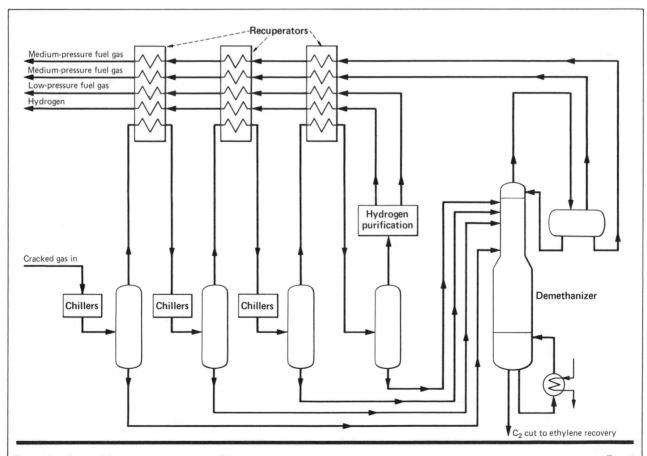

Demethanizer with pure-component refrigerant Fig. 2

boiling at constant temperature, then T_{Ri} is an integral average temperature.

This relationship is very useful because η is independent of T_R for all practical purposes, and depends only on the type of refrigeration cycle, number of flash tempera- tures, pressure drops, and the compressor's polytropic efficiency.

Pressure drop in piping

It is useful to know the relationship for the lost work due to the pressure drop during isothermal flow in piping or equipment:

$$L = MV(T_0/T)\Delta P \qquad (15)$$

This lost work can be added to other items if a detailed study is made for an overall system. For heat exchangers, this term should be added to the lost work due to temperature approaches. In this way, it is possible to pinpoint the relative value of these various losses. Hence, decisions whether to lower the pressure drop and raise the exchanger area (and consequently lower the temperature difference) are put into a more realistic light.

Turboexpander

A turboexpander delivers shaft work and refrigeration, obtained from the cold-material stream leaving the machine. The effectiveness can be given separately for the two contributions. The results are:
Lost work overall:

$$L = (W - W_s)T_0/T_m \qquad (16)$$

Isentropic efficiency of the expander:

$$\eta_w = W/W_s \qquad (17)$$

Characteristics for a mixed refrigerant	Table II

Component	Composition, mole fraction
Methane	0.15
Ethylene	0.50
Propylene	0.35
Total	1.00

Heat of vaporization (−35/−40°C)   389 kJ/kg
Heat of subcooling (−35/−105°C)   170 kJ/kg
Molecular mass 31.5 kg/k-mole

Specific compression power for 1 MW   789.7 kW/MW
of process duty and no recuperators
($\delta = 0$)

For a flow of 100 k-mole/h, the vaporization curve of this refrigerant can be represented by:

Q = $aT^2 - bT + c$
a = 0.00009505, (MW)(K^{-2})
b = 0.02692, (MW)(K^{-1})
c = 1.8626, MW

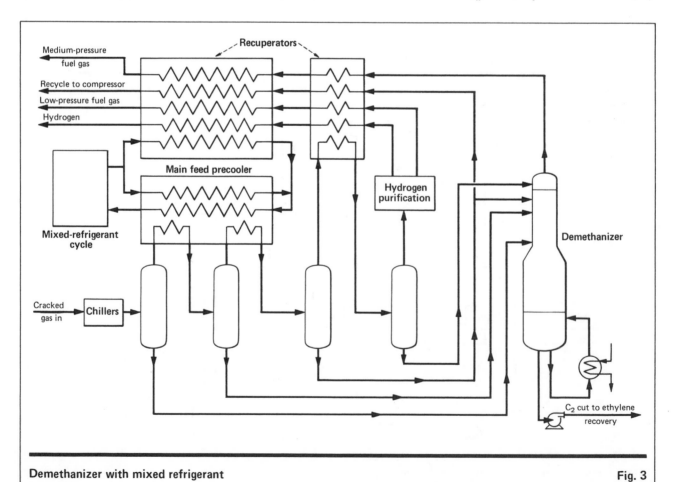

Demethanizer with mixed refrigerant   Fig. 3

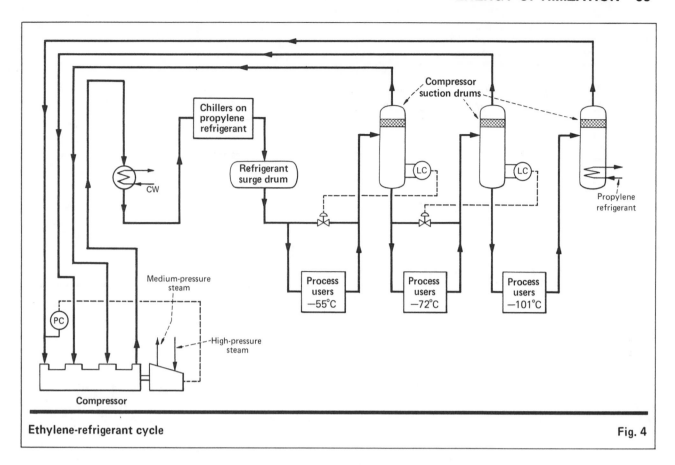

Ethylene-refrigerant cycle

Fig. 4

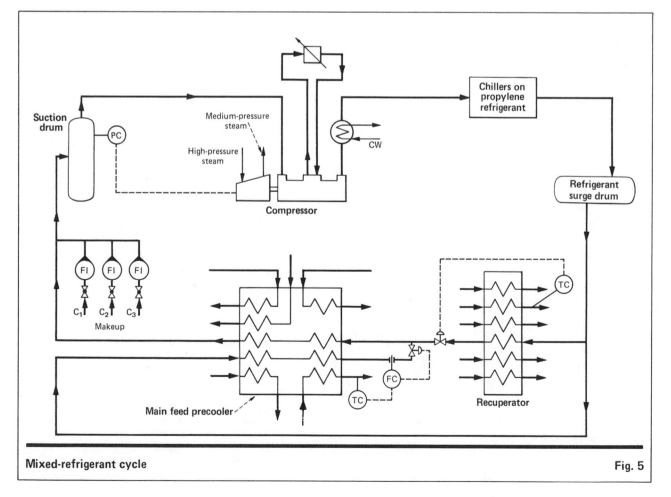

Mixed-refrigerant cycle

Fig. 5

Comparison of ethylene and mixed-refrigerant cycles at same duties and temperatures				Table III
Temperature, T		Heat flux,	Ethylene refrigerant	Mixed refrigerant
°C	K	ΔQ, MW	energy, ΔE, MW	energy, ΔE, MW
−101	172	0.2474	0.1999	0.1626
− 72	201	0.2194	0.1201	0.1060
− 55	218	—	—	—
Totals		0.4668	0.3200	0.2686
$\Delta E/\Delta Q$			0.6855	0.5754
$\dfrac{\Delta E\ (M)}{\Delta E\ (E)}$			0.84	

ΔE (M) represents energy difference for mixed refrigerant.
ΔE (E) represents energy difference for ethylene refrigerant.
Reference temperature, T_o, = 311K

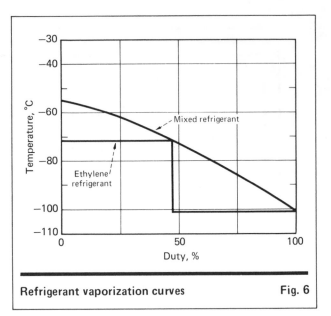

Refrigerant vaporization curves   Fig. 6

Refrigeration-power effectiveness:

$$\eta_R = 1 + (1 - \eta_w)[1 - (T_o/T_m)] \qquad (18)$$

Overall effectiveness, η, in accordance with Eq. (6a) is the shaft power versus input exergy, or:

$$\eta = \eta_w/(2 - \eta_R) \qquad (19)$$

Expander efficiency, η_w, is identical to the isentropic efficiency of the machine. Eq. (17), (18) and (19) clearly show that at low temperatures even a good value for η_w (such as 0.75) results in a modestly low figure for η_R. Once again, this reveals the difficulty of reaching good effectiveness at low operating temperature.

Compressor

Let us use the same symbols for the compressor as for the turboexpander. Here, s denotes exit conditions after an isentropic and adiabatic compression to the same pressure. We find the following relationships. (For the lost work, we will use Eq. (16).)
Isentropic efficiency:

$$\varepsilon_w = W_s/W \qquad (20)$$

Heating effectiveness:

$$\varepsilon_H = 1 + (1 - \varepsilon_w)[1 - (T_o/T_m)] \qquad (21)$$

Overall effectiveness:

$$\varepsilon = \varepsilon_w + \varepsilon_H - 1 \qquad (22)$$

If $T_m > T_o$, $\varepsilon_H > 1$. This means that the real machine has downgraded work to heat, which is not available from the ideal isentropic reference machine. Potentially, some work can be recovered from this heat. Of course, ε is always less than unity.

Heat leaks

Heat leaks, ΔQ, from a stream or equipment can be shown to produce the following lost work:

$$L = -\Delta Q[1 - (T_o/T_m)] \qquad (23)$$

where T_m is the integral average temperature of the

stream or fluid inside the equipment. We can convince ourselves that L is positive:

If $T_m < T_o$, then $\Delta Q > 0$, $L > 0$
If $T_m > T_o$, then $\Delta Q < 0$, $L > 0$

In such a way, it is possible to analyze many unit operations. However, not all can be reduced to general parameters. A rectifying column has to be defined by the incoming and outgoing material streams, so its "exergy balance" must use the enthalpy and entropy values of these streams. This is true whenever concentration gradients or steps are encountered.

Summary of exergy-balance analysis

For each processing step, it is possible to calculate the lost work by applying an "exergy balance."

Presenting a process analysis that shows minimum required work, lost work and actually exchanged work gives a very clear picture of the performance of each section. Examples of such analysis are presented in Ref. [4], [5] and [6].

The effectiveness can be used to check the performance of a processing step with the yardstick familiar to rotating-machine specialists. For very complex processes, effectiveness is not a versatile tool because overall effectiveness does not relate in a simple manner to the effectiveness of each part. On the other hand, the lost work in this case is very useful because the contribution of each part can simply be added up to yield the overall lost work, in quite the same way as with the real shaft-work.

Application to an ethylene unit

The ethylene production unit, as a key element in petrochemical production, is worth investigating in the terms discussed here. Table I summarizes typical values for the basic factors to be considered for energy optimization. In order to analyze such global figures, we can follow the method of Ref. [5], where application of the exergy analysis to the complete unit is discussed, to find areas of low effectiveness. Such a work tends only to highlight evident causes of low effectiveness, and is not

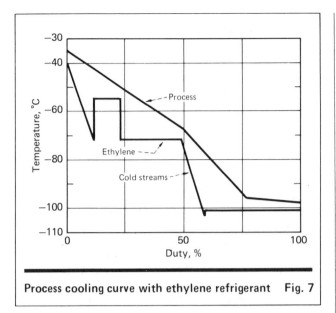

Process cooling curve with ethylene refrigerant   **Fig. 7**

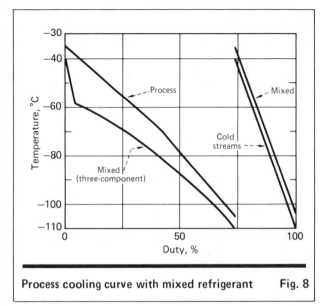

Process cooling curve with mixed refrigerant   **Fig. 8**

very rewarding. On the contrary, for comparing alternative processing schemes, exergy is the best tool to use.

Table I shows that the values of the several factors are not equally satisfactory—recovery being excellent, and yields and selectivity less favorable. For yields, we should, however, consider the kinetic optimum in relation to the technological constraints.

The investment value is in itself not very meaningful. When studying process modifications, we can, however, integrate the investment increments into optimization formulas.

Regarding the specific energy consumption, we notice that the useful energy (the heat of reaction) represents only one-fifth of the total energy input. One-tenth represents compressor power used for cryogenic gas separation. However, considering the conversion from thermal to mechanical energy, close to two-fifths of the energy inputs are really for compressor drivers. The remaining two-fifths of the energy input covers various process heating needs and all thermal losses.

From this general balance, we notice that the cryogenic fractionation, through the required compression power, represents about half of all thermal energy put through the plant. Therefore, improvements in gas separation are very desirable.

Optimization of refrigerant cycles

The refrigerant cycles and the cracked-gas compressor are the support for the cold fractionation throughout the unit. Once a scheme is selected, the problem is to fit the most efficient refrigerant cycle into the process. Ref. [2] provides an example of a study applied to the ethylene refrigerant cycles.

For refrigeration duties above $-35\,°C$, the largest part are constant-température heat exchangers. Hence, a pure-component refrigeration cycle is suitable. Fig. 1 shows a typical propylene-refrigerant cycle in an ethylene plant. By comparison, the gas-cooling section shown in Fig. 2 and 3 requires variable-temperature heat exchanges below $-35\,°C$, and variable-tempera-

ture cold recoveries from off-gases. In this case, a multi-component refrigerant is better suited to the process than a pure-component one.

Traditionally, ethylene has been used as a refrigerant so as to reach the low temperature of $-101\,°C$. It is interesting to study in detail the merits of each type of cycle. Ref. [3] and [4] discuss the process scheme when a mixed refrigerant is used. For this article, we have selected a refrigerant having the characteristics listed in Table II. Fig. 4 and 5 show typical arrangements for an ethylene and a mixed-component refrigerant cycle, respectively.

We begin by first comparing the theoretical refrigeration power that is required for either cycle for exactly the same duty. This corresponds to the curves shown in Fig. 6. Table III summarizes the data obtained. For this service, the mixed cycle has a base power requirement that is 16% lower than that of the ethylene cycle.

Fig. 7 and 8 show the cooling curves for the process, together with the vaporization curves of the refrigerant. We can calculate the effectiveness of these exchangers when operating against one or the other refrigerant, by using the methods described earlier in this article.

We will use the following relationships and data to perform the several computations for this example:

■ Calculation of exergy values for refrigeration.
Ethylene refrigerant:

$$\Delta E = \Delta Q - T_o(\Delta Q / T_R) \qquad (24)$$

Mixed refrigerant:

$$\Delta E = \Delta Q - T_o[2a\Delta T - b\ln(T_2/T_1)] \qquad (25)$$

Process: $$\Delta E = \Delta Q - T_o k\ln(T_2/T_1) \qquad (26)$$

■ Heating and cooling curves.
Mixed refrigerant (see Table II for constants):

$$\Delta Q = aT^2 - bT + c \qquad (27)$$

Process:

$$\Delta Q = kT + Q_o \qquad (28)$$

Ethylene cycle:
 For $-35° > T > -67°C$, $k = -0.1427$
 For $-67° > T > -98°C$, $k = -0.08407$
Mixed cycle:
 For $-35° > T > -70°C$, $k = -0.007726$
 For $-70° > T > -105°C$, $k = -0.005815$

■ Recovery factor.

As shown in Ref. [2], the ratio, δ, of recuperator duty to total process duty is an important factor in cycle op-

timization. This ratio is directly related to the mixed-cycle power as follows:

$$\text{Compression power} = 789.7\Delta Q(1 - \delta), \text{ kW} \quad (29)$$

After performing the several calculations, we obtain the results shown in Tables IV and V.

From this example, we have a means of comparing the merits of various exchanger arrangements. From the cooling curves, we also see that the relatively good effectiveness of the ethylene cycle has been obtained at the expense of very close temperature approaches, down to $3°C$ at the coldest end.

The advantage of a mixed fluid is quite clear when we consider that the mixed refrigerant has a temperature difference with the process fluid of not less than 5 deg, and that this temperature approach can be adjusted in operation by changing the refrigerant composition. The same applies to the cold-gas recuperators. When the cold gases are exchanged against cracked gas, their effectiveness is quite low due to the large temperature difference. By using the mixed refrigerant to recover the cold, a perfectly constant temperature difference of 5 deg can be maintained throughout the exchanger—greatly improving the effectiveness of the recovery.

As another example, we have calculated the performance of the various sections of a 500,000-metric-ton/yr ethylene unit, cracking naphtha. Table VI summarizes the results.

Comparing exchanger effectiveness with and without recuperators, we see that the ethylene cycle is less efficient because the net effectiveness (taking into account the recuperators) is less than the one based on gas cooling (0.76 vs. 0.80, from Table IV). Table VII summarizes the results concerning the refrigerant cycles themselves for the whole system, including the demethanizer.

Analyzing the results of Table VII:

■ Starting from nearly equal total process duty, we first deduct all recuperators in the -35 to $-110°C$ temperature interval. This gives the net refrigeration duty to be supplied by the refrigerants.

■ The recover factor, δ, is a valuable index that is directly related to the compressor power for mixed refrigerants. In relative and absolute values, it is more favorable in the mixed-refrigerant application.

■ Next, we consider the process-side received exergy (line C) in relation to the refrigerant-cycle supplied exergy (line D). The ratio, $-C/D$, indicates the overall exchanger efficiency. Comparing with Table VI, this value is equal for the mixed cycle in both tables, and higher for ethylene in Table VII because the duty for the demethanizer condenser is included. This unit has a very good efficiency, being a nearly constant-temperature heat exchange. Overall, the mixed cycle is still about 13% more efficient.

■ The ethylene and mixed-refrigerant cycles do not supply the total exergy by themselves. Part of it is supplied by the propylene compressor up to the reference temperature of $311°K$ $(38°C)$, as shown on line E. We see that the mixed-refrigerant cycle demands only half the power of the ethylene cycle. This is one outstanding feature of its application, and results in a better load balance between the compressors. This is especially fa-

---

Ethylene cycle, 1 MW of total refrigeration duty (without recuperators)   Table IV

Temperature, T,°C	Heat flux, ΔQ,MW	Enthalpy flux, ΔE_H,MW	Temperature, T,°C	Enthalpy flux, ΔE_C
$-35/-67$	-0.6367	-0.2567	-55	0.1186
$-67/-98$	-0.3633	-0.2310	-72	0.1955
			-101	0.2936
Total	-1.0000	-0.4877		0.6077
		A		B

Exchange effectiveness: $-A/B = 0.80$

Reference temperature, $T_{O'} = 311$ K

Mixed-refrigerant cycle for 1 MW of total refrigeration duty   Table V

Temperature, T,°C	Heat flux, ΔQ, MW	Enthalpy flux, ΔE_H, MW	Temperature, T,°C	Enthalpy flux, ΔE_C, MW
$-35/-70$	-0.5700	-0.2356	$-58/-110$	0.5937
$-70/-105$	-0.4300	-0.2923		
Total	-1.0000	-0.5279		0.5937
		A		B

Exchanger effectiveness: $-A/B = 0.89$

Reference temperature, $T_{O'} = 311$K

Performance of chilling train   Table VI

	Line	Ethylene refrigerant	Mixed refrigerant
Total cooling duty, MW		7.173	10.004
Recuperator duty, MW		2.184	2.605
Net cooling duty, MW		4.989	7.399
Total exergy, process, MW	A	-3.499	-5.281
Total exergy, cycle, MW	B	4.360	5.939
Exchanger effectiveness (without recuperators)	$-A/B$	0.80	0.89
Exergy process (with recuperators), MW	C	-2.445	-3.906
Exergy cycle, MW	D	3.228	4.393
Net exchanger effectiveness	$-C/D$	0.76	0.89

Reference temperature, $T_{O'} = 311$K

vorable for large units where the propylene compressor intake volume is sometimes substantial.

It is possible to show the exergetic efficiency of each cycle (see Table VIII). The low efficiency of the ethylene cycle is due to several factors such as suction pressure drop at low absolute pressure, low intake temperature, and condensing duty concentrated at the $-40°C$ level.

Going back to Table VII, we can attribute a real power to the propylene compressor if we take its overall efficiency to be 0.55. We obtain the values on line F in Table VII. Adding to it the real compressor power for the mixed-cycle or ethylene-cycle compressor, we get the total shaft power, line (F + G), required to supply the exergy (line D). Calculating the overall average efficiency, $D/(F + G)$, we see that both systems are quite similar. The poor efficiency of the ethylene cycle is covered by its smaller contribution overall. However, looking at the overall process efficiency, $C/(F + G)$, the bonus of the mixed cycle, about 16%, is apparent. This advantage can also be seen when we compare the total compressor power, (F + G), showing the mixed-cycle system to represent 80% of the ethylene-cycle system.

Looking at the plant overall, the total compressor-power difference between the systems is smaller, because a recycle stream is treated by the charge-gas compressor in the mixed-cycle scheme. This adds 1.0 MW to the total power. So the ratio for total power between the mixed- and ethylene-refrigerant cycles becomes:

$$(9.04 + 1.0)/11.276 = 0.89$$

This ratio is the overall direct comparison between the two schemes. But a better cold recovery from the demethanizer reboiler, which favors the mixed-cycle scheme, should be taken into account for a global comparison (see Ref. [7]). We see that the mixed-cycle scheme introduces better exergy efficiency at some key points. Due to its better adaptation to the process requirements, it has better potential for improvement.

The example treated here shows how a complicated process system can be decomposed into a sequence of unit operations. The exergetic effectiveness analysis can be applied to each subsection, and even to each equipment item. Due to the definition of lost work, as a balance difference, the contribution of each section to the lost work can be added up to finally arrive at the total for the whole system.

Application to a turboexpander

A turboexpander receives a stream with properties identified by index α. The stream leaving the expander has properties identified by index ω. An ideal isentropic and adiabatic expansion would produce a stream leaving the expander with properties identified by index s. The following relationships hold:
First law of thermodynamics:

$$H_\alpha + W = H_\omega \qquad (30)$$

And, in accordance with Eq. (2):

$$E_\alpha + W = E_\omega + L \qquad (31)$$

For isentropic expansion:

$$S_\alpha = S_s \qquad (32)$$

Performance of refrigerant and power cycles — Table VII

	Line	Ethylene refrigerant	Mixed refrigerant
Process temperature, °C		−35/−98	−35/−105
Total cycle duty, MW	A	10.280	10.004
Recuperators, MW	B	2.184	2.605
Net cycle duty, MW		8.096	7.399
Recovery factor, δ	B/A	0.212	0.260
Energy of process, MW	C	− 4.221	− 3.906
Energy of cycles, MW	D	5.368	4.393
Energy of propylene cycle, MW	E	3.627	1.760
Propylene cycle, MW (real power)	F = E/0.55	6.594	3.200
Real cycle power, MW	G	4.682	5.840
Total compressor power MW	F + G	11.276	9.040
Overall exchanger effectiveness	−C/D	0.79	0.89
Overall cycle efficiency	D/(F+G)	0.48	0.49
Overall process efficiency	C/(F+G)	0.37	0.43
Relative process effectiveness = 0.43/0.37 = 1.16			
Reference temperature, T_o = 311K			

Exergetic-cycle efficiencies — Table VIII

Exergy	Line	Ethylene refrigerant	Mixed refrigerant
Total cycle, MW	D	5.368	4.393
Propylene cycle, MW	E	3.627	1.760
Net ethylene/mixed, MW	D−E	1.741	2.633
Real ethylene/mixed compressor power, MW	G	4.682	5.840
Cycle efficiency	(D−E)/G	0.37	0.45

Example for performances of turboexpander/compressor — Table IX

	Turboexpander		Compressor	
Inlet temperature, °C	−100		38	
Inlet temperature, K	173		311	
Outlet mean-temperature, T_m, K	100		384	
Reference temperature, T_o, K	311		311	
T_o/T_m	3.11		0.81	
Machine efficiency, η_W or ϵ_W	0.75	0.70	0.75	0.70
Refrigeration effectiveness, η_R	0.47	0.37	....	....
Heating effectiveness, ϵ_H	....	....	1.05	1.06
Overall effectiveness, η or ϵ	0.49	0.43	0.80	0.76

But, the technical or isentropic power is given by:

$$-W = H_\alpha - H_\omega \qquad (33)$$

$$-W_s = H_\alpha - H_s \qquad (34)$$

For the ideal machine, Eq. (31) is:

$$E_\alpha + W_s = E_s \qquad (35)$$

Using Eq. (35), Eq. (31) can then be written as:

$$W - W_s + E_s - E_\omega = L \qquad (36)$$

Effectiveness for shaft power, η_w:

$$\eta_w = 1 - [(W - W_s)/(-W_s)] = W/W_s \qquad (37)$$

Effectiveness for refrigeration, η_R:

$$\eta_R = 1 - [(E_s - E_\omega)/(-W_s)] \qquad (38)$$

But: $E_s - E_\omega = H_s - H_\omega - T_0(S_s - S_\omega) \qquad (39)$

Eq. (33) and (34) yield:

$$H_s - H_\omega = W_s - W \qquad (40)$$

Because the exhaust pressures for the real and ideal machines are equal, we know from thermodynamics that:

$$H_s - H_\omega = \int_{S_\omega}^{S_s} T dS = T_m(S_s - S_\omega) \qquad (41)$$

If the specific heat is constant, the integral average temperature, T_m, is calculated as:

$$T_m = (T_s - T_\omega)/[(\ln (T_s/T_\omega)] \qquad (42)$$

Finally: $E_s - E_\omega = -(W - W_s)\left(1 - \dfrac{T_0}{T_m}\right) \qquad (43)$

From Eq. (36):

$$L = (W - W_s)(T_0/T_m) \qquad (44)$$

Now, we can rewrite Eq. (38) as:

$$\eta_R = 1 - \left[(W_s - W)\left(1 - \dfrac{T_0}{T_m}\right)\right]/(-W_s) \qquad (45)$$

$$\eta_R = 1 + (1 - \eta_w)[1 - (T_0/T_m)] \qquad (46)$$

The term $(E_s - E_\omega)$ is the refrigeration power equivalent to the duty between T_s and T_ω. This refrigeration duty is lost because the real-machine exhaust temperature is T_ω rather than T_s. So, η_R represents the refrigeration-power effectiveness measured against the theoretical shaft power, W_s.

Using the effectiveness definition of Eq. (6a), we find the following relationship from Eq. (44):

$$\eta = 1/\left[1 - (T_0/T_m)\left(1 - \dfrac{1}{\eta_w}\right)\right] \qquad (47)$$

Finally, η, η_w and η_R are related as follows:

$$\eta_w = \eta(2 - \eta_R) \qquad (48)$$

For the compressor, we note that all equations, including Eq. (44), hold without any modification. However, for the shaft effectiveness, ε_w, by analogy to Eq. (6b), we will use the definition:

$$\varepsilon_w = 1 - [(W - W_s)/W] = W_s/W \qquad (49)$$

The lost work remains unchanged as per Eq. (36).

Hence, the overall exergy effectiveness, ε, according to Eq. (6b) is:

$$\varepsilon = 1 - [(1 - \varepsilon_w)(T_0/T_m)] \qquad (50)$$

Following an exact analogy with the turboexpander analysis, we can define a "heating" effectiveness, ε_H, as:

$$\varepsilon_H = 1 - [(E_s - E_\omega)/W] = \\ 1 + (1 - \varepsilon_w)[1 - (T_0/T_m)] \qquad (51)$$

Finally: $\varepsilon = \varepsilon_w + \varepsilon_H - 1 \qquad (52)$

If no value is attached to the heat to be recovered from the compressor discharge, then only ε_w is technically meaningful.

A numerical example comparing the turboexpander and the compressor shows the various effects. The final results are tabulated in Table IX. From the table, we see that the amount of cold not recovered from the turboexpander represents a sizable loss. Hence, refrigeration and overall effectiveness are low. For the compressor, the heating to be recovered from the actual machine relative to the ideal machine is low. Here, ε_H is larger than unity because the real machine downgrades work to heat that can be recovered to the extent of 5% or 6% for compressor efficiencies of 0.75 or 0.70, respectively. But this is marginal, so the overall efficiency remains good.

Acknowledgement

This work has been performed as part of the Technip/TechniPetrol development program concerning olefins production. Thanks are due to M. Watrin and D. Gilbourne of Technip for their active support, and to C. Pocini and M. Picciotti of TechniPetrol for their help and advice.

References

1. "Economie d'Energie en Raffinage et Petrochimie," Technip, Paris, 1976.
2. Picciotti, M., Optimize Ethylene Plant Refrigeration, *Hydrocarbon Process.*, May 1979, pp. 157–166.
3. Kaiser, V., Becdelièvre, C., and Gilbourne, D., Mixed Refrigerant for Ethylene, *Hydrocarbon Process.*, Oct. 1976, pp. 129–131.
4. Kaiser, V., Salhi, O., and Pocini, C., Analyze Mixed Refrigerant Cycles, *Hydrocarbon Process.*, July 1978, pp. 163–167.
5. Maloney, D. P., U.S. Dept. of Energy Workshop, Aug. 14–16, 1979.
6. Kenney, W. F., Improving Energy Use: Thermodynamic Analysis for Research Guidance, Paper No. 48a, AIChE Meeting, San Francisco, Nov. 25–29, 1979.
7. Kaiser, V., Heck, G., and Mestrallet, J., Optimize Demethanizer Pressure for Maximum Ethylene Recovery, *Hydrocarbon Process.*, June 1979, pp. 115–121.

The author

Victor Kaiser is process supervisor for Technip, Cedex 23, 92 090 Paris La Defense, France. He heads a group of engineers engaged in process engineering and design of ethylene plants. Previously, he was plant process engineer with Lonza (Switzerland), and then joined Lummus (Paris) as a process engineer. He has a Diplom Ingenieur Chemiker and a Ph.D. in chemical engineering from the Federal Polytechnic Institute (Zurich). He is also a member of the Assoc. Française des Techniciens du Pétrole, and of the Schweizerischer Ingenieur- und Architektenverein.

Section II
THE ENERGY-EFFICIENT PLANT

Energy Conservation In New-Plant Design

U.S. process plants have long been designed for low capital cost, often at the expense of higher operating costs. But with the worsening energy supply, operating costs—hence, energy costs—must be minimized. Here are some of the areas in which design engineers will have to change their ways.

J. B. FLEMING, J. R. LAMBRIX and M. R. SMITH, The M. W. Kellogg Co.

Events of recent years have placed the engineer/contractor in a position to apply energy-conservation technology that, for the most part, has long been known but has been held in abeyance.

Energy conservation measures—make no mistake—add to the first cost of a plant. The desire for low first cost has been, for many years, a powerful deterrent to energy-conservation measures, due both to highly competitive situations in bidding and to low utility costs. Low first cost was considered more important than low operating cost, and the "optimization" or "payout formula" guidelines then used usually brought one to that conclusion.

Almost the reverse has been true for some years in Europe, and plants designed for that area generally have included a greater degree of heat and energy recovery. The ever-increasing costs of crude oil, natural gas and other starting materials—as well as the overriding national interest not to become too dependent upon foreign sources—now promises to reverse this picture in the U.S. All plant designers will now probably be required to include the same or similar energy-conservation design features, in addition to the more-stringent safety features and environmental-protection devices to be required by law (OSHA and EPA).

The 1960s represented a period of relatively stable cost of energy (or its equivalent in utilities). The early 1970s have shown a significant change in these values, and projections more than two or three years from now are very difficult to make. For the purpose of this article, let us consider the tentative price structure (shown in Table I) for fuel and energy, and its effect on new-plant design.

While the values used should be considered typical rather than absolute, the differential will serve to illustrate the need for conservation and recovery of energy.

Table I predicts that the cost of utilities will increase from 50% to 300% over the period shown. During this same period, the projected installed construction cost will increase about 140% (that is, about 2.4 times).

The importance of the price structure of Table I does not stand out until one starts to apply it to the design of a new plant for 1975-80. Instead of comparing natural gas at its 1960 price to the predicted 1975-80 price, one must compare it with the cost of liquid fuels, since natural gas will no longer be available for large industrial uses at any price. The ratio of the costs of energy for this period is 6.0 to 7.5, if we consider substituting liquid fuels for natural gas. Many of the examples in this paper are based on a ratio of 6.0 for energy costs over this 15-yr period using $0.20/million Btu for natural gas in 1960 and $1.20/million Btu for crude oil in 1975-80.

As of this writing, we are concerned that this ratio is too low rather than too high. However, when we consider a cost increase of energy by a factor of 6.0 as compared with a capital cost increase of 2.4, we think the same conclusions will be drawn regarding energy recovery and conservation. It is evident that the capital costs are also rising rapidly at this time, and absolute numbers for more than two or three years in the future are highly speculative.

The energy crisis was not generally recognized until quite recently. Also, EPA has had a great influence during the recent past in emphasizing non- or low-polluting

Originally published January 21, 1974

61

fuels for users of heavy fuel oils or coal. This has greatly accelerated the overall problem of energy costs. We feel that this interplay of fixed energy supply vs. industry, government, and public interest deserves a high level of consideration by all involved. There is a great temptation to use the most economical and desirable sources of energy first and then resort to the more expensive forms. Certainly no consumer will refute this program, but others must consider the influence of such a program on the overall economy.

Energy-Recovery Illustrations

To illustrate some of the major methods that engineers are using to cope with the higher energy costs of the 1975-1980 period, we are presenting nine detailed studies.

These have been selected to illustrate the application of sound engineering to the design of projects, with a comparison of construction and energy costs of 1960 to the 1975-80 period. They include:

- Furnace efficiency.
- Power-recovery systems.
- Hydraulic-turbine power recovery.
- Reflux vs. operating cost.
- Water cooling vs. air cooling.
- Fluid catalytic-cracking unit power-recovery systems.
- Ammonia-plant optimization.
- Ethylene-plant design.
- Cryogenic gas processing.

Furnace Efficiency

This is a very broad subject that has been extensively covered elsewhere. This article will concern itself with points that have received special emphasis because of the energy shortage.

A very informative comparison is made below of a conventional steam-boiler installation, a steam boiler with air preheat, and a hot-oil system. When this comparison is made for a 1960 price structure vs. one for 1975-80, the impact of the energy crisis can be realized even more.

A case study was made of a 400 million Btu/h delivered system that corresponds roughly to a 500,000 lb/h steam system. The predicted installed cost and energy consumed by these systems are given in Table II.

In 1960, the net savings for installing an air preheater on the steam boilers used above would be $16,000/yr vs. $126,000/yr for the 1975-80 period. This is certainly an impressive difference considering that the energy shortage will become even more acute in the future.

An even greater difference can be illustrated by comparing a hot-oil system with a conventional steam boiler. In 1960, the additional cost for the hot-oil system would have been $29,000/yr, whereas in 1975-80 the additional cost will be $358,000/yr.

Hence, the predicted price structure indicates that the use of a steam boiler with an air preheater is much more economic than a hot-oil system. (See Table III.)

This conclusion must be qualified with the assumption that low-level heat can be used economically. While this

Costs of Fuel and Utilities — Table I

	1960-1970, $	1975-1980, $
Fuel, natural gas, million Btu	0.20	0.80
Fuel, light hydrocarbons, liquid propane gas, million Btu	—	1.50
Fuel, No. 1, 2 fuel oil, million Btu	—	1.30
Fuel, bunker "C" crude, million Btu	0.70	1.20
Fuel, coal, million Btu	0.30–0.50	0.70
Steam, 500–750 psig, 1,000 lb	0.50	2.00
Steam, 20–200 psig, 1,000 lb	0.20	0.50
Electricity, kWh	0.010	0.015
Cooling water, 1,000 gal	0.02	0.04

low-level heat is more expensive to recover, it is necessary to continually improve the balance.

Fig. 1 illustrates an application of an air preheater common to two furnaces, in which the preheated air is returned to one of them. This installation provides a very economical unit that allows a great deal of flexibility in the design of the two furnaces.

The point may be raised that hot-oil furnaces are not equivalent to steam boilers. This is certainly true in that hot-oil furnaces are normally designed for high-level heat only. This emphasizes the idea that a high-efficiency furnace essentially improves the recovery of low-level heat. Hence, the use of low-level heat is an important part of the overall plant energy-balance to be discussed in more detail later in this article.

Power-Recovery Systems

A study has recently been made of a power-recovery system for a large chemical plant. The power to be recov-

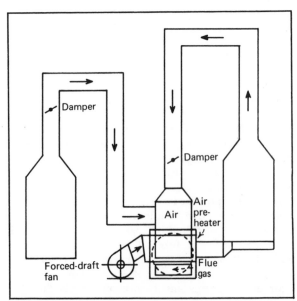

TWO FURNACES with a common air preheater—Fig. 1

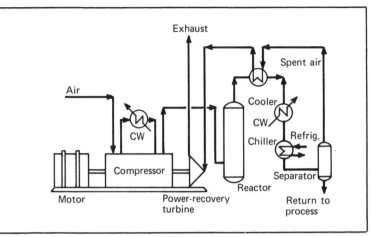

TYPICAL spent-air power-recovery system—Fig. 2

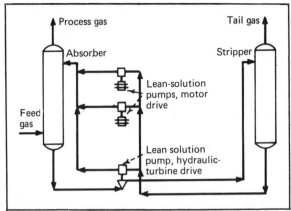

HYDRAULIC TURBINE power-recovery system—Fig. 3

ered is from a spent-air stream that is available at about 70 psig and at 40°F. This stream must be reheated, which can readily be done by heat exchange with other process streams that need to be cooled. This results in a further savings in either cooling water or refrigeration in the process. For the purposes of this study, the reheat temperature was considered to be 200°F, which was enough to allow an expansion-turbine outlet-temperature of 40°F, without condensation.

The power recovered is used to supply part of the energy needed to drive the shaft of the main process-air-compressor, which is motor driven.

A simplified sketch of this power-recovery system is shown in Fig. 2.

The economics of this system are based on a process-air-compressor requiring about 8,000 brake horsepower (bhp), and the associated equipment for a plant of this size. The horsepower recovered by the expansion turbine is estimated to be about 20% of the total, or 1,600 bhp.

The added capital cost of this system over the base case is about $300,000, based on today's price structure. Using 20%/yr, this gives an annual capital cost of $60,000.

The utility savings, based on $0.015/kWh for 1,600 bhp, amounts to $148,300/yr. Therefore, the net annual savings is $88,300. Stated another way, the simple payout on this system is about 2.0 yr.

This system has been simplified to illustrate one possible case. Other variations include: the use of a steam-turbine drive; a two-stage expansion turbine with reheat between stages; and a more complicated exchanger system to recover the cooling available in both the feed to and the discharge from the expansion turbine. These more-complicated cases should certainly be considered, but are very dependent on the overall plant-energy balance, which cannot be discussed in detail in an article of this length.

Hydraulic-Turbine Power Recovery

For more than 15 yr, Kellogg has used hydraulic turbines for the let-down of the rich solution from CO_2-removal systems on large ammonia plants. In these plants,

Fired-Heater Comparison — Table II

(Investment vs. operating cost
based on 400 million Btu/h duty)

	1960	1975–1980
(A) Steam without air preheat		
Investment, $	910,000	2,185,000
20% investment,$thousand/yr	182	437
Fuel at 87% efficiency, million Btu/h	460	460
Cost, $ thousand/yr	761	4,560
Total, $ thousand/yr	943	4,997
(B) Steam with air preheat		
Investment, $	960,000	2,305,000
20% investment,$thousand/yr	192	461
Fuel at 90% efficiency, million Btu/h	444	444
Cost/yr, $	735	4,410
Total, $ thousand/yr	927	4,871
(C) Hot oil		
Investment, $	670,000	1,600,000
20% investment,$thousand/yr	134	320
Fuel at 79% efficiency, million Btu/h	506	506
Cost/yr, $	838	5,030
Total, $ thousand/yr	972	5,350

Fired Heaters — Annual-Operating-Cost Comparison — Table III

(Based on 400 million Btu/hr duty)

	Steam	Steam, With Preheat	Hot Oil
1960			
Cost, $ thousand/yr	943	927	972
Cost, % base	100	98.3	103.2
1975 – 1980			
Cost, $ thousand/yr	4,997	4,871	5,350
Cost, % base	100	97.7	107.1

the CO_2 absorber normally operates at several hundred psi pressure, whereas the CO_2 stripper operates at essentially atmospheric pressure.

Energy is required to pump the lean solution up to the absorber pressure. Much of this energy was lost in earlier plants when the rich solution was let-down to the stripper pressure through a control valve. On the new systems, a steam turbine or motor is used as the primary drive on the lean-solution pump, and a hydraulic turbine is used on the same shaft as a secondary drive (or to drive a separate spare). About 50% of the energy required by the lean-solution pump is supplied by the hydraulic turbine.

Based on a large-scale ammonia plant or SNG (substitute natural gas) plant, the following economics can be developed. Overall, the lean-oil pump requires about 1,600 bhp so that recovered power is about 800 bhp. The installed cost of the system with power recovery is about $100,000.

Using a motor-drive system with electricity at 1.5¢/kWh, the payout of such a system would be $74,000/yr savings for an incremental investment of $100,000, which would give a payout of 1.3 yr.

If a steam-turbine drive were used in place of the motor drive, it would also show a very good payout. However, we would again like to note that choosing a turbine drive rather than a motor drive will depend on the overall energy system, as will be discussed in greater detail later in this article.

Fig. 3 illustrates one method of using a hydraulic turbine. This particular installation includes three 50% (of full load) pumps. One of the pumps is driven by a hydraulic turbine that is powered by the full rich-solution flow. The other two pumps are 50% motor-driven units.

Another variation is the use of two 100% pumps that may be either motor or turbine driven. The shaft of one of these pumps is connected to a hydraulic turbine that provides from 30 to 50% of the driving horsepower required.

Reflux Ratio vs. Operating Cost

The economic balance in fractionating towers will now definitely shift toward the use of more trays and lower reflux ratios in order to save fuel. In order to compare reflux ratio with operating cost, the annual cost of a debutanizer on a large-scale ethylene plant is shown in Fig. 4. This example shows that a reflux ratio of 1.40 would have been selected in 1960, as compared with a reflux ratio of 1.20 in 1975-80. (The overall annual operating cost has increased by a factor of about 2.5 during this period of time.)

We now evaluate each column for large plants, to establish the minimum cost, based on the economic data supplied by the client for his particular installation. In most cases, this is not a straightforward isolated study in which the evaluation is based on steam, electricity, cooling water and installed cost. The particular tower must be integrated into the overall energy balance for the entire plant or complex.

As a rule-of-thumb, a reflux ratio of 1.10 to 1.20 times the minimum is a good number for many installations. Two outstanding exceptions come to mind. For low-tem-

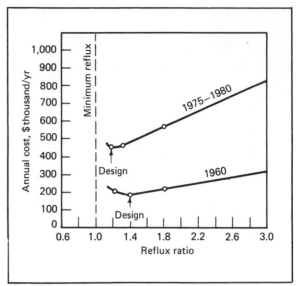

DEBUTANIZER economics, large ethylene plant—Fig. 4

perature and cryogenic applications (where utility costs are exceptionally high, lower reflux ratios in the region of 1.05 to 1.10 may be used. For systems where low-level heat (or recovered heat) is used, higher reflux ratios in the region of 1.20 to 1.50 may prove to be economical. In all cases, direct heat interchange is preferred to indirect heat interchange, such as steam systems or refrigeration systems. This results in the lowest investment cost, since it saves on both hardware and ΔT. However, in many cases, process considerations may make such systems infeasible.

A C_2-splitter study (Fig. 5) is included for a large-scale ethylene plant based on a 1975-80 utility-price structure.

Economics of Air Coolers vs. Water Coolers — Table IV

Case A

q, cooling = 100 million Btu/h	
Installed cost, air cooler, $	200,000
Installed cost, water cooler, $	53,000
Δ Installed cost, $	147,000
Operating cost, air cooler, $/yr	25,000
Operating cost, water cooler, $/yr	172,000
Δ Operating cost, $/yr	147,000
Payout, ($147,000/147,000),yr	1.0

Case B

q, cooling = 15 million Btu/h	
Installed cost, air cooler, $	150,000
Installed cost, water cooler, $	50,000
Δ Installed cost, $	100,000
Operating cost, air cooler, $/yr	17,000
Operating cost, water cooler, $/yr	32,000
Δ Operating cost, $/yr	15,000
Payout, ($100,000/15,000), yr	6.7

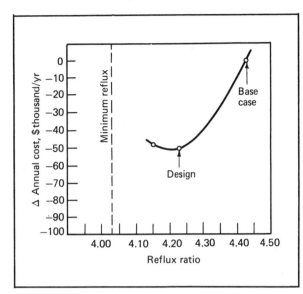

C₂ SPLITTER economics, large ethylene plant—Fig. 5

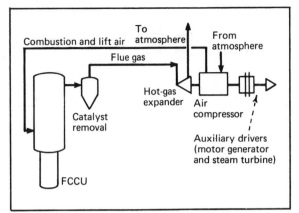

FCCU flue-gas power-recovery system—Fig. 6

For this study, a reflux ratio of 1.05 times the minimum was selected. (Although Fig. 5 shows that an even lower value would be economic, it was thought that 1.05 times the minimum was the smallest reflux that could be properly controlled.)

Air Coolers vs. Water Coolers

Air coolers rather than water coolers are going to be increasingly attractive. At condensing temperatures of 125°F and above, the economics of such systems usually indicates a strong preference for air coolers.

The installed cost of air coolers is basically higher than that of water coolers. When one considers the ΔT, location, motors, and control instrumentation, it is easy to see why installed cost is higher. On the other hand, the operating cost is obviously lower, since the intermediate transfer of the heat, first to cooling water and then to the atmosphere, is avoided.

Kellogg has made a thorough investigation of most large-scale installations, to optimize the use of air coolers vs. water coolers. Two examples are shown in Table IV.

In the two cases shown in Table IV, "A" was switched to an air cooler and "B" was left on water cooling. The final selection is based on many factors, but units with a payout of less than three years would normally be switched to air coolers, and those with a payout of more than five years would be water cooled.

Operating problems that occurred with some of the early air coolers have largely been eliminated and the choice today is based primarily on economics. Air coolers may prove economic at lower operating temperatures, (at temperatures that were not economic 15 years ago).

FCCU Power Recovery

A majority of the fluid-catalytic-cracking units (FCCU) built since 1955 incorporate single-train air-blower installations. The large single-train installation is particularly suited for the application of power-recovery facilities, with their resultant energy savings. Power-recovery facilities in catalytic cracking units were first introduced in 1963 with a 9,000-hp unit at a Louisiana refinery. Since then, over seven additional units have started up in Texas, California, and New Jersey, as well as in Canada and Europe, with power outputs ranging in size from 4,000 to 22,000 hp.

A typical power-recovery system for a 20,000-bbl/d fluid-catalytic-cracking unit is shown in Fig. 6. The gas-expander-driven air-compressor train uses the energy contained in the hot flue-gas from the regenerator to compress the air required by the regenerator. The expander and compressor usually are fairly well balanced in energy availability and consumption; however, depending upon ambient and process conditions, a poten-

||||||||||||||||||||||||||||||||||||||||||||||||||||||||||||||||||||||||||||

FCCU Power-Recovery Economics — Table V

(For a facility handling about 370,000 lb/h of flue gas at 1,300° F)

Air-compressor requirement, bhp	8,200
Expander power recovered, bhp	6,900
Power deficit, bhp	1,300
At Start-up	
a. **Steam**	
600 psig to 150 psig starting turbine (max. rate), lb/h	52,000
150-psig expander seals lb/h	2,500
b. **Electricity**	
Motor generator, kW	2,500
During Normal Operation	
a. **Steam**	
600 psig to 150 psig starting turbine lb/h	750
150-psig expander seals, lb/h	2,500
b. **Electricity**	
Motor generator, kW	300

||||||||||||||||||||||||||||||||||||||||||||||||||||||||||||||||||||||||||||

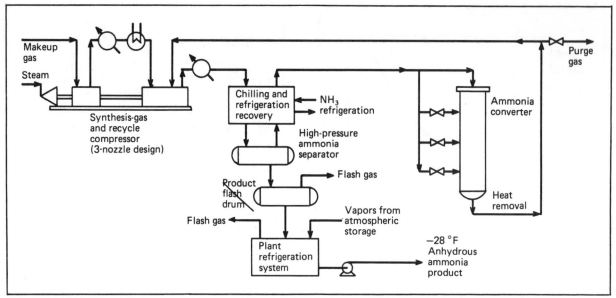

TYPICAL LOW-PRESSURE ammonia-systhesis loop. Product recovery after recycle compression—Fig. 7

tial deficit may exist at times, and a surplus at other times.

These energy differences may be accommodated by the addition of an induction motor-generator. At startup, or whenever there is an energy deficit, this device acts like a motor and supplies the necessary additional power. When the gas expander overdrives the air compressor, the motor generator absorbs power by becoming a generator that supplies power to the refinery electrical system. A small starting turbine may be required to overcome the initial inertia of the train and thus reduce the motor-generator inrush current.

Based on a savings of 6,900 bhp, and assuming an investment differential of $1,500,000, the following economics can be developed:

With electricity at $0.015 kWh the incremental savings in utilities is $640,000, or a payout of 2.35 yr.

Using steam, the payout would be reduced to about 2.0 yr. (but this requires integration into the overall utility balance to get a better value).

Not only does the FCCU power-recovery system show a good payout, it also includes a cleanup system that may be required in any case to meet future EPA requirements.

Table V shows utility quantities associated with the operation of a power-recovery facility handling approximately 370,000 lb/h of 1,300°F flue gas.

Ammonia-Plant Design

The development of modern large-scale ammonia plants has been considered one of the outstanding chemical engineering accomplishments of the last 15 years. To a large extent, this development comprised improvements in both the energy cycle and process efficiency.

In 1960, the manufacture of synthetic anhydrous ammonia was a mature technology that had been developed and practiced worldwide over several decades. The chemistry of ammonia synthesis was well known; thou-

sands of man-years of effort had been expended in perfecting the process, and technical literature in the field was voluminous.

In the face of this apparently mature technology, a coordinated effort comprising laboratory research, pilot-plant development, engineering development, theoretical mathematical modeling, computer technology and machinery development was undertaken. Process engineers started from the basic chemistry, analyzing and reexamining the basic process from the standpoint of total energy requirement (as well as total capital requirement), per unit of product, rather than following the usual practice of optimizing individual process steps.

Study showed that the ammonia process was a surprisingly fertile ground for significant improvement. Because this supposedly mature technology was improved so drastically, it is reasonable to conclude that fertile areas for energy improvement probably exist in other chemical engineering technology.

The new ammonia process has changed the state of the art in several major respects:

■ The synthesis of ammonia is now being conducted at much lower pressure, approximately 150-200 atm, compared to the prior practice of ammonia synthesis at 300 atm or more. (See Fig. 7.)

■ The new process is adapted to the use of low-cost centrifugal compressors, rather than the high-cost reciprocating compressors previously used. The relative inefficiency of centrifugal compression is more than compensated for by the use of an efficient energy-cycle to provide steam-turbine power.

■ Previously unused "waste heat" is now recovered efficiently, and is used to provide the energy for gas compression.

■ A highly efficient steam system has been developed to achieve a significant reduction in ammonia-plant energy requirements.

■ The refrigeration circuit is effectively utilized to liquefy and refrigerate the ammonia product. (See Fig. 7).

These chemical engineering developments, carried out over a relatively short period, have resulted in new ammonia-plant designs that have reduced manufacturing costs by as much as 50% (compared to costs for previously standard production plants).

Fuel and energy costs to produce ammonia were reduced to less than half. The response to this development has been extraordinary since its introduction in 1963—the new process has been adopted for over 50 ammonia plants designed by Kellogg alone. The investment is well in excess of $1 billion.

In the 1960s, when the present ammonia-plant design was evolved, natural-gas price in the U.S. was approximately 20¢/thousand ft³ on the Gulf Coast. U.S. investors generally evaluated incremental investments to conserve energy on the basis of a payout of three years or less. However, also in the mid-1960s, the same new ammonia-plant technology was applied under an entirely different economic framework in Europe, Japan, Africa and Australia. In many overseas locations, the feedstock and fuel, generally naphtha, was priced at 50-60¢/million Btu, approximately 2½ to 3 times the then current U.S. energy costs. In addition, in many overseas areas, incremental investment to conserve energy was justified on the basis of a payout in the range of four to five years. Thus, the design for ammonia plants constructed in the 1960s in Europe and Japan differed considerably from those built in the U.S. These differences may be instructive to us today as the U.S. energy-supply situation approaches (to some extent) that which prevailed in other portions of the world ten years ago. Following are some of the major engineering features (each of which increased plant-investment costs but conserved energy) that were applied to ammonia plants built overseas:

■ Recuperative air preheaters were used to heat combustion-air for the primary reformer, saving 5-10% of the reformer fuel.

■ "Hot carbonate" processes for carbon dioxide removal were adopted, saving 40-50% of the heat required for this process step.

■ Process heat exchangers were generally designed for closer temperature approach. This practice increased capital cost of the exchangers but conserved high-priced energy.

■ More-efficient, but higher-cost, refrigeration cycles were adopted. For example, in U.S. plants it was common practice to use three levels of refrigeration, whereas European plants commonly used four levels.

■ Hydraulic turbines, for power recovery from the expansion of carbon-dioxide-removal solvent, were adopted.

■ Higher-efficiency steam turbines and gas compressors were generally used.

This experience, dating back approximately 10 years, illustrates the effect of an economic framework in which the energy supply/cost relationship is adverse. Consequently, this type of engineering and process modification can be and will be, applied to future U.S. plants. Energy can be conserved over our past practice, but it is important to note that such energy conservation will inevitably tend to increase both the capital cost and the engineering input required for design. Among other engineering variations that have been considered in the past, but not generally adopted either in the U.S. or at overseas sites:

■ Gas turbines for compression power as well as waste heat recovery. Gas turbines have been used in some isolated instances in modern ammonia plants; however, this has been the exception rather than the rule, both in the U.S. and overseas. It is probable that the new economic framework will justify greater use of gas turbines in ammonia plants.

■ Air cooling has not been adopted to any significant extent in ammonia-plant design. Again, however, we anticipate ever-increasing use of air coolers, promoted by energy-supply situations, and environmental and water-supply considerations.

■ The new energy costs projected for the 1970s will undoubtedly encourage use of more-efficient waste-heat recovery, energy cycles, and turbines and compressors.

Ethylene-Plant Design

In recent years, our company has been involved in the planning of many large olefins-complexes, with feedstocks ranging from ethane through heavy gas-oils. Since many of these plants exceed a capacity of 1 billion lb/yr of ethylene, the economics involved have become increasingly important.

First, it was recognized that feedstock was a major contributing factor to olefins-plant economics. To define and optimize these plants for specific feedstocks, a unique pilot plant was especially designed to provide direct scaleup data. Hence, accurate feedstock evaluation and design criteria could be incorporated into the plant design, thereby keeping investment costs down.

Second, each plant was custom-designed to maximize energy conservation. Let us consider some of the principles that go into such a plant design.

Pyrolysis furnaces that employ short-residence-time cracking conditions and adjustable time-temperature relationships provide a capability of handling a range of feedstocks while cracking each feed at near-optimum conditions. Such equipment reduces feedstock consumption and also reduces plant size for a fixed product capacity.

The furnaces are designed for high overall efficiency. Induced-draft fans, economizer coils and air preheat are all evaluated. Efficiencies of at least 90% are usually anticipated owing to the high utility costs outlined earlier in this paper. The use of pyrolysis fuel oil in place of imported fuel gas is also incorporated into the design.

Major consumers of process energy are the three large compressors. (The process-gas compressor is one of the largest.) An overall study is made to optimize compressor-suction pressure versus furnace-effluent pressure. Each 1-lb increase in suction pressure can save over 1,000 hp. This must be weighed against the effect on furnace yields at the increased furnace-outlet pressure and the optimum design-pressures established. A computer program that can simulate various design schemes is used extensively

The dilution steam used in the pyrolysis furnaces is condensed and reused after pretreatment. Generation of

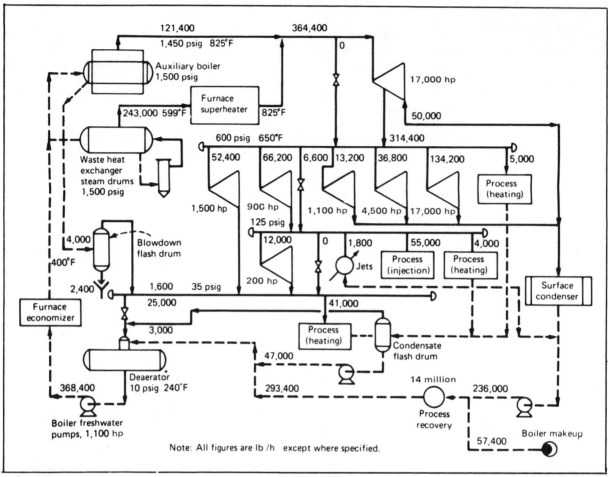

STEAM SYSTEM and typical steam balance associated with a large-scale ethylene plant—Fig. 8

dilution steam is carried out by employing excess low-level heat in the primary fractionator circuit or by condensing turbine-exhaust-steam from an intermediate steam level.

The steam system and the design steam-balance are intimately associated with the process and its flexibility. Turndown and startup requirements must be studied to ensure that the base design can properly handle these situations as efficiently and as easily as possible. A typical "energy system" for a large ethylene plant is shown in Fig. 8.

Design for the separation and final purification of the furnace-effluent products involves the use of several of the principles outlined earlier in this article (for example, the question of reflux ratio vs. the capital cost of extra distillation-column trays).

Ethylene plants employ two major refrigeration systems, each provided with a large compressor. Basic cryogenic practices involving optimization of installed horsepower, heat exchange, and insulation, and reduced heat-leakage, must all be evaluated. "Cold" streams leaving the process are heat exchanged against the main mixed-product stream entering the final fractionation stages in order to conserve this energy. The cooler streams also reduce the need for corrosion-resistant alloys in the process, thereby reducing investment. Heat-exchanger design "approach temperatures" are evaluated against

capital cost and refrigeration savings. The flexibility needed for changes in product composition (as a result of feedstock changes) is included in this overall optimization study.

Reliable vapor-liquid equilibrium data are important as are actual tray efficiency and hydraulic data involved in reactor- and fractionation-system design.

Use of low-level heat from the "hot" end of the plant is evaluated against the use of economizers from the refrigeration cycle, intermediate reboilers, heat-pumps etc.; in order that the overall horsepower and utility requirements can be optimized.

These are only a few examples of the many possible design alternatives that must be taken into account in a process as complex as an olefins-plant design.

Cryogenic Gas Processing

An up-to-date picture of energy conservation would not be complete without consideration of the potential of cryogenic processing for recovery of light components that, for many years, have either been burned as fuel or flared.

Cryogenic gas processing of natural gas or saturated refinery gases can accomplish recoveries that are either impossible, or cannot be done as economically, by other processes. Some cryogenic processes include recovery of

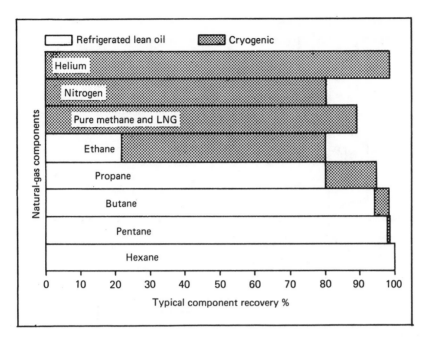

COMPARISON of typical component recoveries in two natural-gas processing methods—Fig. 9

helium, rejection of nitrogen, production of pure methane or liquefied natural gas and recovery of ethane. Fig. 9 is a bar graph that compares the typical economic recoveries obtained in refrigerated-lean-oil plants with those of cryogenic plants.

The refrigerated-lean-oil plant, which is the standard of the industry for the recovery of propane and heavier hydrocarbons, is represented on the graph by unshaded blocks, and the cryogenic plant by shaded blocks. Obviously, the cryogenic plant's recovery exceeds that obtainable by using lean oil.

The general scheme of a cryogenic plant for the recovery of hydrocarbon liquids from natural gas is shown in Fig. 10. Eighty-percent propane or ethane recovery can be achieved by processing condensate at –90°F (propane) or –125°F (ethane). Helium extraction is accomplished by first condensing 80 to 90% of the feed (cooling to approximately –150°F); the 10 to 20% uncondensed vapors are further cooled to approximately –300°F, where all components except helium are condensed.

In the selection of the most economical design for a cryogenic plant to process natural gas, many process and equipment factors must be considered. These include problems of water and carbon dioxide removal, the selection of operating pressures, choice of prime movers, and heat-exchanger design.

The selection of the process method for water removal is generally an economic decision, based on equipment and operating cost; but since plant downtime for defrosting is expensive, reliability considerations often override first-cost considerations.

Fixed-bed desiccant dryer-systems provide bone-dry

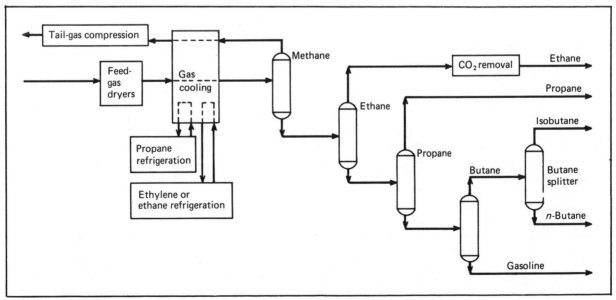

BASIC FLOWSHEET for a typical cryogenic natural-gas-liquids extraction plant—Fig. 10

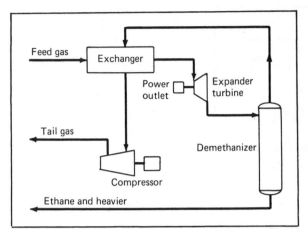

TURBOEXPANDER cycle for cryogenic gas processing—Fig. 11

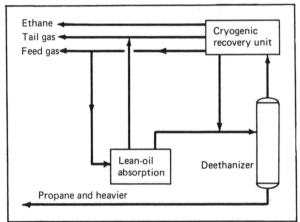

SERIES cryogenic-absorption gas processing—Fig. 12

gas and eliminate operating problems due to icing or formation of hydrates.

In many cases, carbon dioxide will have to be removed, either from the feed gas to avoid freezeup, or from the ethane-mixed-liquids product to obtain the required product purity. Several methods are available for reducing CO_2 content to tolerable levels, but no matter which approach is chosen, consideration should be given to factors that contribute to onstream reliability.

The selection of operating pressure levels in the plant is very important. To reduce equipment size and refrigeration requirements, fractionating towers should be operated at the highest pressure consistent with efficient phase separation. To recompress the demethanizer overhead to tail-gas pressure, the plant should be located at some point along the pipeline that is not expected to exceed that pressure.

Problems arise when an existing upstream extraction plant is temporarily out of operation or the main source of gas is shifted from one field to another.

If this type of situation is not anticipated, the plant can easily be overloaded with liquids (or nitrogen), and the rate of product recovery desired of various pieces of equipment may exceed their capacity. In any case, the product proportions are sure to change.

Surprisingly, electric power has many advantages if it is available. If the utility is connected to a wide-area grid system, the reliability is usually much better than that offered by either gas or steam systems, and electrical systems will certainly be lower in first cost.

If properly located, the utility company will be a purchaser of the fuel gas that would have been consumed by the gas turbines, so there is no loss of pipeline demand. Lost production from plant startup, shutdown, and maintenance is much less with electric power.

Large variations in the cost of refrigeration that occur with changes in temperature place a great premium on process and equipment-design creativity. Intermediate reflux condensers and other design innovations may easily pay for themselves in nitrogen-methane fractionation.

Cryogenic plants generally produce refrigeration by: direct (Joule-Thomson) gas expansion; a turboexpander cycle; a cascaded-refrigeration system (employing propane and ethylene, and sometimes methane and nitro-

PARALLEL cryogenic-absorption gas processing—Fig. 13

gen, as refrigerants); or a mixed-refrigerant cascade.

The turboexpander cycle is usually the best choice if free pressure drop is available—for example, if the tail gas is sent directly as fuel to a large utility. (See Fig. 11.)

The expander cycle is usually more efficient than direct gas expansion, and requires fewer pieces of equipment (and, consequently, a lower capital investment) than a cascade-refrigeration cycle that provides an equivalent amount of refrigeration.

Many natural-gas-processing plants are limited to a nominal 25 to 30-psi pressure differential across the plant, and in some situations tail-gas streams must be recompressed to the inlet-pipeline pressure.

The availability of free pressure drop is quite limited because someone, usually the pipeline-transmission company, has to pay for it through excessive compression or equipment costs. Efficient pipeline engineering will try to avoid such situations.

For ethane recovery, it may appear economically attractive to make modifications to permit 30% recovery of ethane from the lean-oil plant. However, total recovery of ethane may not be sufficient to furnish all the feed needed by one of today's economic-size ethylene plants.

If higher ethane recoveries are required, this can be accomplished either by cryogenically extracting ethane from existing absorber overhead or by splitting the feed into equal parts and processing one part in a cryogenic unit while processing the other part, with full lean-oil rate, in the absorption plant. The lean-oil plant can then extract about 50% of the ethane from the gas going to the absorber, while the cryogenic unit can extract 75 to 80% of the ethane from its portion of the feed.

The main advantages of extracting ethane cryogenically from the existing absorber overhead (Fig. 12) are that very high ethane recovery is possible, and that the existing lean-oil plant continues to operate in its normal fashion.

The main advantage to splitting the feed (Fig. 13) is that the cryogenic unit can be designed for about 50% of the gas throughput, thus saving investment.

Summary

The foregoing specific examples have been selected to typify energy-conservation problems and solutions in plant design today, a situation brought about by the growing energy shortage.

There are a great many energy-saving schemes and methods that have been known for many years but have been considered uneconomical. What with the fast-changing fuel cost, feed supply and legal-ecological situations all old "rules of thumb" and "accepted practices" will have to be critically reexamined when designing new plants.

Many of the "known methods" must be modernized and perfected. And the development of more-selective catalysts to increase product selectivity and reduce feedstock consumption (and therefore energy requirements) will continue apace.

In many processes, the feedstock is also a fuel. Therefore, plant designs that optimize feedstock consumption will indirectly reduce overall energy requirements.

The planning phase of many large new projects has been extended or broadened not only to ensure a more efficient design but also to allow time for government approvals, financing arrangements, etc. Therefore, more time is usually available to make comparative studies and select more-efficient solutions to the plant-design problems.

Energy conservation in plant design must also fit the ecological equation. Thus, the design that requires high circulation of cooling water may exceed allowable water makeup rates. Alternatives within the makeup rates must be reexamined to arrive at the design that conserves the most energy within the required framework

If the plant is part of an overall complex, then overall energy conservation must be examined. It is possible to optimize the efficiency of a plant within its battery limits but impose a greater energy burden on the overall complex.

Designing for energy conservation can be achieved with today's technology. We must not forget, however, that our country looks to us, its engineers, to continue to develop new sources of energy and still-more-efficient processes to reduce the unit energy requirements.

As citizens we can also help by using less energy in our everyday lives and convince those around us to do likewise.

Acknowledgments

The authors of this paper gratefully acknowledge the contributions of the following Kellogg personnel: R. H. Roberts, J. L. Slack, A. Roosov, J. C. Belote, D. B. Crawford, J. E. Wallace, R. E. Daze and J. A. Finneran in the preparation of this article.

References

1. Bing, J. A., "Engineering Energy Conservation Systems for a Grass Roots Refinery," 1972 ASME Petroleum Div. Conf., New Orleans, La., Sept. 1972.
2. Whittington, E. L., others, Striking Advances Show Up in Modern FCC Design, *Oil Gas J.*, Oct. 1972.
3. Brown, C. L., Figenscher, D., Preheat Process Combustion Air, *Hydrocarbon Process.*, July 1973, pp. 115-116.
4. Reed, R. D., Save Energy at Your Heater, *Hydrocarbon Process.*, July 1973, pp. 119-121.
5. Brown, C. L., Kraus, M., "Air Preheaters and Their Application to Process Furnaces," API, 38th Midyear Mtg., Philadelphia, Pa., May 1973.
6. Finneran, J. A., others, Advanced Ammonia Technology, *Hydrocarbon Process.*, **51**, No. 4, Apr. 1972.
7. Eschenbrenner, G. P., Wagner, G. A., A New High Capacity Ammonia Converter, *Chem. Eng. Progr.*, Jan. 1972.
8. Kellogg Information Series, "Better Plants and Economics Brighten Ammonia Picture", *ECN* (*Europ. Chem. News*), Sept. 1971.
9. Quartulli, O. J., Fleming, J. B., Finneran, J. A. Best Pressure for NH₃ Plants, *Hydrocarbon Process.*, **47**, No. 11, Nov. 1968., pp.153-161.
10. Kellogg Information Series, "The Modern Ethylene Plant," Lambrix, J. R., et al., 1972.
11. Slack, J. B., Energy Systems in Large Process Plants, *Chem. Eng.*, Jan. 1972.
12. McCarthy, A. J., Hopkins, M. E., Simplify Refrigeration Estimating, *Hydrocarbon Process.*, July 1971, pp. 105-111.
13. Crawford, D. B., Eschenbrenner, G. P., Heat Transfer for LNG Projects, *Chem. Eng. Progr.*, **68**, No. 9, Sept. 1972
14. Crawford, D. B., others, "Economic Factors in Cryogenic Gas Processing," Natural Gas Processors Assn., Annual Meeting, Dallas, Tex., 1969.

Meet the Authors

J. B. Fleming   **J. R. Lambrix**   **M. R. Smith**

James B. Fleming is Process Manager—Special Projects for The M. W. Kellogg Co. (a division of Pullman Inc.), 1300 Three Greenway Plaza East, Houston TX 77046. Among many other duties as Process Manager, mainly in the field of organic chemicals, he is responsible for the analysis of proposed plants for environmental impact, when required. He holds an A.B. in chemistry from the University of Kansas, a B.S. and an M.S. in chemical engineering from the University of Michigan, and has done graduate work at the University of Delaware, Stevens Institute and Columbia University. He is a member of AIChE, ACS and Research Soc. of America (Kellogg Branch).

James R. Lambrix is Director, Process Engineering Dept., for M. W. Kellogg. This involves administration of the work of the Inorganic, Organic, Refinery, Cryogenic and Technical Data groups. He holds a B.S. (cum laude) in chemical engineering from New York University (degree attained by receipt of The M. W. Kellogg Employee Scholarship Award.) He is a member of AIChE, Scientific Research Soc. of America, Tau Beta Pi and Phi Lambda Upsilon.

Martin R. Smith is Assistant to the Director of Process Engineering at M.W. Kellogg, where he acts as Lead Technologist in charge of Overall Planning. He holds a B.S. from Mississippi State College and an M.S. from M.I.T., both in chemical engineering, and is a member of AIChE, Research Soc. of America, ACS, and is a Licensed Professional Engineer in New York State.

Conserving Utilities' Energy in New Construction

The rising cost of energy makes its conservation important wherever possible. Utilities are a prime target; here are some things you can do.

CHARLES E. SCHUMACHER and BADIE Y. GIRGIS, Bechtel Corp.

Energy consumption and availability are of primary concern to engineers. With the rapidly rising cost of energy must come more efficiency and less waste. New plants already are being designed for higher efficiencies, and more money is being spent on equipment for them (a factor often forgotten when comparing the costs of old and new plants).

Each project differs in its energy requirements. Large amounts may be necessary for material transfer, heat transfer and chemical reactions. This may determine the location of the process, because it may therefore require the cheapest electricity, coal and gas. In addition to the far-ranging political effects of such a requirement, it may determine where you and your family must live if you are to work at the plant.

Energy-System Requirements

When new plants use conventional processes, past performance fairly well predicts future successful operation (although there is, of course, an opportunity to correct problems and improve overall efficiency).

New (prototype) processes demand more attention. However, they offer greater opportunities for unique engineering.

For either, the procedure is the same at the start—set up a utility system. A typical energy and utility balance (Fig. 1) includes all energy input and output items. Whether a complete plant or a small internal unit, a total utility balance brings it all together on a single sheet of paper, and permits rapid evaluation of tradeoffs.

Combinations and flexibilities must be established, with needs and surpluses balanced for an economic conservation of energy. Too little attention has been given to the less-glamorous utility portion of projects. But, more often than not, it is this that determines the success or

Originally published February 18, 1974

failure of the project, since processes are usually repeats of former successes, or are based on highly-proven pilot-plant runs.

Analyzing these utilities is a vital function of any new construction and requires a complete and thorough knowledge of the total plant operations. How often has it been the boiler plant, power station or water supply that delayed plant startup?

The making of heat balances will determine the required energy levels. Often, the source of this energy is predetermined by existing facilities. Heat up to 500°F is generally obtained from steam. Above this, fired heaters or electric heating is used.

To start a balance, several steam-pressure levels that satisfy the process are established. For most larger refineries and chemical plants, turbine drivers are used and their economic operation may set the steam-pressure levels available for the process.

The pressure ratings for piping valves and fittings are a factor in the selection of pressure settings. The 150-psi ANSI carbon-steel flange is rated for 150 psig at 500°F. Many plants use the 150-psig level since it permits 135°F superheat in sustained operations. The same piping materials are used for lower-pressure steam that is set by the end-user's requirements. Higher pressures will use the 300-psi ANSI flange rated for 625 psig at 500°F. It is used for the 200-psig to 500-psig levels, with appropriate allowances for superheat. Above this, the 600-psi ANSI flange rating takes care of the major requirements for pressures above 500 psig that normally occur in chemical plants.

A selection of 20 plants shows an acceptable uniformity of steam-pressure levels. Too low a pressure will give operating problems—40 psig solves most. Large demands are well acknowledged and taken care of because they set the operation. It is the little special heat-users that can

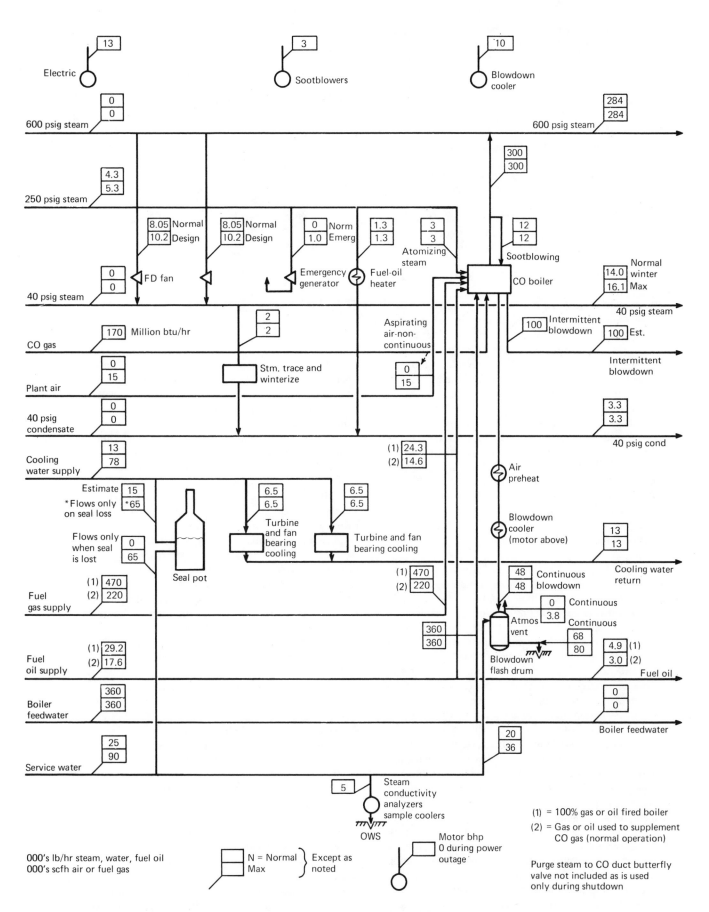

TYPICAL UTILITY BALANCE for a fluid catalytic-cracking unit boiler plant—Fig. 1

be frustrating and challenging. Such systems as Dowtherm, steam recompression, high-temperature water, microwave or electrical heating, and heat pumps are a few of the alternatives that are available for such heating.

Analysis

Once a system is set up and steam-pressure levels are established, one must start to analyze for all eventualities. Here is where some of the new, unproven, rigid and inflexible designs fall out. And here is where it pays to communicate, to see what others are doing or planning.

More loss in efficiency occurs at this stage than any other because of the need to back up and redesign, so as to avoid a patchup later. This is when one selects a turbine that will accept high- and (all) low-pressure steam, rather than selecting an air-cooled enchanger to condense a surplus.

It should be realized that certain pumps and exchangers must run during power outage or the relief system will double in size. You must determine whether the process can take a shutdown if a large synchronous motor is installed. Winter and summer operations must be analyzed. The effects of a unit down that is a big user or producer must be discovered. What about power recovery: Is it available?

Feedback is needed as equipment is purchased because the utility requirements will be fixed by what is bought. The analysis is complex in that there can be double or triple jeopardy. Solution of a complex project, such as the Great Canadian Oil Sands one, required 25 simultaneous utility balances. We must use the computer to best solve these problems.

Selection Basis—Flexible Items

Cost and Payout—The selection of equipment is justified on the basis of original cost plus operating cost during the payout period. Quoted and proven high efficiencies will lead to selections of equipment that is not evaluated simply on low-bid first cost. As the price of energy escalates, selection evaluations will change, and will support higher equipment costs.

Construction estimates must allow for the increase in cost to buy more-efficient equipment. (There is a ratio between plant construction costs and the costs for energy, which could be considered a kind of efficiency factor.)

There are several ways for making the most economic selection of equipment, such as the use of internal rate of return (IRR), net present worth (NPW) or payback period (PBP).* The simplest and most widely used in engineering construction selection is the PBP. The number of years is based on the company's economic policy. At present, for most refinery construction it is four years.

Company policies vary on selection of equipment. Some have insisted upon lowest initial cost because repairs and maintenance can then be charged to operating costs, which is favorable because of the tax structure. However, using the sophistication of a computer, com-

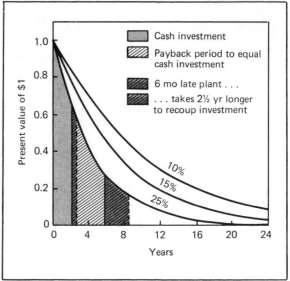

LATE startup has disastrous effect on sales—Fig. 2

panies are developing more-realistic values based on IRR. Profitability calculations should be based on plant operations with a known capacity and service factor. A change in service factor from 80 to 95% has a strong effect on profit, and this is the issue that forces conservatism into plant design.

It is essential that plants are built and brought into operation on schedule. Market conditions can shift rapidly, and a plant six months late coming onstream can have a disastrous effect on sales. Even in a stable market, it is difficult to recoup the losses from a late plant start (a sample explanation is shown on Fig. 2). Early returns far exceed earnings 10 or 20 years later.

Since money depreciates in value, cash flows for long payback periods need to have a present-value-factor evaluation. Despite widespread concern about the energy crisis, few companies are now designing plants with energy conservation as a prominent consideration. The need for quick return on investment limits the number of clients for an energy-saving factory. However, as large new construction is added that has longer payback periods, there will be a shift to the concept of energy savings.

An example of the more recent developments with larger plants is a high-pressure hydrogen facility recently installed in Pascagoula, Miss., for Standard Oil of California. The utility balance includes the following energy savers, for an overall efficiency of 71%:

- Gas-turbine-exhaust waste heat to reformer furnace.
- Waste heat from reformer furnace to steam generation.
- Power recovery from high-pressure-solvent letdown.
- Power recovery from high-pressure-gas letdown.
- Absorption refrigeration using low-pressure steam.
- Low-pressure steam to condensing-turbine-drive for boiler-feedwater circulation.
- Low-pressure steam to condensing-turbine-drive for solvent circulation.
- Steam injection to gas turbine for peak loads.
- Turbine drive with a reduction gear, to permit higher-efficiency turbines.

Because the average refinery overall-efficiency is about 40%, 71% generates savings that justify the added expenditures. Savings to give a daily utility bill of $10,000 support the cost of capital for the 10-15% increase in cost. Payback period is less than a year. (See additional comments in Oil Refiners Squeeze Energy, *Chem. Eng.*, June 25, 1973, pp. 40,41.)

Many plants are built around the concept of energy being supplied by burning a useless byproduct. Even the community garbage dump is up for discussion. But the desire to "do something" may lead the plant into undesirable situations. An example was a polymer plant having a byproduct of no known value, so it seemed reasonable to burn the material for fuel. Since it would dissolve in light fuel oil, the solution looked simple, until it was discovered that costs for such fuel oil plus blending substantially exceeded costs for heavy fuel oil. Sawdust, coke, CO, slops and effluents of all kinds are used now for energy generation. More will be added as EPA requirements are weighed along with disposal-problem costs. Thinking in this area needs to be clear, since there are many opportunities here for "the tail to wag the dog."

New construction should maximize benefits for lowest total costs. If condensate is needed in large quantities, as in a hydrogen plant, it should be made available from large condensing-turbine drivers. In severe climates, installing a complete plant inside a building may offset all the installation and maintenance costs from steam tracing—with added productivity as an obvious benefit.

Reliability

Maximum reliability can be achieved if certain factors are well taken care of during the design, procurement and construction phases, as well as when operating and maintaining units.

These factors are:

System design—A well-studied and -designed system based on latest technology and down-to-earth facts is the keystone of a reliable unit. (This item has already been covered in this article.)

System Flexibility—In an effort to secure uninterrupted operation, particular care should be taken during the design stage to decide on which pieces of equipment will be fully or partially provided with standby units. Normally, equipment subject to wear and tear (pumps, compressors, blowers) or subject to intolerable fouling (filters and certain water coolers) are provided with spare units. Control valves are set up with bypasses to enable their replacement or repair during operation. Spare pumps in crucial services have automatic startup controls. These pumps are usually driven by an alternative source of energy if the primary source is considered unreliable.

A useful practice in the design of net-steam-producing process units is to divide them into a few sections in series, each capable of operating independently, regardless of interruption in the downstream sections. With this arrangement, the steam-generating section, which is normally the front part, will keep on operating and producing steam regardless of any adverse operating conditions on the tail sections. It must be realized that this will occur

at the expense of the process product, which will be disposed of, normally by flaring, for certain periods of time. Therefore, the economics of this mode of operation should be carefully evaluated.

Selection of Equipment—Proper selection of equipment that is expected to perform continuously under various operational and climatic conditions is another critical factor in plant reliability.

Proven equipment having correct metallurgy and being supplied by reputable experienced manufacturers should be obtained, even at a premium in the capital cost. These additional dollars will surely pay dividends in terms of continuous plant operation.

It is of utmost importance that details and small components in complex units should be given the same care and attention, when reviewing the vendors' drawings, as major parts. It is conceivable that a faulty material O-ring on a mechanical seal, or a key on the shaft of a butterfly control valve, can cause the interruption of operation of the whole unit.

Factory Inspection—Factory inspection schedules and procedures should be established by the procurement contractor for every piece of equipment, upon placement of a purchase order. Orderly systematic inspections and tests-witnessing should be carried out, with results relayed immediately to the design office. At this crucial stage of equipment manufacture, all manufacturing problems must be resolved to the satisfaction of the design office.

Bulk standard materials that are manufactured to certain stringent codes and established procedures (e.g., standard piping, valves, bolts and nuts, etc.) do not normally follow this routine.

Construction and Field Inspection—This must be carried out exactly according to the engineering drawings and designs, with particular attention paid to recommended erection procedures of various equipment manufacturers. Details should be given the same care as major parts of the unit.

Plant Operation—The unit should be brought on-stream while following detailed operating instructions, normally prepared by the design office after consultation with the plant's operating management. Operators should refrain from any deviation of the operating instructions until they become familiar with the idiosyncrasies of the unit; then they will be in a position to provide their own input regarding modified operation, after consulting with the design office.

Unit Maintenance—Finally, the reliability of a unit depends to a great extent upon how it is periodically inspected and maintained. The target, of course, is to keep the unit's condition continuously in the same shape as originally built.

Availability of necessary spare parts for scheduled plant turnarounds or unpredicted failures is of utmost importance for unit reliability. Ideally, all spare parts should be available in the plant warehouse, but, on the other hand, an enormous amount of capital can be tied up that way.

The policy for ordering and stocking of spare parts must be set by the operating company during the design and procurement phase. This will be mainly dependent on the geographic location of the plant and the size of operation of the company.

Other Influences

The selection of many of our energy sources is based on the risk involved with their use. Electrical energy in the U.S. has always been available, except for catastrophic occurrences. With brownouts, its use now involves greater risk. Gas and oil have joined the list of growing risk-prone energy sources, as the requirements for clean air accelerate demand over the supply available. Many articles are currently covering these points. What is to be realized is that risk is based on probabilities, and these change. Risk is a moving target that can radically be altered even during the construction period. It is impossible to build a factory without an absence of hazard.

An area of concern is the placement of sophisticated equipment in parts of the world where technical expertise is at a premium. The risk involved in doing so is compensated for by increased equipment reliability and the general improvement in world communications. Though few design companies run a Monte-Carlo-type risk analysis for the selection of equipment, it is available and could become a valuable tool in selection of equipment having long delivery times for plants projected for operations that will not be undertaken until five or more years into the future.

Selection Basis—Fixed Items

The predetermined requirements of a company management, to suit an existing installation or total plant-utility balances, may preclude any choice or flexibility in evaluations. When preplanning for expansions, commitments are made for utilities before contracts are awarded for construction. Contracts so made are based on electric peak demands, uninterruptible gas consumptions, etc. These decisions reduce the degrees of freedom for making the best balance.

Another item of recent development that has restricted freedom is the domination of government regulations. With laws established to specify limits of noise, degree of safety, air and water purities, and with more to come, selections must be based on satisfying such regulations. One offshoot advantage is that the time for management's preparation of such reports should give the designer more time for development work. Members of the public have become keenly aware that engineers and scientists have a dramatic effect upon their lives. The social battles will continue, and engineers will be asked to give answers to questions for which the answers are still obscure.

One of the greatest influences affecting us today is the political stability necessary for rational decisions. It appears we must be more and more self-sufficient, though the contrary is to the benefit of the world community. The U.S. will develop its shale oil. The tar sands will be a major production in Canada, and new oil and gas fields will come into being all over the world because we will demand the energy to give us stability until we develop new sources.

In any plant design, there are concepts built into the philosophy from the start that affect the construction sequence. What the plant can do is fixed in the formative days. This is the first 30% of the project. After this, costs and operations are included in a schedule, and change becomes increasingly expensive and time-consuming. I am not condoning the oft-desired freeze with no changes. What must be, must be. But it must also be an evaluation of nice vs. necessary.

Future Needs

The fact that we build for a 2-4 year payback period and end up with a plant that continues to operate inefficiently for 20-40 years needs to be altered. There is a requirement for an incentive tax structure of some kind, to give advantage for efficient design (a utility guarantee could serve as a basis).

To continue our present methods only leads to lost productivity, and operations that consume our national wealth. This is an easy way to build slums. This inefficiency comes back to all of us in monies required for governmental programs and controls. One of the surest ways to increase our productivity would be to build higher efficiency into the operations of all our new plants while still making sure they operate when and as we wish them to.

References

1. Hayden, J. E., Design Plants to Save Energy, *Hydrocarbon Process.*, July 1973.
2. Hayden, J. E., How to Conserve Energy While Building or Expanding a Refinery, *Oil Gas J.*, May 21, 1973.
3. Slack, J. B., Steam Balance: A New Exact Method, *Hydrocarbon Process.*, Mar. 1969.
4. Chopey, N. P., Oil Refiners Squeeze Energy, *Chem. Eng.*, June 25, 1973.

Meet the Authors

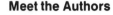

Charles E. Schumacher is a project engineer for Bechtel Corp., 50 Beale St., San Francisco, CA 94119. He has had experience as a project engineer with a variety of refinery, chemical and petrochemical facilities. He has a B.S. in chemical engineering from the University of Cincinnati, an M.S. in chemistry from the Uniersity of Akron, completed the Bechtel Business Management Certification Program at the University of California Business Administration Extension, and is registered as a professional engineer in Ohio, Illinois and British Columbia (Canada).

Badie Y. Girgis is also a project engineer at Bechtel Corp. His experience ▶ has been mainly in the design, maintenance and operation of refinery units. He previously worked for the Anglo-Egyptian Oilfields Co. (subsidiary of Shell Intl.), General Petroleum Co., and Suez Oil Processing Co. of Egypt. He holds a bachelor's degree in mechanical engineering from Cairo University, Egypt.

Energy Conservation In Existing Plants

In this author's company, a plant manager must report the energy expended for each pound of product produced. This has led to the development of numerous ingenious ways of conserving energy, many of which are described here.

J. C. ROBERTSON, Dow Chemical U.S.A.

The "energy crunch" is now well accepted as being real. Shortages of energy have brought about government allocations, and fuel prices are definitely rising and will continue to increase. Further, the producers of energy will apparently be unable to ease the supply problem within the next three to five years.

So, clearly, the near-term solution to our dilemma is energy conservation. We must improve the efficiency of our use of energy.

Actually, this challenge unfolds a new opportunity for the chemical process industries. Because of the higher cost of fuels, the CPI will be able to justify capital projects that will effect energy savings, and these investments will make a profit.

How then do we go about implementing energy-conservation programs in our chemical plants? Perhaps the best way to discuss the subject is to cover what we have

Originally published January 21, 1974

done, and are doing, to conserve energy at our company's plants.

Management Support

Of utmost importance to the success of an energy-conservation program is endorsement by top management. Look at it this way—if management is committed to achieving a goal in saving energy, that commitment becomes of interest throughout the line organization. And it is the line supervisors of the production plants who must see to it that actions are actually taken to accomplish energy conservation.

To illustrate our management's concern for saving energy, let me quote from an address by Dr. Earle B. Barnes, President, Dow Chemical U.S.A., presented before the Instrument Soc. of America in October, 1973. "Last year, we made a 10% reduction in energy used, while increasing yield. Our goal is a 10% reduction this year. Certainly, we will reach a point of diminishing return for our efforts, but we intend to set tough goals each succeeding year."

Consider also the interest of Levi Leathers, our director of operations in the U.S., who has been supporting energy conservation for years. In order to increase awareness of energy consumption on a Btu basis, he initiated a program that called for daily computer reporting of theoretical and actual balances of energy and materials for all unit operations, together with reports of daily production. In the past, material yields were the main measure of efficiency, but now Btus added per pound of product must also be considered. With the computer production-report, each plant superintendent can now monitor his yield and Btu business on a daily basis.

So, at Dow, we do have management support and a continuing program to save energy. We are repairing leaks. We are shutting down that extra motor when it is not needed. We are controlling steam, air and water flows to the minimum required. We are finding situations for cross-exchanging process streams that need heating with streams that need cooling. We are evaluating the replacement of condensing-type mechanical-drive steam turbines with backpressure drives or electric motors. Longer-range plans will include replacement of entire production units with new plants that both have higher yields and consume less energy.

Energy-Conservation Team

What else can be done to intensify efforts to conserve energy? At Dow's Texas Div., management formed a Division Energy-Conservation Team to formulate and coordinate a program. Each major manager appointed one of his manufacturing specialists as his representative on this team. These team members, 20 including a coordinator, hold communications and planning meetings once or twice a month. It may be helpful to discuss some of the team activities.

■ Initially, block-by-block surveys were urged—tracing out the energy streams to find where losses could be reduced or eliminated.

■ The team collected published articles on energy-saving ideas and techniques; copies are made available to those interested.

■ Each team member communicates with key supervisors in his major manager's area. This provides a flow of information and ideas from the team to line supervision and vice-versa.

■ Bulletins are issued every month or two on subjects such as cost of steam leaks; rule-of-thumb methods for the capital expenditure that can be justified for a project with a given fuel savings; tips on air-conditioner checkups to assure efficient operation, etc. Also, four team members are now publishing bulletins that tell about specific energy-saving accomplishments in their areas and give recognition to the individuals responsible.

■ A group in the Maintenance Dept., the "Energy Conservation Ramrods," came up with the design of a decal that features a light bulb having a dollar-sign filament (Fig. 1). The team adopted the symbol for division-wide use, and now the decal can be seen on light

DECAL reminds workers to conserve energy—Fig. 1

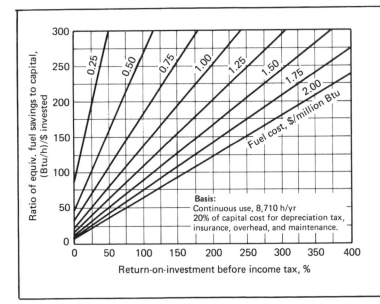

EVALUATION of energy conservation projects—Fig. 2

switches, typewriters, hardhats, etc., all around the plants.

■ Presentations have been made to the team (in some cases, to larger groups) on the energy-saving aspects of such subjects as: steam traps, power-factor correction, lighting, and insulation. At an annual technical meeting of the division, an energy-conservation session of six papers was presented. Supervisor training courses now include a two-hour discussion on energy conservation, presented by the energy-team coordinator.

■ Energy conservation is included in the checklist for approval of capital-project authorizations. On large jobs, the interested energy-team member, the coordinator and the project team, review the flowsheets and energy balances to assure that there is efficient utilization of energy in the design. On small projects, the coordinator confers with the project's supervisor by telephone.

■ The team has also compiled, for management, lists of capital projects for energy conservation that have been

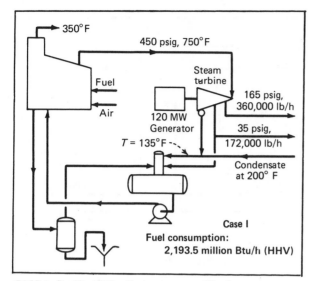

CASE 1: Pre-World-War II steam system—Fig. 3

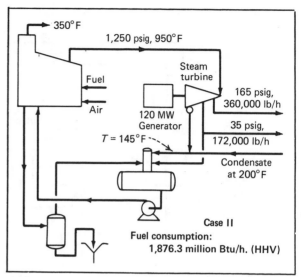

CASE 2: Post-World-War II steam system—Fig. 4

submitted by line supervision. These lists are updated periodically. Savings of energy (electric power, steam, fuel) are expressed as equivalent fuel savings in Btu/hr. The ratio of equivalent fuel savings to capital cost is a convenient indicator for determining the profitability of a project in this period of frequently changing projections of fuel costs. In Fig. 2, this savings-to-capital ratio is related to % return-on-investment (before tax) for various fuel costs ranging from $0.25 to $2 per million Btu. As a simple example, say that you have a $100,000 project that will save 15 million Btu/h. That is a savings-to-capital ratio of 150 Btu/h per dollar invested. Assuming that you project a $1 fuel cost, you can read from the graph a return of 112%. The return on investment, of course, is of prime interest to management in setting priorities on the proposed projects.

Let me describe some of the specific ways we are saving on fuel.

Industrial Power and Steam-Generating Cycles

The power-plant arrangements in the Dow Chemical Co. have changed as more-efficient equipment has become available. Basically, there are three generations of power and steam generating facilities in the Texas Div. Let us call these three generations:

- Case 1—Pre-World War II.
- Case 2—Post-World War II.
- Case 3—Mid-1960s to present.

The process power and steam requirements supplied by one of the company's newest power plants will be used to illustrate how the change in the state-of-the-art during these periods has affected the Division's fuel consumption. The process requirements used in all three cases are:

Power:
  120,000 kW
Steam:
  165 psig   360,000 lb/h
  35 psig   172,000 lb/h

Fig. 3 shows the cycle for Case 1. The first power

plants built in the Texas Div. prior to World War II had boilers generating steam at 450 psig and 750°F to drive automatic-extraction condensing-steam turbines. The fuel required to supply the stated process-utility requirements from a plant of this vintage would be approximately 2,193.5 million Btu/h.

Fig. 4 shows the cycle for the second generation of Texas Div. power plants, which were built after World War II and in the 1950s. These plants were also steam units, but had higher-pressure boilers generating steam at 1,250 psig and 950°F to drive automatic-extraction condensing-steam turbines. The fuel needed to supply the process requirements from these plants would be approximately 1,876.5 million Btu/h.

Availability of the industrial gas-turbines in the 1960s changed the complexion of our units considerably. Fig. 5 illustrates the Case 3 arrangement that is typical of the units built at most of our locations in recent years. The plant consists of industrial gas-turbine - generator sets, steam-turbine units, and waste-heat boilers connected to the gas-turbine exhausts. Fuel requirements for this arrangement are approximately 1,515.8 million Btu/h.

Each step has enabled our company to reduce the fuel

||||||||||||||||||||||||||||||||||||||||||||||||||||||||||||||||||||||||||||||||||

Fuel Saved by Reducing Condenser Load—Table I

	Case I	Case II	Case III
Power, kW			
From process steam	19,700	33,500	12,950
Gas turbine	–	–	93,840
Condensing steam	100,300	86,500	13,210
Total	120,000	120,000	120,000
Steam produced lb/h	1,610.000	1,284,000	660,000
Steam to process	532,000	532,000	532,000
Δ steam	1,078,000	752,000	128,000

||||||||||||||||||||||||||||||||||||||||||||||||||||||||||||||||||||||||||||||||||

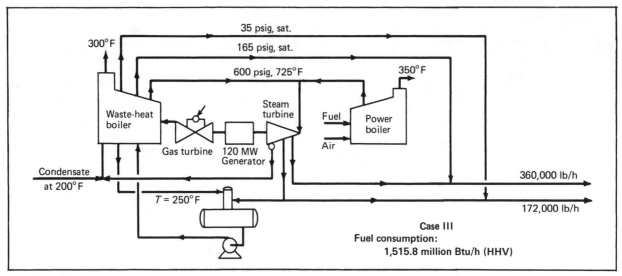

CASE 3: Steam generating systems from mid 1960s to present, incorporate energy-saving features—Fig. 5

required to produce the process' utility requirements. As can be seen from Fig. 6, a reduction in fuel of 14% and 19% has been realized with successive cycle improvements.

How have these reductions been realized? As shown in Table I, the use of high-pressure steam and gas turbines has reduced the condenser load from 100,300 kW in Case 1 to 13,210 kW in Case 3. The fuel improvements were the result of these reductions in condenser loads.

Such improvements can be credited to the "fuel chargeable to power," since they were associated with the power-generating portion of the cycle. As illustrated in Fig. 7, the heat rate for the power produced decreased 20% and 29% with successive cycle improvements.

Looking to the future, we are presently planning modifications for two of our power plants at Freeport, Tex. We plan to add additional gas turbines, waste-heat-recovery boilers and backpressure steam turbines to increase our generating capability by 100,000 kW, while consuming less total fuel than now. We can do this by retiring some of the low-pressure condensing units we built in the early 1940s that are no longer economical at today's fuel costs. In addition to increases in efficiency, we

are significantly reducing the cooling-water requirements and the heat load discharged to the Brazos River.

Chemical-Process-Design Innovations

Process-design innovations must be explored and implemented if significant reductions in plant energy requirements are to be achieved. Application of nonconventional distillation designs can result in lower energy consumption for a given separation. One significant distillation "wrinkle," which can reduce by 25-75% the overall energy requirements for a given separation, is vapor-recompression distillation.

A conventional distillation scheme is illustrated in Fig. 8. The reboiler is driven with steam, hot water, hot oil, or vapors from a refrigeration cycle. The condenser coolant could be air, cooling-tower water, or a refrigerant.

Fig. 9 demonstrates vapor-recompression distillation. The overhead vapors from the column are compressed and then condensed in the reboiler. Heat of compression elevates the overhead vapor temperature sufficiently to provide the needed ΔT to drive the reboiler. The condensed overhead vapors are collected in a reflux drum

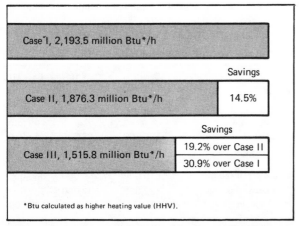

FUEL consumption and savings for Cases 1 to 3—Fig. 6

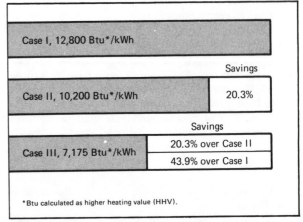

FUEL chargable to power, and savings—Fig. 7

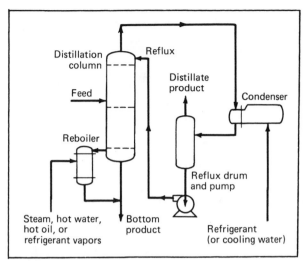

FLOWSHEET of conventional distillation system—Fig. 8

VAPOR RECOMPRESSION distillation flowsheet—Fig. 9

and pressure-controlled (no reflux pump required) to reflux the column. The only energy requirements are for the compressor's driver, which normally would be an electric motor, steam turbine, or even a gas turbine.

Vapor recompression is particularly attractive when the relative volatilities of the components being separated are "enhanced" at reduced pressures, so that for a given distillation separation, less reflux and/or fewer distillation trays are required. Previously to minimize capital, fewer trays would have been designed into a column; but at today's energy costs, the process designer would lean toward more trays and minimum reflux so as to minimize the energy requirements.

Some common systems employing vapor recompression distillation are the separation of ethane from ethylene, and of propane from propylene. However, many other distillation separations, if run at lower pressures, could use vapor recompression to conserve energy. Also worth mentioning—with vapor recompression, the temperature/pressure level of operation is not "tied" to the condenser-coolant availability and eliminates the need for refrigeration in low-temperature applications.

Byproduct Recovery

Many latent opportunities exist in our plants for byproduct recovery, opportunities that conserve energy by conserving valuable raw materials. An example of how a heavy aromatic byproduct stream evolved into a rags-to-riches story, epitomizes such a case.

Here are the steps that occurred:

■ The original plant design called for burning the byproduct in a waste incinerator, which required atomizing steam and fuel gas for combustion. In addition, it was a high-maintenance operation.

■ Later, an outside buyer contracted to purchase the stream, but for a price less than its fuel value. However, overall savings were realized over incineration.

■ More recently, with pending fuel shortages due to the energy crunch, combustible waste streams have been collected to supplement fuel in power-plant boilers.

■ Finally, with the current shortage of aromatics, all

aromatic byproduct streams were evaluated for end-product uses; over 80% of these streams were found to represent a valuable, recoverable end-product.

Fig. 10 shows the original design that was conceived for recovering 80% of the aromatic byproduct as an end-product. Note that it is a very conventional distillation arrangement. The feed is heated by both the product and heavy-fraction streams. The first column removes the heavy fraction, the second column removes the light fraction, and product is removed from the bottom of the second column. The only utility requirements are gas for the fired heater, and motor-power for the air condensers. Use of the hot heavy-fraction to drive the reboiler on the second column is one nonconventional process-design concept incorporated into the scheme. Overall, this was not a bad process design.

However, Fig. 11 illustrates what can be accomplished in a distillation design if the process designer really sharpens his pencil. This design utilizes product side-draw, which simply means selecting an intermediate tray in the distillation column that meets the product requirements. Thus, one column is eliminated and considerable utilities for the reboiler and condenser are saved. However, for this particular process design, that was only the beginning of the story—the pencil was still dull.

By elevating the overhead distillation temperature slightly, the overhead condenser was used to generate 30-psig steam from condensate. This steam was used in the plant to reduce the amount of purchased 150-psig steam.

The heavy bottoms, which freeze at 180°F, could have been burned in an adjacent power-plant boiler, and would have required steam-tracing, storage and pumping. Rather, these bottoms, which had a 5-million-Btu/h heat release, were burned in-plant in an existing gas-fired heater with a 30-million-Btu/h heat load, thus reducing its fuel-gas requirements by about 15%.

Finally, a more-efficient gas-fired heater than was normally specified in then-current plant practice was used for the column reboiler. The plant's existing natural-draft heaters achieve about 65-70% thermal efficiency. However, for this application both a fired heater with forced draft and having 78% efficiency, and one using forced draft together with air preheat that had 88% efficiency, were evaluated. The one having the air-preheat design was selected because it had less than a two-year payout.

The overall "sharp pencil" final design resulted in:

■ Side-draw distillation design reduced energy requirements by 50% over the first design.

■ Steam generation on the column condenser provided 30-psig steam for plant use; its heat content is equivalent to that of the gas required for the fired-heater reboiler.

■ Burning the heavy stream in-plant reduced net plant fuel-gas requirements by about 5%.

■ A high-thermal-efficiency fired heater reduced by 25-30% the fuel-gas requirement over conventional plant heaters.

These results did not just happen—rather, they occurred because management and designers recognized the energy crisis and demanded new, innovative, process designs that conserved energy and, at the same time, made a profit.

Efficient Use of Steam

One area of large potential savings is better utilization of steam. Whenever low-pressure steam is condensed by cooling water or air, roughly 1,000 Btu per pound of steam is wasted. We are searching for, and finding, ways to utilize this energy in low-pressure steam. For example, several of our plants have replaced electric-motor-driven mechanical refrigeration units with absorption refrigeration units, which use 10- to 12-psig steam as the driving force. Previously, this steam was condensed from a 15- to 30-psig system and returned to the power plant as condensate. (The condensate from the absorption units is still returnable to the power plants.) The net effect is to use the energy in the low-pressure steam, thus saving electrical power. One plant in the Texas Div. has saved the equivalent of 20 million Btu/h in this manner, and several others have saved 2 to 5 million Btu/h.

Some of our process-modification projects have accomplished energy conservation and pollution abatement simultaneously. Take the case of a $100,000 capital project at Texas Glycerine 1. One of its towers was extended 14 ft in height. The completed modification also included some design changes, additional trays, and a better mixer. As a result, Glycerine 1 has reduced raw-material costs by more than a quarter of a million dollars per year, eliminated more than 5 million lb of organic compounds going into the waste stream, and reduced the use of steam by 120 million lb/yr.

More Ways To Save Steam

Then there are plants that have made significant savings in energy without spending a dime of capital. For instance, in one plant, 475-psig steam was being throttled to 235 psig because the existing 235-psig system was supposed to be inadequate to provide the full requirement of such steam. Alert plant engineers determined that the line size was adequate; the reason they were not getting full flow was due to the backpressure created by the

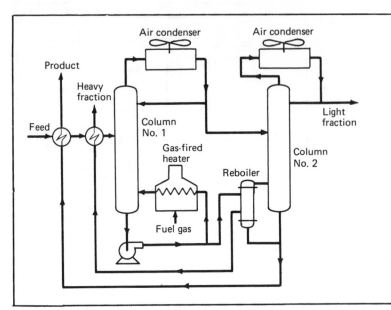
BYPRODUCT RECOVERY, initial design—Fig. 10

steam reduced from 475 psig. The steam-reducing station was shut down and indeed the full requirement of 235-psig steam was made available. The pressure was slightly lower but adequate for requirements. This action resulted in a net savings of 1.5 million Btu/h.

Another plant has several multistage steam jets for the evacuation of noncondensables and inerts from columns. It was found that these jets could handle the noncondensable load adequately with one less stage, so the final-stage steam was turned off. This action saved over a half million Btu/h. This same plant found that it needed only a third to a half the amount of steam used in steam-tracing pipes, and thus saved another 2 million Btu/h.

Tracing is a subject that deserves a bit more discussion. During cold weather, it is commonly used to protect piping and equipment from freeze-ups. Improper operation

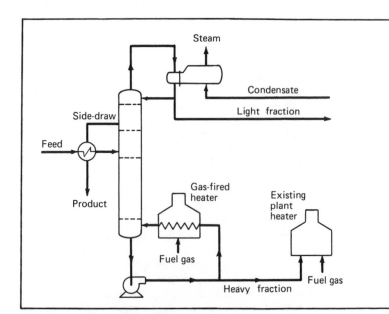
FINAL DESIGN of byproduct recovery system—Fig. 11

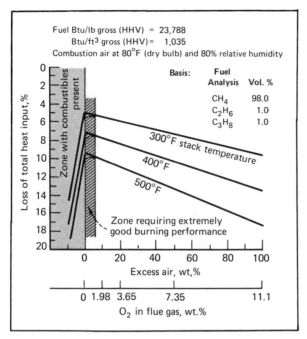

Fuel Btu/lb gross (HHV) = 23,788
Btu/ft³ gross (HHV) = 1,035
Combustion air at 80°F (dry bulb) and 80% relative humidity

Basis:
	Fuel Analysis	Vol. %
	CH_4	98.0
	C_2H_6	1.0
	C_3H_8	1.0

CONTROL of losses in combustion systems—Fig. 12

and maintenance of tracing systems will waste energy. Winter tracing should not be used until necessary and steam should be turned off when the temperature does not require its use. Traced lines should be insulated and traps should be used with all tracers. Preventive maintenance of traps, along with frequent checks to make sure they are not blowing steam, will result in significant steam savings. Also, tracers should be checked for leaks and repaired in the autumn, both to ensure their functioning when needed and to save steam.

Dowtherm SR-1* fluid can be used in place of steam to heat tracers. This type of installation can be economically justified when 15 or more tracers exist. SR-1 is a 50% aqueous solution that freezes at minus 34°F. This makes an SR-1 system practically maintenance-free.

Just one steam trap is needed on an SR-1 system. It is used on the heat exchanger that heats the SR-1 solution in a tank before it is pumped through the tracers. Having fewer steam traps will result in steam savings and lowered maintenance expense. An SR-1 system can achieve steam savings of 1,000 lb/ft of tracers per year.

One of Dow's plants in Midland, Mich., cut its steam consumption by 50% during one winter by converting to a Dowtherm SR-1 heat-transfer-medium system. Although additional electricity was required to pump the Dowtherm, an overall energy savings of about 40% was realized. The additional investment for conversion was almost entirely recovered in the first year.

Condensate Recovery

If condensate is not recovered and returned to the power plant, there is a waste of high-purity water in addition to the heat content of the condensate.

* Dowtherm SR-1 is a tradename registered by Dow for a specially inhibited ethylene glycol.

As an example, one of our large operating plants uses steam engines to drive compressors. The discharge system from these engines contains residual steam-cylinder lube oil that makes the condensate unsuitable for use in the power-plant boilers. This plant installed an oil coalescer in this condensate system that removes the oil (down to less than 1 ppm), and is now returning all the condensate to the power plant. The net savings is 10 million Btu/h.

Another plant recently completed a condensate-recovery project that involved the replacement of two barometric condensers (using cold seawater) on the final stage of two multijet vacuum systems. Approximately 500 lb/h of steam used in each of these systems was lost with the seawater in the condenser, along with minor quantities of a hydrocarbon solvent from the process. The barometric units were replaced by surface condensers, in which the steam condensate (containing the hydrocarbon) is kept separate from the seawater coolant and is collected in a decanter drum, where both condensate and hydrocarbon are recovered. This represents a modest savings of 130,000 Btu/h of energy and eliminates a source of both thermal and hydrocarbon pollution

Minimizing Combustion Losses

Another important area for energy savings is control of combustion. Fig. 12 shows heat losses as percent of total-heat-input versus percent-excess-air for three different stack-gas temperatures: 300°F, 400°F and 500°F. The curves are based on natural-gas fuel of a given composition and on air at specific ambient conditions.

Of course, the higher the stack temperature, the greater the heat losses. For a given stack temperature, the minimum heat losses will occur with zero-percent excess air (theoretically). However, the range of 0 to 5% excess air requires extremely good burner-performance, and one would certainly not want to operate with a deficiency of air (where heat losses increase very rapidly). Furthermore, a choke condition is unsafe and emissions of carbon monoxide and combustibles in the flue gas are objectionable.

So in practice, we operate our power-plant boilers in the range of 5 to 10% excess air. In addition to having percent-oxygen analyzers on individual boiler stacks, we are installing combustibles analyzers. This lets us control each boiler at the proper excess air for complete combustion and optimum efficiency.

Our power-plant boilers incorporate economizers and combustion-air preheaters to reduce flue-gas heat losses. But there are many direct-fired furnaces in our process plants that have high stack-temperatures. Stack heat-recovery equipment (such as air preheaters and heat exchangers for generating steam) that could not be economically justified in the past is now being designed in view of projected higher energy-costs. We are even evaluating incinerators that include heat recovery systems.

The Magnesium Products Dept. in the Texas Div. is burning waste organic-chlorides (from other Dow plants) in acid furnaces. Not only is there a savings in excess of 90 million Btu/h of fuel gas but, also, HCl is recovered for recycle to process, and from waste materials that would otherwise present an ecological disposal problem.

Heat losses through furnace walls are receiving attention. For example, a study made on the lime kilns at the Texas Div. has revealed justification for changing the refractory lining from 6 in. thickness to 9 in. Potential fuel savings amount to $36,000/yr per kiln.

Actually, the whole subject of insulation is being reviewed in light of higher energy-costs. We are revising our specifications for thickness of insulation. We can now afford to insulate at lower temperature levels than previously—for example, condensate-return-lines and storage tanks. And we can justify increasing our effort on maintaining insulation.

Electricity Savings

Energy-loss considerations have always been a part of electrical-system design. The electrical engineer strives for the most economical overall design consistent with good performance, reliability, maintenance, and power costs. Increasing cost of power naturally promotes the use of larger power cables, capacitors for power-factor correction, more-efficient motors, generators, transformers, rectifiers, and other power equipment, to reduce energy losses.

Consider power conductors. In continuous-process plants, the load factors are usually high and, consequently, minimum sizes of conductors for main power feeders and some motor branch-circuits are normally based on power losses rather than thermal limitations, especially where the conductor is aluminum. An exception will apply for copper conductors, due to the higher cost of the metal. Our main feeder sizes range up to 2,000 MCM aluminum, with 3-conductors per phase. (Note: MCM = thousand circular mils. One circular mil = 7.854×10^{-7} in². Hence, 2,000 MCM = 1.572 in².)

Transformer sizes will be determined to a large extent by loss considerations, in addition to initial loads and allowance for load growth. Transformers with low ratios of conductor-to-iron losses are necessary for maximum efficiency where the load factor is high. However, a higher loss-ratio will be preferred for low load-factor cases—for example, where a plant is served by two transformers, each normally operated at approximately 50% capacity, with provision for either to carry the entire plant load. Forced-air-cooled transformers are usually selected for the latter cases, and self-cooled transformers for the high-load-factor cases.

With regard to motors, generators, power rectifiers, and other power-consuming equipment, the same general loss considerations apply as in the previous cases. Some premium cost can usually be justified for the higher-efficiency equipment.

Lighting

For many years, criteria applied to lighting design for new buildings, offices, and process plants were somewhat along aesthetic lines. With cheap energy, operating cost was not a restraint. But now with the need for conservation, we must take steps to make better use of energy required for lighting.

At the Texas Div., lighting energy is now being separately metered in new plants. This was not the practice in the past. Also, high-intensity-discharge lamps are installed in new plants to obtain high efficiency.

New process-plant lighting is designed only to meet applicable industry standards. After the plant is in operation, a light-level survey is conducted and field changes are made to correct deficiencies. This is in contrast to prior days, when overdesign was the rule.

In newer plants, we are using photocell control on outdoor lights to reduce burning time; in older plants, operating personnel are made responsible for turning off lights when not needed. We cannot afford to be lax and leave lights burning when not in use. We are conducting a continuing campaign and post "Turn Lights Off" and "Energy Conservation" decals at light switches.

The use of capacitors in industry to improve electrical power-factors is almost as old as the electric industry. The aims are well known:

- To obtain a more favorable billing charge.
- To increase the kW load of a system through low capital expenditures.

In the past, our applications were usually in large sizes on our 15-kV systems. On our 2.4-kV systems, 300 and 600-kV a.c. banks were scattered throughout the plant to relieve loading problems. Today we are investigating capacitor installations on our 480-V motor centers to relieve system I^2R losses.

Conclusion

In summary, important elements for an effective energy-conservation program are:

- Management endorsement and interest.
- Line-supervision attention to monitoring Btu count and to implementing energy-saving actions and projects.
- An energy-conservation team conducting a continuing program to focus interest in energy conservation and to communicate ideas, techniques and progress regarding energy savings.
- Process Engineering Dept.'s innovations in its designs, both of existing-plant modifications and new plants.

Acknowledgement

The author wishes to express appreciation to his associates on Dow Chemical's Texas Div. Energy-Conservation Team for their contributions to this article.

Meet the Author

Jack Robertson is Senior Technical Specialist in the Utilities Dept. of Dow Chemical U.S.A.'s Texas Div. in Freeport, Tex. For the past year, he has spent full time as coordinator of the Division Energy Conservation Team. His background includes metals research and development, power-plant supervision, and management of a mining and milling operation. He has a B.S. in mechanical engineering from the University of Texas, is a member of the American Soc. of Mechanical Engineers, and a registered professional engineer in Texas.

Changes in Process Equipment Caused by the Energy Crisis

ERNEST E. LUDWIG, Ludwig Consulting Engineers, Baton Rouge, Louisiana.

Energy as a resource has rapidly moved from a position of sufficiency to one of real concern and shortage. A significant point stands out: U.S. industry does not fully appreciate the long-term impact of the energy crisis. In perhaps 80% of today's industrial planning and construction, energy-conserving efforts do not reflect the full seriousness of the situation.

As a result, equipment vendors serving the chemical/petrochemical industries are generally responding only to the specific problems being presented. Only in limited instances is there a completely innovative approach with significantly improved equipment for a conventional problem.

Process-equipment vendors usually respond to requests for specific performance by modifying or applying their conventional line. If the process engineer is not innovative, the equipment will fall into the usual "mold." So he is an important part of the creative process for conserving energy. To reinforce his position, management must insist that energy conservation is a real target, which requires accountability if not attained.

Implementing Conservation

Necessary improvements to processing steps must be translated into the appropriate hardware or equipment. For example:
- Increased use of catalysts.
- Process optimization via instrumentation and equipment designed for the purpose.
- Process integration, rather than independent self-sufficient process units.
- Improvements in metallurgy.
- Equipment-efficiency improvements.

What can be expected? Where such energy conservation is in effect, the savings are estimated to be upward of 30%. Alfred Waterland, Manager of Energy Management Services for Du Pont Applied Technology Div., predicts savings of as much as 50%.*

*Speech to Chemical Marketing Research Assn., New York, Nov. 12, 1974.

Originally published March 17, 1975

The predominant forms of energy as applied by the chemical process industries (CPI) are:

	Costs	
	Current	Est. 1980
Electric power, kWh	$ 0.015	$ 0.025-0.05
Steam, 1,000 lb	0.80	2.50-4.00
Natural gas, MMBtu	0.75	1.80-4.00
Oil, bbl	11.00	14.00-20.00

Unfortunately, the problem is not simply one of cost but of availability, and this changes the approach to many equipment-performance problems. A shortage of natural gas for fuel after 1985-1990, and the doubt that synthetic natural gas will be available, requires a significant shift in equipment planning. Design philosophy must recognize that the *cumulative impact* of 1 kWh saved from one pump and 20 kWh from a compressor drive, etc., is not only economical but essential.

The Impact of Energy in Processing

The two obvious energy forms, heat and power, are utilized in many ways in process plants. Two of the more prominent energy-intensive operations are:

Fluid Pumping—Often the heights to which liquids must be pumped are taken for granted, and receive no specific consideration during the planning of buildings, structures, etc. In the same way, processing steps involving high backpressures need scrutiny to reduce the head and corresponding energy requirements. By evaluating alternate situations, more-efficient pumps can often be selected. The equipment manufacturer's role here is to call possible efficiency differences to the attention of the design engineer.

Heat Recovery—Significant energy savings are often possible in heat recovery. Here, the interchange of heat levels can include; either fluid-to-fluid for process and/or utility streams, or recovery from process or utility operations.

Reduction in actual heat requirements—by use of ef-

85

ficient reformer furnaces by means of catalyst conversions and tube metallurgy—can be significant, sometimes as much as 5-15%.

The selection of boiler type and capacity has a real impact—15 to 25%—on the utilization of fuel for firing.

Equipment-Industry Trends

Heat Exchange—The conventional heat-exchange fabrication industry has received the most impressive initial impact of conservation work, indicating that perhaps the most obvious and sometimes easiest installation is being sought out. The trend to some extent includes:

■ Heat cross-exchange on fluids having low temperature differentials, thereby necessitating large surface areas. Previously, this application would have been termed marginal or submarginal.

■ Integrated heat exchange between systems that are independent process-wise, but close enough physically to make a heat recovery feasible. In this way, plants provide heating (or cooling) to streams that had been utilizing steam or cooling water independently.

■ Equipment designs to accomplish the maximum amount of heat transfer with the lowest practical pressure drop.

■ Air coolers/condensers, which are in great demand as the industry seeks to achieve cooling/condensing operations that were previously carried out with water. There is an energy saving in the pumping portion of water circulation. (This also satisfies environmental protection regulations in the thermal pollution classification.) Greater emphasis is being placed on marginal cooling levels, thereby requiring large air-cooled bundles in multiple banks. Temperature applications as low as 100-115°F. are being considered for areas where the climate will allow an acceptable and dependable ΔT. Previously, 120-125°F. was about as low a temperature as economics would justify.

As the costs for motive energy increase, there will usually be a strong position for air coolers, because of expenses that develop with time against inadequately maintained cooling-water circuits.

■ Higher thermal efficiencies for fired process heaters to provide the very basis for some process operations. Manufacturers are now confronted with designing and building smaller units to serve as high-temperature heat sources, replacing high-pressure steam. At the other end of the spectrum, larger units are being designed to accommodate the economics of single high-capacity systems. These units demand increased investigations of tube metallurgy, creep, and high-temperature stresses.

Fuel types are more varied, and the ability to switch or use mixed gaseous-liquid fuels is more frequently involved in the design criteria. Dual burners are now being designed into the firing systems.

Heat recovery in many forms—i.e., preheating, superheating and trimming (representing many different streams)—is involved. This makes the design balance more complicated and reflects in the attention needed during operation. Waste-heat recovery in the form of steam, using packaged boilers from 10,000 lb/h to 100,000 lb/h and even larger in multiple units, is in great demand. Heat that previously was wasted to the air is recovered to the lowest level practical just above the flue-gas dewpoint. Usually, this is 500-600°F., but it could go lower in the proper equipment arrangement.

Heat-recovery efforts have boosted many furnace systems from 83-85% efficiency to 89-93%, which represents significant savings in fuel consumption and in operating costs. Obviously, there is a capital cost for the premium equipment; however, most CPI companies go after the fuel saving and do not feel obligated by the conventional rules for return and payout on the incremental premiums.

■ Packaged steam boilers, which are being produced at a record rate to satisfy the increasing demand for applications previously foreign to such heat recovery. They are being made smaller, even 5,000 lb/h, to fit specific situations, as well as larger (100,000-300,000 lb/h) for applications formerly not considered package situations. Also, the operating pressure levels are increasing from the 250-psi range to 1,000 psi.

Of perhaps most significance is the strange conglomeration of chemicals and hydrocarbons—liquids and gases-being burned as the fuel source with only minimum maintenance of load by fuel oil or natural gas. Some of the materials are chlorinated hydrocarbons, phenols, solvents of all types, chlorine, pharmaceutical wastes, etc. These systems operate satisfactorily after due care has been given to the design and the selection of materials of construction. Of course, life expectancy and maintenance should be less favorable than for conventional clean-fuel units.

Compression Equipment

The key point in compression equipment is the increase in number of vapor-recovery applications. Part of this relates to EPA regulations for atmospheric pollution control, but approximately 50% relates to the economy of increased investment to reduce operating costs.

The compressor manufacturer sees a demand to accommodate small flow conditions at anywhere from vacuum to medium pressure. Many of these applications require new energy demands.

The high demand has created shortages in forgings and castings, and has forced a consolidation of models/sizes into a limited number of standards, for which the manufacturer attempts to develop an inventory of assembly parts.

Due to the loads on foundries, the call for highly specialized alloys creates delivery problems. Consequently, some manufacturers have stopped offering particular designs and materials of construction.

Refrigeration Systems

The power consumption of new refrigeration systems is being examined in greater detail to achieve the lowest energy level. This contrasts with the previous practice of selecting the lowest capital cost. Such refrigeration systems involve normal compression condensation with evaporation, plus interstage flashing and process heat exchange.

The compressor manufacturer may provide many case studies for the design engineer; however, once the system is fixed, this joint effort dissolves, and the balance of the equipment is designed separately.

The use of refrigeration for vapor recovery has increased markedly in the past two years. Many applications require reciprocating compressors in the 20-500-hp range. Problems have arisen in deliveries of castings and forgings for these units, as there are not acceptable compromises in the selection of materials of construction.

According to F. B. Neyhart of Carrier Corp., centralization of refrigeration generating plants is developing, with such benefits as:

- Efficiencies of larger systems, both for the refrigerant compressor and the electric-motor drive train, including gears where applicable.
- Efficiencies of operating manpower.
- Power savings from the use of larger-than-nominal heat exchangers for high suction temperatures and lower condensing temperatures. In one installation, 15% power reduction was accomplished by using 9,000-ton heat exchangers for a 4,000-ton refrigeration application. This trade of extra capital dollars for power savings is becoming reasonably common among the progressive and energy-sensitive CPI.

Process designers are more frequently providing refrigerant-system designers with more economic data, to allow the latter to better recommend a system. Absorption refrigeration is receiving increased consideration in situations where waste heat is available, since the resultant energy saving can be significant, compared to a mechanical compression system.

Materials of Construction

The materials of construction—metals and nonmetals—used in process equipment are truly caught in a squeeze play. The demands of high-temperature heat recovery and higher pressure boilers, along with heat recovery from more-corrosive fluids, have created significant shortages in the more popular alloys. For example, high-temperature reforming is encouraging investigation of many previously untried or unproven alloys, and the recovery of heat from chlorinated and other corrosive process streams has extended metallurgical capabilities to new limits.

Many designs are being constructed with liners of plastics or rubber to protect vulnerable metals; and in application of 1,500-3,000°F, poured, cast, or troweled-in-place refractories are applied with due consideration for presence of acidic or basic components in the contacting streams.

Scrubber towers in many systems are receiving a double demand: recovery of heat, and neutralization of corrosive vapors. To avoid delay in deliveries of the more popular metals, plastic or rubber linings are applied over carbon steel protected by acid-resistant brick linings with appropriate mortars. In other applications, a buttering of acid/alkali-resistant mortar directly to the plastic or rubber linings has proved successful. Although some of these techniques have been applied for many

years, recent energy demands plus environmental requirements have encouraged their use. Unfortunately, brick and ceramic packing for towers is also in very short supply.

Energy Losses in Existing Plants

Although existing plants might be considered stuck with their installed equipment, they are taking steps to detect and evaluate potential and real energy losses. The larger companies have established corporate taskforces, assigning them to reduce energy consumption by from 10 to 25%. Smaller companies leave such assignments to the local plant.

Those companies with most success in energy savings have followed the approach of studying the overall picture, rather than local or isolated situations. Because of the relationships between electrical power and steam or waste hot fluids or liquids, the maximum benefit requires an energy-flow balance throughout the plant and with adjoining process or utility-related plants.

It is well known that random steam leaks throughout a plant, due to insensitive plant supervision or maintenance, can account for thousands of pounds of steam, which can be translated into a waste of natural gas or fuel oil. In many instances, low-pressure steam that has been vented can be recovered by condensation and the condensate pumped back for reuse. In other instances, the low-pressure steam can be recompressed to a higher pressure for more-effective use. And another approach involves reexamining the levels of steam throughout a plant and perhaps adjusting some of these for a more beneficial use without low-pressure waste. The flashing of pressurized condensate can also provide useful steam for specific applications.

A common source of steam losses involves the misapplication and poor operation of many steam traps. In closed systems, plant personnel cannot fully realize how much semi- to continuous blowby exists. The answer to this problem is not to hastily replace the trap with a new one of the same size or design, but rather to carefully examine the operation of the equipment that the trap serves and to attempt to select a trap better rated for the system characteristics.

Detection of high-temperature heat losses is often made rather easy by just touring the plant and examining the temperature readings for the various streams. In order to avoid rash and illogical process recommendations, it is necessary to lay out the heat balance for all interrelated streams plus other available streams, in order to determine whether (a) more-complete heat recovery can be achieved and (b) the quantities will, in fact, balance when considering their respective enthalpies.

Rearrangement of streams is often involved and can pose significant physical problems. Often, improved heat recovery requires parallel or series sets of heat exchangers put into the system. For those rather straightforward situations where high-temperature flue gases or air or acceptable process vapors are being released to atmosphere, it may become only a matter of adding a waste-heat recovery unit, which may be anything from a preheater for another stream to a waste-heat boiler for steam generation. The magnitude and temperature level of the waste heat determine just how much can be done process-wise. Of course, physical space may pose challenging problems in accomplishing the process heat recovery.

Consistent improvement efforts and the recovery of heat to the lowest practical working level have reduced fuel consumption by some plants 5-20%. At times, the mixing of compatible combustible waste streams with boiler fuel has allowed particular installations to cut natural gas or fuel oil by 50%. Many plants have hydrocarbon-rich streams, including some with hydrogen, that can effectively be utilized.

A very common source of heat loss is inadequate and poorly applied insulation. The economics of heat-loss insulation is changing rapidly. Thicker layers of insulating material, or the use of high-quality material, is often justified.

Identification of power losses also requires the preparation of an integrated power "flow/use" diagram. With this as a base, actual measurements of power consumption can be made. Perhaps the most common waste of power is the pressure reduction of liquid or vapor streams. Some processes do attempt to recover such pressure energy by specialized power-recovery equipment. The gas-production industry has used gas power-recovery turbines for over 50 years.

Process-plant streams vary widely in magnitude of flowrate, which poses real problems in the selection of power-recovery equipment. For gas/pressure letdown, there are small and large power-recovery turbines that can be used to drive pumps or compressors. For liquid systems, hydraulic-recovery turbines are available. Although these have not been in widespread process uses, they can now serve in this capacity, following proper application analysis.

Electrical energy is frequently wasted by the selection of (a) inefficient driven equipment, and (b) small equipment that requires small electric motors that are inefficient compared with large units. Electric motors are most efficient when operating at high speeds; for example, an induction motor of 200 hp is about 90.5% efficient at 514 rpm and full load, whereas the same motor at 1,800 rpm is about 92% efficient. If the corresponding motor were selected as a synchronous one, the efficiencies would jump to 92.6% and 93.6%, respectively. When speed changes from the motor to the driven equipment require the use of gears, the overall system efficiency falls into the 75-85% range.

Unfortunately, as energy becomes less available. the mere fact that these existing plants are built with energy-intensive equipment may in some instances cause a plant to shut down. In effect, it will not be able to compete for the available energy. Recognizing this predicament, many plant managers are initiating programs to reduce all forms of energy consumption. Other plants that try to stick with their present situation may very well be in serious trouble later on. With such plants, equipment modifications will not be a matter of payouts in the true sense of economics.

Existing Equipment Modifications

Unfortunately, some existing equipment is very difficult to modify to reduce energy consumption. This type of equipment is usually designed to suit a set of, or range of, operating conditions, and if these are changed, the performance becomes worse. Examples are: shelf-type dryers, rotating kilns or drum dryers, certain scrubbers/separators. In order to achieve worthwhile improvements for such equipment, specific tests and reevaluation must be made, because no "standard" modification can be expected to produce fixed results.

Rotating equipment such as fans or blowers can often have wheels or impellers replaced or modified to a more efficient configuration. Some gear drives for such equipment can be replaced by gears with more-efficient ratios. Old steam-turbine drives can usually be replaced with modern, more-efficient units.

High-horsepower equipment mixing liquids and suspensions can usually effect some power savings by replacing small high-speed units with one that is larger and lower speed; however, this requires careful evaluation to ensure equivalent process performance. The gear drives on such units can often be replaced by units of improved efficiency.

Several streams can be combined into one pump to allow operating in an improved efficiency range; and inefficient pumps can be changed to more-efficient ones, as for example, a centrifugal pump for an axial or turbine pump.

Compressor replacements are usually more difficult to justify unless the horsepower is large and the present equipment's efficiency is relatively low. Of course, each incremental improvement may contribute to keeping the plant operating and in business, so it should not be ignored. Old reciprocating units may be improved upon by physical changes by cylinders, or by replacement by more-efficient centrifugal units.

Meet the Author

Ernest E. Ludwig is owner of Ludwig Consulting Engineers, 5615 Corporate Blvd., Baton Rouge, La. 70808, which he has operated since 1969. Prior to that, he was vice-president for manufacturing, engineering and industrial relations at Copolymer Rubber and Chemical Corp., project and process manager for a polyethylene plant for Rexall Chemical Co., and general works manager of its Odessa, Tex., polyolefin facility, and process design manager for Dow Chemical Co. Well known for his books and papers, as well as his activities in many societies, he is a registered professional engineer in 6 states, and holds BS and MS in chemical engineering from the University of Texas.

Techniques for saving energy in processes and equipment

In an attempt to gain perspective on the effects of the energy shortage we surveyed, earlier this year, all of the firms known to us to be offering technology for license or sale, asking what processes had been altered to achieve significant reductions in energy consumption.

The following four articles are a consequence of that survey. The first three were selected for the energy-saving technology they reveal, as well as their significance in terms of energy consumption. The fourth tells how to save energy—and money—by including energy-consumption in bid evaluation. These articles are:

Originally published July 4, 1977

Balancing energy costs against equipment costs

The rise in energy costs has changed plant-design concepts. Experience with ethylene plants suggests modifications that might be used elsewhere.

S. B. Zdonik, *Stone & Webster Engineering Corp.*

☐ A careful study of over 100 different points of energy consumption in a typical large ethylene plant has yielded enough modifications to provide at least a 20% reduction in the energy consumption of plants designed within the last three years. Since the earlier plants consumed on the order of 11,000 − 15,000 Btu of equivalent fuel per pound of ethylene, depending on the feedstock, a 20% saving represents an appreciable reduction in the cost of manufacturing ethylene and its coproducts.

The problem is that of balancing fuel savings against the additional investment for equipment and auxiliaries. Many energy-saving modifications or additions that formerly promised too little in return for the investment required are now being considered because of the rapid rise in energy costs relative to equipment costs. Thus, an investment of approximately $2-3 is required to save $1 per year in fuel cost in producing ethylene. For each energy-saving feature studied, only those providing less than a two- to three-year payout were considered.

Process steps for energy reduction

An optimum balance between energy reduction and investment increase can only be realized through an analysis of the overall ethylene process. A flowsheet for a typical ethylene plant is shown in Fig. 1, which contains the basic operating sections, including cracking, primary fractionation, cracked-gas compression, gas treating, gas drying, demethanizer-feed cooling, hydrogen/methane separation, cold fractionation, hot fractionation and gasoline hydrotreating. Steam and refrigeration systems, although not shown, must also be considered in the search for energy-saving possibilities.

In this process, the Stone & Webster USC furnace cracks a range of hydrocarbon feedstocks to produce an ethylene-rich stream. The primary fractionator separates most of the heavy-molecular-weight fractions (fuel oil in Fig. 1) from a lighter portion containing hydrogen, plus heavier components through cracked gasoline.

The gasoline is condensed, and the C_4 and lighter portion is then compressed to a pressure that will permit separation by distillation, using refrigeration where required. Four stages of compression are frequently applied, with interstage cooling. Condensed liquids are recycled from the intercoolers to prior stages of compression. Acid gases (mostly CO_2 resulting from pyrolysis, and H_2S) are removed by scrubbing the compressed cracked

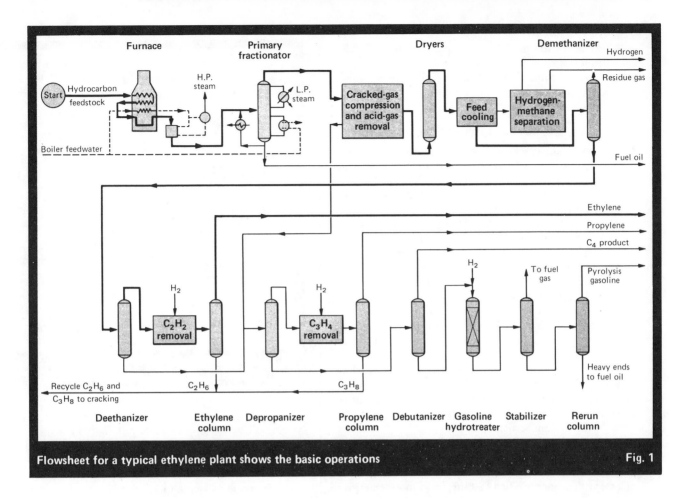

Flowsheet for a typical ethylene plant shows the basic operations   Fig. 1

gases with caustic, usually prior to the fourth stage of compression. The discharge of the final stage of compression is dried to a low water-dewpoint.

In the cold-distillation section, hydrogen and methane are separated. Subsequent separations are usually in the order of increasing molecular weight (i.e., ethylene, ethane, propylene, propane, etc.). To meet the rigid product specifications for polymer-grade ethylene and propylene product, C_2 acetylene, methyl acetylene and propadiene must be removed. These highly unsaturated compounds are most conveniently eliminated by hydrogenation. The net overheads from the deethanizer and the depropanizer are catalytically hydrogenated to convert the contained C_2H_2 and C_3H_4, respectively, in these overhead streams.

The debutanizer net overhead containing butadiene is sent to battery limits and usually becomes the feed to a butadiene recovery unit.

The bottoms from the debutanizer ($C_5 - 400°F$) is the raw pyrolysis gasoline and is usually sent to a gasoline hydrotreater for conversion of dienes, if a stable high-octane gasoline is desired. If BTX aromatics are to be recovered as separate products, the olefins must be converted, and a BTX heart cut produced, prior to aromatics extraction and product fractionation.

Energy-saving modifications

In this overall process scheme, six modifications have been selected for illustration because they not only represent possibilities for energy savings in ethylene plants but also have some application to other energy-intensive processes. These six are:

■ Increasing the thickness of furnace insulation.
■ Reducing the interstage pressure drops of the cracked-gas compressor.

Intermediate reboiler on deethanizer takes heat from quench water   Fig. 2

■ Increasing the number of temperature levels in the refrigeration systems.
■ Using intermediate reboilers on distillation columns.
■ Reducing operating reflux ratios.
■ Using turbo-expanders.

Increasing the furnace insulation

Ethylene-plant cracking furnaces are well known for their extremely high operating temperatures. The radiant tubes are supported vertically in a box lined with high-temperature insulation and refractory bricks. These high temperatures are achieved by the use of radiant-type burners installed in the walls facing both sides of the radiant reaction tubes. Cracking furnaces capable of producing 100 to 150 million lb/yr of ethylene are being designed. The fired duty for one of these furnaces can range from 90 to 150 million Btu/h (LHV).

It can be appreciated that even a 1% reduction in heat loss can result in a significant fuel economy.

Because of the high firebox temperatures, the heat loss from an ethylene-plant cracking furnace is typically more than that from other types of process furnaces. In the past, the high cost of firebrick and block insulation, with low fuel cost, did not mandate high furnace efficiencies. However, changes in energy costs have in some cases made it economical to double the thickness of the brickwork in the radiant section.

Our studies have shown that such modifications can save as much as 1% of the furnace fuel costs, with a payback of less than three years on the additional investment. Furthermore, we find that the use of more-extensive and more-expensive types of insulation for furnaces, reactors, high-temperature ductwork, heat exchangers, etc., is generally worth reexamination.

Reducing interstage pressure drops

Considering that the compressor requirements for ethylene manufacture are roughly 0.50 hp per lb/h of ethylene product, and that the horsepower of a given compressor is roughly proportional to the pressure differential, the complex interstage cooling/condensing and recycling systems of the usual ethylene plant offer a prime target for energy reduction.

Our studies have shown that the lower compressor brake-horsepower resulting from larger interstage piping more than balances out the added investment for reducing interstage pressure-drops through equipment and lines. Typically, the payback period is less than two years, from energy savings of up to 2% of the total plant fuel-cost.

Increasing levels of refrigeration

Most ethylene plants utilize both ethylene and propylene refrigerants to provide feed cooling and to condense overhead reflux for the low-temperature distillation towers. Usually, the ethylene refrigeration system will operate at two temperature levels, and the propylene system at three or four.

However, some of the cooling services could be adequately provided at temperature levels intermediate to, or higher than, those normally afforded by these two systems. Increasing the number of levels reduces the total brake horsepower required for refrigeration compression. Three levels, instead of the usual two, for the ethylene

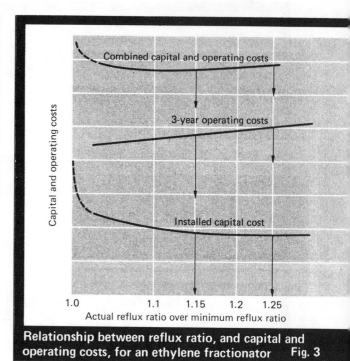

Relationship between reflux ratio, and capital and operating costs, for an ethylene fractionator   Fig. 3

system will permit approximately 0.3% reduction in total plant-utility costs for a payback period of less than three years. Increasing the levels of the propylene system to five permits a utility cost reduction of about 1.7% for a payback period of less than 1.5 years.

Use of intermediate reboilers

The deethanizer column in Fig. 1 separates C_2s as overhead, from a C_3s-plus as bottoms product. Normally, the deethanizer is reboiled with low-pressure steam, with the overhead being cooled by refrigeration. By adding an intermediate reboiler near the middle of the tower, in order to handle part of the reboiler duty at a lower temperature (Fig. 2), quench water can be substituted for the steam. This not only saves the steam, but also reduces cooling-water requirements for a combined fuel saving of approximately 1%. The payback period for the added equipment is much less than one year.

Reducing operating reflux ratios

A generally accepted rule-of-thumb for distillation-column design has been to use a factor of 1.25 times the minimum reflux for establishing the operating reflux and determining the theoretical number of trays required to effect a separation. In the past, numerous studies indicated that this factor gave the minimum overall cost for a wide variety of distillations within the framework of fuel costs at the time.

However, increased energy costs have required that reflux ratios, particularly in many ethylene-plant columns, where the reflux duty is provided by refrigeration instead of cooling water, be reexamined. Ratios of actual reflux to minimum reflux are now more economical in the range of 1.1 − 1.2, with the cost of additional distillation trays being offset by the reduction in refrigeration requirement.

The relationship between total cost and reflux ratio for

the ethylene tower is shown in Fig. 3. The approximate reduction of 0.5% in total fuel cost, made possible by reducing the reflux ratio, usually provides payback periods of less than one year. In general, a careful examination of reflux ratios for all types of distillation towers, particularly of those employing refrigeration, is recommended.

Use of turbo-expanders

From 8 to as much as 65 lb of methane are produced for each 100 lb of ethylene, depending on the feedstock. Normally, this methane is used as fuel in the cracking furnace, where it is required at pressures on the order of 50 psig. Since the demethanizer in an ethylene plant frequently operates at a pressure of around 475 psig, the large methane stream offers an opportunity for recovering considerable pressure energy.

Thus, when the $-145°F$ net overhead vapor from the demethanizer of a 1-billion-lb/yr ethylene plant using naphtha feedstock is preheated through heat exchange with the deemethanizer feed and then passed through a turbo-expander, the overall fuel saving is 0.5 – 1.0%. In this system, the methane is expansion-cooled in the turbo-expander to serve as a substitute for ethylene refrigeration and then passes through a compressor driven by the turbo-expander prior to delivery to the fuel-gas system.

The total heat-exchange duty of the demethanizer system remains the same, and all the required compressor power is supplied by the turbo-expander. However, the system's ethylene refrigeration duty is decreased significantly. The extra capital costs for this modified system include those for the turbo-expander/compressor, the exchanger modifications, and a bypass loop. The payback period is less than two years.

In summary, the foregoing energy-saving modifications can be used with a variety of other processes. Each plant should be considered separately, since variations in throughput, type of feedstock and product distribution will often have a determinng effect on the economic justification of a specific modification.

The author

S. B. Zdonik is Manager, Process Dept., Stone & Webster Engineering Corp., P.O. Box 2325, Boston, MA 02107, where he is responsible for technical coordination of the process design activities of several offices of the firm's Process Industries Group on a worldwide basis. Author of more than 40 technical papers, and a P.E. in Massachusetts, he has been elected a fellow of AIChE. He holds an S.B. and S.M. in chemical engineering from Massachusetts Institute of Technology.

Optimizing the ICI low-pressure methanol process

By using the exothermic heat of methanol synthesis to generate high-pressure steam, this popular process can economically handle synthesis gas from a wide range of sources.

A. Pinto and *P. L. Rogerson*, *Imperial Chemical Industries Ltd.*

☐ Total world capacity of plants licensed to use the ICI low-pressure methanol process now comes to about 5.7 million short tons/yr, representing 80% of the world's new methanol capacity built since 1967.

When this process was first introduced in 1966, the operating pressure was that adopted for the original plant built by ICI—750 psig. But the efficiency of this methanol synthesis pressure was subsequently questioned as the cost picture changed.

Beginning in 1969, a critical examination showed that, with the energy and capital costs prevailing in the U.K., a higher synthesis pressure would bring significant advantages. Also, as part of a policy of continuing research, a

newly formulated methanol synthesis catalyst was developed, one that was more suited to operation at higher pressures.

Consequently, a pressure of 1,500 psig was selected for a new ICI methanol plant, which was brought onstream in 1972. Use of this higher pressure affords advantages that accrue largely from a reduction in capital investment and a simpler engineering design. At a methanol synthesis pressure of 1,500 psig, the "carbon efficiency" of the synthesis loop is about 17% higher than at 750 psig. Hence less natural-gas feedstock is required.

The total capital and operating costs for a methanol plant depend on the efficiency with which CO and CO_2

are converted to methanol through reaction with H_2. The efficiency of this conversion is here defined as "carbon efficiency," which is:

$$\frac{(\text{moles of methanol produced}) \ 100}{(\text{moles of CO} + CO_2 \text{ in the synthesis gas})}$$

As the carbon efficiency increases, the requirement for reformed synthesis gas is reduced and, therefore, reformer size and capital cost. Also, loop equipment is simplified at the higher pressure. This leads to an overall lower capital cost, compared with a similar-sized unit designed to operate at 750 psig.

However, the 1,500-psig operation incurs a greater requirement of horsepower. Fig 1 shows how the total compression-power requirement varies with methanol synthesis pressure for a production of 1,000 short tons/d of refined methanol from natural gas, assuming a fixed suction pressure and constant machine efficiency. Although the flow of synthesis gas decreases with increasing synthesis pressure, the power required to compress the gas increases substantially and is not compensated for by the reduction in flow. Thus the total compression horsepower increases, and this requires more steam generation for turbine drives, and additional fuel for generating the steam.

Recovering reaction heat

On the other hand, the synthesis reaction liberates about 38,000 Btu/lb-mole of methanol. It is possible to recover approximately 70% of this heat at a high temperature and to use this high-grade heat to generate steam in several different modes (Fig. 2), for which the maximum steam pressure lies between 350−650 psig.

The most effective and economic method for recovering this high-grade heat, however, is to use it for preheating high-pressure boiler feedwater, in order to generate steam

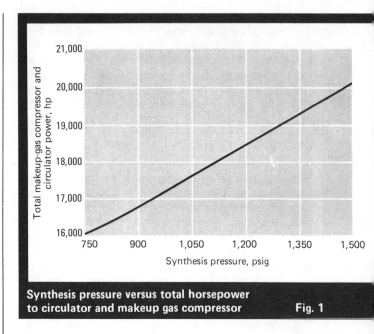

Synthesis pressure versus total horsepower to circulator and makeup gas compressor   **Fig. 1**

at 1,500 psig. This high pressure affords a much higher thermodynamic efficiency when letting the steam down through turbines, resulting in lower total steam needs.

Fig. 3 shows how this boiler-feedwater-heating system is integrated into a typical methanol plant based on reforming natural gas or naphtha. Since the process sources of high-grade heat are limited, however, all of the power requirements cannot be met by backpressure turbines exhausting to low-pressure steam users. Instead, the increased power requirements of operating above 750 psig must be met through the use of condensing turbines, which reduces the overall efficiency of the process because of their low thermal efficiency.

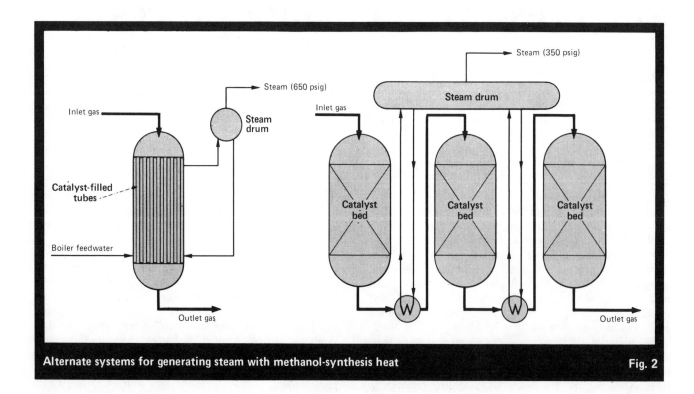

Alternate systems for generating steam with methanol-synthesis heat   **Fig. 2**

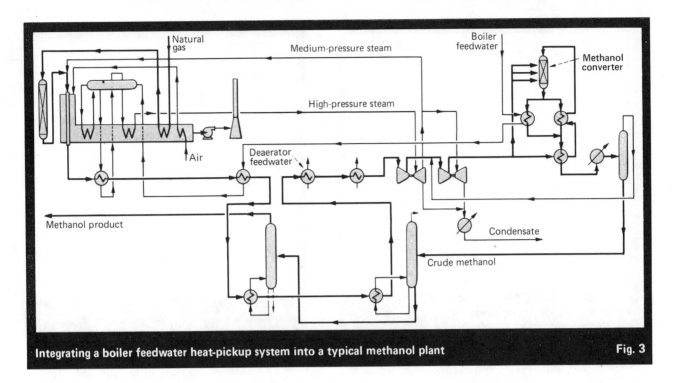

Integrating a boiler feedwater heat-pickup system into a typical methanol plant   Fig. 3

Fig. 4 shows the improvement in thermal efficiency obtained by raising the maximum operating steam pressure from 900 to 1,500 psig. The two steam systems considered are:

Case I: 1,500-psig, 860°F high-pressure steam
400-psig medium-pressure steam
21-psig low-pressure steam

Case II: 900-psig, 860°F high-pressure steam
400-psig medium-pressure steam
50-psig low-pressure steam

Also, Case I uses gas-heated reboilers, condensing turbines, and the maximum heat recovery from the methanol synthesis loop; Case II uses the same conditions, but without condensing turbines. It is apparent in Fig. 4 that a maximum saving occurs, in both cases, when the pressure of the methanol synthesis reactor is operated at 900/1,200 psig.

Combining gas preparation/synthesis

The above-described synthesis loop and heat recovery system adopted for the ICI process have the necessary

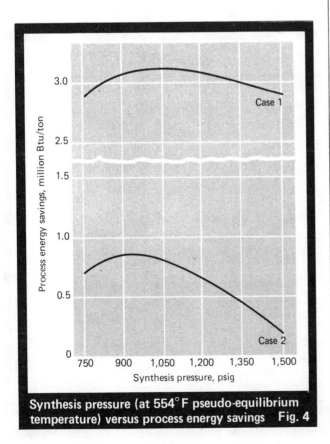

Synthesis pressure (at 554°F pseudo-equilibrium temperature) versus process energy savings   Fig. 4

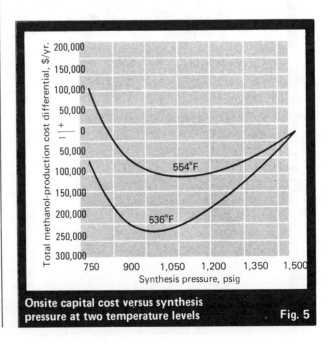

Onsite capital cost versus synthesis pressure at two temperature levels   Fig. 5

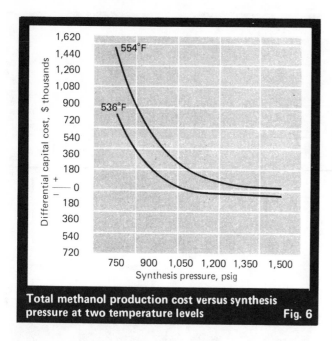

Total methanol production cost versus synthesis pressure at two temperature levels   **Fig. 6**

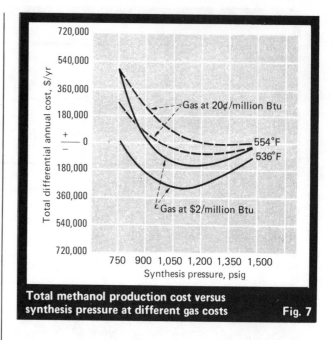

Total methanol production cost versus synthesis pressure at different gas costs   **Fig. 7**

flexibility and impose few constraints on the total design of a methanol plant, which may include various processes for preparing synthesis gas. Depending on the feedstock available and the type of synthesis-gas process, the methanol synthesis loop is designed for a wide range of conditions so as to achieve optimum economy in capital and operating costs.

An increase in the CO/CO_2 ratio of the gas increases the carbon efficiency of the synthesis, since methanol is made directly by reaction with CO and only indirectly

from CO_2. Also, an increase in pressure and decrease in temperature will increase the equilibrium concentration of methanol in the gas leaving the synthesis reactor (le Chatelier's principle), which in turn leads to a lower recirculation rate in the methanol synthesis loop and a corresponding reduction in synthesis-gas requirements for a given gas composition.

Such conditions are approached differently by the two principal processes for providing synthesis gas: steam reforming and partial oxidation.

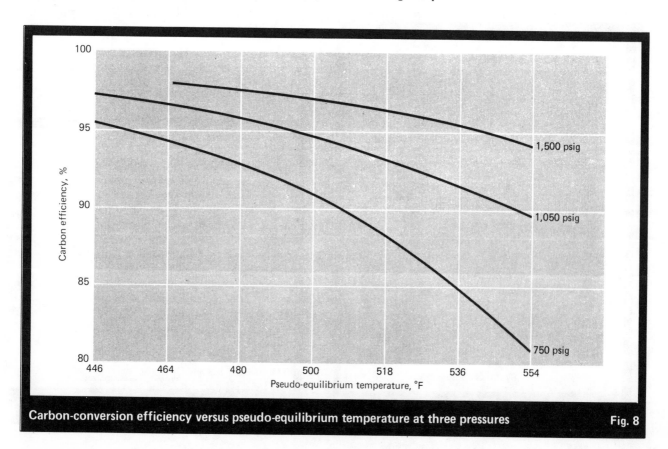

Carbon-conversion efficiency versus pseudo-equilibrium temperature at three pressures   **Fig. 8**

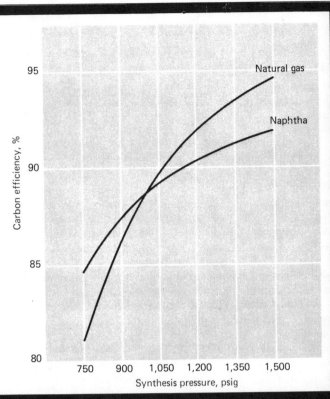

Carbon efficiency versus synthesis pressure for natural gas and naphtha feedstocks   Fig. 9

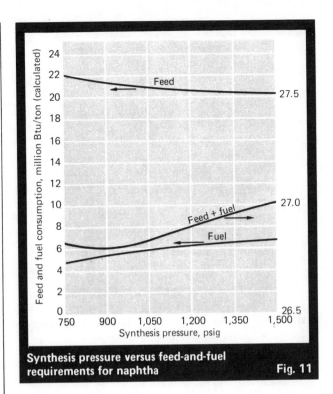

Synthesis pressure versus feed-and-fuel requirements for naphtha   Fig. 11

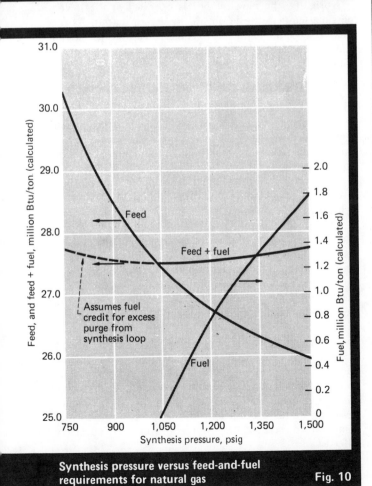

Synthesis pressure versus feed-and-fuel requirements for natural gas   Fig. 10

Synthesis gas by steam reforming

The distribution of capital costs among individual units of a steam-reforming methanol plant are:

Unit	% of total cost
Reformer/gasmaking	45
Compression	25
Synthesis	15
Distillation	15
	100

In this context, an increase in plant pressure improves the carbon efficiency and reduces the size of the reformer, as well as the cost of the higher-pressure equipment of the methanol synthesis loop. A reduction in the pseudo-equilibrium temperature of the synthesis reaction also improves carbon efficiency and reduces reformer size. Fig. 5 shows the effect of pressure and pseudo-equilibrium temperature on plant costs for natural gas- and naphtha-based plants with 1,000 short tons/d capacity. Total production costs, depreciating the capital over a three-year period, are shown in Fig. 6 and 7.

In addition, the CO/CO_2 ratio of the reformed gas can be increased by increasing the reforming temperature and reducing the steam ratio. The design reforming temperature should be as high as economically possible, usually in the range of 1,560 − 1,615°F. The design steam-to-carbon ratio in the reformer should be kept as close to the safe minimum as possible; this is normally 2.5 − 3.0, depending on the feedstock.

However, a reduction in steam ratio increases the methane level in the reformed gas. Methane behaves as an inert in the synthesis loop, and affects the thermal efficiency of the process. The optimum methane level varies with the type of feedstock and the synthesis reactor's operating conditions.

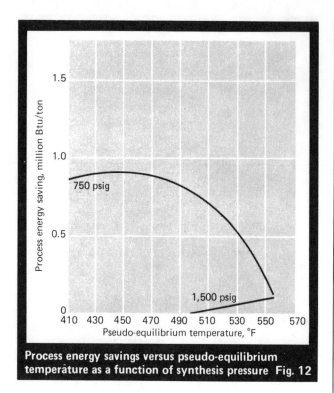

Process energy savings versus pseudo-equilibrium temperature as a function of synthesis pressure Fig. 12

losses have been considered in the preparation of these figures.

Synthesis gas by partial oxidation

Synthesis gas may also be produced by a partial oxidation process using heavy fuel oil or coal as feedstock. The synthesis gas produced is rich in carbon, which must be removed in the form of CO_2 in order to achieve the stoichiometric gas composition. The gasification conditions are such that a synthesis gas is produced containing a high CO/CO_2 ratio conducive to high carbon efficiencies over a wide range of pressures in the synthesis loop. For partial-oxidation plants, there is no advantage in increasing the synthesis pressure above 750 psig, since carbon efficiency is increased only from 97% to 99% as pressure increases from 750 to 1,500 psig.

Synthesis temperature

From a thermodynamic point of view, the lower the temperature in the methanol synthesis reactor, the better the carbon efficiency (Fig. 8). However, the designer also has to consider catalyst consumption, the size of the synthesis reactor, and the life of a single catalyst charge. Catalyst volumes are chosen to give a reasonable life up to four years. The designer can reduce these volumes, still maintaining the same process conditions. However, a smaller catalyst volume means more frequent changes and slightly more catalyst usage, because higher temperatures are experienced by the catalyst.

Normally, the new catalyst is operated at the lowest possible temperature to obtain the design output. As the catalyst ages, the temperature is increased until the design temperature is reached. Because the catalyst temperature is lower than design in early life, a better carbon efficiency is achieved and worthwhile savings can be made.

A higher average carbon activity can be maintained by changing the catalyst before it reaches the normal design temperature; or new catalysts can be operated at a higher temperature than needed to achieve design capacity. In this manner, a close approach to the methanol conversion equilibrium may be achieved; and provided that the exit temperature is not too high, the pseudo-equilibrium temperature is reduced and a similar improvement achieved in carbon efficiency.

The effects of reducing the pseudo-equilibrium temperatures are shown in Fig. 12 for two cases, 750 and 1,500-psig synthesis pressure. This study assumes there is a reduction in heat recovery directly proportional to a decrease in pseudo-equilibrium temperature below 554°F. Accordingly, the process savings at 1,500 psig diminish, because the small improvement in carbon efficiency does not compensate for the loss in heat recovery. At 750 psig, by contrast, a significant improvement in savings reaches a maximum at 490°F, and then diminishes as the reduction in heat recovery becomes significant.

In the case of steam-generating synthesis loops, it is not possible to take advantage of higher thermal efficiency resulting from a reduction in the pseudo-equilibrium temperature because the steam pressure is adversely affected.

In summary, a wide range of practical process conditions (for which experience exists) show that the design of the methanol synthesis loop has a significant effect on the

Once reformer temperature and steam ratio have been fixed to maximize the CO/CO_2 ratio, the only remaining variable affecting the methane level is pressure. The optimum reforming pressure is also influenced by the power balance of the overall plant, and by the relative costs of single-stage versus two-stage synthesis-gas compressors. For natural-gas-based plants, it has been found that the optimum reforming pressure is that at which the combustion heat value of the methanol flash gas (Fig. 3) and the purge from the synthesis loop (Fig. 3) are equal to the combustion heat requirements for reforming and steam generation. From a practical point of view, the actual reformer pressure must be slightly lower than this theoretical optimum.

Fig. 8 has been prepared to show the relation between synthesis conditions and the size of the reformer. The effects of pressure and pseudo-equilibrium temperature on carbon efficiency are indicated for the synthesis converter. It is apparent that: (1) as the pressure increases, the carbon efficiency increases; (2) as the pseudo-equilibrium temperature increases, the carbon efficiency decreases; and (3) as the synthesis pressure increases, the carbon efficiency becomes less dependent on the pseudo-equilibrium temperature existing within the methanol synthesis reactor.

Also, the carbon-to-hydrogen ratio in the feedstock has a marked effect on the way in which carbon efficiency varies with synthesis pressure. Fig. 9 shows that this efficiency is much more dependent on synthesis pressure for a natural-gas feedstock, relative to a naphtha feedstock.

Fig. 10 and 11 show the process feed and fuel requirements at various pressures for a 1,000-short-tons/d plant based on natural gas and naphtha. The combined feed-plus-fuel is at a minimum at pressures of 800−950 psig for naphtha and 900−1,200 psig for natural gas. No

economics of producing methanol. These are the basic conclusions:

1. The optimum synthesis pressure for methanol plants based on steam reforming varies with energy costs. At a cost of \$2/million Btus, the optimum pressure is 1,050 − 1,250 psig; at 20¢/million Btus, this optimum is 1,200 − 1,500 psig. These options apply to plants up to about 1,200 short tons/d; in order to maintain single-stream operation for larger plants, the synthesis pressure has to be increased to 1,500 psig.

2. The high CO/CO_2 ratio of the synthesis gas produced by a partial oxidation process allows the synthesis loop to achieve very high carbon efficiencies, in the range of 97 − 99% both at and above 750-psig synthesis pressure. Because of this, the optimum pressure for this type of plant is about the same as the gasification pressure, so that synthesis-gas compression can be avoided.

3. Plants designed for lower pressures can be operated at minimum cost by reducing the pseudo-equilibrium temperature during the early life of the catalyst. However, the beneficial effects are negligible for plants operating at about 1,500 psig, because the overall carbon efficiency is already high at this pressure.

The authors

Alwyn Pinto is Process Development Manager in the Projects & Engineering Dept. of ICI, Agricultural Div., P.O. Box 1, Billingham, Cleveland TS23 1LB, England. He holds a degree in chemistry and a post-graduate diploma in chemical engineering.

Peter Rogerson is a member of the Catalyst and Licensing Dept. of ICI, Agricultural Div., Billingham. He is responsible for licensing ICI's Low Pressure Methanol process and formaldehyde technology. He is a graduate of London University, where he earned his B.Sc. and Ph.D. in chemistry.

A formaldehyde process to accommodate rising energy costs

A single modification can affect almost every aspect of this relatively simple process.

C. W. Horner, Reichhold Chemicals, Inc.

☐ At a seminar held April, 1977, we had an opportunity to review the importance of energy conservation with licensees from all over the world of our Formox formaldehyde process. Our conclusion has been that the driving force behind recent changes has been not so much fuel costs as increasing pressure toward economy.

Formaldehyde, which is a large-volume commodity chemical, consumes an energy-intensive feedstock, methanol, which has been proposed as an alternative automotive and heating fuel because of its ease of transportation. Furthermore, formaldehyde manufacture does not require either fired furnaces or large compressors, as do other fuel-consuming processes such as ammonia.

Consequently, the rapidly rising costs of energy have been apparent to formaldehyde makers primarily in the form of rising feedstock prices. One might expect, therefore, that recent modifications to our formaldehyde process

would be directed toward obtaining the maximum amount of byproduct energy from reacting the feedstock, without any sacrifice in yield. This is precisely what has happened.

A comparison of yield, and consumption/production of chemicals and utilities, between 1973 and the present is shown in the table.

The outstanding change in these numbers is the increase in steam generation, from 410 lb of 150-psig steam to 684 lb of 300-psig steam per 1,000 lb of formaldehyde solution. Considering that today some of our licensees value steam at over \$4 per 1,000 lb, this steam represents an important byproduct. It results primarily from increasing the generating pressure of waste-heat boilers. However, this is combined with other modifications for the overall effect. The principles involved might be applicable to other processes as well.

Processing problems

In making formaldeyde, methanol is oxidized with air, with formaldehyde recovered from the resulting mixture of gases by absorbing it in water (Fig. 1). Although simple, the process involves a number of problems. The methanol must be vaporized; air and vaporized methanol must be proportioned so as to avoid their explosive limits; and the proportioned mixture must be fed through the reactor and subsequent absorber. The reaction must be controlled so that the oxidation neither proceeds all the way to carbon oxides nor leaves too much unreacted methanol in the effluent. The considerable heat of reaction must be removed; and the absorber must be designed to optimize heat and mass transfer with respect to energy demand.

Resolutions to these problems are interrelated. Part of the waste gases leaving the top of the absorber are recycled to mix with fresh air and methanol feedstock (see Fig. 1). By thus reducing the concentration of oxygen in the mixture, it is possible to increase the concentration of methanol from 7.0 mols per 100 mols of air to $9.0 - 12.0$ mols per 100 mols of gas without forming an explosive mixture. This reduces the volume of reaction gases $17 - 37\%$, thus reducing the horsepower for the compressor, and the pressure drops through both the reactor and the absorber. Since the waste gases leaving the top of the absorber contain the unwanted byproducts, the recycled gas affects the equilibrium in the reactor so as to improve yields. Furthermore, since the waste gases comprise an atmospheric pollutant that must be incinerated, recycling part of those gases reduces that problem.

The reactor consists of a multiple-tube unit, having tubes that are filled with a catalyst composed of molybdenum and iron oxides, and having a heat-transfer fluid around the tubes. Reaction temperatures are on the order of 550°F, which is high enough to generate 300-psig steam and still have ample temperature difference for heat transfer.

This, in turn, permits application of a patented thermosyphoning system on the reactor heat-transfer fluid, whereby the temperature is controlled in the reaction tubes without use of a circulating pump. The heat is

Comparing chemicals and utilities consumption/production

Item	Per 1,000 lb of CH_2O as 37% solution	
	1973/74	Present
Expected yield	91.95%	91.95%
Catalyst	0.050 lb	0.050 lb
Cooling water, 85°F	5,380 gal	5,380 gal
60°F	1,860 gal	—
50°F	—	746 gal
Export steam, 150 psig	410 lb	—
300 psig	—	684 lb
Electric power	43.4 kWh	35.9 kWh

recovered in waste-heat boilers at two pressure levels. The higher level, up to 20 atm, provides steam for use by letdown turbines, which drive cooling-water pumps, etc., and still exhaust steam at a pressure suitable for process heating.

The formaldehyde absorber has been redesigned to minimize pressure drop on the process side, while the cooling is on a tray-to-tray basis internal to the column, so that recirculation of formaldehyde solutions through side-arm coolers is not required. Cooling-water pressure drop through these coils is optimized on a countercurrent basis, both with regard to temperature and pressure drop, by manifolding externally and internally.

The gaseous emission, already minimized by the recycle, is burned to eliminate atmospheric pollution through the use of a self-sufficient catalytic incinerator, which requires energy only for startup or abnormal operating conditions.

Formaldehyde's future

We and our licensees feel that conservation of energy will certainly continue to be important. Just as important to us will be conservation of materials. Formaldehyde has several alternatives for feedstock. It has been made directly from methane, and a return to this feedstock is possible. It

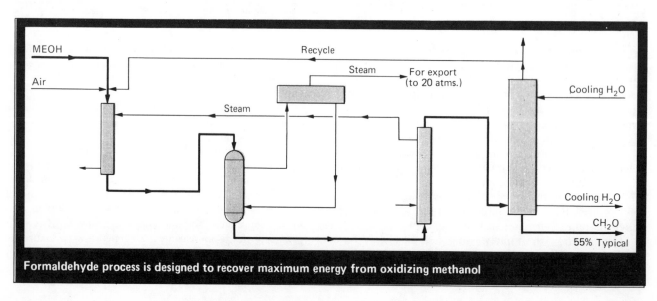

Formaldehyde process is designed to recover maximum energy from oxidizing methanol

can also be made by using a combination catalyst that converts synthesis gas, perhaps produced from coal, through methanol to formaldehyde, *in situ*. More likely will be the growth of methanol as a means of conveniently transporting energy, with a concurrent use as a raw material for formaldehyde.

Also, the conservation of capital is as vital as the conservation of energy. Although the factors of scale have been pushed to near their limits with the current process, progress was reported at the seminar on processes that, while retaining the energy efficiency of the present, would also lead toward large-scale plants that could conserve capital.

The author

C. W. (Chuck) Horner is Manager—Process Engineering Licensing at Reichhold Chemicals, Inc., RCI Building, White Plains, N.Y. 10602, where he is responsible for technical information regarding any prospective licensing for RCI processes worldwide. He joined Reichhold in 1953 and was appointed Chief Process Engineer in 1955, continuing in that position until becoming manager of technical licensing in 1975. Before joining RCI, he worked for General Electric, Westvaco and St. Regis Paper. He is a graduate in chemical engineering from Rensselaer Polytechnic Institute.

Equipment-purchasing policies that save energy

Most chemical process equipment can be designed to an economic optimum of installed cost plus operating cost. Here is how one major corporation obtains equipment having that design.

Enrique J. Armstrong, Union Carbide Corp.

☐ In 1973, Union Carbide Corp. started an energy conservation program to reduce its dependence on fuels and hydrocarbon feedstock materials, hold down operating costs against rapidly escalating energy costs, and cooperate with the national effort toward saving energy.

By October 1976, that program had resulted in a new policy, under which the Purchasing and Material Services Dept. now buys equipment according to energy consumption as well as purchase price and performance. Since the purchase of equipment takes up an appreciable percentage of UCC's worldwide sales revenues, posted at $6.3 billion in 1976, this policy is an important facet of the corporation's operation.

While the energy crisis has inspired this new policy, implementation causes the policy to function primarily as a cost-cutting measure. In fact, it has resulted in documented savings of over $750,000 for the first year alone. Since much of this amount is in operating costs, the savings accrued over the years will be many millions.

Basis for analyzing energy consumption

To control costs, the corporation must assure that every piece of purchased operating equipment be as energy-efficient as possible within economic parameters. To do this, it has been necessary to:

■ Provide suppliers with a basis for making an accurate cost evaluation of energy-efficiency improvements in their equipment.

■ Give the corporation's engineering and purchasing departments the information necessary to evaluate bids with respect to energy consumption. Such information must permit an economic comparison and optimization for equipment having a higher initial cost and lower operating cost versus that having lower initial cost and higher operating cost.

■ Supply the purchasing department the information needed to negotiate, when necessary, a liquidated-damages clause directly related to the vendor's quoted efficiency numbers.

Energy-conservation clauses

All of this information, both for suppliers and for corporate departments, is provided by special "Energy Conservation Clauses," which have been adopted for worldwide use by the corporation. These clauses, in effect, tell the suppliers or equipment manufacturers that Union Carbide Corp. will use energy efficiency as a major factor in the evaluation of new major equipment, and also informs them of the energy rates that will be used in the evaluation of bids.

While the clauses are used primarily for the procurement of "major equipment," defined as costing at least

$50,000 and consuming energy equivalent to about one third of the operating cost, the clauses are also used for less-expensive equipment whenever such use appears justified. The less-expensive equipment is individually screened to see if any advantage can be gained by using the clauses.

The actual clauses are formalized, and their implementation is described in numbered paragraphs for corporate personnel. However, their broad intent can be described according to three categories: language, information, and enforcement.

Language

The language of the energy-conservation clauses normally follows the example clauses (see Boxes—right), which are written in general terms for broad application. However, this language may be modified to suit particular instances. The clauses are written directly into the forms used to make requests for quotation, as well as purchase-order forms.

Blank spaces for energy rates are included in the clauses. These are filled in by the purchasing department according to information supplied by the engineering department.

Information

The clauses are worded so as to give UCC's engineering and purchasing departments a choice between least energy consumption, lowest initial cost, and the lowest overall cost, with the initial cost depreciated over a number of years. The energy costs, which are stated in terms of dollars per quantity of energy used, are the same as those energy costs that were used for the justification of the project.

Of course, all safety and engineering specifications must be met by all proposals, and the energy-conservation clauses cannot be used by vendors as an excuse to alter the mechanical and operating specifications of their equipment.

Enforcement

A key section of the clauses requires that suppliers undergo a joint "verification of equipment efficiency" performance test. Accordingly, an integral and major part of the energy-conservation policy consists of a stern warning to would-be suppliers that they must prove energy-efficiency claims for their equipment or be liable for corrective action or even rejection of the equipment and refund of its purchase price.

This warning is communicated to potential suppliers by means of the first clause (see Box) of the "Request for Quotation" sent out for the equipment. The selected vendor is again informed by the enclosure of a second clause (see Box) in the final purchase order or contract sent out for the equipment.

Enforcement is accomplished through a "Performance Warranty" clause in the purchase order. Further enforcement may be achieved through the use of a "Liquidated Damages" clause in the purchase order. In the event that such a clause is considered, this will be stated in the request for quotation.

Verification of the equipment's quoted efficiency is done by a performance test mutually agreed on by buyer and seller. The requirement for this test is announced in

---

Energy conservation clause for use in requests for quotation — Box I

All pieces of equipment and their auxiliaries specified herein shall be evaluated as to their efficiency in terms of energy conservation, as well as safety, price, durability, maintainability, performance, quality and delivery.

Seller is to propose the most *energy efficient* design to satisfy the enclosed specifications. Seller shall also propose the most *economical* (lowest initial cost, as comprising all elements stated in the paragraph above) design to satisfy the enclosed specifications, should it be different from the most energy efficient design.

The energy consumption of the equipment is to be specified in appropriate terms (i.e., MBtu/h, klb/h steam, kW, hp). The percentage of accuracy of the efficiency data submitted is to be stated. Fuel and/or any other energy input consumption is to be specified. Energy consumption of the specified equipment shall be evaluated at the rate (s) of $____ per ____ for the purpose of bid comparison.

Verification of equipment efficiency shall be done by a Performance Test as mutually agreed upon by Seller and Buyer, to be performed at a time mutually agreeable to Seller and Buyer, usually within the warranty period.

In the event that said equipment should fail to meet the quoted efficiency, Seller agrees to modify the equipment, at its own cost, to comply with the quoted efficiency. These modifications shall be performed within a reasonable time after Seller receives written notice from Buyer that said equipment failed the Performance Test.

In the event the equipment fails to comply with the quoted efficiency after a third Performance Test, Buyer shall have the rights as specified in the attached Performance Warranty clause.

Energy conservation clause for use in purchase orders — Box II

Verification of the quoted efficiency as stated in Section No.____ of this contract shall be done by a Performance Test as mutually agreed upon by Seller and Buyer, as specified in Section No.____ of this contract.

Energy consumption of the equipment shall be rated at $____ per ____ for the purpose of evaluation and/or liquidated damages.

In the event that said equipment should fail to meet the quoted efficiency, Seller agrees to modify the equipment, at its own cost, to comply with the quoted efficiency. These modifications shall be performed within a reasonable time after Seller receives written notice from Buyer that said equipment failed the Performance Test.

In the event the equipment fails to comply with the quoted efficiency after a third Performance Test, Buyer shall have the rights as specified in the attached Performance Warranty clause.

the request for quotation, and then the test is detailed in the purchase order as to type, mode and timing. If a test can be identified for the equipment among the standards of the ΛSME, IEEE, etc., that test shall be stated in the request for quotation and purchase order.

In the event that the equipment fails to meet the quoted efficiencies, the supplier will:

■ Modify the equipment to comply with the quoted efficiencies at the supplier's own cost and within a reasonable time.

■ Or—if after the modifications are completed, the unit still fails to meet the quoted efficiencies—accept an equitable downward adjustment of the purchase price of the equipment according to the rates shown in the clauses.

■ Or, accept the return of the equipment (f.o.b. place of installation) and refund all payments theretofore made by Union Carbide.

Limitations

Using the energy-conservation clauses is not without potential disadvantages, including: (1) higher initial costs to obtain lower long-term operating costs, (2) a possible reduction in the number of suppliers due to the unwillingness of some to incorporate energy performance as part of their bids, and (3) the possibility for longer lead-times needed to obtain the extra information and to evaluate the bids.

Some vendors may be reluctant to bid because of the requirements that they guarantee efficiency numbers. Union Carbide realizes that some few suppliers may not readily want to participate in the energy-conservation effort, but no important supplier is expected to refrain from bidding for this reason.

When equipment is purchased by UCC for a customer, as for a licensed process, UCC will urge the customer to adopt the energy-conservation clauses, but the customer has the option of using or not using those clauses.

In conclusion, Union Carbide is committed to a continuous reduction of its energy requirements in support of the national energy-conservation effort. The proper evaluation of energy usage in the purchase of major equipment is an important tool for accomplishing this objective. This does not mean that energy efficiency is the only criterion for evaluating a piece of equipment. It does mean, however, that energy efficiency will be among the major considerations.

The author

Enrique J. Armstrong is Corporate Purchasing Agent, E.C.E. Group, at Union Carbide Corp., 270 Park Ave., New York, N.Y. 10017. Except for one and a half years in the Army, including a tour in Vietnam, he has worked for UCC since 1964, advancing through the positions of Electrical and Instrument Engineer, plant Chief Electrical Engineer, and plant Energy Systems Dept. Head, to his present post. Λ member of the Inst. of Electrical and Electronic Engineers and the Instrument Soc. of America, he holds a B.S. in electrical engineering from the University of Puerto Rico.

Starting point for other processes

At first reading, the foregoing articles appear to present a tangled network for the practicing engineer to unravel for energy-saving modifications applicable to his own plant—but a closer look should reveal a reliable starting point.

This is the so-called "back-pressure," or "let-down" steam turbine, which converts part of the energy of high-pressure steam to mechanical energy, then discharges low-pressure steam for process uses. In the ethylene process, such turbines drive pumps and compressors, then provide process steam or reboiler heat. In certain types of methanol plants, limited possibilities for using low-pressure steam act to reduce the possibilities for the back-pressure turbine, and thus, possibilities for energy reduction. And a principal economy of a formaldehyde process can be byproduct steam at a pressure suitable for letdown turbines.

Moreover, there are few pieces of process equipment for which energy economy can play a greater role in bid comparisons than for the back-pressure steam turbine.

Comparison is made on the basis of guaranteed "water-rate," that is, the number of horsepower per pound of steam that can be delivered by the turbine, as it operates between two given steam pressures. The differences between water-rates claimed by different turbine manufacturers can pay back the price differentials in a matter of months.

Usually, the engineer will find that the horsepower required by a turbine is fixed. A lower water-rate thus means more steam through the turbine. As long as the discharged steam can be used, the turbine is part of an energy cycle with an efficiency typically over 90%. When the discharged steam cannot be used, it must be condensed in a so-called "surface condenser." This gives flexibility, but an energy efficiency on the order of 35%.

Consequently, the most efficient energy cycle is obtained by starting with available users of low-pressure steam, and then buying a back-pressure turbine having a water rate that will balance those users against horsepower requirements.

Choosing equipment for process energy recovery

Today's energy and raw material costs mandate increased power recovery from process streams, which increases capital costs. Here are pointers on coping with the problem.

M. J. Rex, *Davy Powergas Ltd.* *

☐ Cost increases of raw materials and energy have caused the balance between capital and operating costs to shift away from lowest investment that is consistent with reliable equipment to designs that maximize the use of energy released within the process.

The rapid growth in single-stream plant capacity over the past 10 years has brought a corresponding increase in process streams (which were previously vented through valves) that can now be considered viable sources of energy. Moreover, the availability and acceptance of reliable expansion engines with efficiencies of over 80% can allow process engineers to consider higher pressures, knowing that a considerable proportion of the compression energy can be recovered.

Changing economics will cause designers to run acid-gas scrubbing systems at higher pressures (changing from chemical to physical absorption), and hydraulic turbines will become standard features.

A primary need for energy recovery by mechanical equipment is a high potential. The use of low-potential waste heat from boiler flue gases at a few inches of water gage pressure, for example, is not economical.

Possibly the most established area for energy recovery is gas-expansion equipment used in cryogenics. The modern air-separation unit for the production of liquid oxygen is a fine example of how plants can be designed to minimize energy consumption.

Unfortunately, in many operations, the liquids and gases arrive at the machine as a process fluid, rather than as a utility, and special attention has to be given to phase separation within the machine. Also, design features must be incorporated to accommodate the corrosive and erosive aspects of the fluid. Finally, the new

* This article is adapted from a paper titled "Equipment for Power Recovery," presented at a meeting on Energy Recovery in Process Plants, held by The Inst. of Mech. Engrs., 1 Birdcage Walk, Westminster, London SW1H 933. The paper will be published, along with others from the meeting, by IME.

Originally published August 4, 1975

energy-recovery machines represent improvements nonessential to the process; equipment failure that shuts down the whole process unit cannot be tolerated.

Wheel or runner; reciprocating or rotating

The equipment to be considered for energy recovery falls into two distinct classes, depending on whether the process fluid is liquid or gaseous. Equipment that handles liquids can further be classified as Pelton wheel or Francis runner, whereas equipment handling gases can be classified as either reciprocating or rotating.

For high-head, low-flow applications, a Pelton wheel turbine (see box, p. TT) can be used with one or more nozzles; but for most applications, a Francis-type runner (a runner is an impeller in reverse) is more suitable. This latter comes in two forms, one being a hydraulic turbine specifically designed as such, the other being a centrifugal pump running backwards in (perhaps) a strengthened casing. Diagrams and dimensions from one manufacturer of the turbine are shown in the box.

For hot or high-pressure gas streams, turbo machines are now made to handle the low flows and high expansion ratios that previously required reciprocating expansion engines or units of the free-piston type. These turbo machines can be further divided into two distinct types, the inward-flow radial turbine, which is similar to the Francis turbine (see box), and the axial-flow type, which can be either impulse or reaction.

Hydraulic turbines

It is possible to reduce the energy consumption of pumping regenerated absorbent in a CO_2-recovery unit as in Fig. 1 (F/1) by as much as 60%, if a suitable hydraulic turbine, or "letdown" turbine, is installed. On the CO_2 unit for a large ammonia plant this pumping power can be as much as 1,000 kW.

Manufacturer's data on Pelton wheels and Francis runners

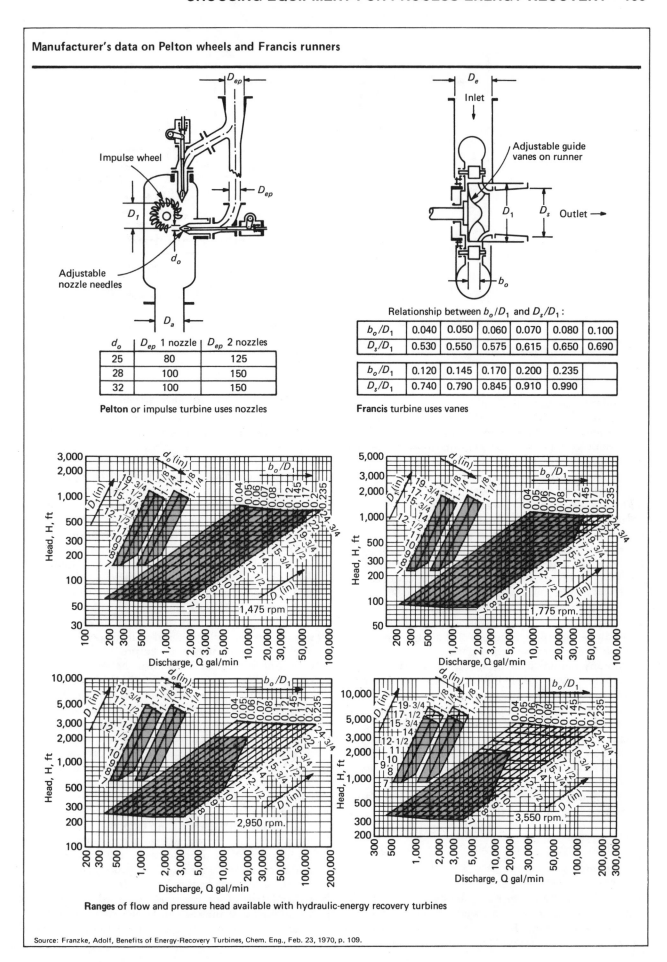

d_o	D_{ep} 1 nozzle	D_{ep} 2 nozzles
25	80	125
28	100	150
32	100	150

Pelton or impulse turbine uses nozzles

Relationship between b_o/D_1 and D_s/D_1:

b_o/D_1	0.040	0.050	0.060	0.070	0.080	0.100
D_s/D_1	0.530	0.550	0.575	0.615	0.650	0.690

b_o/D_1	0.120	0.145	0.170	0.200	0.235
D_s/D_1	0.740	0.790	0.845	0.910	0.990

Francis turbine uses vanes

Ranges of flow and pressure head available with hydraulic-energy recovery turbines

Source: Franzke, Adolf, Benefits of Energy-Recovery Turbines, Chem. Eng., Feb. 23, 1970, p. 109.

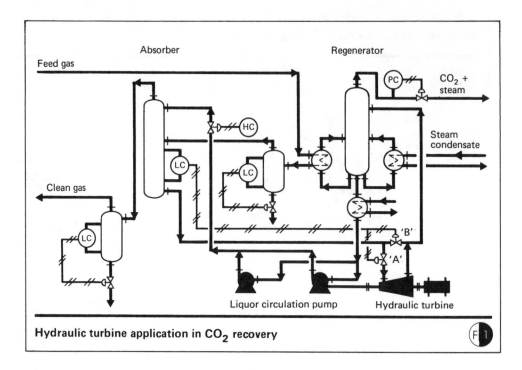

Hydraulic turbine application in CO₂ recovery

Similar economics are possible in a number of operations involving high-pressure liquids, but several other factors have to be taken into account, such as: equipment delivery time; supplier's familiarity with the process fluid; relative costs of types of equipment available; corrosion-resistance experience with the materials of construction; reduced-throughput operation; control requirements; mechanical-design features; method of coupling power-recovery equipment for driven units; maintenance.

Beyond such considerations, the choice between a Pelton wheel and a Francis runner is indicated in F/2. The number of nozzles around the Pelton wheel can be increased to four, but beyond approximately 800 m³/h, the single Francis runner becomes more economical and mechanically simpler. Capacity of the Francis turbine can be increased by means of a double-flow back-to-back arrangement of runners.

In many applications, a Francis runner is the automatic choice, but the engineering contractor and prospective user are then faced with the problem of choosing between a reverse-running pump and a hydraulic turbine. In the past, the tendency has been for U.S.-based companies to favor reverse-running pumps, while European companies have used hydraulic turbines. It is likely that the future will see increased use of hydraulic turbines.

A comparison of typical percentages of design output at reduced flows provides the following:

	Percentage of Design Power				
Flow, %	100	80	60	40	20
Turbine	100	80	60	37	15
Reverse-running pump	100	55	22	−5	−15

In addition, the power recovered by means of a turbine can exceed that by a reverse-running pump by 10% or more. Although the pump is less expensive, it offers control problems and higher operating costs.

Turbine-design considerations

The runner of the Pelton wheel rotates in a gas-laden atmosphere, so the design of the chamber and the discharge line must be such that foam or liquid cannot build up quickly.

Turbine casings must be designed to withstand the highest pressure in the upstream system, or relief valves must be placed in the discharge line. When hot solutions are handled, the turbine should be supported about the centerline, with special attention paid to methods of retaining alignment.

Many of the process fluids suitable for power recovery will form crystals when cooled, and it is important that turbine casings be completely drained and flushed with a clean fluid when shut down.

The rotating assembly must be robust and preferably running below its first lateral critical speed in air. Adequate thrust bearings must be provided, and growing operating experience has shown the value of installing tilting-pad bearings on Francis-type turbines.

Hydraulic turbines, with adjustable inlet vanes, are available from several manufacturers. The vanes improve the part-load performance of the turbine but increase its complexity. A decision to use these vanes must be determined according to the process fluid and the time spent at part load. If the process fluid is clean and not liable to form crystals, adjustable vanes become attractive. Also, they can be coupled to a controller for level control of the preceding vessel.

Leakage, both internal and external, requires careful attention. The Pelton wheel has no internal sealing problems, but the shaft-to-casing seal can be difficult, due to the lack of a liquid on the process side.

Where internal seals are necessary (runner to casing) either labyrinth or ring seals are used. It must be remembered that the gas can be released within the turbine at any point of pressure reduction, and that the turbine may run dry. These problems are overcome by combinations of: increased seal clearances, introduced clean, compatible flushing-fluids, and overrunning-clutch couplings.

Leakage at the shaft/casing interface can be controlled in several ways: single-face-type mechanical seals, with or without flushing medium; double mechanical seals; soft-packed glands with water cooling and lubrication. Each user and contractor appears to have a preference among these alternatives, depending on experience with pump seals.

The recovered power can be used to drive a pump or an electric generator. Coupling a hydraulic turbine to an electric generator requires careful attention to overspeed correction, since the electrical load can be shed very quickly, and the consequent rapid acceleration of the turbine will overspeed the generator rotor.

The coupling to a pump can be a conventional flexible coupling or an overrunning-clutch coupling. Assuming that the system consists of turbine-pump-motor, in that sequence, the use of a clutch coupling allows the motor to run the pump at full speed with the turbine stationary. Such a turbine can be commissioned and brought up to speed, at which point the clutch automatically engages and delivers power from the turbine to the train. If flow to the turbine ceases, the clutch automatically disengages, and the turbine comes to a halt, leaving the pump to the motor.

Consideration must also be given to the size of the helping motor. Should it be considered purely for makeup power, or for full power? Each installation must be assessed for the effects of turbine failure and operation on a motor sized for partial or full power.

Gas-expansion units

There are many potential applications for gas-expansion units in process plants. For example, purge gases are removed from ammonia and methanol converters at pressures of 100-300 bars, to be subsequently used as fuel at 3-4 bars. Such purge gases can be reheated several times for increased power recovery and more-acceptable discharge temperatures, as for example in a large methanol plant (F/3).

When a large pressure drop is available and the flow is small, the radial-flow turbine is usually preferred. In order to obtain reasonable efficiencies, speeds of 10,000-100,000 rpm are required. Such speeds impose problems on the designer of the turbine, but the corresponding problems of trying to harness the output of the turbine and convert torque at these high speeds are even more severe.

Efficiencies of small units are comparatively low, with small, light, rotating assemblies being less stressed, because of their small size. As the size of the unit increases, so do the efficiencies, but also do the mechanical problems.

Construction of inward-flow radial turbines is fairly standard. The turbine wheel is either cast or machined from a solid forging, while very small units have the turbine wheel machined integral with the shaft. The wheel is of the overhung type, carried by a substantial shaft designed so that operating speeds are well away from the critical speeds. Casings are likely to be radially split and may be fabricated, cast or forged, de-

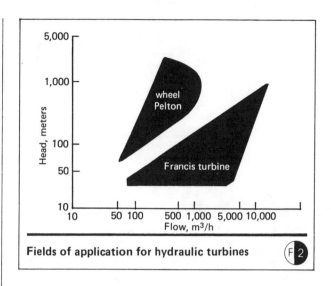

Fields of application for hydraulic turbines   F 2

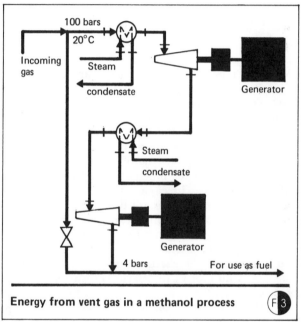

Energy from vent gas in a methanol process   F 3

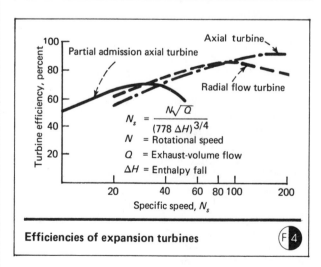

$$N_s = \frac{N\sqrt{Q}}{(778\,\Delta H)^{3/4}}$$

N = Rotational speed
Q = Exhaust-volume flow
ΔH = Enthalpy fall

Efficiencies of expansion turbines   F 4

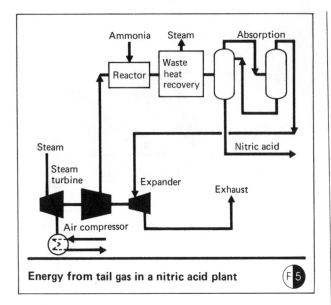

Energy from tail gas in a nitric acid plant (F 5)

pending on the pressure. Gas inlet is via nozzles, and complete or partial admission units are available, as are units with variable-inlet nozzles for efficient part-load operation. Bearing design is perhaps the most important element to reliable continuous operation.

Very small units, using clean dry gas as the bearing medium, are available for oxygen, hydrogen and helium liquefaction plants; however, it is usual to use journal bearings of the multipad type for general process applications. These journals should be selected only after a careful analysis of the dynamics of the rotating system. Leakage of process gas, either into the bearing housing or out into the atmosphere, is prevented by labyrinth seals at the back of the impeller and possibly on the shaft, together with mechanical seals and a flush of clean dry gas compatible with the process.

Axial-flow turbines cannot handle the very low flows handled by a radial turbine, and are expensive at pressure-differentials above 40 bars (F/4).

Process-plant experience with axial-flow turbines has been mainly in nitric acid plants, where the pressure into the turbine is over 4.5 (F/5). This has led to the impression that these turbines are not economical at lower pressures. However, although such turbines become large and expensive, they can still be economical whenever hot gases that are at more than 1 bar above atmosphere are available.

Axial expansion can be through single- or multistage wheels of a rotor that is either overhung or supported between bearings. According to size, the blades may be added to a disk, or machined integral with the disk or rotor. Since axial turbines usually handle tail gases, which often contain acid-gases or solids, special attention is given to materials of construction. Casings are normally split horizontally for easy access to the rotor, and can be fabricated or cast.

The shaft/casing interface seal is usually easier than with radial-flow turbines, since the axial machine is larger, operates at lower speeds, and has more space for buffer-gas chambers, mechanical- and labyrinth-seals.

Whenever possible, axial turbines should be used to drive the compressor that generates the pressure, with supplementary power coming from a helper such as a variable-speed steam turbine or gas turbine. Driving electrical generators leads to such problems as: added gearboxes, with higher costs, lower reliability and sources of vibration; fixed-speed operation, although small changes in the process fluid have marked effects on generator output; possible sudden load loss, making it necessary to protect the turbine with rapid-closing trip valves; the need for a speed-control governor.

By contrast, harnessing the expander to a compressor generally permits direct drive, gives a compact unit, affords stability in the speed characteristics, and provides an ever-present means of absorbing the expander output. However, the startup of a closely coupled compressor-expander train requires careful study to assure that adequate power is available for accelerating and sustaining the unit until useful energy is recovered by the expander.

Trends

Francis-runner turbines capable of continuous operation at high load-factors are well established. However, there is still a tendency among operating and maintenance people to neglect running and repairing such units unless necessary.

The Pelton wheel seems to offer advantages when high-pressure boiling liquids saturated with gas are being expanded, and it should find more application in the future.

An interesting development by one manufacturer combines pump and turbine into a single casing, thus reducing the number of bearings, seals, couplings, etc.

Applications for the gas expander are broadening, as it gains acceptance and proves its reliability. The radial machine will probably be developed and enlarged, so that units developing 6,000 kW and more are available. The expansion turbine will find applications at high pressures when pressure letdown is required, and at low pressures in such instances where large quantities of low-grade heat are released to the atmosphere.

Optimization studies of energy-recovery from high-pressure hot gases are needed to ascertain whether a combination of hot-gas expander plus waste-heat boiler is not a better combination than the single waste-heat boiler conventionally installed.

The author

Mervyn James Rex is Chief Engineer and Associate Director of Davy Powergas Ltd., 8 Baker Street, London W1M 1DA, where he is responsible for all engineering aspects, except process, associated with the company's projects. He joined Davy Powergas as Senior Mechanical Engineer in 1971, after having designed steam turbines and centrifugal compressors for A.E.I. Company Ltd. and having selected and applied rotating equipment for Air Products Ltd. and Humphreys & Glasgow Ltd. A Chartered Engineer, Member of the Institution of Mechanical Engineers, he passed a 5-yr engineering apprenticeship with British Thompson Houston Co. Ltd., Rugby, England.

Heat Recovery in Process Plants

Rising costs and shortages of fossil fuels emphasize the urgent
need for lessening the demand of primary fuels through the optimum
recovery and use of heat from elevated-temperature process streams.

J. P. FANARITIS and H. J. STREICH, Struthers Wells Corp.

The process designer involved with recovering heat to improve the economics of the process has an unlimited range of applications and techniques to consider. The increasing cost of primary fuel has broadened the range of heat-recovery applications that can be economically justified.

Several factors will influence the recovery of heat from a process:

1. Competitive market conditions on most products make it essential to reduce processing costs.

2. The cost of fuels keeps rising.

3. Limited fuel availability is already causing plant interruptions.

4. There are restrictions on using some of the lower-cost fuels because of environmental pollution.

5. Increasing emphasis is being placed on minimizing thermal pollution.

6. Increasing amounts of elevated-temperature flue-gas streams are becoming available from gas turbines, incinerators, etc.

7. New, sophisticated heat-recovery equipment is providing increased recovery efficiencies, and converting the recovered heat to high-pressure steam in the 600 to 1,500-psig. range, where the economic value of the steam is significantly higher.

The process designer has a number of alternatives to consider for recovering energy and/or heat. The specific method selected will depend on (a) type of energy available for recovery, (b) pressure or temperature level of the available energy, and (c) specific requirements of the process under consideration. This article will devote itself

Originally published May 28, 1973

only to the recovery of heat from elevated-temperature streams. Some of the methods used to recover heat are listed in Table I. These will be individually discussed as to advantages and disadvantages plus typical applications.

Steam Generation

Recovery of heat from elevated-temperature streams by the generation of steam is one of the oldest, easiest to engineer, and most widely used methods. This is probably the reason for the term "waste-heat boiler" evolving into a standardized term of reference for any type of heat recovery, including that in which no change of phase occurs in the coolant.

The purpose for which the generated steam is to be used will in large measure dictate the economics of the system, the type of heat-recovery equipment to be used, and whether or not steam generation is the proper heat-recovery technique.

The principal uses of steam generated by a heat-recovery system are:

• For process heating. In this application, the steam will probably be generated at pressures of 125 to 650 psig.

• For power generation. In this application, the steam will probably be generated at pressures of 650 to 1,500 psig., or higher, and will likely require superheating.

• For use as a diluent or stripping medium in a process. This is a low-volume use.

The use of steam generation as a means of recovering heat from high-temperature streams provides a number of advantages:

1. It generally results in a relatively compact heat-recovery installation because of the high rate of heat transfer associated with the boiling of water.

2. It usually will result in the lowest initial installation cost of any type of heat-recovery system.

3. It generally will incur fewer operating problems when applied to the cooling of high-temperature streams, since the high heat-transfer rates secured with boiling water will maintain metal temperatures close to the boiling-water temperature.

4. It will provide a rapid response rate.

5. It will permit some adjustability in heat-removal capacity, by raising or lowering the steam-side operating pressure within the design limitations of the equipment.

6. It does not require the close coordination between

Methods for Recovering Heat—Table I

☐ Generating steam.

☐ Preheating boiler feedwater.

☐ Preheating combustion air.

☐ Superheating steam.

☐ Preheating a process feedstream, or heating a process stream at some intermediate point in the process.

☐ Heating circulating heat-transfer media, which are then used to provide process heat.

☐ Preheating air for use in applications such as direct-contact dryers.

☐ Using flue-gas streams at elevated temperatures for process applications such as direct-contact dryers.

☐ Evaporating process streams.

☐ Providing space heating and utility steam.

Guidelines for Fire-Tube Steam Generators—Table II

1. Usually limited to steam pressure of under 1,000 psig., although technology is extending this to 1,850 psig.

2. Elevated-temperature streams being cooled may be liquid or gas.

3. Adaptable to cooling of elevated-temperature gas streams operating under pressure. Cooling of streams at pressures of 500 psig. and higher is not uncommon.

4. Most efficient when process-gas streams having a reasonable heat-transfer film coefficient are being cooled.

5. Generally limited to process-stream flowrates that can be handled by shop-assembled units.

6. Can handle clean or highly fouling elevated-temperature streams. Generally less susceptible to fouling and easier to clean on the high-temperature side than a water-tube design.

7. A fire-tube unit will generally be less expensive in applications where either fire-tube or water-tube design may be used.

8. In high-temperature service (1,000 to 1,800 F. inlet-gas temperatures), the inlet or hot tubesheet of a fire-tube unit is highly vulnerable.

9. Generally requires a higher pressure drop on the high-temperature streamside than a water-tube design.

10. Natural-circulation and forced-circulation designs are available.

Guidelines for Water-Tube Steam Generators—Table III

1. Can be designed for any steam pressure, including supercritical.

2. Wider range of designs is available than in fire-tube units, including the use of extended-surface tubes as well as bare tubes.

3. Ability to use extended heat-transfer surface makes this design more efficient than a fire-tube unit when cooling gas streams that have poor heat-transfer characteristics or low allowable pressure drop.

4. Designs are available for handling low-pressure elevated-temperature gas streams at much higher flowrates, since water-tube designs may be field assembled in large sizes.

5. Water-tube designs do not lend themselves to cooling elevated-temperature gas streams with highly fouling characteristics. Even high-density bare-tube designs do not lend themselves to efficient soot blowing.

6. Water-tube designs lend themselves more readily to supplementary firing, where such additional heat input is desired, than fire-tube units.

7. Water-tube units can be designed for low friction losses (2 to 10-in. water column) on the high-temperature side while handling high flue-gas flows, and still provide efficient heat transfer.

8. Water-tube units are generally less susceptible to mechanical failure as a result of malfunction than fire-tube units.

9. Natural-circulation and forced-circulation designs available.

10. Certain water-tube designs may incorporate steam superheater and other auxiliary service coils beyond the primary function of steam generation. Included in a single housing can be superheater, generator and economizer coils to provide a high-efficiency heat-recovery unit.

process-stream flows and temperatures and feedwater flowrate, as required by other heat-recovery techniques.

Disadvantages of steam generation as a means of recovering heat:

1. A steam-generation system must operate at fairly high pressure to ensure economic justification. Current trend is for operation at the 650 to 1,500-psig. range, with higher operating pressures being evaluated.

2. Steam generation cannot cool elevated-temperature streams through as wide a range as other heat-recovery techniques because most of the heat is recovered primarily by vaporization of water at constant temperature corresponding to the system operating pressure.

3. A high-pressure waste-heat boiler generating steam requires high-quality feedwater, comparable to that required for a fired boiler. The cost of water treatment may significantly reduce the economic advantages steam generation might have over alternate heat-recovery techniques.

4. Steam generation has limited flexibility in utilizing the recovered energy. In certain types of installations, there is simply no use for the steam generated through heat recovery.

Heat-recovery steam generators or waste-heat boilers may be either of the fire-tube or water-tube design, depending on a variety of factors. The process designer considering a steam-generator heat-recovery system must be familiar with the advantages and disadvantages of both designs. Improper application of either will likely result in a more-expensive-than-necessary installation and/or operating problems. Application guidelines for the two types of steam generators are given in Tables II and III.

The process designer must first establish whether heat recovery by means of steam generation is the most desirable for his application, and must then evaluate the available steam-generation arrangements. With guidelines previously outlined, the process engineer should be able to select the design that will result in the highest economic return and have minimum operating problems.

Preheating Boiler Feedwater

A preheating system for heating a large volume of feedwater through a reasonable temperature range will have a low initial cost and prove highly efficient in the recovery of heat. Here, heat may be recovered from flue gas or from process liquids and gases. Flue gas is the usual source of heat for this application.

Either water-tube or fire-tube heat-recovery designs may be used. Most of the guidelines for evaluation previously described are also applicable. Since heat is usually recovered from flue gases, an extended-surface water-tube type of coil is the normal design. In installations where small heat loads are involved and where the heat is to be recovered from a process stream, shell-and-tube heat exchangers are frequently used.

In feedwater preheating, the process designer should consider the following:

1. Flue-gas streams can become extremely corrosive when they are cooled below the dewpoint, particularly when traces of sulfur are present in the fuel. If the tube-wall temperature is below the dewpoint, condensation will occur even though the main-body gas temperature is still above the dewpoint. On occasion, semiconcurrent flow may be used to prevent condensation, even at the sacrifice of some of the log-mean temperature difference, and with the need for additional heat-transfer surface.

2. When heat is recovered from a process stream in a shell-and-tube exchanger, the stream is generally cooled over a long temperature range. With the process stream flowing through the tubes, a severe temperature differential across the multipass tubesheet can develop and result in operating problems.

3. It is generally not economical to use a high-temperature gas stream above about 800 F. for feedwater preheating. Greater economic value can be obtained by using higher-level heat for steam generation or the heating of process streams.

Preheating Combustion Air

Preheating combustion air is one of the oldest and most widely used means of recovering heat from elevated-temperature gas streams. This technique has primarily been used in central-station power plants where large volumes of flue gas are available, and where plant thermal efficiency has always been of primary importance. Air preheaters for central stations are of specialized design because of the requirement to exchange heat between two streams having equally low heat-transfer characteristics. Rotating-element metallic preheaters are the principal type, but there are some shell-and-tube air preheaters used on smaller boilers. Some special extended-surface exchangers are also in this service.

Air preheaters have been used in conjunction with process heaters on a relatively small number of installations where high temperatures at the process inlet have not permitted use of convection sections.

Preheating of combustion air is an effective means of recovering heat from high-temperature flue-gas streams, but it can only be justified in a limited range of applications. Some of the disadvantages are:

1. It generally will require a relatively high energy expenditure in the form of horsepower for the induced-draft and/or forced-draft fans.

2. There is a relatively high initial cost because of the proportionately high heat-transfer surfaces.

3. The recovered energy must be used in conjunction with a combustion system.

Steam Superheating

Adding 50 deg. (F.) or more of superheat to high-pressure steam significantly adds to the dollar value of the steam when it is to be used for power generation. Elevated-temperature streams can frequently be used to superheat steam, thereby appreciably increasing the efficiency of a process.

Both water-tube and fire-tube waste-heat boilers can be provided with auxiliary superheaters to handle the steam that they generate. The water-tube design, which recovers heat from the exhaust of gas turbines, is particularly adaptable to using a steam superheating coil. Process furnaces also lend themselves to the superheating of steam in the convection section.

The one precaution to be exercised in the design of steam superheater coils in heat-recovery installations is that the steam must have a relatively low solids content, preferably under 1 ppm. Deposit of solids on the superheater tubes can lead to rapid tube failure when the elevated-temperature stream is over about 1,000 F.

Preheating Process Streams

Preheating of a process stream by an elevated-temperature flue gas or process stream generally will result in the most favorable economics of any form of heat recovery. This technique directly reduces the primary energy input required by the process, as well as eliminates or minimizes the size of the equipment required to transfer the primary energy to the process.

Feed-to-effluent heat exchangers have been used by refineries for many years to conserve heat. Any catalytic process involving high exothermic-reaction heat can very effectively conserve energy by exchanging heat between the feed and effluent.

Process-stream preheating is generally performed in water-tube coils or in shell-and-tube exchangers. Water-tube coils, usually of the extended-surface type, are used when heat is being recovered from flue gases. Shell-and-tube heat exchangers are generally used when heat is being exchanged between two process streams.

Heat transferred to a process is invariably more valuable than heat converted into steam. The process designer must explore all possibilities for heating process streams by recovering heat from elevated-temperature streams before considering other types of heat recovery. Among the considerations involved in this technique:

1. There is some hazard in tieing in an elevated-temperature gas stream from one processing unit into a process stream in a different unit. A shutdown in the one unit for any reason can force a shutdown of the second. From this standpoint, feed-to-effluent heat exchange within a single processing unit is ideal.

2. The source of the elevated-temperature stream and the point of use of the coolant stream must be relatively close to permit effective utilization of this system.

3. Preheating process streams from gas-turbine exhaust, especially with auxiliary firing, offers a means of heat recovery that can produce excellent economic returns.

4. The system must have a way to control the exit temperature of the process stream being heated.

Circulating Heating System

An effective method of recovering heat from elevated-temperature streams is to use a circulating heating medium as the coolant in an intermediate step, and then to transfer heat from the medium to other process units such as tower reboilers, evaporators, etc. This permits distribution of the heat recovered from a single elevated-temperature stream to multiple units that may be widely separated.

A heat-transfer medium other than steam has these advantages:

• Heat can be transferred to the process at high temperature levels without the high pressure associated with steam. This can sharply reduce the cost of equipment.

• Organic heat-transfer fluids can transfer heat at temperatures up to 750 F., a significantly higher temperature than can be attained with saturated steam.

A circulating heating medium to recover heat provides the designer with a highly versatile tool for optimizing the design of a process unit. Vapor-phase or liquid-phase heat transfer can be used, or a combination heating system can be provided with certain heat-transfer media.

Heat-Recovery Equipment

The primary equipment for heat recovery is broadly classified as either fire-tube or water-tube type, depending on whether the elevated-temperature stream is flowing on the inside or the outside of the tubes. There are some specialized designs of heat-recovery equipment that do not fit into these two categories. However, such equipment is outside the scope of this article.

WASTE-HEAT boiler for heat recovery—Fig. 1

A typical fire-tube waste-heat boiler with separate steam drum, and with interconnecting risers and downcomers, is shown in Fig. 1. This design is widely used in process-plant heat-recovery applications. The separate steam drum permits the generation of higher-purity steam and better circulation through the bundle than can be achieved with a design incorporating integral steam-separation space above the tubes.

A fire-tube unit presents a highly versatile and low-initial-cost design for heat recovery. Some of the advantages and disadvantages of this design have been outlined under steam generation. In any fire-tube unit, certain features must be incorporated in order to minimize potential operating problems:

1. The inlet channel should generally be refractory lined, since most process applications involve inlet temperatures of 1,200 to 2,150 F.

2. The inlet or hot-end tubesheet is the single most vulnerable area of a fire-tube boiler. Maximum success has been achieved with relatively thin, $\frac{5}{8}$ to $1\frac{1}{4}$-in., flat flanged tubesheets having a generous corner radius to provide some expansion flexibility. Other features that have contributed to greater reliability of the hot tubesheet are: (a) strength welding of the tubes to the tubesheet, (b) refractory lining of the tubesheet, (c) insulated inlet ferrules to reduce the transfer of heat to the tubesheet ligaments, and (d) a wide tube spacing to ensure adequate tubesheet cooling area.

3. Heat flux in the inlet portion of the unit must be carefully analyzed to ensure that film boiling conditions leading to rapid tube burnout do not occur. This is especially important when cooling streams that have a high hydrogen content, because such streams have excellent heat-transfer characteristics. The combination of high heat-transfer rates and a high temperature differential can produce dangerously high heat fluxes in the inlet portion of a fire-tube steam generator. With carefully engineered fire-tube units, heat fluxes on the order of 140,000 to 160,000 Btu./(sq.ft.)(hr.) have been successfully achieved.

4. Design of the risers and downcomers is important for successful operation of a high-temperature fire-tube process cooler. The design must meet two criteria: (a) risers and downcomers must be sized and located to compensate for the much heavier steam generation that

WATER-TUBE heat-recovery coil for direct-fired heater will be used as a natural-circulation steam generator—Fig. 2

occurs in the hot inlet end of the unit, and (b) design must ensure high-velocity circulation of water across the hot tubesheet, and adequate steam removal so that there is no possibility of steam blanketing the tube sheet.

5. An internal bypass arrangement is necessary when outlet temperature of the process stream must be closely controlled.

The least complex of the water-tube designs used in heat recovery consists of a rectangular hairpin coil having tubes connected by 180-deg. return bends. This design is invariably used with gaseous elevated-temperature streams operating at low pressures. The tubes are frequently of the extended-surface type in order to compensate for the low heat-transfer rates characteristic of flue gases. A wide range of materials can be heated in this type of coil including steam, water, process liquids and gases.

The basic water-tube design of a heat-recovery unit, as shown in Fig. 2, is a typical direct-fired-heater convection section. This can serve as a complete heat-recovery unit or be incorporated into other units. It is limited to use with low-pressure, gaseous, elevated-temperature streams.

Another water-tube design frequently used in heat recovery is the two-drum boiler, as shown in Fig. 3. This may be used with elevated-temperature streams at any temperature level, but these streams must generally be at low pressures, usually under 2 psig. Extended surface can reduce the number of tubes. The design can be used over wide capacity ranges but must be in vaporizing service, with water generally, but occasionally with hydrocarbon fluids. It can be installed as a separate unit or as part of an integrated heat-recovery train.

A water-tube heat-recovery boiler is widely used in ammonia plants. It features conventional shell-and-tube construction, with modifications required to handle the high process-stream temperature and pressure, plus high generated-steam pressure. Special features include:

• The shell is internally refractory lined for protec-

TWO-DRUM boiler with extended surface tubes—Fig. 3

tion against inlet-gas temperatures of 1,500 to 1,800 F. Some designs include a water jacket to protect the shell from overheating.

• The bundle is of the U-tube, bayonet tube, or other construction, to provide maximum differential thermal expansion between the shell and tubes.

• High velocity is maintained in the tubes with either natural or forced-water circulation, so as to attain high heat fluxes without tube damage.

• Control of the process-gas outlet temperature within narrow limits can be achieved by bypassing a

WATER-TUBE steam generators recover heat from exhaust of gas turbines. These units include steam superheater, steam generator and economizer coils to produce steam at 650 psig. and 750 F. —Fig. 4

MODEL of cascaded water-tube unit generating steam and preheating process streams by recovering heat from high-temperature flue gas leaving a reforming furnace. Steam drum is at top of structure and one of the two fire-tube steam generators, recovering heat from the process stream, is shown at the grade level—Fig. 5

portion of the partially cooled process gas from an intermediate point in the shell.

• The units are normally designed for steam pressures in the 1,500 to 1,650-psig. range, and can be designed for even higher steam pressures.

This water-tube heat-recovery unit has primarily been used on hydrogen units in an ammonia plant but has the versatility for other applications where process gases in the 500-psig. range at elevated temperatures must be cooled. The principal limitations are that the unit must be kept within shop-fabricated sizes, and that it is non-competitive economically unless steam pressures over 1,200 psig. are desired.

The dramatic increase in gas turbines as prime movers for electrical generators, compressors and pumps has accelerated development of a high-capacity heat-recovery unit. Fig. 4 shows a typical installation.

A gas turbine produces high volumes of exhaust gases at temperatures ranging from 750 to 975 F., and with an oxygen content of about 17%. Thus, there are not only large volumes of gas available for cooling but the high oxygen content permits auxiliary firing at high thermal efficiency to further increase the available heat. In order to handle this combination of conditions with efficient heat recovery, the multiple-coil water-tube heat recovery unit was developed. This design may also find other applications where large volumes of low-pressure flue gas are available.

Features of the highly versatile design include unlimited size and capacity, and very low gas-side friction losses, from 2 to 12 in. water column.

These units can handle any type of coolant stream. Many have multiple-duty coils. A typical unit might preheat a process stream, superheat steam, generate steam, preheat boiler feedwater, and produce low-pressure steam for deaeration.

The design lends itself to either natural- or forced-circulation-type steam-generator coils using extended-surface or bare tubes, or any combination of the two. Furthermore, the ability to provide either single-stage or dual-stage auxiliary firing provides added versatility and control adaptability.

Steam can be developed at high pressure and superheated for power generation. Alternately, lower-pressure steam can be generated for use in process heating, eliminating the need for a separate fired boiler.

Applications for Heat Recovery

Any flue-gas stream available at elevated temperatures (600 F. or more) is a prime candidate for heat recovery. Typical applications include high-temperature process furnaces, incinerators, glass furnaces, high-temperature direct-fired dryers, and steel-mill furnaces. Fig. 5 shows extensive multiple-coil heat recovery as applied to a high-temperature process furnace. In this instance, a portion of the heat is recovered by the process feedstream, and the remaining heat is recovered by generation of steam. The expanding use of incinerators has opened up a completely new field for the application of heat recovery. With exit flue-gas temperatures of 1,200 to 1,800 F. from typical incinerators, heat recovery can reduce the costs of waste disposal.

Any process that requires the feed material to be heated to elevated temperatures in order for catalytic or thermal conversion to occur has the problem of cooling the effluent gas. A typical example is a plant where natural gas or liquid hydrocarbons are cracked to produce hydrogen. Cooling of the effluent stream is necessary in order to prepare it for further processing, and the magnitude of available heat is such that efficient heat recovery is essential in order to improve the economics of the process. A similar but more critical application is the recovery of heat from ethylene-furnace effluent, where the process stream must be cooled rapidly through a critical range in order to prevent continuation of the reaction.

Many catalytic reactions are highly exothermic, and the temperature of the effluent stream is substantially higher than that of the feedstream. The exothermic heat may have to be totally or partially removed in the reactor, and the effluent stream will also generally require cooling. The exothermic reaction heat can prove to be a valuable source of additional energy when it is properly converted by a heat-recovery system. A special extended-surface tube, developed to remove heat from the catalyst bed, uses heat-transfer salt as the coolant, and subsequently generates steam while cooling the molten salt.

Economics of Heat Recovery

The broad range of applications for heat recovery makes it virtually impossible to provide specific guidelines for economic evaluations. Such evaluations must be based on the cost of primary energy, depreciation rates, tax benefits and all the other factors that enter into an economic analysis. However, certain basic guidelines for heat-recovery applications are almost universally applicable:

1. Economic value of the recovered heat should exceed the value of the primary energy required to produce the equivalent heat at the same temperature and/or pressure level. An efficiency factor must be applied to the primary fuel in determining its value compared to that obtained from heat recovery.

2. An economic evaluation of a heat-recovery system must be based on a projection of fuel costs over the average life of the heat-recovery equipment.

3. Environmental pollution restrictions may force the use of a more costly fuel.

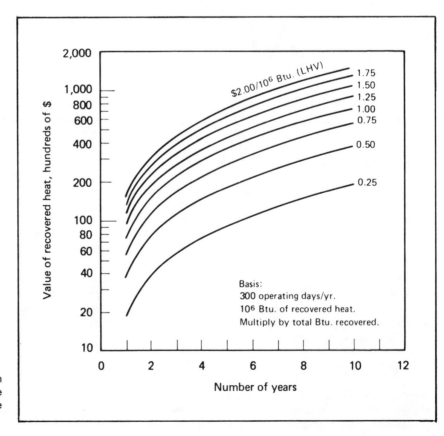

CHART yields value of one-million Btu. of recovered heat. This value is based on the projected average costs for primary fuel—Fig. 6

Equipment Cost for Heat-Recovery Units—Table IV

System Type	Approximate Cost $/10⁶ Btu.
1. Fire-tube steam generator, 650-psig. design, with external steam drum and interconnecting piping.......	1,670
2. Water-tube shell-and-tube steam generator, 1,650-psig. design, with external steam drum and interconnecting piping............	2,170
3. Gas turbine waste-heat boiler with superheater, generator and economizer coils, including field erection..	1,500
4. Water-tube steam generator for incinerator.................	1,400

4. Many elevated-temperature process streams require cooling over a long temperature range, regardless of whether or not heat recovery is used. In such instances, the economic analysis should credit the heat-recovery installation with the saving that results from eliminating the nonheat-recovery equipment that would normally have been provided.

5. Fuel availability might have an impact on the ability of a plant to maintain operations. It is possible that extensive heat recovery in a plant may reduce primary-fuel requirements to the point that plant operation can be maintained, where it otherwise might be faced with interruptions during periods of fuel shortage.

Fig. 6 represents a plot of the value of the recovered heat when evaluated against a range of fuel costs. This value has been developed for a heat-recovery rate of 10^6 Btu./hr. over 300 operating days per year. The value for any other heat-recovery rate may be determined by direct ratio. Similarly, the value of the recovered heat may be determined directly for any period from 1 to 10 yr., and by direct ratio for any other time interval. Fuel values have been plotted in $/10^6 Btu., based on the lower heating value (LHV) of the fuel. Conversion efficiency from the lower heating value of the fuel (which represents available energy) to the useful heat actually recovered from the fuel must be applied.

An example will illustrate how to use Fig. 6. Heat is recovered at the rate of 150 million Btu./hr. in the form of 650-psig. steam superheated to 750 F. The projected average primary-fuel cost over a 12-yr. evaluation period is $0.75/10^6 Btu. LHV. Expected thermal efficiency of a conventional power boiler to produce steam at the equivalent pressure and temperature is 86%, based on the LHV of the fuel.

From Fig. 6, the value of recovered heat for one year at a recovery rate of 1-million Btu./hr. and a fuel cost of $0.75/10^6 Btu. LHV is $5,400.

The total value of the recovered heat would be:

$$\$5,400 \times \frac{150 \times 10^6 \text{ Btu./hr.}}{1 \times 10^6 \text{ Btu./hr.}} \times \frac{12 \text{ yr.}}{1 \text{ yr.}} = \$9,720,000.$$

Since the power boiler has a thermal efficiency of 86%, the equivalent cost of the primary fuel would be:

$$\$0.75(1/0.86) = \$0.873/10^6 \text{ Btu.}$$

Hence, the total value of the recovered heat if purchased as primary fuel would be:

$$\$9,720,000 \times \frac{\$0.873}{\$0.75} = \$11,314,000.$$

This is a startling amount to anyone not thoroughly familiar with the large savings available through heat-recovery systems. For a plant producing 1,000 tons/day of product, the use of heat recovery per this example could reduce cost of the product by about $3.14/ton. While not all installations will produce results as spectacular as the example, it must be recognized that properly applied heat recovery can produce significant savings.

Any economic evaluation of heat-recovery equipment will require a reasonable estimate of the cost of the equipment. Here again, there are so many different types of heat-recovery equipment and so many alternate designs that it is difficult to provide cost guidelines. The cost figures in Table IV are given for some of the principal types of heat-recovery units, but the user must recognize that they are very approximate, and are based on relatively large-sized units: 50,000 lb./hr. steam or more.

This article has attempted to cover the field of heat recovery from process and flue-gas streams so as to provide some broad guidelines. Increased interest in energy conservation because of rising fuel costs and serious fuel shortages already has forced, or will force, the process designer to become heat-recovery conscious.

Meet the Authors

J. P. Fanaritis is executive vice president of Struthers Wells Corp., Warren, PA 16365. Mr. Fanaritis joined Struthers Wells in 1941 upon graduating from the University of Pittsburgh with a B.S. in chemical engineering. He has been actively involved in the design of distillation equipment, direct-fired heaters, and heat-recovery equipment.

H. J. Streich is manager of the special products department at Struthers Wells Corp., Warren, PA 16365. He has been involved in the design of direct-fired heaters and heat-recovery equipment, including high-temperature applications involving liquid metals. He has a B.S. in chemical engineering from Case Institute of Technology.

Energy Recovery In a Petrochemical Plant

Maximizing energy recovery and increasing yields of hydrocarbon feedstocks lowers operating costs and has environmental advantages, because exhaust gases and liquids are consumed within the plant.

J. A. BARLOW, BP Chemicals International Ltd.

In the engineering of acid plants in petrochemical works, a major aspect of the design and selection of equipment is to recover as much energy as possible, as well as to optimize hydrocarbon use. Energy is recovered from processes in the form of steam and electricity, and by expanding high-pressure process off-gases—mainly nitrogen and other inerts—for process-stream cooling.

Waste heat is recovered from gas-turbine exhausts, exothermal reactions, high-temperature off-gases, and by burning of acid residues in water-tube boilers.

To provide an example of how the above can be done, the operation of one of the acid plants at the Hull Works of BP Chemicals International will be described. Here, the energy recovered is used within the plant, with any excess being exported for general use in the Works. Located in Yorkshire, England, the plant is part of the largest complex of its kind in Europe.

Each plant within the complex was designed to economize fuel and maximize energy recovery by:

- Maximizing overall efficiency through an economi-

cal arrangement of compressors in the gas-turbine-driven air-compressor units.
- Generating steam from gas-turbine exhausts.
- Recovering heat from exothermic reactions.
- Generating electricity from high-pressure off-gases.
- Using the "cold" produced by the expansion of the high-pressure off-gases.
- Recovering heat in the form of steam from the incineration of heavy-ends acid residues.

Process and Equipment

The basis of the process is the continuous oxidation of a light-hydrocarbon feedstock in stainless steel reactors. Gas-turbine-driven compressors provide the air for oxidation purposes, and the exhaust from the turbines is passed through a waste-heat boiler, which generates steam. This steam is reduced in pressure and is used for heating the reactor during startup, as well as in the distillation section of the plant.

As the reaction mixture leaves the reactor—reduced in pressure and separated into various acids in a common distillation section—the heat of reaction is removed from

This article is based on a paper presented by the author at a conference held by The Institution of Mechanical Engineers (1 Birdcage Walk, Westminster, London SW1H 9JJ, England) in January 1975, from which the complete proceedings of the conference can be obtained.

Originally published July 7, 1975

Steam Generation and Usage — Table I

Production and Usage	Pressure	Unit Quanity
Production		
Gas turbines	High, reduced to medium	100
Reactors	Low	135
Waste-heat boilers	Low	67
Total		302
Usage		
Distillation	Low	112
	Medium	100
Evaporators and heaters	Low	10
	Medium	5
	High	0.5
Tracing, ejectors, etc.	Medium	4
Miscellaneous	Low	3.5
Total		235

Overall Steam Situation

Kind of Pressure	Production, Units	Usage, Units	Balance, Units
Low	202	125.5	76.5 export
Medium	100	109	9 import
High	—	0.5	0.5 import

the reactors by tube heat-exchangers. The steam is separated in a steam drum (Table I).

The off-gases from the top of the reactors—which contain mainly N_2, CO and CO_2, together with vapors from hydrocarbons and some mixed acids—are passed through waste-heat boilers, where more steam is generated.

After the off-gases have the acid and hydrocarbon removed, they pass through turbo-expanders, from which the maximum quantity of electricity is recovered, without reheating. As the gases exhaust from the expanders at about –60°C, they become the primary coolant in the latter stages of the process.

The heavy ends and acid residues produced from the process are burnt in an acid-residue boiler, where steam is generated at 44 atmospheres (atm).

Each reactor is supplied with process air from gas-turbine-driven compressors, as shown in Fig. 1. For the largest unit, air at a rate of 850,000 lb/h—under atmospheric conditions—is taken into the axial compressor via inlet filters, and discharged at 4 atm and 220°C.

To reduce the number of compressors needed to produce high-pressure air, the axial compressor is overrated in relation to combustion-air requirements. Air for the process is taken from the outlet of the axial compressor and then cooled and compressed in two radial compressors to the required pressure.

Turbine controls are designed to allow the machines to be operated independently of the plant by having a silenced air blowoff system on the roof of the compressor house. The same controls allow the machines to be run at constant speed; variations in flow are controlled by adjusting the blowoff valves on the roof.

Steam and Electrical-Energy Generation

Since the air compressors can be run as separate units, the plant can be started up independently of any other steam source. Once the reactor is in production, steam is generated by the exothermal reaction, as well as by the off-gas, waste-heat-boiler recovery system. Additional steam is also generated by burning acid residues in water-tube boilers. These were originally of the conventional coal-fired type but were converted to burn acid residues.

Steam generation and usage naturally depend upon the size and throughput of the plant, but the proportions remain approximately the same. If it is assumed that the gas-turbine steam production is 100 units, Table I outlines the steam generation and usage at the plant.

As can be seen, when the plant is operating at the design rate, a maximum of 76.5 units of low-pressure steam are exported to other plants in the complex. In addition, an extra 33 units of high-pressure steam are exported from the acid-residue boiler to the Works power station to operate turbines that drive electric generators, as well as to the low-pressure steam system of the complex.

The amount of electricity required to start the plant on one reactor—together with its auxiliary equipment such as water coolers, reactor feeders, condensate pumps, etc.—can be obtained from either the Works power station or imported from the national grid.

When one reactor is operating at the design rate, the turbo-expander alternator—driven by the reactor off-gases—generates enough electricity to balance plant usage. When a second reactor is started, the plant again becomes an importer of electricity, but turns into an overall net exporter when a second turbo-expander alternator is online.

Refrigeration

In the latter stages of the process, low-temperature cooling is essential for economy and environmental considerations. Such cooling can be accomplished by utilizing the low-temperature off-gas that is obtained by expanding the process gas through turbo-expanders, rather than spending extra energy for cooling purposes.

Capital expenditures are therefore reduced, by allowing off-gas condensers to be cooled by the process gas rather than by some other medium. This extra cooling allows the process to operate at a higher efficiency, thus providing plant operations with a higher yield on hydrocarbon feedstock.

Selecting the Prime-Mover Drive

The choice of drive for the main air compressors was made after a detailed analysis involving various alternatives, capital and operating costs, and plant flexibility and reliability. The main compressor-drive alternatives considered were: (1) electric motors, (2) gas turbines, and (3) steam turbines. When considering the economics of each scheme, two of the main controlling features that emerged were steam requirements and cost of electrical generation.

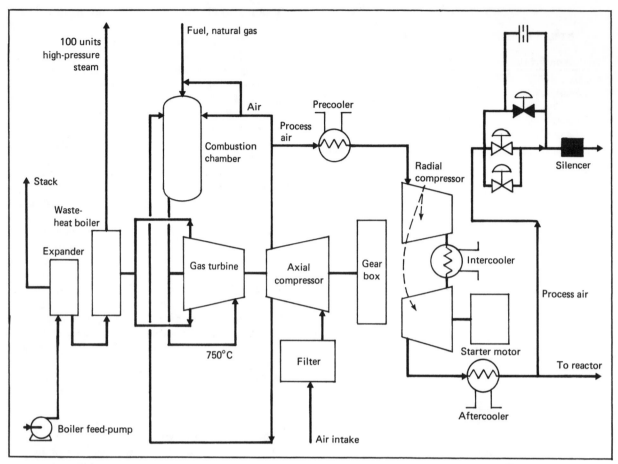

PETROCHEMICAL-PLANT FLOW DIAGRAM of a gas-turbine-driven air compressor and accessories—Fig. 1

With an electric-motor-driven compressor, an additional boiler would be required to produce process steam. Electricity cost would therefore be high. On the other hand, the controls would be simpler and probably more reliable. Steam-turbine-driven compressors would need an even larger boiler—operating at high pressure—to cater for the process steam, and steam for the turbine drive. So, even though the gas-turbine-driven compressors had the highest capital cost, running costs were the lowest due to the economics of generating steam from exhaust gases.

If the capital cost of the electric-motor scheme is considered to be 1.0, then a simplified comparison of cost for the other schemes (including operating costs) is provided in Table II.

The Gas-Turbine System

A standard Sulzer gas turbine was chosen as the prime mover for the compressors. The design was modified to enable the axial compressor to be used for the dual purpose of providing combustion air and process air. But because the compressor was to provide excess air over that needed for combustion purposes, a bypass was installed between the axial-compressor outlet and the gas-turbine outlet to prevent choking and surging during the startup of the gas-turbine sets.

Because the fuel for the turbines of the early plant was

oil, corrosion of and deposition on the turbine blades was experienced. The systems were therefore converted to natural-gas firing. Since then, the corrosion and deposits on the blades have disappeared.

A modification was also needed to the control system to ensure that the gas turbine did not shut down when the air to the reactor was cut off for process reasons. The effect of isolating the air from the reactor was that the turbine control responded to the reduction in air flow, allowing additional fuel to the combustion chamber to

Cost Comparison of Prime Movers — Table II

Expenses	Electric Drive, Units	Gas-Turbine Drive, Units	Steam-Turbine Drive, Units
Capital costs	1.00	1.69	1.31
Operating costs			
Fuel, electricity	1.75	1.25	—
Steam	2.00	1.31	3.19
Installation, maintenance, etc.	0.31	0.50	0.50
Total	**4.06**	**3.06**	**3.69**

compensate. The result was that the machine shut down automatically when the gas-turbine inlet temperature was too high.

Corrosion of the intercoolers was another major problem. Water from the cooling bundles entered the air stream, causing excessive fouling and erosion of the first stage on the high-pressure radial compressors. To overcome this, the materials of construction were changed from copper to cupronickel. The life of the tube bundles has definitely been increased, but further improvements are being considered.

Energy Recovery

When a reactor is operating at the design air rate, a substantial quantity of heat is generated. The heat content of the air from the gas turbine increases this total heat a small amount. The reactor heat-exchanger removes approximately 50% of this heat by generating steam, and the process waste-heat boilers remove another 25%. A further 6% is recovered by the turbo-expander alternator in the form of electricity. A simplified energy balance is shown in Fig. 2.

Heat-Exchanger Problems

Because the heat sources are highly corrosive, special precautions are needed to ensure that immediate action is taken should a leak develop between the product and the steam. In addition, special fabricating, design and welding techniques are required when manufacturing or maintaining the equipment.

During plant shutdowns for scheduled maintenance, all heat-exchange equipment is carefully monitored, using ultrasonic thickness measurements, visual examinations and radiography. And in-between major shutdowns, the shell of the reactors and the off-gas piping leading to the process waste-heat boilers are regularly monitored, using ultrasonic thickness measurements.

All waste-heat boilers at the complex were designed as tubular heat exchangers, with the tubes, tube-plates and end-boxes fabricated of corrosion-resistant materials. The shell of the waste-heat boiler—which contains either the water or the steam—is ordinarily constructed of carbon steel.

Because of the process environment and the operating pressures, the heat-exchanger design selected is that of a double tubeplate with an annulus. This has the advantage of establishing whether leaks occur from the steam or process side of the exchanger.

The tubes in the front tubeplate are seal-welded and expanded, whereas those in the rear tubeplate are only expanded. An allowance is made on the thickness of the tubeplates to enable them to be reused when the exchangers are rehabilitated, because tube-tubeplate attachment failures represent a major problem.

The size of these exchangers (which weigh about 25 tons each) and their position within the structure—together with other large heat exchangers in similar locations—calls for special lifting equipment. A crane is

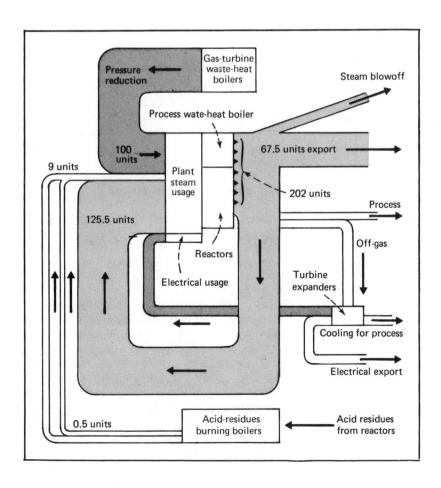

ENERGY BALANCE (electrical, steam, heat, etc.) of the total system—Fig. 2

mounted as a permanent feature on top of the plant structure, some 250 ft above ground level.

Turbo-Expander Problems

For startup and control purposes, the control system for the turbo-expander is independent of the gas flow and pressure. This enables the electrical load to be varied to suit factors other than process needs, and allows the outlet temperature to be adjusted to suit process-cooling requirements. The pressure at the inlet to the turbo-expanders is reduced by means of a 13-stage expander, to yield an outlet temperature of about –60°C.

Piping layout takes into consideration the fact that the turbo-expander could shut down for reasons other than normal process operation (i.e., mechanical failure, axial trip, overspeed, loss of oil pressure, etc.). Such flexibility is achieved by fitting a pressure-controlled bypass around the machine, which opens when the turbo-expander shuts down. Although under these conditions some of the cooling obtained by the expansion of the gases through the turbines is lost, the process can still operate successfully, but at a lower overall efficiency.

The main problems encountered with turbo-expanders (in relation to their effect on the process) have been associated with control-valve instability. Unless the machine is perfectly stable when the alternator is synchronized, plant upsets do occur.

One other problem that has been experienced is that of fluctuating electrical output, which occurs because the control valves on the inlet to the turbo-expander determine the pressure on the reactor. If this back-pressure increases due to the closing of expander inlet-valves, the bypass valve around the expander opens; this allows the gas to bypass the machine, which causes electrical-output variations.

This bypassing of the gas around the machine also affects the control of the process downstream of the expander. This happens because the gases bypassing the machine are at a higher temperature (due to the loss of expansion), which reduces the cooling on the latter stages of the process. The result is a loss of yield and efficiency. When this happens—and the machine is taken out of line for maintenance—the reactor throughput is reduced to obtain a thermal balance on the system.

Modifications to the control system have been necessary to minimize valve movement, due to the valve's design and to pulsing of the control oil. Since the inlet pilot valve was completely redesigned, the stability and control of the system have been greatly improved.

Managing a Plant Startup

Although high-pressure process air is not required until the reactors are charged and pressurized, the gas turbines are commissioned to full working pressure, with the air blown off. The waste-heat boilers thus produce the steam needed for heating the reactors.

The main problem with the startup of the gas turbines has been the ignition of the main burners. Once the main flame has been established, the units can be exporting steam at a quarter rate in approximately 20 min. All auxiliaries are checked out during this period. A detailed vibration analysis of the entire gas-turbine and compressor units is made.

When the reactor has been pressurized and heated, air is carefully introduced into it, and the reaction is started. The heating steam is gradually reduced, while the condensate level in the reactor steam-drum is raised, as steam is generated within the reactor system. The control of reaction temperature determines the amount of heat removed.

As the reactor system starts to export steam, the steam from the gas turbine is available for heating the first of the distillation columns. Further heating can be applied to the columns as the reactor steam generation increases and a second gas turbine is started up.

When the second reactor is ready for startup, there is insufficient steam available from the operating reactor and the two gas-turbine waste-heat boilers. To supply the additional steam, the acid residues from the first reactor are burned in a water-tube boiler. The steam thus generated is sufficient to supply the needs of the second reactor and the remainder of the distillation columns. Eventually, when a plant unit is operating at design rate, it becomes a net exporter of steam.

As a reactor reaches approximately 60 to 70% of its design rate, the turbo-expander is synchronized and brought online. When a reactor is at design rate, the electrical output from it approximately balances the plant usage for the one running reactor. When the second reactor is started and the expanders are synchronized, the plant becomes a net exporter of electricity.

General Conclusions

Since the commissioning of the plants, no major problems have arisen with the energy-recovery equipment. Actually, the design quantities for steam generation and usage have improved, which has resulted in slightly more steam being available for export and reuse.

As mentioned at the beginning of the article, there are other acid plants in operation at the Hull Works. Maintenance work is planned so that only one plant is offstream at any one time. This allows the steam and electrical export from the running plants to be used for startup purposes.

These acid plants show how, with careful design and well-practiced operation, the maximum recovery of energy in the form of steam, electricity and refrigeration can successfully be achieved. Initial capital costs are usually higher, but operating costs are lower, especially in this present era of high fuel costs.

Major problems are usually associated with more-difficult maintenance and more-complex startup problems. The latter are caused by the dependence of one section of the plant on the energy produced in another section; thus, process plants within the Works may become dependent on the operation of the exporting plant.

Acknowledgment

Permission to publish this article has been granted by The British Petroleum Co.

Conserving fuel by heating with hot water instead of steam

Hot-water systems for unit processes can effect fuel savings of 20% over systems using steam. Such reduced costs, plus increased reliability and less maintenance, will accelerate the use of hot-water heating during the coming years.

William M. Teller, William Diskant and *Louis Malfitani,*
American Hydrotherm Corp.

☐ Heating with high-temperature water under pressure was in use in Germany in the 1920s in the 250–420°F range. Apparently, engineers at I. G. Farben had trouble obtaining uniform temperatures on presses that were heated with steam. The presses were therefore heated with the hot water from the drum of the boiler, by circulating it at fairly high velocity through the presses. The water was returned to the boiler, which also produced steam for other applications.

While such changes were made to obtain a better product, it was quickly realized that a heat saving of up to 50% could also be obtained. In contrast to the U.S., fuel costs were a sizable manufacturing item in Europe. Within a very few years, high-temperature water (HTW) was adopted by many plants as a substitute for steam, with fuel savings amounting to 20–50%. HTW refers to water at about 250–420°F (300 psig). Mechanical problems caused by high water pressures above 420°F make this temperature the practical upper level.

At present, conditions in the U.S. are not unlike those in Europe in the twenties. Conversion from steam heat to HTW is quite attractive. Conversion costs from steam to hot water can usually be paid off in not more than two years for systems rated at about 20 MM Btu/h.

Economic studies made 25 years ago in the U.S. proved that, despite cheap fuel, substantial savings could be realized if HTW heating systems were used in multibuilding complexes.

Since 1950, most American air bases have been required to install HTW district heating in preference to steam systems. University campuses, hospitals and industrial-building complexes also started to follow this trend. Large segments of the chemical process industries have also benefited from such conversions, by having hundreds of HTW systems supplied for process heating.

The smaller systems, ranging from 5–15 MM Btu/h are only marginally more economical to operate than steam, but they are still favored because they provide much more accurate and uniform temperature control. Because all these factors have become increasingly critical in the U.S., it is timely to review the advantages of installing HTW systems or of converting existing steam-heating systems to hot water.

Disadvantages of steam heating

Apart from fuel savings, the following situations contribute to heat losses in steam-heating systems: open vents on condensate receivers; flashout losses and leaks in steam traps; and boiler blowdown. These losses involve at least an additional 5% in fuel costs. Another disadvantage is boiler deterioration, caused by instantaneous changes from low fire to peak loads. This is practically eliminated by HTW systems, because their "flywheel" effects dampen those of peak loads.

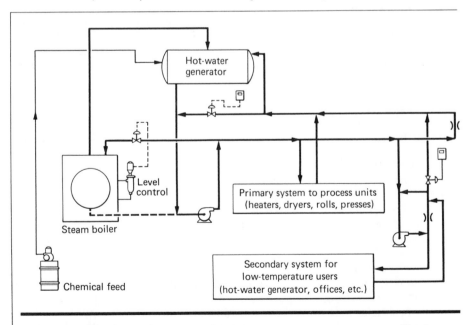

Hot-water generator produces water from steam Fig. 1

Originally published June 21, 1976

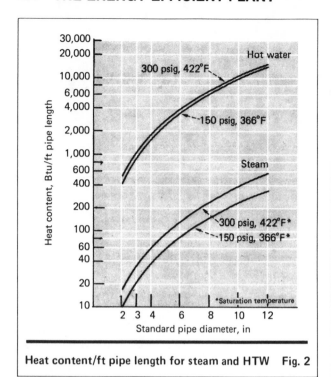

Heat content/ft pipe length for steam and HTW   Fig. 2

A hot-water system, on the other hand, is a closed circulating loop, with only very minor losses from leakage at valve stems and at pump stuffing boxes. Inasmuch as the amount of makeup water required is only a small fraction of that required for a steam system, this eliminates the need for blowdown, thereby doing away with a source of considerable heat loss.

Condensate lines are usually subject to corrosion, because mineral-free water, combined with atmospheric oxygen, becomes very corrosive. Theoretically, a properly maintained hot-water-cascade system, derived from condensing steam, will last indefinitely. The water volume increase in a hot-water-cascade system supplies makeup water to the boilers without flashing, thereby eliminating steam-condensate flashout losses.

Condensate collection systems usually flow by gravity, so all lines are pitched in the direction of the receiver. Forced-circulation hot-water lines are entirely independent of the building layout. They may, for instance, frame plant windows or doors to prevent cold air leaks.

Hot-water system advantages

For new installations, total capital investment is about the same for both steam and hot-water systems. However, the savings involved in fuel costs and maintenance make the payout period for a new hot-water system shorter than for conversion of an existing steam system.

Many plants use their steam boilers for both process and space heating. Cascade heaters can generate up to 350°F water from 150-psi steam (or 400°F from a 250-psi boiler). This temperature is adequate for rolls, presses, extruders, evaporators, conveyors and reactors. Steam-pressure reducing valves are not necessary to maintain the different temperature levels required by each machine.

Plants that have steam boilers can convert to hot-water heating simply and quickly by installing direct-contact water heaters in, or adjacent to, the boiler room. When boilers must be replaced because of age, hot-water generators can be installed as indicated in Fig. 1.

Direct-contact, hot-water heat exchangers (cascade heaters) not only can convert steam into hot water but can also serve as heat reservoirs. Such cascade heaters absorb sudden peak loads and make it possible to operate the boilers at fairly constant load levels. (Continuous changing of loads is one of the main factors that shorten boiler life.)

A central HTW system that supplies the higher temperature levels for process equipment, as well as low temperatures for space heating, is simple and uncomplicated, and has high distribution efficiency. A continuously circulating closed loop discharges HTW at constant temperature from the cascade heater (at only a few degrees below the saturation steam temperature), returning to the heater after releasing any required heat into separate secondary loops.

Each loop is individually controlled, either manually or automatically, to maintain the various temperatures at which the process equipment, the space heating, and the utility hot-water system operate. Cooled water from the secondary loops discharges into the return side of the primary loop and from there back to the cascade heater.

Since the only heat extracted is that required for the various services, distribution efficiency of the overall system approaches 95%. A typical system is shown schematically in Fig. 1.

The film heat-transfer coefficient, $Btu/[(h)(ft^2)(°F)]$, of hot water flowing through a pipe is constant at all points around its periphery for any given cross-section, varying only with velocity and temperature. For process applications requiring extremely close temperature control, the circulating rate through a secondary loop can be designed to limit the difference between inlet and outlet temperatures to as little as $\pm2°F$.

On the other hand, the film heat-transfer coefficient of condensing vapors flowing through a horizontal pipe is higher in the top region and lower in the bottom, due to the forming of a condensate film along the bottom. The increasing accumulation of condensate between the inlet and the outlet of the pipe further reduces the overall inside-film heat-transfer coefficient. It has been demonstrated by actual tests that the condensate film thickness on the bottom of a pipe is five times thicker than that at the top.

The greater heat capacity of hot water over steam at equivalent saturation temperatures and the narrower pipelines required are other advantages. A steam line must be many times larger than an equivalent hot-water line, which need only be one or two sizes larger than the condensate line required in the steam system.

As to heat capacity, steam yields up about 800–960 Btu/lb when it condenses, while HTW can transfer only 100–150 Btu/lb. This, however, is no basis for comparison. Consider, for example, a 1-ft-long, standard 6-in-dia. pipe filled with 150-psig steam. This pipe will hold 0.073 lb of steam with a heat capacity of roughly 90 Btu. This same pipe with water at 366°F (saturation temperature corresponding to 150 psig) will hold 11

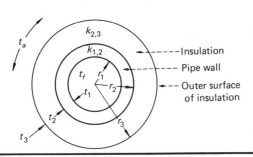

k = thermal conductivity, Btu/[(h) (°F) (ft)]
t_f = fluid temperature, °F
t_a = ambient temperature, °F
t_f, t_1 and t_2 are temperatures
  that have approximately the
  same value

Steam 10-in-dia. pipe, 2-in insulation	Stream conditions Condensate 5-in.-dia. pipe, 1-in insulation	High-temperature water 10-in.-dia. pipe, 2-in insulation
t_f = 365°F	t_f = 212°F	t_f = 340°F (average)
t_a = 70°F	t_a = 70°F	t_a = 70°F
D_1 = I. D. of pipe, 10.02 in (say 10.00 in)	D_1 = I. D. of pipe, 5.04 in	D_1 = I.D. of pipe, 10.02 in (say 10.00 in)
D_2 = O.D. of pipe, 10.75 in	D_2 = O.D. of pipe, 5.56 in	D_2 = O.D. of pipe, 10.75 in
D_3 = pipe plus insulation dia., 14.75 in	D_3 = pipe plus insulation dia., 7.56 in	D_3 = pipe plus insulation dia., 14.75 in
$k_{1,2}$ = 30 Btu/[(h) (ft) (°F)]	$k_{1,2}$ = 30 Btu/[(h) (ft) (°F)]	$k_{1,2}$ = 30 Btu/[(h) (ft) (°F)]
$k_{2,3}$ = 0.035 Btu/[(h) (ft) (°F)]	$k_{2,3}$ = 0.033 Btu/[(h) (ft) (°F)]	$k_{2,3}$ = 0.034 Btu/[(h) (ft) (°F)]
h_a = surface coefficient of heat transfer outside insulation, 1.5 Btu/[(h) (ft²) (°F)]	h_a = 1.5 Btu/[(h) (ft²) (°F)]	h_a = 1.5 Btu/[(h) (ft²) (°F)]
t_3 = temperature at surface, 102°F	t_3 = 96°F	t_3 98°F
L = length of pipe, 1,000 ft	L = length of pipe, 1,000 ft	L = length of pipe, 2,000 ft

Steam, condensate and HTW conditions for length of pipe plus insulation   **Fig. 3**

lb/ft, which contains about 3,800 Btu. Hence, the ratio of absolute heat-storing capacity is 42 to 1 in favor of water. A graphical comparison of heat contents for different pipe diameters is provided in Fig. 2.

An HTW-distribution system acts as a heat accumulator due to its capacity to store heat. This may be likened to an energy reservoir, which can accommodate sudden heat demands without loss in temperature. Steam systems often suffer a temperature drop when shock or peak loads occur, which cause a drop in the boiler pressure. But because of the heat storage, HTW generators need not be sized for maximum peak loads. Steam boilers require such sizing to prevent pressure losses and accompanying temperature losses. Such smaller pipe diameters for HTW systems are economically important when heat is transported over long distances. Smaller pipes also reduce insulation costs.

One argument given for preventing many installations from switching to hot water is that steam lines are needed for certain production methods, and that such steam would be difficult to obtain from hot water. This is a fallacy, because obtaining steam from high-temperature water can be done inexpensively.

Comparison example

To compare the advantages of high-temperature water versus steam process heating, assume that 50,000 lb/h of steam at 150 psig are to be produced to deliver heat to equipment 1,000 ft away. Assume further that the saturation temperature of the 150-psig steam = 360°F; specific volume = 2.75; enthalpy of evapora-

tion = 857 Btu/lb; enthalpy of saturated vapor = 1,195.6 Btu/lb; ambient temperature = 70°F; velocity = 5,000 ft/min; and density of water at 340°F = 56 lb/ft³. (See Fig. 3 for further information.)

Steam system

Area required = (50,000 × 2.75)/(5,000 × 60)
                     = 0.46 ft², or 66 in²

For a 5-lb pressure drop, a 10-in pipeline would be required, plus a 5-in condensate line. Assume also that the 10-in line would have a calcium silicate insulation 2 in thick, and that the 5-in line would have a 1-in insulation of the same material.

Pipeline heat losses:

To calculate the heat losses, this equation* is used:

$$q = \frac{\pi(t_f - t_a)L}{\left(\frac{1}{h_f D_1}\right) + \left[\left(\frac{2.3}{2k_{1,2}}\right)\log\left(\frac{D_2}{D_1}\right)\right] + \left[\left(\frac{2.3}{2k_{2,3}}\right)\log\left(\frac{D_3}{D_2}\right)\right] + \left(\frac{1}{h_a D_3}\right)}$$

where h_f = surface coefficient of heat transfer on the inside wall of the pipe; h_a = surface coefficient of heat transfer on the outside of the insulation. Usually, the first term of the denominator can be neglected because its effect is minimal.

The surface temperature, t_3, and the surface coefficient, h_a, must be computed by trial and error. Here, only the final computation for the heat losses are provided because the intermediate steps would be too long.

*Kern, Donald Q., "Process Heat Transfer," Chapter 2, p. 19, McGraw-Hill, New York (1950).

Therefore:

Loss in 10-in line =

$$\frac{\pi(365-70)(1,000)}{\left(\frac{2.3}{2\times30}\right)\log\left(\frac{10.75}{10.00}\right)+\left(\frac{2.3}{2\times0.035}\right)\log\left(\frac{14.75}{10.75}\right)+\left(\frac{1}{1.5(14.75/12)}\right)}$$

$$= 183,200 \text{ Btu}$$

Loss in 5-in line =

$$\frac{\pi(212-70)(1,000)}{\left(\frac{2.3}{2\times30}\right)\log\left(\frac{5.56}{5.04}\right)+\left(\frac{2.3}{2\times0.033}\right)\log\left(\frac{7.56}{5.56}\right)+\left(\frac{1}{1.5(7.56/12)}\right)}$$

$$= \underline{78,100 \text{ Btu}}$$

Total heat loss for steam system = 261,300 Btu

Amount of condensate = 183,200/857 = 214 lb

Amount of heat delivered = (50,000 − 214)857
= 42,666,600 Btu/h

Flashout heat losses:

If the produced flash vapor is not used when the condensate is flashed out to atmospheric pressure in the return line and condensate receiver, the losses due to flashout will equal the enthalpy of the saturated water at 365°F minus the enthalpy of the saturated water at 212°F, or 338.5 − 180 = 158.5 Btu/lb.

To produce 857 Btu of latent heat per pound of steam, the boiler must supply 1,195.6 − 180 = 1,015.6 Btu/lb (assuming the condensate is returned to the boiler at 212°F). Therefore, condensate losses due to flashout = (158.5/1,015.6)100 = 15.6%. In addition, an approximate 5% loss occurs due to leakage of steam and condensate, plus blowdown losses, bringing the total losses up to 20%.

Total heat required:

Inasmuch as 80% of the condensate (40,000 lb/h) would be returned to the boiler, the enthalpy of the feedwater to the boiler, including makeup water, is:

40,000 lb condensate at 212°F (180 Btu/lb)
= 7,200,000 Btu

10,000 lb makeup H_2O at 50°F (18 Btu/lb)
= 180,000 Btu

Total heat to boiler = 7,380,000 Btu

The boiler must therefore produce (50,000 × 1,195.6) − 7,380,000 = 52,400,000 Btu/h.

Assuming 75% boiler efficiency, the adjusted total amount of energy needed for steam heat is 52,400,000/0.75 = 69,867,000 Btu.

Water system

To deliver 42,666,600 Btu to the equipment, assume a 40°F temperature drop between flow and return. The amount of water needed would then be 42,666,600/40 = 1,066,700 lb/h. If we assume a velocity of 10 ft/s, the pipe area needed would be 1,066,700/[(3,600)(10)(56)] = 0.529 ft², or 76.2 in².

To arrive at this pipe area, the equation $Q = AV$ is used, where Q = flow, ft³/h; A = area, ft²; and V = velocity, ft/h. Thus:

$$\frac{1,066,700 \text{ lb/h}}{(3,600 \text{ s/h})(10 \text{ ft/s})(56 \text{ lb/ft}^3)} = 0.529 \text{ ft}^2$$

This area requires a 10-in pipe.

The flow and return lines would require 2,000 ft of 10-in line, with a 2-in calcium silicate insulation. If the flow temperature = 360°F, and the return temperature = 320°F, the mean temperature would be 340°F. Then the heat loss would be:

$$\frac{\pi(340-70)(2,000)}{\left(\frac{2.3}{2\times30}\right)\log\left(\frac{10.75}{10.00}\right)+\left(\frac{2.3}{2\times0.034}\right)\log\left(\frac{14.75}{10.75}\right)+\left(\frac{1}{1.5(14.75/12)}\right)}$$

$$= 326,800 \text{ Btu}$$

Heat delivered to equipment = 42,666,600 Btu

Total heat required = 42,993,400 Btu

Assuming a boiler efficiency of 77%, the total adjusted amount of energy required from the hot-water system would then be 42,993,400/0.77 = 55,835,600 Btu/h.

Steam versus hot-water system

Heat required by steam system = 69,867,000 Btu

Heat required by hot-water system = 55,835,600 Btu

Difference = 14,031,400 Btu

Saving of high-temperature-water system over steam system = (14,031,400/69,867,000)100 = 20%.

The authors

William M. Teller has been associated with American Hydrotherm Corp., 470 Park Ave. South, New York, NY 10016 for over 28 years. Formerly, he was vice-president and partner of the corporation; now he only works part-time. He is a graduate in textile engineering from the University of Florence, Italy, and worked for three years in Austria for Caliqua. Thereafter he held various positions as a textile-machinery and photogrammetric-equipment designer.

William Diskant is Executive Vice-President, American Hydrotherm Corp., 470 Park Ave. South, New York, NY 10016, with which he has been associated since 1956. Previously, he worked for Bechtel Corp. and was an instructor at the College of the City of New York, from where he received a B.M.E. degree. He also holds an M.M.E. degree from New York University. A registered professional engineer in five states, he belongs to the American Soc. of Mechanical Engineers, the American Soc. of Heating, Refrigerating and Air Conditioning Engineers, and the International District Heating Assn.

Louis Malfitani is Vice-President, American Hydrotherm Corp., 470 Park Ave. South, New York, NY 10016, with which he has been associated since 1952. Since 1965 he has been chief of the High Temperature Water District Heating and Cooling Dept. He has a degree in mechanical engineering from the College of the City of New York, and is a member of the American Soc. of Mechanical Engineers and the International District Heating Assn.

Recovering energy from stacks

Many plant officials complain about the high cost and the scarcity of fuel. Yet the stacks of their plants are usually hot, which means energy is being wasted. Here is a way many companies could save money.

Howard Summerell, H. M. Summerell Co.

☐ It is quite possible that your plant can meet its responsibilities to the environment and still survive economically. Unexpected benefits—even profits—reveal themselves in lower equipment costs when efforts are made to avoid energy waste.

The most economical approach to energy and environmental conservation appears to be a combined or multilevel one, when possible. You can save money and materials by combining heat recovery with air-pollution control. In the past, low fuel costs made it uneconomical to install heat-recovery equipment in many industrial applications, such as on stacks of direct-fired furnaces. Today, shortages and the higher cost of fuel turn the spotlight on fuel economy.

When we recover a large part of the sensible-heat energy from an air stream, we find that the resulting drop in temperature causes a dramatic reduction in the size and cost of air-pollution control equipment. Thus, we save not only in equipment and operating costs (due to lower horsepower requirements), but also save scarce and expensive fuel. Recovered heat can be used to reduce fuel requirements by preheating primary combustion air; to preheat air going to another process; and to heat buildings.

Plant examples

Consider the following problem:

A company has several tobacco dryers, each discharging 12,480 acfm (actual ft³/min) of tobacco-dust-laden air at 365°F. If an engineer considers the air-pollution problem alone, the outlay for dust collectors and blowers will be about $22,260 (Fig. 1 and 2).

However, if he also considers the heat-recovery problem, certain pleasant surprises are in store for him. After considering heat-recovery equipment of various

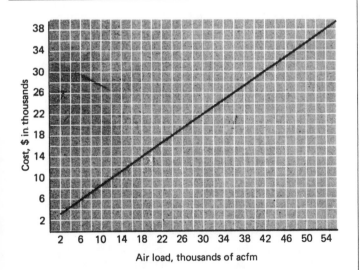

Graph is based on current prices and several sizes of Mikro-Pulsaire dust collectors, manufactured by MikroPul Div., United States Filter Corp., Summit, N.J. Cost includes reverse-pulse, dry-filter collector at 10/1 filter ratio, filter bags, controls, rotary air-locks, erection, screw conveyors, made with standard materials of construction.
Note: When temperature goes above 275°F to a maximum of 425°F, add 10% to cost.

Cost of reverse-pulse, dry-filter collectors   **Fig.1**

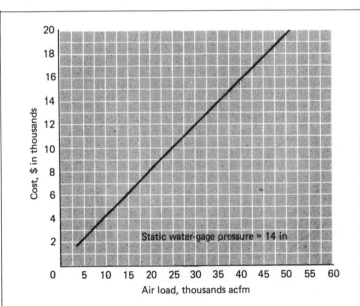

Graph is based on current prices and several sizes of centrifugal blowers made by Aerovent, Inc., Piqua, Ohio.

Cost of centrifugal blowers with various acfm   **Fig. 2**

Originally published March 29, 1976

types, he decides in favor of the heat pipe,* which was developed by NASA and others. Since this is a truly passive device requiring no additional power input, it is most desirable in an energy conservation program. This super heat-conductor is compact, requires only a small pressure drop, has no moving parts, and offers efficiencies up to 75%.

After selecting an exchanger with 60.7% efficiency, the plant engineer now finds the exhaust reduced to 9,100 acfm at 144°F (Fig. 3), and the cost of the dust collector and blower reduced to only $11,500. Thus, cost savings of 48% ($10,760) are effected:

Air location	Exhaust, acfm	Temperature, °F	Reverse pulse collector cost,$	Centrifugal blower cost, $	Total cost, $
Before exchanger	12,480	365	17,160	5,100	22,260
After exchanger	9,100	144	7,500	4,000	11,500
Savings					10,760

Most reverse-pulse, fabric dust collectors operate below 425°F, which is the temperature limit of Nomex (high-temperature nylon) filter medium. If air must be filtered above this temperature, the expense of exotic materials and design must be considered. In our sample problem, wet scrubbers could be chosen, but these have high operating costs, and require slurry disposal—which adds a water-pollution problem.

By dropping the operating temperature to 144°F, a much smaller collector and a blower of standard materials of construction could be used, thereby saving plant space and equipment outlay.

By recovering heat from the high-temperature exhaust, additional benefits result by the savings of $38,190 during the first year of operation. These savings are obtained by recovering 1.913 million (MM) Btu/h, and by using No. 2 fuel oil at 30¢/gal, 70% combustion efficiency. Since fuel costs are $3.10/MM Btu, the total gross savings are: (1.9 MM Btu) × ($3.10) × (24 h) × (360 d) = $50,890. The cost of heat-recovery equipment is as follows: heat-pipe exchanger (8-coil row, 60.7% efficiency, 8,000 scfm at $1.15/scfm) = $9,200; centrifugal blower = $3,500.

Thus, the gross annual savings minus the cost of equipment = $50,890 – ($9,200 + $3,500) = $38,190, which is the net first-year heat-recovery savings.

Thus, this company has realized first-year savings on combined air-pollution control and energy recovery of $48,950 ($10,760 + $38,190) on an investment of $24,200 ($11,500 + $12,700), plus installation costs!

Considering another example, assume that a plastics plant has several monomer dryers exhausting 11,000 acfm at 650°F. Since the air stream meets the state code, no pollution problem exists. However, a close look at this application reveals that the dryer manufacturer's recommended exhaust of 11,000 acfm is, typically, much too high. This could be safely reduced to 8,500 acfm, thereby reducing the exhaust blower cost from $4,700 to $3,500 (Fig. 2). The reason for this is that power and pressures vary directly with air density, even though the actual weight of air is not changed.

* Described in *Chem. Eng.*, Aug. 19, 1974, pp. 89-91.

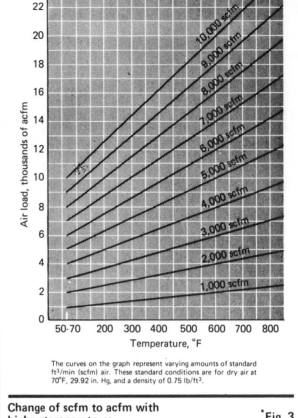

The curves on the graph represent varying amounts of standard ft³/min (scfm) air. These standard conditions are for dry air at 70°F, 29.92 in. Hg, and a density of 0.75 lb/ft³.

Change of scfm to acfm with higher temperatures   *Fig. 3

If heat-recovery equipment is installed so that 1.785 MM Btu/h are saved, the gross annual fuel savings amount to $47,780. These savings are accomplished by burning No. 2 fuel oil at a cost of $3.10/MM Btu, or (1.785 MM Btu) × ($3.10) × (24 h) × (360 d) = $47,780.

Equipment costs would be $8,600 for a heat-pipe exchanger (8-coil row, 63.6% efficiency, $2.15/scfm), and $2,300 for a centrifugal blower. Thus, total equipment costs would be $10,900 (installation not included). Subtracting this amount from the gross annual savings yields net savings of $36,880 for the first year.

Recovered-heat misapplication

The first impulse of many engineers is to use the recovered heat to heat up the plant. Such an application, however, is primarily a seasonal problem, which lasts less than six months out of the year. If the recovered heat is instead returned to the process for 24 h/d, 360 d/yr, tremendous savings become apparent.

Payback calculations are of overriding interest and should be carried through the entire plant-life expectancy of 20 years. They should reflect not only fuel savings, but also eliminated equipment.

At first glance, the expense of meeting state air-pollution codes may appear unbearable, but such costs can be made much less burdensome, as indicated here.

* © Copyright 1976 by Howard Summerell

The author

Howard Summerell is president of his own firm, H. M. Summerell Co. (P.O. Box 8646, Richmond, VA 23226), which is a manufacturers' representative concern that specializes in air-pollution control, industrial air handling, and energy conservation. He has a B.S. degree in mathematics from Davidson College, in North Carolina, and is a member of the American Soc. of Heating, Refrigerating and Air-Conditioning Engineers, Instrument Soc. of America and the Technical Assn. of the Pulp and Paper Industry.

Strategies for curtailing electric power

This system of energy management saves up to 25% of the power costs for industrial customers and adds to the effective capacity of existing power plants. Here are guidelines to its implementation.

Edmund A. Perreault and *Paul J. Prutzman*, *Air Products and Chemicals Inc.*

☐ A novel form of electrical power usage, known as voluntary peak shaving (VPS), is being employed to reduce energy costs. With VPS electrical service, the customer makes a decision of when, for how long and to what extent he must reduce operations in order to avoid taking more than a minimum amount of electricity during the electric company's peak demand hours for the year. The VPS contract provides a substantial price incentive for the customer to curtail operations during that annual peak.

At Air Products and Chemicals, Inc., VPS curtailment strategies have been developed through the combined efforts of management, operators and designers, and entrusted to a special taskforce. These strategies consist of statistical guidelines indicating the optimum times for plant operations to be cut back in order to avoid the utility's annual peak hours.

Plants at New Orleans, La., Burns Harbor, Ind., and North Baltimore, Ohio, utilize the VPS techniques to be described. These plants have an aggregate annual demand of about 60 MW, with the curtailment portions offering potential savings of $1.5 million/yr.

A look at the daily load curve (Fig. 1) for a typical utility will aid in understanding the benefits achieved from VPS service. The demand of large round-the-clock industry rises slightly at 9:00 a.m. and again at 1:30 p.m., but its total load profile is flat. When the remaining industrial load is added, the effects of a single-shift operation at the smaller plants are readily seen. This peak, while somewhat more prominent, is not sharp, and the total load profile is still relatively flat.

Adding general-service customers and miscellaneous categories, such as street lighting, to this industrial base produces a daily load curve more closely approximating the familiar form. The peak now occurs during the normal business hours, in the early afternoon (1:30 p.m.)

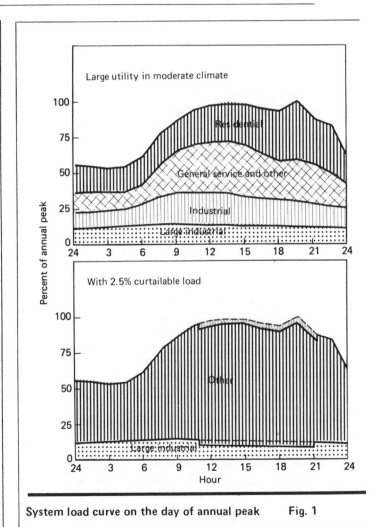

System load curve on the day of annual peak   Fig. 1

Originally published May 23, 1977

when the sun is near its zenith. A second, lower, peak occurs in the evening when the lighting load increases.

As the last increment, the residential load, is added to the daily load curve, the peak shifts dramatically. In the example, it moves out of normal business hours to a time when people returned home from work, turned on the air conditioning or heat, and are preparing the evening meal. This peak can occur at different times and different seasons for different utilities, and can change from morning to mid-afternoon to evening in different years for the same utility. The timing of its occurence is due to the life style of the population.

Now, if there were certain manufacturers that could reduce operations during the times when the utility load were approaching its maximum, the resulting load curve would look like that in Fig. 1. As shown, the load for the large industrial class—and thus the system total load— drops at 11:00 a.m. and comes back up at 9:00 p.m. If timed correctly, this reduction in industrial loads could reduce the utility's peak load for the year. Since a utility's total need for generation capacity is predicated primarily on the annual peak load of the system, such a peak-load reduction can reduce the need for expensive generating-plant additions.

VPS is effective because some manufactured products can be stored, whereas electricity cannot. The great fluctuations in demand for electricity cause the generating plants to be used only about half of the time. Thus, as VPS is extended, utilization of the existing generating plant rises, since no capacity is needed to provide industrial service at the time of the annual peak load. VPS customers take power when the utility operates below 100% of its maximum demand, and consume (without adding to capacity requirements) added kilowatt-hours for which the utility can recover its fixed costs, thus lowering the average cost to all customers.

What does the customer get in exchange for these benefits to the utility? A chance to reduce his electricity costs. Successful avoidance of the utility's annual peak load can result in savings of as much as 25% of the power costs to an ordinary high-load-factor industrial customer.

The need to know power demand

Once a utility and its customer have agreed as to the applicability of VPS service, and have established an appropriate rate schedule (approved by the local regulatory body), the customer must implement a curtailment strategy to maximize the benefits available under the rate schedule, i.e. curtail operations at the time of the utility system's annual peak, yet minimize the hours of lost production. To do this, the customer must become intimately familiar with the utility's peak power patterns, in order to anticipate future peaks.

It is appropriate to ask, "What is an industrial concern doing trying to forecast electric-power peak demands?" Without being presumptuous, the simple answer is that, for our specific purpose, we think we can often do the job best. In most cases, we are asking a question that is more important to us than to the utility. Knowing the value of the annual peak one year in advance will not greatly influence the operations of most utilities, since time is too short to make capacity additions. Peak demands five or ten years into the future are more likely to interest utility-

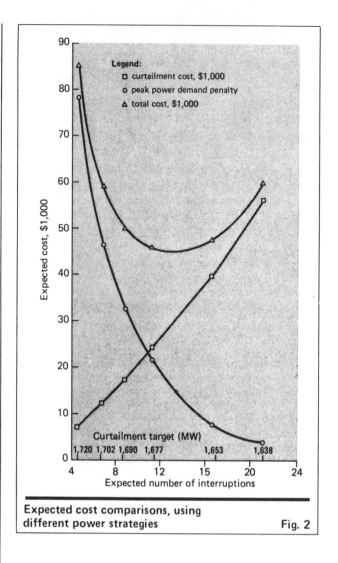

Expected cost comparisons, using
different power strategies   Fig. 2

company forecasters. Our management feels more comfortable with a forecast developed with a specific purpose in mind.

Also, forecasting methods can depend on the purpose for which the forecast is to be used. The utility is more likely to select a technique that will not underestimate future power demands, because of the need to avoid shortfalls in capacity planning. The curtailable customer, on the other hand, must be very concerned with overestimating the peak, since that could lead to a long off-peak season of paying peak-demand charge penalties while wondering whatever happened to the peak that was forecast for the previous on-peak season.

Justifying the cost of the analyses and techniques appears easy when power-bill savings of $250,000 to $750,000 per plant can be achieved; but such justification becomes more difficult on closer inspection. In theory, we could close our plant for the entire peak season and reap the full power-demand savings, but we would obviously lose all of our customers, since providing sufficient storage is not economical. In practice, we usually arrive at strategies requiring 10 to 20 plant interruptions a year and having a 95% or better chance of avoiding peak charges. This requires an estimate of costs versus benefits, in calculating the savings associated with curtailment strategies.

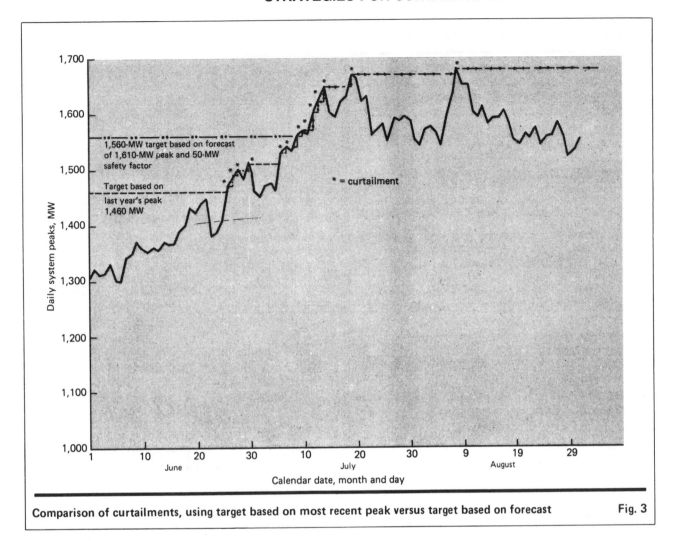

Comparison of curtailments, using target based on most recent peak versus target based on forecast   Fig. 3

A strategy for curtailment

The general approach involves three basic steps: forecasting, simulation and selection.

A variety of forecasting techniques have been employed over the years with varying degrees of success. Levels of sophistication have ranged from simple straight-line regression models to Box-Jenkins or complex econometric methods.

To develop a curtailment target, one must combine forecasts, data on daily and seasonal load profiles, demand measurement, and cost details, so as to provide a complete picture for use in selecting a strategy.

We have chosen a simulation model as the mechanism for incorporating all these factors into a consistent cost-measurement framework. The simulation approach is technically straightforward and, therefore, easily explained. Simulations require relatively few assumptions—definitely a positive factor in winning confidence and support for the analysis.

Finally, evaluation of alternative curtailment strategies must be considered. Fig. 2 presents the results of a set of simulation runs for a range of curtailment targets. For each target, five years of load profiles were simulated 200 times. In each trial, a value was drawn from a distribution of forecast errors to provide an actual peak-load value.

Within each trial, values of load change and load-measuring error were selected from distributions for each day of the peak season. The results in Fig. 2 are the expected values of the cost of curtailments and the peak-demand power penalty. These costs can be compared with a charge of just over $300,000 for operating at normal demand levels during a power peak.

As can be seen from Fig. 2, the total cost curve has a reasonably wide flat section between 10 and 15 curtailments, in which the expected total costs vary only about $5,000. In selecting among these alternatives, it is worth noting that the expected peak-power demand penalty results from the combination of a rather low incidence of occurrence (1.7–25.8%)with a very high cost. Since the expected curtailment costs have much less variability, a good argument can be made for selecting the curtailment target with the least uncertainty when the total expected costs are similar.

One point should be emphasized: both the analysis and the implementation of the strategy depend on utility load data. In the analysis activity, we are usually able to work with data that have been compiled carefully and declared official; in implementing a policy, we often face tradeoffs between accuracy and timeliness. The most accurate data are only available after meter errors and other biases have been removed after the fact, while curtailments must be

An example of curtailment costs for an air separation plant   Table I

Unit curtailed	Average hours production lost	Remaining load (kW)	Associated costs	
None	–	10,000	–	
Liquefier	6	4,000	Energy to idle main air compressor	$300
Main air compressor	10.5	0	Power to restart	$250
			Labor to restart	$54
Any	–	–	0 – 5% lost production	$20/hr lost production
			5 – 10% lost production	$40/hr lost production
			> 10% lost production	$80/hr lost production
			(on a 30-d average)	

made on real-time information. This is compensated by introducing safety factors in the curtailment policy to allow for the risk of not recognizing the peak when it occurs, and the companion problem of unnecessary curtailment.

Forecasting

Forecasting is done to avoid premature shutdown due to temporary peaks. Given the availability of reasonable load data, our first step is to forecast the coming season's peak demand. A good forecast is one of the most important factors in achieving savings with a strategy. Our strategy, as provided to plant operating management, consists essentially of an indicated electrical load at which plant consumption should be curtailed. In some cases, it is supplemented by curtailment targets that apply only at certain times of day or for certain pieces of equipment; but primarily, we are concerned with that single overall curtailment target.

Fig. 3 compares those curtailments that are required by a policy of curtailing when the previous peak is reached, versus those required by a target of 1,560 MW based on both a forecast of 1,610 MW for the season's peak and a safety factor. The safety factor in this example is 50 MW and is subtracted from the forecast to compensate for historical forecast errors. In both cases, the target was adjusted upward when the actual demand exceeded the then current target.

For utility contracts in which the peak-demand-charge calculation is based on the peak load of the last twelve months, use of a forecast-based policy is even more important. If the peak of one summer exceeds that of the previous summer, the preceding example holds; but in the case of a year-to-year peak decline, a policy based on the prior peak could result in not curtailing on the new lower peak. While, at first glance, the idea of a lower peak from one year to the next seems remote, such an incident did occur in 1970 even before OPEC introduced many Americans to the finiteness of energy supplies. Now, with a greater consciousness of energy conservation and more-intense load-management efforts, year-to-year peak

declines must be given serious consideration.

Our early work with time-series curve-fitting techniques was fairly successful in picking up the long-range trends but was somewhat erratic in its year-to-year accuracy. This, therefore, is an approach that would be useful to a utility in determining long-range generation requirements but would not account for short-term fluctuations very well.

We recognized that there were two avenues of improvement open to us. In the first place, we had not been including any economic factors that could influence electrical consumption. Secondly, since we were concentrating on the peak loads occuring in the summer months, we had been discarding nearly a year's worth of the most current load data.

We found that both of these factors could be accommodated by means of a single technique. A number of analyses were conducted to measure relationship between economic variables and loads, but without notable success. The most significant problem apeared to be the national scope of available economic data as compared to the very limited regional nature of the utility's native demand. Because the level of regional disaggregation required of the economic variables did not exist, we borrowed from a Federal Reserve Board practice of using electric power consumption to measure components of industrial production. Although this appeared to bring us full circle, by declaring electricity demand to be a good indicator of economic activity that subsequently drives electricity demand, the procedure is not misleading. It really points out that electricity demand, a readily available data stream, is one of the best measures of economic activity.

Using electricity demand as a proxy for economic activity does, however, introduce another problem: we have no forecast of this proxy for the coming season. At this point, introduction of the load data from the non-peak season proves useful. The techniques of seasonal adjustment analyses allow the utility's load data for both peak and non-peak seasons to be decomposed into seasonal and trend-line components.

We, therefore, have modified our forecast horizon from predicting a peak based on previous peak data at least a year old, to predicting the next two- or three-month trend in line-load growth, using data from the months immediately preceding the peak season, and then applying the relatively stable seasonal factors to this underlying load trend. We have shifted the jumping-off point of our forecast from about twelve months before the peak to about two months. This ten-month advance has proved especially valuable in dealing with the recessionary periods of the 1970s when a ten-month period could see substantial changes in the business cycle.

We have employed this forecasting methodology since 1970, a year in which it successfully forecast an unexpected downturn in peak demand at one utility. Since then, many factors related to electricity demand have changed, and have necessitated continuing analysis of forecasting approaches and performance. Throughout these continued studies, however, this basic approach has remained the key element of our work.

Simulation

Developing the simulation model requires close cooperation between the analyst and the plant operations and distribution personnel. First, the contract with the utility must be understood fully and modelled accurately. Computer languages are well suited to modelling the ifs, ands, and buts of a contract, but care must be taken that all contract aspects are linked in the proper relationship.

Energy-intensive operations, such as caustic-chlorine, phosphate, and industrial-gas manufacture, are best suited to operate under VPS service conditions. Plants that use curtailable electric service must be able to reduce their demand on the electric system with relative ease and rapidity, without causing damage to their own equipment thereby.

Discussion with plant operations staff personnel are necessary to determine how the plant's power requirements can be curtailed, and what implications the curtailments will have on production rates. In processes where equilibrium conditions must be reached before productive operations can begin, it is important to recognize the costs of operating the plant at full power during the period between the time of startup and the time that specification product is made. The possibility of curtailing power demand only to an idling level from which a rapid restart can be made should be considered. This procedure can sometimes be employed to avoid a large percentage of the peak demand charge, while the impact on production is greatly reduced from that of a full curtailment.

Usually, the cost of lost production is hard to quantify. Depending on the person queried, it will vary from full unit-price to nothing. Not surprisingly, the accurate value is usually somewhere in between. If a plant is in a sold-out condition, then the lost-product cost is equal to the selling price less out-of-pocket production costs. In the typical multi-facility-commodity environment, it is usually possible to sustain some lost production at essentially no cost. As constraints are reached, however, it becomes necessary to supply product from alternative sources, thus generating a dislocation cost. Ultimately, lost sales may result.

It has been our practice to work with the marketing and distribution staffs to attempt to quantify these effects. Our most effective vehicle for this part of the analysis has been a linear programming model of our production distribution network. Using this sourcing optimizer, we can see the results of gradually decreasing the product supply at one source and having the demands satisfied by plants with succeedingly higher production costs. Development of such a sourcing model is no minor task in itself, but if such a model exists, it should be exploited.

An air separation plant can provide a good example of three types of curtailment costs: idling, restart, and lost production.

Table I shows a plant with two possible levels of curtailment. At the first level, the liquefaction portion of the plant is shut down, but the main air compressor is kept running to maintain process stability in the separation part of the plant, as well as temperatures in the distillation column. In this operational mode, the plant is not producing any liquid oxygen or nitrogen for sales but is consuming electricity. The advantage of this mode of operation is that the plant can be restarted with a minimum of time and effort. In the partial-curtailment mode, with the main air compressor idling, production is halted for only about half an hour longer than the electrical curtailment.

When full curtailment is necessary, the main air compressor can be shut down, bringing all process flows to a halt. Restarting from this level is a complex and time consuming process, requiring operation of the plant at full power demand for several hours until process balances are restored, and usually resulting in overtime labor costs. When the entire plant is curtailed, production generally cannot resume for four to five hours.

The lost production hours are particularly important because when a plant is heavily loaded the cost of lost product increases with the frequency of the curtailments. Cost ranges for various levels of lost production in Table I represent the increasing difficulty of supplying customers via alternate sources.

The final input required for our simulated operation is the utility's profile. By using several years of daily peak observations, and adjusting them for our current peak forecast, we introduce the historical peak that is characteristic of the utility's demand into the simulation. If, for instance, the utility system tends to have an abrupt peak, the results of the simulation will show that relatively few curtailments are necessary to avoid it.

On the other hand, a utility having a stable (usually non-temperature-sensitive) demand will require many curtailments to avoid the peak. While some people may raise the objection that "history never repeats itself," we have found no better indictor of demand variability.

Although the simulation model well represents the curtailment costs and patterns, other real-world complications must also be considered. The most important are the short but significant time required for an orderly curtailment rather than just "pulling the plug," and the possible uncertainty in the measurement of the utility's native load on a real-time basis.

The time lag between the decision point and the actual curtailment is compensated for by using distributions of load growth in the interval prior to historical peaks. Our target must now include a safety factor for this load growth, as well as a forecast error.

The most troublesome reality is the error that can exist between actual load and the concurrent measurement of that load as reported to our plant. We have experienced this error in three ways.

In the least serious case, the reported load is declared official by the contract, and used by the utility to calculate the bill.

The two other types of error have caused more than a little concern. In one case, the reported load is the result of telemetry from all points in the utility's transmission network where energy flows in or out. In theory, this can be exact but in practice we have experienced errors in excess of 5%. Our ability to forecast has, on occasion, been more accurate than the ability of the utility to provide us with information on the current load.

In the other case, the load has been extrapolated from selected key interchange points in the utility's interchange system. Our experience with this approach has been satisfactory, but the estimated load has been found sensitive to changing load patterns and the selected monitoring points, and has led to the possibility for even larger errors that must be recognized when the curtailment strategy is implemented. In recognizing these factors, it is necessary to analyze the potential errors they may introduce in order that compensation for any consequent problems may be incorporated into the curtailment strategy.

VPS decision

How do a manufacturer and his utility supplier decide whether this type of service would be mutually beneficial? Table II is an example of the capital budgeting decision related to the comparison of generating- and manufacturing-plant costs. (Air separation plants are used as an example.) Recently published market values for oxygen and nitrogen are used in this analysis. Line 2 shows the amount of lost production or sales forgone, as determined on the basis of 30 curtailments per year at a five-hour average curtailment and four-hour average restart time—a production loss of about 3% of the total plant production for the year.

The average power use for these energy-intensive products is shown on a per-ton basis in Line 3. The value of the lost production on these figures is based on $27.60 per kW-yr for liquid oxygen and nitrogen. Since these values are less than the usual carrying charges for generating plants (omitting the cost of transmission capacity), it is probably beneficial to curtail these plants during the periods that affect the generating capacity of the utility. The industrial plants would forgo about 3% of their annual production, making unnecessary the generating and transmission investment that would be required to serve them at the time of the system peak. These firms must be compensated by a reduction in power cost in excess of the value shown in Line 4. Similar results may be obtained for a large number of other products.

The figures shown are for illustration only. Since each particular plant and utility situation is different, specific circumstances must be investigated. For example, the energy cost for restarting the plant has not been included, since it varies with each utility and customer facility. In addition, if the energy-intensive manufacturing plant has a peak customer demand in the same season as the utility peak demand, use of an annual price applicable to a steady year-round customer account may not represent the

Generating-plant and manufacturing-plant costs: capital budgeting comparison	Table II
Item	Liquid cryogen
Market value per ton	$92
Lost production, tons/yr	270
Power requirements, kWh/ton	900
Value of lost production, $/kW-yr	$27.60
Annual cost of generating plant, $/kW	$50

value of production lost at that plant during the peak period.

The main point of Fig. 3: an approximation of the total potentially curtailable demand may be made by similar calculations for a number of industries assuming that production lost due to curtailment can be valued at the equivalent selling price (to a customer whose demand for the plant's output follows a pattern similar to the pattern of production loss due to curtailment).

In the case shown, the consuming industrial plant has a higher annual capital cost per kilowatt of demand than the utility plant's annual capital cost per kilowatt of generating capacity.

We have unfortunately not been able to provide a magic formula that will work in all combinations of industries and utilities. But we hope that this description of our analytic process, emphasizing interdepartmental implementation, will offer some comfort as one considers the feasibility of similar programs.

Most utilities can offer curtailable service, but many have been reluctant to implement curtailable rates because they have little knowledge of their customers' ability to reduce electrical load in a short period of time to implement strategy with any degree of success. We need better communication between the industrial consumer and the utility, to assure them that such cooperation is indeed possible and has resulted in successfully operating curtailment arrangements in many instances. Promulgation of such assurance appears to be well worth the effort, for VPS applied on a national scale would considerably extend the present electrical generating capacity and go a long way toward relieving some aspects of the present energy shortage.

The authors

Edmund A. Perreault is Manager of Utilities for the Cryogenic Systems Div. of Air Products and Chemicals, Inc., Allentown, PA 18105. His responsibilities include negotiation, administration and budgeting the utilities required for operating the company's industrial gas plants. In this position, he has delivered sworn testimony before state regulatory agencies and participated in hearings before both houses of Congress. Earlier, he was an industrial engineer in the company's nuclear-power-plant gas-treatment systems group. He holds a B.S. in mechanical engineering, an M.S. in manufacturing engineering and an M.B.A., all from Boston University.

Paul J. Prutzman is manager of the Management Sciences Section of the Management Information Dept. of Air Products and Chemicals, Inc., where he was previously Manager of the Operations Research Section. He joined Air Products in the Career Development Program in 1967. he holds a B.S.E.E. (1966) from Lehigh University, as well as an M.S. in Management Science (1968) under an Ogden Corporation Fellowship. He is a member of the National Assn. of Business Economists and the American Institute of Industrial Engineers.

Energy-Saving Ideas . . . Here Are the Winners

The panel of judges in CHEMICAL ENGINEERING'S energy-saving-ideas contest (*Chem. Eng.*, Nov. 12, 1973, p. 253) has picked 7 entries as most deserving of publication—and $100 cash awards.

Winning ideas, presented on the following pages, range from a complex scheme for disposing of wastewaters in an evaporative-cycle gas turbine, to the simple step of closing a seal-vent in a compressor.

We sponsored the contest to recognize practical, energy-saving methods that might otherwise not receive wide attention. By disseminating these methods, and by focusing on the positive aspects of energy conservation, we hope to make a small contribution to help solve what has become a worldwide problem.

Not all of the winners completely conform to a strict interpretation of the contest criteria: proof of practicality, full and concrete data, novelty, and sufficient detail to permit duplication elsewhere. On balance, however, the judges singled out the ideas published here as being most deserving of recognition.

Panel of Judges

E. L. Ekholm is manager of energy planning, Bechtel Corp., San Francisco, Calif. He has been with the company eight years, as manager of process engineering in Houston and manager of business development in Atlanta. Mr. Ekholm is a 1946 graduate of Georgia Institute of Technology, with a B. S. degree in chemical engineering.

Robert E. Lenz has recently taken early retirement from Monsanto Co. to become a consultant in engineering management. He spent 34 years with the company, holding a number of positions in research, engineering and development. Most recently, he was director of corporate energy planning. Mr. Lenz graduated from Washington University (St. Louis) with a B. S. in chemical engineering.

E. O. McBride is a partner in Atkins, McBride, and Owen, consulting chemical engineers in Houston. He has had extensive experience in refining and petrochemicals manufacture with Humble Oil and Refining from 1939 to 1956 and Tenneco Oil from 1956 to 1970. He is a chemical engineering graduate of the University of Oklahoma.

Originally published September 2 1974

Versatile Oil Burner Offers High Efficiency and Reduced Emissions

LORNE C. LAMBERT, Gulf Oil Canada Ltd.

The Vortometric burner, which we developed together with the Ontario Research Foundation, offers substantially reduced emissions plus fuel savings of at least 5% and reduced operating and maintenance costs. It burns a wide range of liquid and gaseous fuels—from the lightest to the heaviest industrial residual—in a manner comparable to natural gas.

Typical process heaters are similar in their operating principle to a once-through boiler. A pump moves fluid through heating coils in a combustion chamber and then to distillation columns or towers. It has been usual practice to fire process heaters with gas, oil, or both, using natural draft to supply combustion air. The enclosures are designed to operate at a slight negative pressure. However, since draft is normally controlled manually with a stack damper, firebox pressures frequently are either positive or highly negative, with massive amounts of excess air present.

This causes corrosion of steel structures and firebox casings, and deterioration of refractories. Overfiring to maintain capacity, which accelerates damage, is the ultimate result of problems arising from poor combustion control.

Our new burner achieves essentially complete and smokeless combustion while burning a wide range of liquid and gaseous fuels at any practical level of excess air. with minimum operator attention and reduced mechanical maintenance. In addition, a significant decrease of polluting emissions results when the system is designed for and operated under low-excess-air conditions.

The benefits of low-excess-air techniques, as first practically demonstrated in Germany, have been well publicized. Fig. 1 shows how combustion deposits can be dramatically reduced by cutting back on air relative to the customary 15% excess. Today there is little doubt about the desirability of operating at the lowest practical level of excess air. That is, as close as possible to stochiometric combustion, usually described as combustion with no more than 3% excess air.

Efficient combustion of liquid fuel requires near per-fect proportioning of air and fuel. The Vortometric burner has high performance because it is an integrated design incorporating separate systems for fuel flow, atomization, air-fuel mixing, vaporization and recirculation. None of these systems impedes the functioning of any other as is the case, for example, in burners where steam and fuel are premixed for atomization.

Theory of Burner Operation

A basic concept is that combustion is an exothermic process in which heat results from the rapid oxidation of a combustible material. Since the generally accepted theory is that oxygen and fuel molecules combine only in the vapor state, liquid fuels must be vaporized. This process is obviously facilitated by atomizing the fuel to the

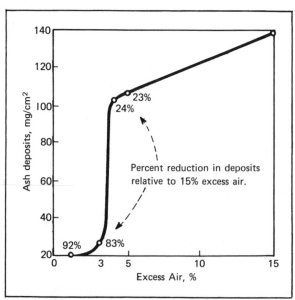

LOW EXCESS AIR reduces combustion deposits—Fig. 1

Originally published September 2, 1974

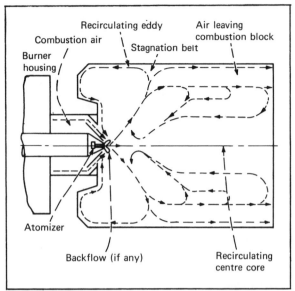

TYPICAL FLOW PATTERNS in Vortometric burner—Fig. 2

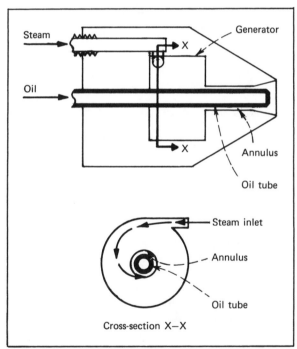

ATOMIZER is key component of Vortometric burner—Fig. 3

smallest possible droplet size. Heat for vaporization can then be provided externally by a continuously burning pilot, or internally by the impingement of other flames. But the most immediately available source is the burning combustible itself.

Convective and radiant heat for vaporization can be provided effectively by recirculating burning gas to the atomizing area of the burner. This can be done in several ways: by producing a vortex in the flame—the basis of the Vortometric system—or by placing baffles, known as diffusers or bluff bodies, upstream of the atomizer in the path of the air flow.

Combustion is sustained and stable when the heat sup-

ply provided or developed in the reaction zone is greater than the outflow of heat to the furnace. When burning liquid fuels, the time required for the complete combustion cycle is critically dependent on the vaporization rate, which, in turn, is a function of fineness and the consistency of atomization.

If sufficient oxygen is mixed intimately with the combustibles at the necessary temperature, there is no smoke or sooting. However, if conditions of mixing and temperature are less than ideal, some degree of thermal cracking occurs, producing carbon and hydrogen that burn with a smoky flame. Also, some of the reactions are endothermic and reversible at certain stages. Cooling the combustion gases before the components are completely burned, under a localized or overall air shortage, can favor the formation of elementary carbon instead of carbon dioxide.

At any temperature, there is a preferential order of forming oxides of any compound, according to its oxygen affinity. If there is not enough oxygen to complete all the reactions, the oxides having the lowest free energy will form. Sulfur trioxide and vanadium pentoxide have a high level of free energy. This means that with low excess air, it is possible to complete the combustion reactions of carbon and hydrogen with minimal formation of these oxides.

Similarly, the amount of nitrogen oxides in the fuel-burning zone can be minimized in three ways: by decreasing the excess air, by reducing the high-temperature residence time, and by recirculating combustion gases to lower temperature.

The Vortometric System

In the Vortometric system, the three functions required of an oil burner are accomplished by an air volute—the fuel atomizer that is the heart of the unit—and a combustor in which burning takes place. A vortex within the flame locates and controls the reaction zone between fuel and oxidizer. Air entering through a volute provides the energy to generate the vortex. The system requires a total air pressure of 6 in of water at the volute for maximum heat release. There is no need for secondary air.

The air stream rotates and swirls as it flows into the combustor through a circular nozzle connected to the volute (Fig. 2). The stream continues spinning within the combustor, and while expanding it generates a vortex that produces a negative pressure gradient on the center axis. A large portion of the hot flue gas is drawn into the center core, where it flows toward the atomized fuel and continuously supplies heat for evaporation and ignition.

The most important part of any burner is the atomizer, which in the Vortometric is mounted coaxially within the air volute. It is generally agreed that conventional atomizers operate reasonably well, provided they get fairly frequent cleaning and mechanical maintenance. However, they are not capable of good performance with low excess air, and without an effective servicing program even conventional burning efficiencies are seldom realized.

The oil gun in such atomizers premixes steam and oil at pressures from 50 to 140 psig, and ejects the resulting emulsion through a series of holes at the outer tip. The major atomizing effect occurs when the emulsion expands from the high internal pressure within the nozzle to the lower pressure of the firebox—in other words, the differential pressure across the outlet orifices.

Atomization is poor if the oil is too viscous, unless compensated by an increase in steam pressure. Also, since steam pressure is usually maintained at a constant 15 to 20 psig above oil pressure, atomization quality varies as the oil pressure changes with different loads. Another inherent problem is a minute accumulation of thermally decomposed or coked oil around the periphery of the outlet holes. This requires considerable routine attention or, alternatively, acceptance of indifferent performance.

Multistage Atomizer

The Vortometric atomizer (Fig. 3) is a multistage device utilizing shear forces in the first atomizing stage, and vibrational or accoustic energy in the final stages. Fuel oil flows through the center pipe to the tip, where it is ejected from a circle of holes drilled obliquely or radially to the axis. Fuel pressure ranges from 2 psig at low load to 50 psig or slightly higher at high load, with fuel viscosity at 150 to 500 SUS. Fuels have been burned at 1,200 SUS—equivalent to bunker No. 6—at only 140°F. Although it is important to have uniform viscosity for good flow control in any burner, this atomizer performs well over fairly wide variations in viscosity.

The atomizing fluid (steam is the most efficient because of its expansive energy) flows to the gallery of a circular chamber labelled "Generator" in Fig. 3. It enters the outer periphery of the chamber through one or more tangential inlets, which are analogous to the nozzle block of a steam turbine. The fluid spirals inwardly at continuously accelerating angular velocity toward the center axis of the generator. It leaves through an annular space formed between the outer surface of the fuel pipe and the internal surface of a cylindrical hole in the atomizer nose cone, while spinning at close to sonic velocity. The fuel, which is ejected from the fuel pipe in a number of radial jets, impinges on the atomizing stream at a certain point of confluence. At this juncture, the multiple stages of atomization start.

Since only low-velocity fuel oil flows through any of the atomizer passages, wear of fuel ports is virtually zero. This means that nozzles seldom need to be replaced, and that flow control is predictable and reliable. Furthermore, since atomization is unaffected by the load changes, a wide turndown or turnup ratio—of as much as 20:1—is available. This permits almost any degree of modulating control without degrading combustion efficiency.

The third component, the combustor in which the reaction takes place, is a refractory-lined cylinder that receives atomized fuel and combustion air. Ignition is a simple procedure, using a pilot torch or high-intensity spark ignitor inserted in the throat area.

Flame stability is outstanding over a range from rich to very lean mixtures. This is possible because of the internal recirculation of hot flue gas and radiant heat of the

Typical Flue-Gas Analyses

Fuel	2.0 Million Btu/H Test Unit			4x20 Million Btu/H S/T Gulf Canada
	Bunker No. 6			Bunker No. 6
Load, %	100	85	32.5	100
Excess air, %	2.9	2.4	4.5	4.2
CO_2, %	15.6	16.2	15.3	15.5
CO,* %	ND	ND	ND	ND
O_2, %	0.6	0.5	0.9	0.8
N_2 + Inerts, %	83.8	83.3	83.8	83.7
SO_2, ppm	1,100	1,232	1,020	760
SO_3, ppm	4	4	NM	11
NO_x, ppm	307	197	407	62
Hydrocarbons as hexane, ppm	1	NM	NM	NM
Particulates†	0.18	0.30	NM	0.18–0.3
Bacharach rating	4–5	3–4	2–3	3

*CO not detected (ND) below 0.01%
†gm/m³ of flue gas at STP
NM-Not measured

combustor. This radiant heat also provides a wide margin of safety should flameouts occur as a result of interruptions in fuel supply, steam, etc. Operating experience with a high-alumina refractory brick lining indicates a minimum service life of two years.

Verifying Performance

We obtained the first real verification of projected performance of the Vortometric system during tests on a direct-fired heat exchanger and in the boiler system of our tanker, the S/T Gulf Canada. Typical flue-gas analyses from these tests are in Table I. Note that with the smaller unit the amount of nitrogen oxides decreases with decreasing excess air. Also, there was a much lower concentration of NO_x in the tests on the S/T Gulf Canada.

The results are not comparable because these combustion systems are entirely different and burn dissimilar fuels. The small burner could not be expected to achieve as high a rate of gas recirculation as those installed on the ship, a factor of considerable influence on NO_x production. The analyses of nitrogen oxides in the flue gas from the boilers of the Gulf Canada were very encouraging because the system used 400°F preheated air, and produced a very hot flame that normally favors maximum NO formation. The Vortometric system has powered the ship more than 175,000 miles with a minimum of operating attention and maintenance.

Similar improvement, but more dramatic because of a long history of traditional design, has been achieved with the petroleum-type heater. Reported combustion rates have exceeded those obtained with natural-draft burners by a factor of 10 to 15. An improvement of about 40% is indicated by the use of high-intensity, forced-draft burners over conventional burners due to an increase of the average transfer rate and a simultaneous lowering of the maximum point transfer.

It has been demonstrated that Vortometric burners can significantly increase the efficiency of existing furnace designs. Perhaps more interesting, however, is the very real possibility that by designing furnaces for the short flame length and high-radiant-heat characteristics of the Vortometric burner it may be possible to reduce capital costs of furnaces by as much as 25%.

In addition to a potential reduction in capital costs, there is the saving in operating expense that would benefit many other operations. We know, for example, that improved radiative transfer has contributed impressive results to the metals-casting process. Remelt and alloying times have been reduced with a 40% decrease in rejections, all at reduced fuel expense.

Lorne C. Lambert is a heat-transfer specialist for Gulf Oil Canada Ltd., P. O. Box 460, Station A, Toronto, Ont. He is involved in engineering design, standardization, project administration and engineering economics. Mr. Lambert has been with Gulf since 1951, and holds a B.Sc. degree in mechanical engineering from Queens University, Kingston, Ont.

Energy-Saving Turbine Uses Wastewater To Raise Output

ROGER WYLIE, Exxon Chemical Co. U.S.A.

Some 70% of the energy obtained from burning fuel in a gas- or liquid-fired gas turbine leaves with the exhaust gases. This loss must be partially converted into useful energy in order to make the gas turbine an economical power-generation machine.

The energy in the exhaust gases can be used to supply low-level process heat, either directly or via steam generation. This operation gives maximum thermal efficiencies and thus should be selected whenever possible. Unfortunately, in most manufacturing processes, the need for mechanical energy usually far exceeds the need for low-level heat. Furthermore, this option cannot be applied at power-generation stations, remote compressor stations, or with drivers for transportation equipment. For these reasons, engineers have devised other methods to convert some of the low-level heat into mechanical energy.

For example, the regenerative-gas-turbine cycle, in which compressor air is preheated by the exhaust gases, can be used anywhere. This cycle has found applications in widely different fields such as process drivers, power-generation facilities, and ship, train, truck and automobile propulsion.

Combined gas and steam turbines also provide an attractive energy-conversion process. Waste-heat boilers operating at different pressure levels have been installed

Operating Conditions and Efficiencies

Compressor	
Temperature, °F	80.0
Pressure, psia	14.56
Suction filter ΔP, in of water	4.0
Water vapor in air, wt %	1.6
Air flow, million lb/h	1.74
Compression ratio	9.1
Adiabatic efficiency, %	86.0
Polytropic efficiency, %	90.7
Adiabatic Flash Separator	
Approach to wet bulb, °F	5.0
Blade Cooling, % of Flow	3.0
Regenerator	
Pressure drop, air/exhaust	5 psi/6 in of water
Efficiency, approach, °F	70.0
Heat loss, %	1.5
Firing Temperature, °F	
Compressor air cooling	1,850
Separator vapor cooling	2,200
Combustor Efficiency, %	99.5
Expander	
Mechanical efficiency, %	99
Adiabatic efficiency, %	—
Polytropic efficiency, %	86 first, 88 second and third
Back pressure, in of water	6.0
Generator Efficiency, %	98
Cycle overall efficiency (LHV), %	
Firing 1,850°F	47.6
Firing 2,200°F	53.6

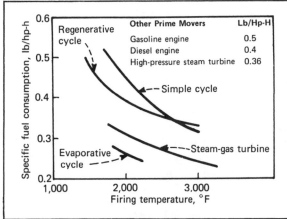

FUEL CONSUMPTION for various prime movers—Fig. 1

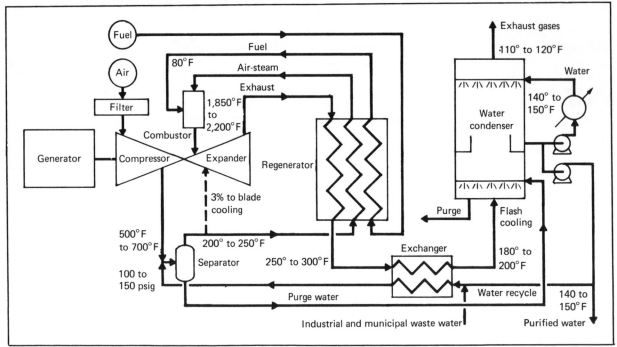

FLOWSCHEME for using evaporative-cycle gas turbine with industrial or municipal wastewaters—Fig. 2

in series in the exhaust stream from a gas turbine. Steam at various pressures is routed to the appropriate stage of a multistage condensing turbine.

This combined operation is more complicated than the regenerative cycle, and a source of cooling water must be provided. On the other hand, it is more efficient than the regenerative cycle, and the presence of the steam turbine increases the output of mechanical energy from the package. Several gas turbines can exhaust into the same system of waste-heat boilers, which simplifies the heat-recovery operation. This option is frequently selected by utility companies when they install heat-recovery facilities along with gas turbines.

Lately, articles are beginning to appear about a lesser-known cycle, the evaporative-cycle gas turbine, which is interesting because of its inherent high efficiency. In this cycle, heat from the turbine exhaust evaporates water, and the resultant steam expands through the turbine for additional power. The data in Fig. 1 show how the evaporative cycle outperforms other prime movers.

To understand the advantage of the evaporative cycle, consider that the expansion stage of a gas turbine converts about one-half of the thermal energy in the high-pressure gases leaving the combustors into mechanical energy. About 55% of this energy is used to drive the air compressor, and the remaining 45% is available for useful work. This results in a leverage effect: any change that slightly decreases the energy required by the compressor, or slightly increases the output of the expander, significantly boosts the turbine's net output.

Steam injection has been used for a number of years to raise turbine output. However, consumption of once-through high-quality steam in large quantities can become prohibitive. With the evaporative cycle, large quantities of steam (or water vapor) from low-quality water streams can be injected into the turbine.

A novel application of the evaporative cycle is presented in Fig. 2. This differs from previously published evaporative cycles in that (1) a recycle water stream is provided rather than using once-through water, and (2) industrial or municipal wastewaters are used rather than high-quality water.

Although there would be minor differences in the process conditions depending on the gas turbine design, the ranges shown in Fig. 2 should cover most major manufacturers' equipment. Air would be filtered and compressed in an axial compressor to 110-176 psia, and then cooled adiabatically with water from 550-700°F to about 215-250°F. Cooling water could be a recycle stream or purge water from process units. The cooled air would be separated from the unvaporized water, washed with clean water if necessary to remove solids, and routed to the regenerator. Here, the air and water vapor streams would be preheated by exchange with the expander-turbine exhaust gases.

The fuel to the gas turbines would also be preheated with the same exhaust gases to above its autoignition point. This would prevent any possibility of a flameout in the combustors when the fuel entered the air-water vapor stream. The high concentration of water vapor in the combustors would result in improved burning, particularly with hydrogen-deficient fuels, and would thereby minimize trace quantities of pollutants.

Water vapor in the hot gases from the combustors would increase the volume of gases flowing through the expander about 15%. This would result in a net power gain from the gas turbine of about 35%.

The exhaust gases from the expander would be cooled by heat exchange in the regenerator and water preheater. These gases would then be adiabatically flash cooled with the purge-water stream from the high-pressure separator. If the fuel contained more than trace quantities of sulfur, chemicals could be added in this stage to remove pollutants.

The vapors from the flash-cooling stage could be routed directly to the atmosphere or sent to a dehumidifier for recovery of most of the water, which could be recycled to the water preheater whenever wastewater was unavailable.

The concentrated purge water from the evaporative cooler would be processed before disposal. In some cases, enough organic materials would be present so that this stream could be economically dried in a fluid-bed dryer operating at elevated pressures. The combustion products could then be used as fuel for the gas turbine.

More Power, Higher Efficiency

A portion of the air from the axial air compressor would cool the expander blades and nozzles. Published data from gas-turbine manufacturers indicate that firing temperatures could be increased from 1,850°F to about 2,200°F if cool air, rather than the hot compressor-discharge air, were available. Such an increase in firing temperatures would give 30% more power and about 7% higher efficiency.

Overall, an evaporative-cycle gas turbine having a compression ratio of 9.1 and a firing temperature of 1,850°F would have a thermal efficiency of about 47% (LHV). Its power output would be about 35% larger than that of a simple-cycle machine with the same air flow. At a firing temperature of 2,200°F, the cycle efficiency would be about 54% and the power output would be about 75% larger than the original simple-cycle machine. At the 54% efficiency, a plant operating on a limited fuel basis would be able to produce around 40% more power than would be possible with high-pressure steam turbines (supercritical steam boilers).

The large amount of water in the turbine exhaust gases would allow condensing most of it at temperatures higher than those normally found in steam-condensing turbines. In locations lacking cooling water, extended-surface air coolers could be employed to cool a recirculating water stream that would be used to condense the water from the turbine exhaust gases.

Because water in the exhaust gases could also be condensed with ambient air, it might be possible to use the evaporative cycle on moving equipment. This cycle's inherent high efficiency would decrease fuel consumption about 35-40% compared to the low-pressure-steam turbines, diesel engines, and regenerative gas turbines that propel ships and trains.

Ecological Advantages

The proposed cycle would be very useful in disposing of industrial wastewater streams containing ammonia, cyanides and oxygenated hydrocarbons such as methanol, ethanol, isopropyl alcohol, isobutyl alcohol, phenol, acetone, etc. These compounds are often present in waste streams from petrochemical and steel plants, and decomposing them to harmless materials is quite expensive by conventional means. On the other hand, their end-products in the gas turbine—CO_2, H_2O and N_2—are quite harmless.

Electric utilities could help many of their major customers by installing the proposed cycle. The system in Fig. 2 could process about 0.24 to 0.27 gal of wastewater per kWh of power. With a slight sacrifice in efficiency, the cycle could be modified to handle 0.5 gal of wastewater per kWh. If all electric power producers had been using the evaporative cycle in 1972, they could have converted 400-800 billion gal of wastewater for reuse—five times the 1972 water requirements of industries and municipalities in the Houston - Galveston Bay area.

Engineering Considerations

Ambient conditions affect the operation of any gas-turbine cycle. The efficiencies calculated for the evaporative cycle described here are representative of what would be expected under average summer conditions at a Gulf Coast location, allowing for realistic inlet and outlet pressure losses. Table I lists pertinent data on the various component parts of the gas-turbine cycle. We tried to make these figures as consistent as possible with published data from turbine manufacturers.

Several innovations could be made to further improve the cycle's efficiency. For example, a supply of high-quality recycle water would make it very easy to install direct intercooling stages on the compressor. As mentioned earlier, raising the firing temperature significantly increases the efficiency of any turbine cycle. The Federal Government has given several grants for engineering studies on gas turbines with firing temperatures of 2,500 to 3,500°F.

With existing technology, the evaporative cycle should make it possible for the first time to convert at least one-half the energy in fossil fuels into mechanical energy.

See page 143 for author's biography.

Air Preheaters Can Save Fuel In Many Furnace Installations

ROGER WYLIE, Exxon Chemical Co. U.S.A.

Large process furnaces and steam boilers have been equipped with air preheaters for a number of years. These units realize fuel savings of 5 to 20%, with about 10% being average. Higher fuel savings are normally obtained on high-heat-flux furnaces, although when these furnaces are fitted with waste-heat boilers, air preheaters reduce the amount of byproduct steam produced. This must be made up by steam from other process facilities, which lowers the net energy savings. However, a 20% gross savings in high-quality fuel may at times be equally

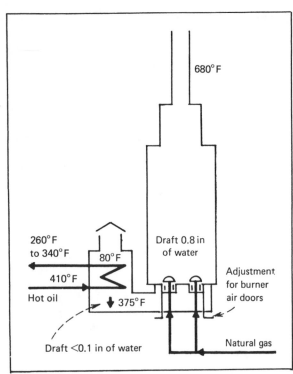

PREHEATER-COOLER can operate on natural draft—Fig. 1

Distribution of Energy

	Fuel Fired, %	
	Without Air Preheater Stack Temp. 710°F	With Air Preheater Stack Temp. 680°F
Stack gases	16.6	15.4
Heat losses through walls	1.4	1.5
Lower versus higher heating value	10.0	10.0
Air preheat	—	−6.6
Process heat	72.0	79.7
	100.0	100.0
Fuel fired, % of without preheater	100.0	90.3

———— Energy Saved, % ————	
Heat input to air	60
Stack gas losses	30
HHV versus LHV of fuel	10
	100

as important as a 10% net savings in plant fuel.

Rotary or tube air-to-flue-gas heat exchangers require both forced-draft air fans and induced-draft exhaust fans. Power costs for operating the fans are roughly 20% of the dollar savings from lowered fuel consumption.

Recently, extended-surface preheaters have been installed on some furnaces. On gas-fired units, these preheaters can be designed for natural-draft operation. On oil-fired furnaces, only forced-air blowers are required, which consume 15 to 30% of the power needed to run equivalent rotary air-preheaters. This means that extended-surface preheaters, which have nominal installation costs and low power consumption, can be justified today on even small furnaces.

Types of Preheaters

Three types of extended-surface air preheaters are available commercially. The preheater-cooler, the simplest method of installing an air preheater on a furnace, has limited application because it requires a hot process stream that is being cooled with air or water. In a typical installation (Fig. 1), a hot oil stream that otherwise would be cooled with recirculating water or air, travels through an extended-surface heat exchanger in the air duct to a reboiler furnace. If this exchanger is oversized, the temperature of the preheated air is relatively independent of flowrate changes in the hot oil.

It is economical to design this type of exchanger for an air-side pressure drop as low as 0.1 in of water, which makes it possible to use natural draft on gas-fired furnaces. If fans are required to atomize fuel, it would probably be most economical to design the exchanger with an air-side pressure drop of 0.5 to 1.5 in of water.

Adding air preheating to an existing furnace transfers 10 to 20% of the furnace's heat duty from the convection section to the radiant section, resulting in more-efficient transfer of heat into the oil. Stack temperatures should decrease because the flow of flue gas through the convection section is reduced. Pertinent temperature and pressure data for an air preheater installed on a gas-fired furnace operated by Exxon Chemical Co. U.S.A. are shown in Fig. 1. The distribution of energy obtained from the fuel with and without the preheater is listed in Table I.

Complete Air Preheater

Fig. 2 shows a flowscheme including pressures and temperatures for a second type of air preheater. In this system, a portion of the furnace charge passes through extended-surface heat exchangers located in the air duct to the burners. After the air flowing to the burners cools this stream, it travels to an exchanger above the convection section of the furnace. Stack gases are cooled to the desired temperature by controlling the amount of liquid routed through the two exchangers. The volume of the slip stream is related to its specific heat and the quantity of flue gas. Maximum energy recovery is obtained when the system meets the following conditions:

$$\Delta T_{\text{air}} \geqslant \Delta T_{\text{liquid}} \geqslant \Delta T_{\text{flue gas}}$$
$$\text{or}$$
$$(A)(Cp) = (L)(Cp) = (FG)(Cp)$$

where A is air flow in lb/h, L is slip-stream flow in lb/h, FG is flue-gas flow in lb/h, and Cp is the respective specific heats in Btu/(lb)°F.

The heat exchanger in the air duct usually would be

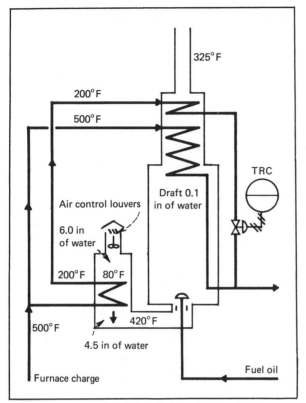

NORMAL USE of extended-surface preheater—Fig. 2

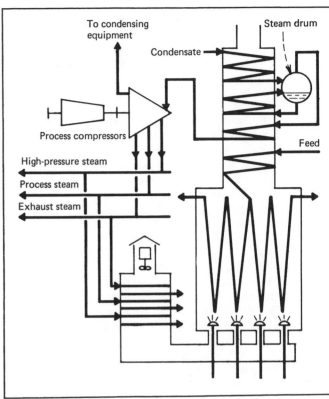

AIR PREHEATER combined with steam extraction—Fig. 3

constructed of carbon-steel tubes with extruded-aluminum fins, similar to the fin-fan air coolers that are used widely in process units. The heat exchanger in the convection section would be constructed with methods similar to those used in building the convection section of the furnace. With oil firing, solid-coiled fins have advantages. They should be at least 0.1 in thick, no more than 1.0 in high, and be placed two to three per inch. Steam-soot blowers should be provided for cleaning the flue-gas exchanger.

Induced-draft flue-gas fans and the stack-gas bypass around the preheater, which are needed in case of mechanical failures in a rotary-type preheater, are not required with extended-surface air preheaters. Also, the low-pressure-drop exchangers operate with smaller forced-draft fans. Because of these factors, the power requirements for extended, surface preheaters are 15 to 30% those of rotary types. In large process plants and refineries, power savings could be as much as 5,000-10,000 kW if preheaters were installed on all furnaces.

Extraction-Steam Air Preheater

The third type of extended-surface air preheater (Fig. 3) can be installed on large furnaces where an economizer produces high-pressure steam. This type of economizer is usually found on high-heat-flux furnaces where only a small portion of the energy in the stack gases can be recovered by heating process streams. Normally, the high-pressure steam expands through large process turbines, and multiextraction nozzles supply steam for various plant requirements. Some of the mechanical energy

for driving process compressors and other equipment is obtained this way (the remainder comes from condensing-steam turbines).

Furnace air is preheated by condensing steam in a series of extended surface exchangers, each operating on successively higher pressure steam. This system realizes the normal fuel savings for air preheaters and in addition increases the amount of high-efficiency mechanical energy obtained from the extraction turbines. Usually, mechanical energy obtained via steam-extraction turbines requires one-third to one-half as much fuel as an equal amount of energy obtained from condensing-steam turbines.

Other Considerations

Using the furnace charge as the preheater liquid in the second type of system discussed simplifies the equipment but does increase the chance of fuel entering the air ducts. The amount of hydrocarbon leakage that would cause an explosion in the air ducts would be roughly one-half the furnace fuel rate. For safety, alarms indicate any sudden change in fuel rates, and hydrocarbon detectors installed in the main air duct show small leaks if they develop.

Heat Research Corp. has recently been issued a patent covering the second type of air preheater. No patents are known to the author that may cover the other types.

Roger Wylie is an engineering associate in the engineering technical dept., Exxon Chemical Co. U.S.A., P. O. Box 4004, Baytown, TX 77520. He has been with the company since obtaining a B. S. degree in chemical engineering from Rice University in 1944. Mr. Wylie holds patents in several fields; three major units in the Baytown Refinery and Chemical Plant are based on his process developments.

Grate Cleaner in Pelletizing Plant Cuts Energy Requirements by 20%

OLIVERIO MACIAS A., Las Encinas, S.A., Mexico

By installing a very simple grate-bar cleaner, our pelletizing plant during 1974 will save 9.12 million kWh in electrical energy and the equivalent of 262 billion Btu in fuel oil. Both these figures represent reductions of more than 20% from previous levels.

We operate a straight horizontal-grate pelletizing machine that produces pellets from magnetite iron ore. The machine has a grate area of 1,940 ft², a designed grate factor of 1.89 tons/(ft²)(24 h), and an installed fan capacity of 7,200 hp.

In 1971, our actual grate factor was 1.78 tons/(ft²)(24 h), using 6,000 hp from the fan motors. We were consuming 26.45 kWh/ton in electrical power and 693,000 Btu/ton in fuel. At this time, we were faced with the need to increase production about 35%.

In this type of process, fan capacity is normally the bottleneck in raising production. Thus, in order to avoid a major modification, we had to improve fan efficiency.

The pelletizing machine has 103 pallets, each approximately 5 x 10 ft. These hold the grate bars, which tend to plug, causing a reduction in the quality and quantity of production. It was obvious we needed a method for cleaning the grate bars. However, we could not find a commercially available cleaner that had high efficiency.

After many tests, we designed our own cleaner, which is 100% efficient. That is, it removes 100% of the plugged material so we can utilize all of the void area in the grate bars for air circulation. The cleaner, which cost about

IIIIIIIIIIIIIIIIIIIIIIIIIIIIIIIIIIIIIIIIIIIIIIIIIIIIIIIIIIIIIIIIIIIIIIIIIIIIIIIIIIIIIIII

Grate Cleaner Boosts Production, Saves Electricity and Fuel Oil

	Grate Factor, Tons/(Ft²)(24 H)	Electricity, kWh/Ton	Fuel, Btu/Ton
Before	1.78	26.45	683,000
After	2.43	20.93	534,600
% change	36.5	−20.1	−22.3

IIIIIIIIIIIIIIIIIIIIIIIIIIIIIIIIIIIIIIIIIIIIIIIIIIIIIIIIIIIIIIIIIIIIIIIIIIIIIIIIIIIIIIII

$1,000, is a simple device consisting of three-tooth wheels on a rotating shaft that can be raised and lowered to remove material from between the bars. We also improved the cleaning operation by changing the cross-section of the grate bars from a rectangle to a truncated cone.

As shown in the table, the grate cleaner cut electric power to 20.93 kWh/ton (a 20.1% reduction) and fuel to 534,600 Btu/ton (22.3% less). We are producing 28% more than design capacity with the fan motors still at 6,000 hp. Operating costs are lower by $0.098/ton, which will save us $162,000 annually.

Oliverio Macias A. is superintendent of the pelletizing plant at Las Encinas, S.A., Apartado 130, Colima, Col., Mexico. He joined the company, which is part of the Hylsa group, in 1967 as a production assistant, and later worked in the quality-control department. He is a mining and metallurgical engineering graduate from the University of Guanajuato.

Closing Vents on Compressor Seals Prevents Loss of Hydrocarbon Gases

HOWARD B. HILE, Cooper-Bessemer Co.

Centrifugal compressors handling hydrocarbon gases are predominantly fitted with oil-buffered seals—either the bushing or carbon-ring type. These provide positive sealing by supplying oil at a pressure higher than the gas side. The small amount of oil lost through the seal is drained and recovered in a trap.

In some instances, even the smallest amount of oil leakage into the gas stream is objectionable. To eliminate such inner leakage, a small amount of gas is vented from each trap. Where possible, the vented gas is recovered in a lower-pressure system, but often the most economic solution is to flare or vent the gas. Expressed as a percentage of the flowing stream, the vent rate appears as a negligible quantity. But when evaluated on an annual basis, this hydrocarbon loss may carry an impressive price tag. So, why not simply close the vent?

There are two precautionary checks one should make

before turning off the vent line for the seals:

■ Be certain that the inboard and outboard (or discharge and inlet) traps of a given compressor and its associated reference piping are isolated from each other. Otherwise, end unbalance will cause an excessive oil leakage into the gas stream. (End unbalance is a difference in reference-gas pressure between the inlet and discharge. It typically runs between 0.1 and 10.0 psi.)

■ Each seal's oil trap must be connected to the compressor with a 1.5-in minimum line having a continuous slope. Or, each trap should have an equalizing line back to the compressor-seal reference port. While the vented trap has a forced flow of gas, without these provisions the unvented trap will be prone to "gas bind," causing all of the inner-seal leakage to find its way into the process gas stream. (See illustration, p. 60)

Historically, only a few companies have consistently

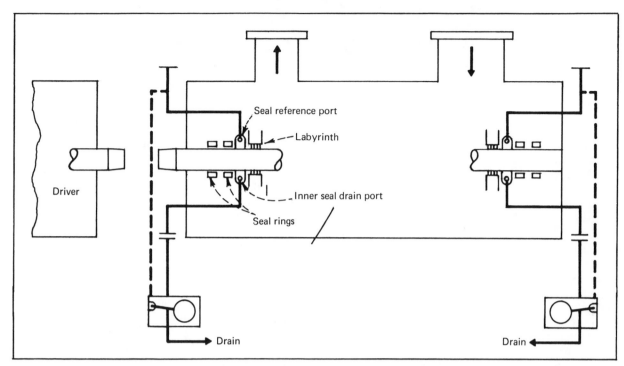

COMPRESSOR-SEAL VENTS may be closed to avoid hydrocarbon losses if inboard and outboard traps are isolated.

specified that the shaft seals of rotating machinery should have unvented traps. From the supplier's viewpoint, the specter of excessive oil leakage into the gas stream looms large by comparison to the small, readily evaluated gas-vent rate. Rarely is the supplier of the equipment in a position to know whether oil leakage to the gas stream of up to 1 gal/d will be acceptable.

In practice, many processes will readily tolerate such a leakage rate, but due to standards—either those of the user or the system supplier—the vented-trap system is most frequently installed. Today, the increasing value of hydrocarbons provides the stimulus for reviewing existing plant operations and revising specifications for new systems.

Howard B. Hile is a project engineer with Cooper-Bessemer Co., Mount Vernon, OH 43050. He is currently involved in design and application of centrifugal compressors and gas turbines, and has held similar positions with Dresser Industries and Monsanto Co. He has a B.S. degree in chemistry from Bucknell University and is a registered professional engineer in Ohio.

More-Careful Control of Ventilation For Enclosed Operations Saves Energy

BRUCE A. JENKINS, Hercules Inc.

Enclosed operations involving explosives or toxic materials require ventilation with single-pass air-handling equipment to maintain necessary temperature limits. The units ordinarily run continuously, maintaining building temperatures within close tolerances.

In an effort to conserve energy, we changed the adjustable-pitch pulleys on the fan-drive motors to lower the fan speed and reduce delivered air quantities. To determine the extent of reduction possible without adverse effect on temperature requirements, we took readings and established temperature profiles within the building as air quantities were reduced.

By reducing the air flow to several units in one building by 2,400 ft³/min, we saved $3,015 in annual heating costs (10,052 gal of oil at $0.30/gal) and $253 in electrical energy (19,500 kWh at $0.013/kWh).

We made further savings by taking into account that actual operations usually occur at intervals, e.g., one shift per day or several shifts per week. This permits use of a pneumatic or electric selector switch to place the air-handling unit on standby, allowing the building temperature to fluctuate within wider limits.

We are implementing this control modification in several processing buildings, which have a combined air flow of 100,000 ft³/min. We expect that the yearly savings will come to $32,550 for heating (108,500 gal of oil at $0.30/gal) and $1,050 for electricity (81,120 kWh at $0.013/kWh).

Bruce A Jenkins is a senior facilities engineer for Hercules Inc., Allegany Ballistics Laboratory, P. O. Box 210, Cumberland, MD 21502. He previously was an air-conditioning engineer for Du Pont, and served as chief engineer on various ships. Mr. Jenkins is a graduate of the U.S. Merchant Marine Academy with a degree in marine engineering.

Heat-Transfer Fluid Conserves
30 Billion Btu in Tracing System

WALTER SEIFERT, Dow Chemical U.S.A.

A heat-transfer system using ethylene glycol in place of conventional steam tracing is saving 30 million lb/yr of steam in one of our organic chemical units at Midland, Mich. This is equivalent to an energy saving of 30 billion Btu/yr.

The unit has 160 tracer runs, each approximately 200-ft long, for a total of 32,000 ft of tracing. When using steam tracing, tests showed that an average of 23.3 lb/h of steam was consumed for every 100 ft of run. In addition, quite a bit of steam was being lost from the system, a common fault of steam tracing systems.

With a closed circuit containing a heat-transfer fluid (Dowtherm SR-1, an inhibited ethylene glycol), only 8 lb/h of steam is needed for every 100 ft of run. The steam is used only to heat the transfer fluid.

We are also obtaining a substantial saving in maintenance costs because with conventional steam tracing the lines often freeze in winter. A 50% aqueous solution of the heat-transfer fluid freezes at −34°F, which means the system is virtually maintenance-free.

The product made in this facility is sensitive to heat. The heat-transfer fluid eliminates the danger of heat damage because to maintain the product at 248°F, the tracing medium need be only 18 deg higher. In contrast, steam tracing with 150-psig steam produces a temperature of 365°F. A pocket of the heat-sensitive product laying in the line could be discolored by prolonged exposure to this high temperature.

The facility uses heat-transfer fluid at two temperatures—176°F and 248°F. The higher temperature is used for lines containing the product, which freezes at 140°F. The fluid at 176°F traces lines containing water, some process materials and 50% caustic. Holding the caustic below 150°F permits transporting it in iron pipe without having trouble from caustic embrittlement.

Cost of the shell-and-tube exchanger for heating the fluid is roughly equal to the expense of steam traps, check valves, thermostatic air vents, drains and fittings in a regular steam-tracing system. Only one steam trap, on the heat exchanger, is required with the fluid. It is relatively easy to convert from a conventional system to one using a heat-transfer fluid.

Walter F. Seifert is technical service and development manager for heat transfer products, Dow Chemical U.S.A., Midland, Mich. He joined Dow in 1960, and has held a number of engineering and special technical positions in research and development. Mr. Seifert has a B.S. degree in chemical engineering from the University of Colorado.

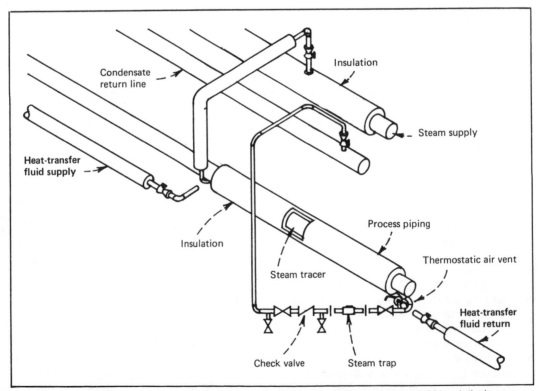

CONVERSION from conventional steam tracing system to one using heat-transfer fluid is relatively easy.

Energy-Saving Contest Runners-Up

Fuel Saving in Chemical-Plant Power Stations

LALLO GASPARINI, Franco Tosi S.P.A., Legnano, Italy

For a chemical-plant power station that is fairly large and must supply steam (more than 150 tons/h) and electrical power for plant usage, there are several means of obtaining the maximum electrical energy possible with the least amount of fuel consumption. These means involve: increasing the boiler pressure and temperature, increasing the feed-water temperature, and the heating of makeup water.

Assume that the steam needed for a particular plant is at low pressure—although the boiler pressure may be higher—and that by flowing through a back-pressure turbine, it generates electric power. Taking into consideration only the losses related to electrical-energy generation, the overall heat rate is approximately 1,000 kcal/kWh (4,000 Btu/kWh).

||||||||||||||||||||||||||||||||||||||||||||||||||||||||||||||||||||||||||

Energy Requirements for 100 Tons/H of Low-Pressure Steam and 25,000 kW of Electric Power

Power-Station Type	Recovery and Condensing Systems	Straight Recovery
K value, kWh/ton	150	250
Recovered energy, kWh	15,000	25,000
Additional condensing energy, kWh	10,000	0
Fuel for steam generation, kcal/kWh x 10^6	60	60
Fuel for energy recovery, kcal/kWh x 10^6	15	25
Fuel for additional condensing energy, kcal/kWh x 10^6	25	0
Total fuel, kcal/kWh x 10^6	100	85

Assumptions: boiler efficiency = 0.93; low-pressure steam enthalpy = 600 cal/kg; makeup-water enthalpy = 40 cal/kg; heat rate for recovered energy = 1,000 cal/kWh; heat rate for additional energy = 2,500 cal/kWh.

||||||||||||||||||||||||||||||||||||||||||||||||||||||||||||||||||||||||||

Originally published September 30, 1974

This heat rate may be compared with the one typical of a condensing, power-generating station, which involves condenser losses dictated by the Carnot cycle. For units of 50-100 MW, it is usual to consider 2,500 kcal/kWh (10,000 Btu/kWh).

Consider the ratio of electrical energy (E) to the low-

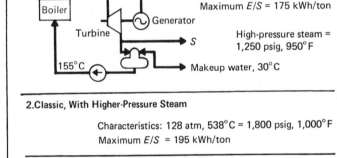

1. Classic
88 atm, 510°C
Boiler — Turbine — Generator → E → S
155°C — Makeup water, 30°C
Maximum E/S = 175 kWh/ton
High-pressure steam = 1,250 psig, 950°F

2. Classic, With Higher-Pressure Steam
Characteristics: 128 atm, 538°C = 1,800 psig, 1,000°F
Maximum E/S = 195 kWh/ton

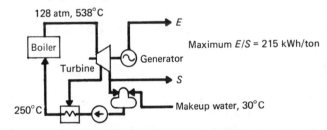

3. As Step 2, With Higher Feed-Water Temperature
128 atm, 538°C
Boiler — Turbine — Generator → E → S
250°C — Makeup water, 30°C
Maximum E/S = 215 kWh/ton

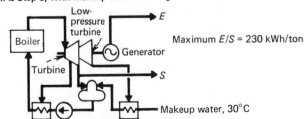

4. As Step 3, With Makeup Water Heating
Boiler — Turbine — Low-pressure turbine — Generator → E → S
Makeup water, 30°C
Maximum E/S = 230 kWh/ton

STEP-BY-STEP EVOLUTION of straight-recovery plant.

147

pressure steam (S) consumed as a basic parameter relating the fuel consumption (F) for both E and S. The symbol F_0 is to be considered the fuel consumption for steam generation only. If E/S increases, F must also increase at a rate of 1,000 kcal/kWh, up to the maximum feasible recovery value, K. If even larger E/S values are required, F will increase at a rate of about 2,500 kcal/kWh because the additional power must be generated with a condensing cycle. The table shows a comparison between the two types of plants (recovery and condensing, and straight recovery).

When the plant boiler is large enough to be compa-rable to boilers in utility power stations, it may be convenient to use the same boiler type, with equal steam pressure and temperature (127-140 atm, 540°C, or 1,800-2,000 psig, 1,000°F).

The feed-water temperature may also be heated by steam flowing through the high-pressure turbine. Makeup-water heating can also be accomplished in the deaerator by steam flowing from an additional low-pressure turbine, which generates power at a heat rate equal to the one in the recovery system. The accompanying figure shows the steps that can be considered and the maximum E/S ratios that are attainable.

Recovery of Compression Heat

A. J. HERMANS, Pittsburgh, Pa.

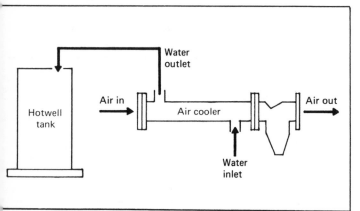

CIRCULATING WATER recovers additional heat.

In most power plants serving industrial complexes, a sizable quantity of both compressed air and boiler-water makeup are used. The heat that is generated by compression of air can be saved by introducing it into the boiler makeup water.

The accompanying sketch shows how this water circulates through an intercooler or aftercooler, thereby picking up the extra Btu. This heated water then goes into the hotwell tank.

For instance: if 100,000 gal water/d circulate through a cooler at an air compressor, the difference of the inlet and outlet temperatures could be 22°F. The sensible heat picked up by the cooling, makeup water is then:
(100,000 gal/d)(8.33 lb/gal)(22 Btu/lb) = 18×10^6 Btu/d.

Electrostatic Precipitator Overkill

J. COOPER and G. G. SCHNEIDER, Enviro Energy Corp., Encino, Calif.

With the advent of stricter air-pollution control laws, engineers specifying electrostatic precipitators have developed a trend of combining the worst possible process conditions to assure clients that installed plants will meet government requirements under almost any set of circumstances. The possibility of alternate sources of raw materials or fuels has also forced the listing in specifications of broad ranges of chemical and physical characteristics, as well as expected dust loadings, gas volumes and temperatures.

In recognition of an industrial history in which many precipitators failed to meet guaranty conditions, manufacturers have adopted more and more conservative approaches in their offerings.

Reducing drift and face velocities to handle the most difficult dusts, retention time has been increased in parallel, creating very large equipment. Power sets have increased in size and number to improve efficiency through sectionalization, and to achieve the desired "sparking condition" indicative of maximum power input.

To ensure reliability, some installations are equipped with extra fields (chambers) as a safety margin against short-circuiting from wire breakage, overfilled hoppers, or some other malfunction that would take a field out of service. Maintaining contract efficiency with a chamber out of service is common in the steel, cement and paper industries.

Anticipating stricter SO_2-emission requirements, a Northeast plant sized its equipment for low-sulfur fuel. The high-sulfur coal on hand, however, made it impossible—when the precipitator was started—to test the ash-handling system for the rear hoppers without shutting off the two front fields. During this period, with 40% of the precipitator shut off, the stack was visually clear. Early in the operation, because of rapper and control problems, the precipitator was working with almost half the fields out of service. Again, the stack was perfectly clear. Six identical precipitators exist at this plant.

Operating Example

Consider the specified conditions for a typical, large, pulverized-coal plant: gas volume, 2 million ft³; inlet dust-loading, 4 grains (gr)/ft³; outlet dust-loading, 0.02 gr/ft³; efficiency, 99.5%; sulfur content, 1%.

Manufacturers conservatively select drift and face velocities from experience with such installations. Within economic bounds, they try to provide as much sectionalization as possible. Let us examine some alternate operating conditions that could occur: inlet-dust loading, 2 gr/ft³; outlet-dust loading, 0.02 gr/ft³; efficiency, 99%. Theoretically, the precipitator could operate with 8 of the 64 fields shut down. Energy savings would be about 400 kVA, plus perhaps 100 kVA for idle associated auxiliaries.

If the gas were 1,800,000 ft³, this 10% volume reduction—affecting face velocity and retention time—should increase precipitator efficiency sufficiently to allow operation with 8 of the 64 fields shut down. And if sulfur content were 2%, the marked change in dust resistivity would yield a drift velocity almost twice the design value. The unit would probably meet its required efficiency with 24 fields out of service. This would represent a reduction of power of about 1,500 kVA.

As mentioned before, transformer-rectifier sets and field sizes are selected in anticipation of driving the field to sparking levels. Automatic voltage controls are installed to maintain the best peak-to-average voltage-ratio possible. For many years, engineers have provided more power than was theoretically needed for energizer sets as part of the "manufacturer's margin."

Economically, there is a very small cost differential between an 800 mA and a 1,000 mA set for the same kV. Historically, where power deficiencies have occurred because of small sets, it has been the result of insufficient

mA/1,000 ft² of dust-plate surface. As a result, in many installations the transformer-rectifier sets operate at maximum rating; there is no sparking. If this condition exists in several of the fields in series, there is an opportunity to reduce energy usage with an insignificant reduction in precipitator efficiency.

Auxiliaries, such as rappers, hopper vibrators, heating cables and the ash-removal system itself are usually operated in accordance with the instructions of the manufacturer, within control limitations and the judgment of his service engineers and plant operators.

At one installation, the rappers ceased to function because of a control problem. There was no deterioration of stack appearance for long periods of time, except for objectionable puffs, when large accumulations of dust dislodged themselves from the collecting curtains and fell into the hoppers. However, with a new control component, the rappers were restored to service utilizing the preset frequency and cycle. This resulted in far more rapping than apparently was necessary.

The cure for rapping deficiency has been to replace the original rappers with larger ones, even though many engineers have recognized that the problem may lie in how the energy of the rapper is lost in transmission through casings and suspension systems. Perhaps the "brute-force" approach must be weighed against other design changes to capture energy savings.

The plant previously mentioned—which was designed for operation on low-sulfur coal—could probably operate in compliance with the law, under full-load conditions, with about 60% of its installed precipitator capacity. Since there are six precipitators at the plant, further reductions could be possible over weekends, when the load is reduced.

Optimizing auxiliaries requires additional testing. In this area, operators may find it necessary to introduce additional flexibility through alteration of existing controls. It is possible that the addition of a modest amount of equipment would allow for better sectionalization of the operation of the auxiliary equipment from inlet to outlet fields. The information gathered at full and partial loads would be used to establish new operating procedures. Power now being used to operate equipment at higher levels of efficiency than required could be conserved and would find its way to the grid.

Utilizing Byproduct Hydrogen*

PETER DE ANGELIS, Sobin Chemicals, Inc., Orrington, Me.

Our plant is a mercury-cell, chlorine-caustic plant that started operations in 1967. Because of potential pollution problems and the severe winter conditions of our area, our two 15,000-lb/h boilers were designed to operate on No. 2 fuel oil.

*Although the utilization of byproduct hydrogen gas from chlorine caustic plants has been practiced by large installations, this is a reminder to other companies of energy savings that can be accomplished.

One of the byproducts of our manufacturing process, hydrogen gas, was vented to the atmosphere because there was no inplant use for it; plant steam is used primarily for heating. Because of the long winters in Maine, our use of No. 2 oil was about 500,000 gal/yr. Therefore, it was decided to use the heat value of the vented hydrogen by designing a system to collect the gas, compress it to a pressure of 5-10 psig, and pipe it to one of our two boilers as fuel.

The total installed cost was $35,000. During the first year of operation, our usage of fuel oil was reduced by 300,000 gal; present use is about 150,000 gal. The system has worked so well that we are presently converting the second boiler to burn hydrogen.

Since hydrogen is a very clean fuel, our boiler has re-quired very little maintenance. Our major emphasis on design was safety, so a number of protective devices were installed. No problems were encountered with our insur-ance underwriters.

At oil prices when the system was proposed, the unit payout was less than 2 years after taxes. At today's oil price, the payout on our second boiler installation will be 1.5 years after taxes.

Sonic Fuel Atomization

DAVID M. BRINCKERHOFF, Clayton, Mo.

Ultrasonics can be used as a tool to superatomize (acoustically cavitate) liquid particles. The resulting in-crease in surface area of the smaller particles is propor-tional to the cube of the reduction in diameter of the pre-existing particles: $\Delta A = 1/\Delta d^3$.

The payoff in fuel savings when this idea is applied to combustion systems is that it allows the delivery of a fine and uniform mist. This, in turn, reduces the need for ex-cess air in the secondary combustion stage, which results from a less than stoichiometric ratio in the primary stage.

Nitrogen oxides are lower, as well, because of the uni-formity of mist particles and the lack of significant flame eddies that allow a significantly less compromised com-bustor design.

Sonic nozzles operate at low liquid pressures, and mul-tiple orifices can be used for highly viscous fluids. Other advantages include consistency over wide flow ranges, and versatility. From coarse to ultra-fine particles (1-5 micron droplets) can be produced by air-pressure vari-ation.

Symbiosis To Reduce the Energy Crunch

PAUL N. GARAY, Consultant, Moraga, Calif.

Symbiosis is defined as the closed union of two dis-similar organisms, usually when this is advantageous to both. This definition may be applied to a feasible part-nership between a utility organization and an industrial plant.

A typical utility plant may generate power at a fuel rate of some 10,000 Btu/kWh. Fuel chargeable to power generation in an industrial plant is as low as 4,000 Btu/kWh when all the exhausted or extracted steam from a steam turbine is used for a given process. Such turbine-steam-generated power is often referred to as byproduct power.

Sometimes, the process cycle can be arranged to pro-vide byproduct power in excess of the plant's require-ments. Appreciably less fuel will then be needed than the 10,000 Btu/kWh, which is typical for a utility. It is in these cases that a profitable partnership between the util-ity and the industrial plant can be realized.

To illustrate the potential benefits resulting from such joint ventures, assume that Refinery X, located in a small town, generates some 12,000 kW of power for plant use, and requires about 500,000 lb/h of steam for process use. The electrical requirements of the town—supplied by a generating station several miles distant—amount to about

10,000 kW. The refinery operates on a 10-day-on, 4-day-off schedule. Thus, a good probability exists that the town can be supplied with refinery-generated power at a heat rate of 4,500 Btu/kWh. Furthermore, transmission losses inherent in the utility supply would be almost completely eliminated. An analysis of three possible

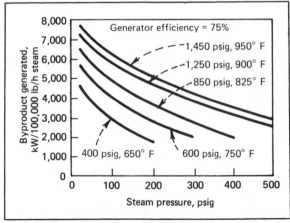

BYPRODUCT POWER and steam generation relationship.

cases shows the economics of power production and fuel use:

1. In this instance, the refinery produces its own process steam, but both refinery and town purchase some 10,000 kW each from the utility's network. The total heat input to the combination is 918×10^6 Btu/h.

2. Here, the refinery produces its own power—10,000 kW—as byproduct of process heat requirements. The town purchases 10,000 kW from the utility system. The total heat required in this particular instance is 860×10^6 Btu/h.

3. This is the most favorable case, in which the power for the town and the refinery—20,000 kW—is generated by the refinery. The total heat input for the town-refinery

system is now reduced to 801×10^6 Btu/h, which is equivalent to 13% reduction in fuel use.

In the situation of this small town and refinery (not considering transmission losses), a possible conservation of 221 bbl/d of fuel is readily possible.

An industrial plant can be tailored to suit its anticipated power needs, plus the external load. The curve shows the relationship between byproduct power and steam generation. The net effect, aside from the saving of fuel, would benefit both partners: the utility grid would be enlarged, and power would be available at low cost. The industrial plant would improve its operating situation by having an adequate operating, as well as standby, power source.

Fuel Savings Through No-Vent Systems

M. W. STOUT, David Stout & Sons, St. Louis, Mo.

No-Vent equipment is a much more efficient means of conserving energy than are steam traps. Essentially, No-Vent installations consist of a vertical motor mounted on a bracket, so that special packing can be inserted in the extra-deep stuffing box to permit stainless-steel impeller shafts to rotate easily. A manifold removes all excess air

or other non-condensables from the hot condensate line automatically and completely, so that corrosion and entrapped air are eliminated.

The advantages of returning process-steam condensate to the boiler without throttling and venting flash steam to the atmosphere are shown in the graph. Hot-condensate return markedly reduces boiler load, especially at higher pressures. Curve D shows the proportion of condensate flashed to steam when it is throttled to atmospheric pressure. If this steam is not utilized, the proportion of energy lost is shown by Curve A. Curves C show the increase in boiler load caused by replacing the lost, flashed steam with makeup water at temperatures of 40 and 100°F, and assuming the remaining condensate is returned at 212°F. Curves B show boiler-load reduction by returning hot condensate under pressure to the boiler, and not replacing it with makeup water at inlet temperatures of 40, 70 and 100°F.

Assume 30,000 lb/h of steam are condensed at 205 psig and 390°F. If steam costs 80¢/1,000 lb, you spend $24/h. No-Vent fuel savings amount to 19.6% of this $24/h, or a savings of $4.70/h. Better than 40 years experience has shown that there is often enough pipe scale and foreign matter in steam-trap competitive systems to double the $4.70/h No-Vent savings. The payout with No-Vent fuel savings with 205-psig steam pressure is so fast, that the loss due to steam-trap leakage is entirely neglected in figuring out payouts.

No-Vent systems require no steam traps, steam vents, orifice plates, steam-pressure regulators, or steam-reheating in the entire circuit. Some 1,500 installations, plus 40 years experience with them, has shown that this will provide large fuel savings.

If you divide the cost of the unit—$2,860—by $4.70/h savings, the payout is 609 hours of operation. If your system operates 24 h/d, payout time will be 25.4 days, even if your present steam traps are perfect and no leaks of live steam exist whatsoever.

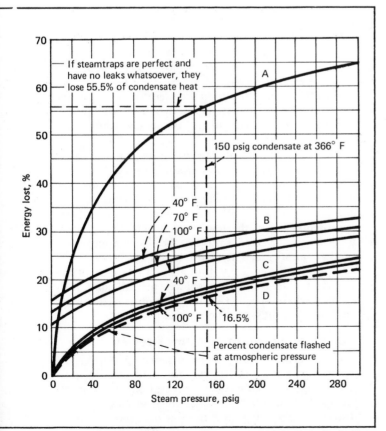

ADVANTAGES of returning steam condensate to the boiler.

How to optimize the design of steam systems

As energy and equipment costs increase, efficient steam systems become more important in the overall economics of process plants. Here is a method for modeling and optimizing an industrial steam/condensate system by using linear-programming techniques.

James K. Clark, Jr. and Norman E. Helmick, Fluor Engineers and Constructors, Inc.

☐ An iterative linear-programming (LP) algorithm can be used to minimize net costs in steam-distribution systems. These costs represent everyday operating expenses, or annualized costs including operating expenses and capital investment.

The LP technique will determine optimum values for the process-design variables, so as to achieve minimum cost. Typical variables include steam-header temperatures, turbine rates, letdown flowrates, makeup and desuperheating water flows, cooling/heating requirements and deaerator pressure.

When alternative designs are feasible, LP will also determine the optimum system configuration, based on selection between electric-motor or steam-turbine drivers, and selection and sizing of optional superheaters and condensate coolers.

The LP procedure is more flexible than methods based on solving a system of equations for heat and material balances. In addition, for a desired set of conditions with a specified objective, a final solution is obtained in a single computer run—thus eliminating the numerous solutions required by the case-study approach. Published work on the subject has emphasized methods based on the solution of heat and material balances with computer programs having little or no optimization capability [1,2,3,5]. A linear-programming

optimization approach has been proposed by Bouilloud [4], but it requires that all enthalpy values be known and that all thermodynamic calculations be done prior to the LP solution. It is up to the user to make repetitive trials based on different thermodynamic assumptions. Because of these requirements, the model is limited in what it can optimize.

When enthalpies are introduced as problem variables, direct LP solution becomes impossible due to the nonlinear nature of the energy-balance equations. The iterative technique described in the following sections gets around this difficulty by solving successive linear approximations of the nonlinear steam-balance equations. With each new solution, the linear approximations are improved. The final solution is achieved when all approximations are within a small tolerance of the actual nonlinear equations. Since each LP solution is optimal with respect to its linear approximations, the final solution must also be optimal.

Linear programming: a brief review

Linear programming is one of several mathematical techniques, known collectively as "mathematical programming," that attempt to solve problems by minimizing or maximizing a function of several independent variables. LP is the most widely used of these

Originally published March 10, 1980

methods, and is one of the best for analyzing complex industrial systems. Typical applications include determining the optimum allocation of resources (i.e., capital, raw materials, manpower and facilities) to obtain a particular objective such as minimum cost or maximum profit for the project.

In practice, optimal allocation of resources must be determined under conditions where there are alternative uses of resources and where physical, legal and managerial constraints must be met. Constraints take the form of upper or lower bounds that must be placed on resources or on operating conditions. For example, the availability of a raw material may be limited, or a process unit may have a maximum throughput. Other examples include the blending of petroleum products where sulfur contents, gasoline octanes, fuel-oil viscosities, etc., must meet upper or lower bounds to make salable products.

The LP procedure will maximize or minimize a linear objective function of the form:

$$\sum_{j=1}^{n} c_j x_j \qquad (1)$$

where the x_j are unknown problem variables, and the c_j are constant coefficients. The variables must be con-

strained by equality or inequality relationships, having the form:

$$\sum_{j=1}^{n} a_{ij} x_j \lesseqqgtr b_i, \ i = 1,2 \cdots m \qquad (2)$$

where the x_j are the same problem variables as in Eq. (1), a_{ij} are constant coefficients, and b_i are right-hand-side constants. The m relationships indicated in Eq. (2) are called the constraint set. The objective function and the constraint set, together, form an LP model. Relationships indicated by Eq. (2) can be upper bounds (maximum constraints), lower bounds (minimum constraints), or equalities. The relationships are linear.

When developing a steam-system LP model, the engineer must derive the heat- and material-balance equations that relate the problem variables (i.e., process flows and enthalpies) to one another. In this article, most of the constraint set will consist of just such equations. However, it is also necessary to specify all of the upper and lower bounds that apply to the problem. Most of the inequalities will identify maximum and minimum temperature bounds. Finally, all LP solution codes require that all x_j be nonnegative (i.e., $x_j \geq 0$). This property helps to ensure feasible heat and material balances.

A number of computer codes have been developed to "solve" LP models. Many of these codes are available through the major computer manufacturers and service companies. Solution codes determine a set of values for the x_j that maximizes (or minimizes) the objective function while satisfying the relationships contained in the constraint set. Much has been written about the simplex algorithm that is used by all LP codes—such as the mathematics, options, and interpretation of results. These topics are beyond the scope of this article. For further information, see Refs. 6 and 7.

LP applied to steam systems

The constraint set that describes a typical steam system consists largely of the following four relationships:

1. Equations for the material and energy balances for each process unit and equipment item—this includes all headers, boilers, superheaters, turbines, and heat exchangers, flash drums, deaerators, etc.

2. An equation for each power demand (i.e., steam turbine), and each process-energy demand.

3. Upper and lower bounds for all independent variables.

4. Equalities and upper/lower bounds that are problem-dependent.

By way of example, let us consider the following steam header:

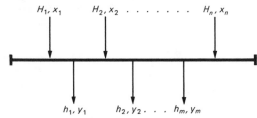

The x_j are input flows to the header. For each x_j, there is a corresponding enthalpy variable, H_j. (There is no loss of generality if some of the x_j or H_j are known, fixed values.) The y_i and h_i represent the header output flow and enthalpy variables, respectively. A material balance is expressed by Eq. (3), and the energy balance by Eq. (4).

$$\Sigma y_i - \Sigma x_j = 0. \tag{3}$$
$$\Sigma h_i y_i - \Sigma H_j x_j = 0. \tag{4}$$

Eq. (3) and (4) are not sufficient to describe the status of the header. The specific values of h_i depend upon the header pressure and state (liquid or vapor), as well as temperature. Thus, we must know the temperature, pressure and state in order to use a thermodynamic correlation to compute the enthalpy. Conversely, if the pressure, enthalpy and state are known, the temperature can be calculated. Throughout this article, header pressures are taken as known, constant values. Further, the output enthalpy variables, h_i, must all have the same value that we shall designate as H, e.g., $h_i = h_2 \cdots = h_m = H$.

Let k_1 be the enthalpy of saturated steam at the header pressure. It is apparent that if we wish to ensure a vapor state for all steam flows exiting the header, we must append the inequality:

$$H \geq k_1 \tag{5}$$

Process or equipment limitations normally require adding an upper bound on the header steam enthalpy. If k_2 is the maximum value, then the following inequality is also necessary:

$$H \leq k_2 \tag{6}$$

With the assumption of known header pressure, the relationships given by Eq. (3), (4), (5) and (6) are sufficient to describe the mass and energy balances.

Let us now consider a backpressure steam turbine, as shown:

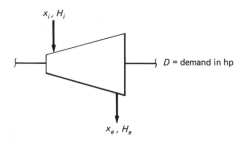

where H_i and x_i represent the input steam enthalpy and flow; H_e and x_e are the exhaust steam enthalpy and flow. The material balance is:

$$x_i = x_e \tag{7}$$

The energy balance is also the horsepower-demand equation:

$$H_i x_i - H_e x_e = D/C \tag{8}$$

In Eq. (8), C is a constant that converts the horsepower demand to Btu/h, and accounts for mechanical efficiency. Since the input steam is from a header with a known pressure, and the output enters a header with a known pressure, the exit enthalpy, H_e, can be expressed as a function of the input enthalpy, H_i:

$$H_e = \alpha + \beta H_i \tag{9}$$

The constants α and β in Eq. (9) will be derived later. Eq. (9) is problem-dependent and is only required if the header supplying the steam has a variable temperature (and hence, enthalpy). Further examples of problem-dependent constraints include bounds on fuel availability, boiler blowdown rate, and deaerator pressure.

Types of objective function

The objective function for a steam-system LP can range from very simple to quite complex. The simplest consists of a single variable. It represents either the steam-generation rate (in excess of that generated by process-heat recovery), or the total fuel consumed. In either case, the objective function is minimized.

An alternative objective function minimizes total operating expense. It includes costs for fuel, boiler-feedwater makeup, electric power, cooling-water makeup, and catalysts and chemicals for water treating. (The example that will be discussed in this article has as its objective the minimization of operating expense.) This type of objective function should be used by operating companies when the goal is minimum operating expense for an existing system.

A more comprehensive objective can be defined by including operating expense, and the cost of capital recovery plus return on investment for major equipment. For a new steam system in the design stage, this objective function represents the total variable cost of the system and, of course, is minimized.

The three types of objective functions described here are typical. Selection of a particular objective function depends on the purpose of the study.

The nonlinear problem

The relationships given by Eq. (4) through Eq. (8) may contain nonlinear terms such as Hx, where H represents an unknown enthalpy, and x an unknown flowrate. The constraint relationships must be linear because the simplex algorithm (used by the LP solution codes) cannot handle nonlinear terms.

To get around this problem, it is possible to replace the nonlinear terms with single variables, or with linear approximations, in order to form a linear model that can be solved by using LP methods. The approximations require estimates of coefficients. When the estimates are correct, the errors introduced by the approximations become insignificant. The method to be described uses iterative estimation of coefficients, and subsequent problem solution, until a stable set of coefficients is found that satisfies maximum error requirements. The LP solution that utilizes this last set of coefficients is the solution that we are seeking.

In subsequent sections of this article, we will describe the linearization technique and the iteration strategy for obtaining the desired solution by means of an example of a typical steam system.

Modeling a steam system

A process flow diagram for a typical steam-distribution system is shown in Fig. 1. This system provides steam at three pressure levels: 850, 250 and 40 psig. Header temperatures can range from saturation to maximum superheat values. The optimum value for each header temperature will be determined by the LP procedure.

Process steam requirements and/or heating demands are indicated for each pressure level. Normally, such demands are given in constant Btu/h or lb/h. The diagram also shows that each pressure level has fixed sources of steam, which the system must accept.

The low- and intermediate-pressure levels each have one source where the volume and heat content are fixed. However, the temperatures of two other sources, going to the high- and intermediate-pressure headers, respectively, can be increased by superheaters. The latter units are shown by dashed lines to indicate optional equipment items. The LP procedure will determine the optimum temperature (fuel usage) for these superheaters. A zero fuel rate to a superheater simply means that the optimal solution does not include that superheater.

Power requirements

The four horsepower demands shown in Fig. 1 are designated by the symbol T/M. This indicates that the power demand can be satisfied by either a steam turbine (T) or a motor drive (M). The dashed lines for these units indicate that the proper selection will be made by the LP procedure. Three different types of tur-

bines have been included: backpressure, condensing, and condensing with extraction. In some cases, there may be a choice among turbine types for a given horsepower load, as well as a motor alternative. Here again, the LP procedure can choose the alternative required to minimize annual operating expense.

It is possible that a combination of turbine types, or both a turbine and a motor, will be selected to satisfy a horsepower demand. If the demand represents a single driver, such a solution is not physically possible, and it becomes necessary to make several case studies to select the optimum configuration. For each case study, all alternatives, except a single power source, are set to zero activity for each demand. Thus, if a power demand has been satisfied by both a motor and a turbine, two case studies are required. The solution with the lowest cost is optimum.

For the case where the horsepower requirement represents multiple drivers, combinations are acceptable. Individual driver types are selected by using the relative amounts of alternative energy sources. For example, if a power requirement represents ten drivers, and the LP solution indicates a 40%/60% split between turbines and motors, then four drives would be turbines, six would be motors.

Condensers, coolers and desuperheaters

Four of the five condenser/coolers shown in Fig. 1 are indicated by dashed lines; again, the LP algorithm will determine whether a given condenser/cooler is required for an optimum configuration. A zero cooling-water rate indicates that the subject cooler is not needed.

We notice that no desuperheaters are shown in Fig. 1 for the intermediate- and low-pressure headers. However, the diagram does show letdown steam and condensate going to each of these headers. This is all we need, since the LP procedure will determine the amounts of steam and condensate necessary to meet the mass and heat balances for a header. An LP solution that requires condensate to be added to a header indicates the need for a desuperheating station. If there is more than one steam flow that can be desuperheated, the design engineer has the option of selecting which streams to desuperheat. Again, this is an example of how the LP model helps to determine the optimum system configuration.

Deaerator pressure

A significant variable for the system is the deaerator pressure. It is a system variable because the enthalpies of the vent steam and the boiler feedwater reflect saturated values at the deaerator pressure, and these enthalpies are variables in the LP model. In practice, only one enthalpy, that of the boiler feed, is carried as a model variable. A linear correlation has been developed to represent the enthalpy of the vent steam as a function of the boiler-feed enthalpy over a practical pressure range.

Since boiler feedwater goes directly to all boilers and desuperheating stations, the enthalpy of the water has an important effect on the overall system design. The LP procedure chooses the optimum value for the deaerator pressure, as part of the final optimal solution.

The mathematical model of a steam system (such as

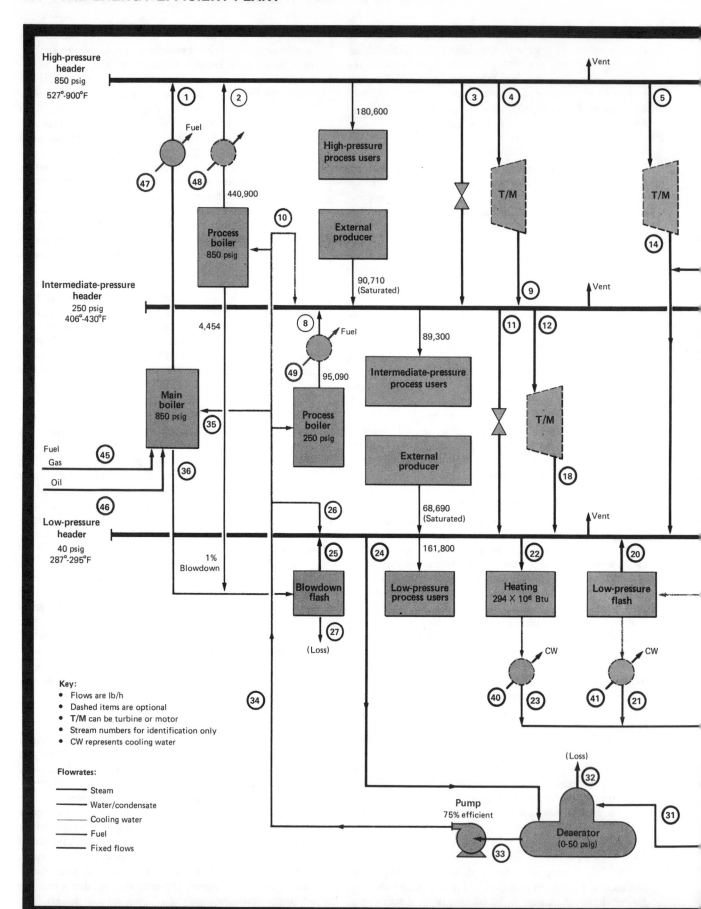

Steam balance for a typical steam-distribution system will be optimized by linear-programming techniques

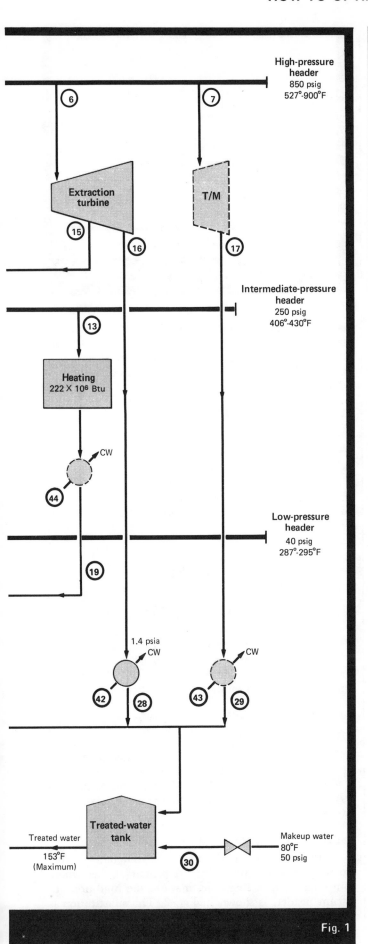

High-pressure
header
850 psig
527°-900°F

Extraction
turbine

T/M

Intermediate-pressure
header
250 psig
406°-430°F

Heating
222 × 10⁶ Btu

CW

Low-pressure
header
40 psig
287°-295°F

1.4 psia
CW

CW

Treated-water
tank

Treated water

153°F
(Maximum)

Makeup water
80°F
50 psig

Fig. 1

Fig. 1) will include equations for heat and material balances for each element contained in the system. Eq. (3) through (6) are typical for a steam header. These equations will be developed further, along with additional examples. But let us first consider linearization.

Linearization

We have noted that Eq. (4) shows terms of the type Hx, where H is the steam enthalpy in Btu/lb, and x is the flowrate in lb/h. In the general case, both H and x are problem variables, and their product is nonlinear. Where possible, such terms will be replaced by a simple variable substitution. (An example of variable substitution will be discussed as part of the equations that represent a primary boiler.) In most cases, the product, Hx, must be replaced by a first-order Taylor-series expansion (see p. 325 of Ref. 8). For example:

$$Hx \approx H_0 x + x_0 H - H_0 x_0 \qquad (10)$$

In Eq. (10), H_0 and x_0 are the Taylor-expansion coefficients, and are the best available estimates of the true values of H and x. Values for the coefficients must be estimated initially, and then reevaluated by successive LP solutions until a tolerance test can be met. The computer algorithms that perform the tolerance testing will be discussed in a later section, as will be the method for making initial estimates for the coefficients.

A similar linearization technique is used when nonlinear terms are encountered in the objective function. This usually occurs when the capital costs of equipment items are represented as functions of problem variables.

Let us consider two examples that will illustrate linearization by Taylor expansion.

Steam header

For the first example, let us analyze the equations for a steam header (as shown in the illustration on p. 118). The material balance is given by Eq. (3). The enthalpy balance is now expressed as:

$$H\Sigma y_i - \Sigma H_j x_j = 0 \qquad (11)$$

Using the concept of Eq. (10), we can write the linear form of Eq. (11) as:

$$\Sigma H_0 y_i + (\Sigma y_{i0})H - \Sigma H_{j0}x_j - \Sigma x_{j0}H_j = \\ H_0\Sigma y_{i0} - \Sigma H_{j0}x_{j0} \quad (12)$$

In Eq. (12), each symbol that carries a zero subscript is a Taylor-expansion coefficient. In addition to the heat-balance and material-balance equations, it is necessary to limit the temperature, and hence the enthalpy, of steam leaving the distribution header (header pressure is fixed). We have noted that the minimum enthalpy will correspond to the saturation temperature as in Eq. (5), and the maximum enthalpy will be determined by process or equipment requirements [Eq. (6)].

Eq. (3) and (12), along with Eq. (5) and (6), are sufficient to describe the steam header. Note that the upper and lower bounds, similar to Eq. (5) and (6), are required for all system enthalpies that are independent variables.

Each steam header in Fig. 1 contains a flow labeled "vent." This flow represents steam or condensate that is thrown away to maintain valid header heat-and-material balances. Such a variable can be included in Eq. (3) and (12) as one of the y_i, i.e., one of the output flows. However, these variables deserve special consideration. Let v represent vent flow for the header shown on p. 118. Eq. (3) and (11) become:

$$v + \Sigma y_i - \Sigma x_j = 0 \qquad (13)$$

$$Hv + H\Sigma y_i - \Sigma H_j x_j = 0 \qquad (14)$$

If the system is well designed, the optimum value of v will be zero, in most cases. Therefore, when applying the Taylor expansion to the product Hv, we have:

$$Hv = H_0 v + v_0 H - H_0 V_0 \qquad (15)$$

$$v_0 = 0 \qquad (16)$$

$$Hv \approx H_0 v \qquad (17)$$

Substituting these relations into Eq. (14), and expanding the nonlinear terms, gives the final heat-balance equation:

$$H_0 v + \Sigma H_0 y_i + (\Sigma y_{i0})H - \Sigma H_{j0} x_j - \Sigma x_{j0} H_j =$$
$$H_0 \Sigma y_{i0} - \Sigma H_{j0} x_{j0} \qquad (18)$$

Eq. (13) and (15) are the final set of heat- and material-balance equations for a typical header. A vent variable should be added to the equations for each steam header and condensate loop. For a well-designed system, all such variables will have a zero value. If more heat energy is available from process sources than can be used by the system demands, this will be indicated by nonzero vent values. For such cases, it may be wise to review the process-design and equipment specifications to determine whether the heat sources can be economically reduced or the system design modified.

Backpressure turbine

For the second example, let us consider the backpressure turbine (as shown in the illustration on p. 118). The LP module for this in matrix format is:

	x_i	x_e	H_i	H_e		RHS
(I)	-1	1			$=$	0
(II)	$-H_{i0}$	H_{e0}	$-x_{i0}$	x_{e0}	$=$	$x_{e0}H_{e0} - x_{i0}H_{i0} - D/C$
(III)			Z_1	-1	$=$	Z_2

where:

$$Z_1 = \left[1 - \rho_t\left(\frac{\partial \Delta h_s}{\partial H_i}\right)_p\right]$$

$$Z_2 = \rho_t\left[\Delta h_{s0} - \left(\frac{\partial \Delta h_s}{\partial H_i}\right)_p H_{i0}\right]$$

Row I in the matrix array is the material balance; Row II, the enthalpy balance; and Row III relates the output enthalpy, H_e, to the input enthalpy, H_i: C in the right-hand side (RHS) changes the horsepower demand to Btu/h. Row III expressed as an equation becomes:

$$\left[1 - \rho_t\left(\frac{\partial \Delta h_s}{\partial H_i}\right)_p\right]H_i - H_e =$$
$$\rho_t\left[\Delta h_{s0} - \left(\frac{\partial \Delta h_s}{\partial H_i}\right)_p H_{i0}\right] \qquad (19)$$

The development of Eq. (19) follows. Let Δh_s be the

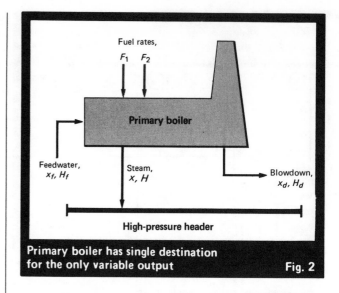

Primary boiler has single destination for the only variable output

Fig. 2

isentropic enthalpy change from inlet conditions for temperature and pressure to exhaust pressure. The actual work available is less than for an ideal reversible process. Hence, it is necessary to introduce a thermodynamic efficiency, ρ_t, defined as:

$$\rho_t = \frac{\Delta h_s - \int T ds}{\Delta h_s} = \frac{\Delta h}{\Delta h_s} \qquad (20)$$

The outlet enthalpy, H_e, is given by:

$$H_e = H_i - \Delta h = H_i - \rho_t \Delta h_s \qquad (21)$$

Since the exhaust pressure is fixed, the value of Δh_s depends only on the inlet conditions. Let Δh_{s0} be the value of Δh_s for the latest LP solution. Then Δh_s can be expressed:

$$\Delta h_s = \Delta h_{s0} + \left(\frac{\partial \Delta h_s}{\partial H_i}\right)_p (H_i - H_{i0}) \qquad (22)$$

$$H_e = H_i - \rho_t\left[\Delta h_{s0} + \left(\frac{\partial \Delta h_s}{\partial H_i}\right)_p (H_i - H_{i0})\right] \qquad (23)$$

Eq. (19) then follows from Eq. (23).

The partial derivative $(\partial \Delta h_s / \partial H_i)$ can be approximated numerically by using the values of H_{i0} and Δh_{s0} from the latest LP solution:

$$\left(\frac{\partial \Delta h_s}{\partial H_i}\right)_p \approx \frac{\Delta h_{s0} - \Delta h'_{s0}}{H_{i0} - H'_{i0}} \qquad (24)$$

The values of $\Delta h'_{s0}$ and H'_{i0} are computed by making a small temperature change (from the value of the last LP solution, which gave H_{i0} and Δh_{s0}), and calculating H'_{i0} and $\Delta h'_{s0}$ from the thermodynamic correlation for steam.

Substituting a variable

In this example, we will use substitution of a variable to remove a nonlinear term. A primary boiler whose output goes to a single destination, the high-pressure steam header, is shown in Fig. 2. The substitution is possible because the output has a single destination.

The diagram shows two fuels, F_1 and F_2, whose flowrates are expressed in lb/h. Associated with each fuel is a net heat of combustion, ΔH_1 and ΔH_2; and a boiler efficiency, e_1 and e_2. The blowdown rate, x_d, is a constant fraction, f, of the feedrate, x_f. Finally, F_1 has an upper bound of K, lb/h. The material-balance equations for boiler feed and steam are:

$$x_f - x_d - x = 0 \qquad (25)$$
$$f x_f - x_d = 0 \qquad (26)$$

The heat-balance equations must include heats of combustion for the fuels:

$$H_f x_f + e_1 \Delta H_1 F_1 + e_2 \Delta H_2 F_2 - Hx - H_d x_d = 0 \quad (27)$$

In Eq. (27), H is the enthalpy of steam entering the high-pressure header, and x is the flowrate. The minimum value of H corresponds to saturated conditions for the header pressure; its maximum value corresponds to the maximum amount of superheat that can be tolerated. H_d is the enthalpy of the blowdown, which is equal to the saturated liquid enthalpy at the pressure of the boiler. Since there is but a single destination for x, we can substitute $Hx = E$ into Eq. (27) to give:

$$H_f x_f + e_1 \Delta H_1 F_1 + e_2 \Delta H_2 F_2 - E - H_d x_d = 0 \quad (28)$$

where E represents the total energy delivered to the high-pressure header in Btu/h. Now, we note that E, x and H are related by:

$$E/x = H \qquad (29)$$

Further, let K_m be the minimum enthalpy value for H that corresponds to saturation conditions; and let K_h be the maximum value. Since H is an independent enthalpy variable, we must have the inequalities:

$$H = E/x \geq K_m \qquad (30)$$
$$H = E/x \leq K_h \qquad (31)$$

Finally, we have noted that the fuel rate, F_1, has an upper bound of K:

$$F_1 \leq K \qquad (32)$$

After expanding the term $H_f x_f$ in Eq. (28), and rearranging terms in Eq. (30) and (31), the final equations for the boiler model in matrix form are:

	x_f	H_f	x_d	x	F_1	F_2	E		RHS
(I)	1		-1	-1				$=$	0
(II)	f		-1					$=$	0
(III)	H_{f0}	x_{f0}	$-H_d$		$e_1\Delta H_1$	$e_2\Delta H_2$	-1	$=$	$H_{f0}x_{f0}$
(IV)				K_m			-1	$\leq$	0
(V)				$-K_h$			1	$\leq$	0
(VI)					1			$\leq$	K

Rows I and II are direct expressions of Eq. (25) and (26). Row III is Eq. (28) with the term $H_f x_f$ expanded, where H_{f0} and x_{f0} are the Taylor-expansion coefficients. Rows IV and V are Eq. (30) and (31) after rearranging, and Row VI corresponds to Eq. (32).

If steam from the boiler were supplying two, parallel, high-pressure systems, the substitution $Hx = E$ could not be made. This is simply because it is not possible to force the total heat, E, and mass, x, to be distributed in the same proportions at both destinations linearly.

The substitution can be, and should be, made when a

Modular LP models for a steam system   Table I

1. Steam boiler — variable flow and enthalpy
2. Process waste-heat boiler — variable Btu rate
3. Stand-alone superheater — variable Btu rate
4. Process user — fixed Btu demand
5. Process heater — fixed Btu demand
6. Steam header — variable temperature
7. Condensate flash — adiabatic, variable steam and Btu input
8. Deaerator — variable pressure
9. Backpressure turbine — fixed horsepower demand
10. Condensing turbine — fixed horsepower demand
11. Extraction turbine with condenser/cooler — fixed horsepower demand
12. Condenser/cooler — stand-alone
13. Pumps
14. Treated-water storage tank

boiler (or any process equipment item) has only one variable output and the output has only one destination. Upper and lower bounds, similar to Eq. (30) and (31), are usually required, as in the boiler example.

The complete model

The three examples that we have discussed are typical of what is required for obtaining linear equations to model process equipment. Space limitations preclude the development here of all the equations for the system shown in Fig. 1. A condensing turbine can be modeled by adding equations representing condensing and aftercooling to the model of the backpressure turbine. An extraction turbine can be modeled as two turbines in series with steam removal between them. A complete list of all mathematical models required for the system shown in Fig. 1 is given in Table I.

Input data and model definition

When solving a steam-balance problem, the first task is to define a model representing the desired configuration and operating requirements. This is accomplished by using input-data modules for each of the process units in the steam system. Table I shows the units for which data modules have been developed.

Each module is defined by completing a set of "fill in the blank" data sheets. Specific inputs for a given unit module include operating data required to define the unit's performance. This information is used to generate the constraint and heat- and material-balance equations for the unit. Examples of operating data would be a turbine's required horsepower, a process boiler's duty, or the exit temperature range for a superheater. Costs or revenues associated with the unit's operation or connecting process streams are also given.

The input-data sheets also identify the process streams connecting the units, along with the physical state of each stream. Data for these streams includes information defining their initial flow, pressure, enthalpy and stream type. Several stream types are possible: a stream may be either fixed or variable flow, and/or fixed or variable enthalpy, depending upon the process units connected to it.

Input data for process units Table II

Boilers	Main	High-pressure process	Intermediate-pressure process
Rate, lb/h		440,900	95,090
Pressure, psig	850	850	250
Superheat range, °F	827-900	827-900	406-430
Blowdown rate, %	1.0	1.0	0.0
Heat-transfer efficiency, %	80.0	77.0	77.0
Fuel-gas heating value, Btu/lb	20,800		
Fuel-oil heating value, Btu/lb	17,300	17,300	17,300
Maximum fuel gas available, lb/h	20,000	0	0

Steam headers	High	Intermediate	Low
Pressure, psig	850	250	40
Maximum temperature, °F	900	430	295
Minimum temperature, °F	527	406	287

Turbines	H.P.-I.P.	H.P.-L.P.	I.P.-L.P.	Ex. #1	Ex. #2	H.P.-Vac.
Horsepower required, hp	11,170	13,555	2,217	19,080	7,241	2,633
Thermodynamic efficiency, %	75.9	68.55	65.51	86.67	78.9	67.8
Inlet pressure, psig	850	850	250	850	40	850
Exhaust pressure, psig	250	40	40	40	1.4 psia	1.4 psia
Condenser subcool, °F					0	0

Process heating	I.P.	L.P.	L.P. flash
Duty, million Btu/h	221.622	294.024	
Optional cooler:			
Minimum exit temp, °F	287	90	90
Maximum exit temp., °F	406	287	287
Maximum C.W. Δt, °F	70	70	70
C.W. inlet temp., °F	80	80	80

Process users/external sources	H.P.	I.P.	L.P.
Consumption, lb/h	180,600	89,300	161,800
Generation, lb/h		90,710	68,690

Treated-water tank	
Makeup temperature, °F	80
Makeup pressure, psig	50
Maximum exit temperature to deaerator, °F	153

Deaerator	
Maximum pressure, psig	50
Minimum pressure, psig	0
Minimum vent flow, lb/h	500
Feedwater pump efficiency, %	75

H.P. = High pressure, I.P. = Intermediate pressure
L.P. = Low pressure, Vac. = Vacuum, C.W. = Cooling water
Ex.#*n* = Extraction turbine, Stage *n*

In the example for the steam boiler, the flows designated as feedwater and steam have both variable flowrates and enthalpies. However, blowdown has only a variable flowrate, because the enthalpy is that of saturated water at the operating pressure of the boiler. When a stream is defined as having both variable flow and enthalpy, two Taylor-expansion terms and a constant are created to replace the single term that represents total energy in a heat-balance equation, according to Eq. (10).

Using the modular approach, any combination of units may be pieced together and solved. The user has great flexibility in modifying the configuration or operating data. Units may be added or deleted, and system designs may be altered by manipulating the input-data deck.

Solution strategy and technique

The system used to solve steam-balance problems consists of two computer programs. The first is an executive program that monitors and controls the solution strategy. The second can be any standard LP solution package. Problems are solved by repeated executions of, first, the executive program, and then the solution code, until specified convergence criteria are satisfied.

Our primary interest in this article is with the executive program because it makes all evaluations, decisions and model adjustments during the solution process. Its functions include: initial model construction from the input data, analyzing the LP solution output, checking for convergence, modifying the expansion coefficient values between iterations, and generating the LP-formatted input for each new iteration. The executive program includes correlations for generating thermodynamic properties of steam and water [9]. These are used in calculating the equation coefficients and right-hand-side values.

The purpose of the executive program strategy is to achieve a solution to the linearized model such that all Taylor-expansion coefficients are good approximations of their corresponding actual variable values. This is accomplished by using the recursive procedure as outlined in Fig. 3. As indicated, the solution strategy is divided into two phases: initialization and recursive-solution.

Initialization phase

The purpose of initialization is to obtain a consistent set of Taylor-expansion coefficients. These are estimated values of steam/condensate flowrates and enthalpies. To be consistent, they must satisfy the heat- and material-balance equations and the upper/lower bounds of the system model.

To begin initialization, all Taylor-expansion flow coefficients are set to zero, while the Taylor-expansion enthalpy coefficients are set to chosen "typical" values (as determined from input data). In addition, the enthalpy variables are fixed at the same values chosen for the coefficients. All such values must be within the ranges defined by the upper and lower bounds that have been specified for the enthalpy variables. When completed, the above steps create a linear model having constant enthalpy values.

The linear model is then solved, and the values obtained for the steam and condensate flow variables become the initial Taylor-expansion flow coefficients for a second model. The initial Taylor-expansion enthalpy coefficients are simply the fixed values previously noted. Since all flow coefficients are initially zero, no convergence checking is done after the first LP solution.

To complete the initialization phase, the fixed status imposed on each enthalpy variable is removed, and the bounds are restored to their initial upper and lower limits specified by the input data. The system is now ready to obtain an optimum solution.

Recursive phase

After obtaining a consistent set of Taylor-expansion coefficients, the recursive phase begins. The second model is put in LP format by the executive program (using the Taylor-expansion coefficients just obtained) and is then solved by the solution code. The LP output is then interpreted, and the existing Taylor-expansion coefficients are compared with the corresponding flow and enthalpy solution values.

After each recursive pass, the executive program reads the LP solution output and checks for convergence of the Taylor coefficients. Fig. 4 outlines the convergence algorithm used. Both enthalpy and flow coefficients are checked against an allowable maximum fractional change, ε. In addition, flow coefficients may

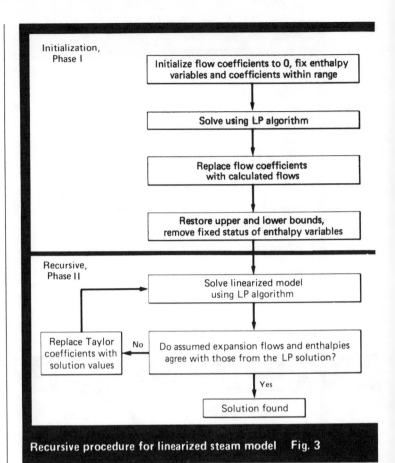

Recursive procedure for linearized steam model   Fig. 3

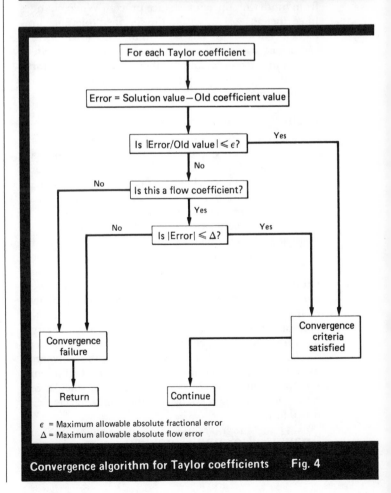

ε = Maximum allowable absolute fractional error
Δ = Maximum allowable absolute flow error

Convergence algorithm for Taylor coefficients   Fig. 4

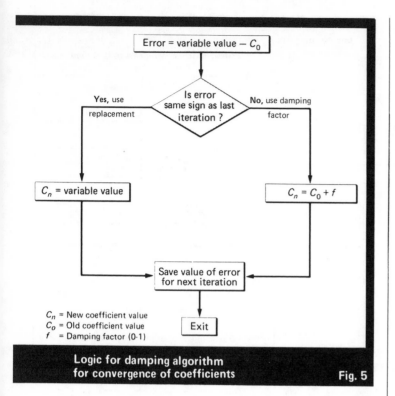

C_n = New coefficient value
C_o = Old coefficient value
f = Damping factor (0-1)

Exit

**Logic for damping algorithm
for convergence of coefficients**

Fig. 5

Comparison of problem specifications for different mathematical modeling systems		Table III
Variable	**H & MB**	**LP**
Main-boiler superheat	Fixed	Upper and lower limit
Steam-level pressures	Fixed	Fixed
Steam-level temperatures	Fixed/calculated	Upper and lower limit
Process-generation superheat	Fixed	Upper and lower limit
Condensate subcool	Fixed	Upper and lower limit
Treated-water temperature	Fixed	Upper limit
Deaerator pressure	Fixed	Upper and lower limit

H & MB = heat-and-material balances
LP = linear programming

Utilities consumption and operating expense	Table IV

Utilities consumption*

Cooling water, thousand lb/h	5,138
Power, kW	1,034
Fuel gas, thousand lb/h	20.0
Fuel oil, thousand lb/h	20.6
Makeup water, thousand lb/h	279

Utilities costs

Cooling water	$ 10.00/million lb
Power	$ 0.03/kWh
Fuel gas	$ 1.21/million Btu
Fuel oil	$ 2.18/million Btu
Makeup water	$100.00/million lb

Annual operating costs	$ 9.96 X 10^6/yr

*Optimal solution values

satisfy convergence by meeting a maximum allowable error, Δ, in total flow. The total-flow tolerance is added to prevent streams having relatively small flowrates from hindering the convergence process. When changes in coefficients are all within the specified tolerance limits, a final solution has been found. If, however, any coefficient fails the test, all coefficients are recalculated, placed in the model, and another recursive pass is initiated through the program for the linearized steam model (see Fig. 3).

Stability of convergence

In certain cases, some variables may not follow a stable convergence pattern. They tend to fluctuate around the final solution value, delaying or even preventing convergence. This is especially true in the case of letdown flows between pressure levels, where small changes in header enthalpies can cause large changes in desuperheater-water or letdown-steam flows.

To solve this problem, damping logic (illustrated by Fig. 5) was included in the coefficient replacement procedure. As long as a coefficient is changing in a consistent direction, direct replacement of the coefficient with variable value is used. If the direction of the coefficient change reverses, the replacement is made by using the old value plus a specified fraction of the total change. This fraction, f, is specified as input data.

The sample problem that will be discussed converged in seven to eleven iterations, regardless of the problem's starting assumptions for header enthalpies, where ε was set to 0.0001, Δ = 5 lb/h, and the damping factor, f, was 0.667.

Analysis of example problem

The steam-system design (outlined in Fig. 1) was optimized by using the recursive LP procedure. Variable

flowrates for the steam system are color-coded as follows:

Color	Flow
Red	Steam
Blue	Water/condensate
Green	Cooling water
Orange	Fuel

All known (i.e., fixed) flowrates are colored black. Numbers beside each stream in Fig. 1 are for identification. A description of the problem is given on p. 119.

The numerical input data used for the LP solution are shown in Table II. These were obtained by first specifying a set of typical data that would be required to obtain a solution by using the traditional method of solving a set of simultaneous heat- and material-balance (H&MB) equations. The variables to be determined by the LP optimization were then removed from this list and replaced with a variable range. Each variable removed represents a degree-of-freedom that the LP procedure uses for obtaining the minimum-cost solution. Data in Table II are for illustrative purposes only, and do not represent operating data from an actual plant.

A comparison of the differences in problem specification between the LP and H&MB methods is shown in Table III. Note that ranges rather than fixed values are shown for most variables in the LP case. If desired, these variables can also be fixed, but since our purpose is that of minimizing costs, every effort should be made to give the LP algorithm the greatest possible flexibility. This means that variables should not be fixed unless required by equipment specifications or process design. As a result, less-specific input data are required for the LP procedure than for the H&MB method. Further, much of the required LP data is easily obtained from process and equipment requirements (loads, sources, and temperature limits), or from experience and engineering judgment (minimum steam temperatures).

The objective function of the example minimizes operating expenses. In this case, it includes the cost of fuel, makeup water, cooling water and power. The unit cost values are shown in Table IV.

The example LP-problem generated a linearized model containing 86 rows and 119 columns. Seven iterations (initialization + six recursive LP solutions) were required to obtain an optimum solution, using a coefficient-damping factor, f, of 0.667, and convergence tolerances, ε, of 0.0001, and, Δ, of 5 lb/h. The starting values, specified for the initialization phase, for this problem are those shown in Table II. Table V shows the results of the LP-problem solution for all steam, water and fuel flows. Steam-header vents are all zero as expected, and are not shown. The stream numbers in Table V correspond to those shown in Fig. 1. Utilities and total operating expense are given in Table IV.

Some points concerning the LP solution, which define important design requirements, include:

■ No electric motors were selected. All drivers are steam turbines.

■ An optional superheater with a duty of 6.3×10^6 Btu/h was added to the high-pressure process boiler (exit temperature 720°F). The intermediate-pressure superheater is not required.

■ Both the intermediate- and low-pressure headers indicated minimum enthalpies of saturated steam.

■ Treated-water tank temperature went to its maximum value of 153°F.

■ There is no letdown between the high-pressure (H.P.) and intermediate-pressure (I.P.) headers.

■ The desuperheating-water flow of 27,200 lb/h to the intermediate-pressure header indicates that a desuperheater is required on the H.P.–I.P. turbine exhaust. All other steam flows to the intermediate-pressure header are at saturated conditions.

■ The low-pressure (L.P.) header requires letdown steam and condensate. This indicates that the letdown steam plus the two turbine exhausts should be combined and introduced into a desuperheating station.

■ Deaerator pressure is at its minimum value of 0 psig.

■ Optional subcoolers were included on the low-pressure process condensate (3.5×10^6 Btu/h) and the low-pressure flash condensate (46.4×10^6 Btu/h). Since both coolers handle low-pressure condensate, the total duty of 49.9×10^6 Btu may be divided between the two streams (at the option of the design engineer).

Results of linear-programming calculations for the problem				Table V
Steam/ condensate, Stream no.	Flow, lb/h	Pressure, psig	Temperature, °F	Enthalpy, Btu/lb
1	430,677	850	900	1,455
2	440,900	850	720	1,347
3	0	850	810	1,400
4	276,609	850	810	1,400
5	186,587	850	810	1,400
6	207,729	850	810	1,400
7	20,052	850	810	1,400
8	95,090	250	406	1,200
9	276,609	250	564	1,298
10	27,179	850	213	183
11	59,325	250	406	1,200
12	70,234	250	406	1,200
13	270,729	250	406	1,200
14	186,587	40	362	1,216
15	106,969	40	287	1,167
16	100,760	1.4 psia	115	984
17	20,052	1.4 psia	115	1,066
18	70,234	40	287	1,120
19	270,729	250	406	381
20	36,520	40	287	1,175
21	234,209	40	90	58
22	319,933	40	287	1,175
23	319,933	40	276	245
24	53,510	40	287	1,175
25	2,535	40	287	1,175
26	4,383	850	213	183
27	6,270	40	287	256
28	100,760	1.4 psia	115	83
29	20,052	1.4 psia	115	83
30	279,070	50	80	48
31	954,024	0	153	121
32	500	0	212	1,150
33	1,007,034	0	212	180
34	1,007,034	850	213	183
35	435,028	850	213	183
36	4,351	850	527	521

Cooling water, Stream no.	Flow, gpm	Duty, million Btu/h	Fuel, Stream no.	Flow, lb/h	Duty, million Btu/h
40	100	3.5	45	20,000	332.80
41	1,325	46.4	46	7,600	105.18
42	7,265	90.77	47	8,026	111.01
43	1,578	19.72	48	4,974	66.30
44	0	0	49	0	0

Note: Stream numbers correspond to those on Fig. 1

■ The cooler on the intermediate-pressure condensate is not required.

To illustrate the advantages of the LP procedure, let us consider solving the minimization problem by using an H&MB program. In order to find the combination of conditions that minimizes operating expenses, many case studies would be required.

Each solution for a case study would represent the results based on an assumed set of fixed values for the problem variables given in Table III. If each of the nine variables [boiler superheat, header temperatures (3), process superheat, condensate subcool (2), treated-water temperature, and deaerator pressure] were investigated at only three levels, a possible 3^9 or 19,683 combinations would be required. If one assumes that many cases can be eliminated, based on past experience and engineering judgment, case-study optimization would still require a prohibitive amount of time and effort. In addition, the case studies use discrete values of the variables, which greatly increases the chance of missing the true optimum solution.

Even if a great many case studies are made, success in finding a near-optimum solution depends upon the sensitivity of the objective function (in this example, operating expenses) to changes in values of the operating variables. The greater the sensitivity, the greater the chances of not finding the optimum, or even a near-optimum, solution. By comparison, the LP algorithm uses continuous variables and will always find the true optimum solution for any problem where such a solution exists.

Applications and benefits

The LP method can be used as an operations-planning tool for existing steam systems to minimize operating expenses. This is especially true when one or more power demands can be satisfied by either electric motors or steam turbines. Maximum savings can be realized when there is significant flexibility in steam-header temperatures and deaerator pressure.

Incremental investments for existing steam systems can be studied by the LP method. Additions to an existing system can include new primary boilers, superheaters, desuperheaters, condensate coolers, turbine drives, etc. By including the total system in the LP model, the indicated savings due to an investment will not disappear because of unforeseen effects in parts of the system that are not directly associated with the new investment. Furthermore, we can be sure the LP procedure will adjust all aspects of system behavior to obtain maximum benefit from a new investment.

The greatest potential benefits of the LP method are in design engineering. For preliminary design, it is sufficient to minimize total steam or fuel consumption. For detailed design, the objective function should include all operating expenses and the daily capital-recovery plus return on investment of major equipment items. Principal benefits are derived from the ability of the method to choose the major design-parameter values that produce a truly optimized system. The design parameters include:

1. The ability to choose between turbine and motor drives, and/or between several turbine options. The incremental cost of primary-steam generation is included in such decisions because all steam boilers are part of the LP model, and because the total system cost is minimized.

2. Steam-header and condensate temperatures are problem variables and are part of the optimum solution. The design engineer need not estimate values for these variables to satisfy heat- and material-balance requirements. The procedure will select optimum temperatures so that all heat- and material-balance requirements are satisfied.

3. All inequality constraints, such as maximum or minimum temperatures, Btu demands, flowrates, etc., will be satisfied. This is a unique property of mathematical programming procedures. If the LP model is feasible, the LP-solution procedure will find a feasible and optimal solution. A lower-cost design is not possible, given the same problem definition.

4. Letdown-steam rates and desuperheating requirements will be determined by the LP solution. The design engineer need not "approximate" the requirements and recalculate all system heat-and-material balances to see whether the system is still feasible.

5. Cooling-water requirements and heat-exchange duties are determined as part of the LP solution.

6. Deaerator pressure is a system variable and becomes part of the optimum solution.

It is apparent that problem definition is more flexible for the LP method than for the traditional heat-and-material-balance simultaneous-equation approach, when the same level of economic sophistication is included. The reason is that estimates for heater temperatures, motor vs. turbine decisions, letdown flows, deaerator pressure, etc., are not required. The design engineer is able to concentrate on the demands of the overall system, its availabilities, and its upper and lower bounds.

References

1. Dodge, R. D., Gordon, E., Hashemi, M. H., and LaRosa, J., A Steam Balance Program Which Also Handles Combined Power Cycles, *Chem. Eng. Prog.*, July 1978.
2. Ruggerie, M. T. V., Automate Steam Balance, *Hydrocarbon Process.*, July 1977.
3. Slack, J. B., Steam Balance: A New Exact Method, *Hydrocarbon Process.*, Mar. 1969.
4. Bouilloud, P. H., Computer Steam Balance by LP, *Hydrocarbon Process.*, Aug. 1969.
5. Arnetin, R., and O'Connell, L., What's the Optimum Heat Cycle for Process Utilities?, *Hydrocarbon Process.*, June 1968.
6. Dantzig, G. B., "Linear Programming and Extensions," Princeton University Press, Princeton, N. J., 1963.
7. Gass, S. I., "Linear Programming Methods and Applications," McGraw-Hill, New York, 1958.
8. Zangwill, W. I., "Nonlinear Programming, A Unified Approach," Prentice-Hall, Englewood Cliffs, N. J., 1969.
9. Schmidt, E., "VDI Steam Tables," Springer-Verlag, New York, 1963.

The authors

James K. Clark, Jr., is a principal process engineer and systems analyst for Fluor Engineers and Constructors, Inc., 3333 Michelson Drive, Irvine, CA 92730. He joined Fluor in 1974 and has worked with and developed computer programs for various processes. Previously, he spent six years with Union Carbide Corp. as a senior process control engineer. He has a B.S. in chemical engineering and an M.B.A. from the University of Southern California.

Norman E. Helmick is chief of systems engineering for Fluor Engineers and Constructors, Inc., Irvine, CA 92730. He has over 29 years experience in chemical engineering, operations research and systems analysis, including extensive experience in developing linear programming models. He has a B.E. in chemical engineering from the University of Southern California and is licensed as a professional chemical engineer in California. He is a member of ACS, The Institute of Management Science, and Operations Research Soc. of America.

First Steps in Cutting Steam Costs

How to install primary flow-sensing devices in steam pipelines provides the means for accurately measuring steam flow in the process plant.

DONALD L. MAY, Monsanto Co.

Steam for process heating and turbine drives plays a vital role in today's chemical process plants. The rising cost of fuel for steam boilers has made it necessary to take steps wherever possible to reduce steam usage. Before this can happen, it is necessary to install adequate instrumentation to measure steam flow to each process. Then process revisions that are intended to reduce steam consumption in the plant can be properly evaluated for their effectiveness.

Several methods for steam-flow measurements can be set up in a chemical plant to initiate a cost-reduction program. It is not practical to achieve high accuracy (±¼%) at each flow location. The methods shown here are economical and will provide ±1.5 to ±2% accuracy at each flowmeter. Because of the averaging effect of the ±2% errors at each flowmeter, the readings from these instruments can be added up to provide an accounting system of total steam flow with an overall accuracy of ±5%.

Other aspects to be discussed here for these procedures include the effects of temperature and pressure changes on steam-flow measurements, cost-saving ideas, details of installations, and potential problems with installations.

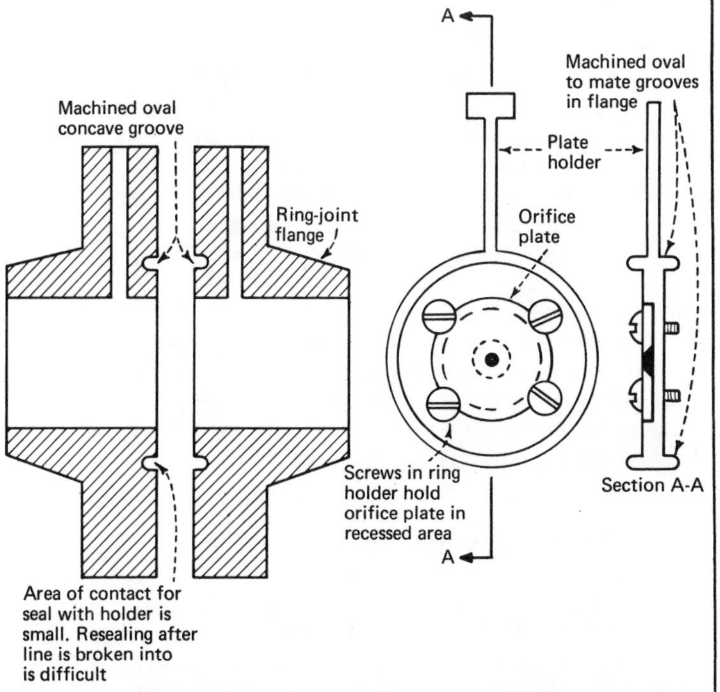

RING-JOINT flanges support orifice—Fig. 1

Originally published November 12, 1973

Temperature and Pressure Effects

The accuracy obtainable from an orifice installation in steam is directly related to how constant the steam temperature and pressure are maintained. The formula for the correction factor is:

$$w_a = w_r \sqrt{v_d / v_a} \qquad (1)$$

where w_a is actual flow, w_r is recorded flow, v_d is specific volume at design temperature and pressure, and v_a is specific volume at the actual temperature and pressure.

Long pipelines, inadequate insulation and large variations in flow (0 to 100,000 lb./hr.) may cause the pressure and temperature at the point of flow measurement to be quite different from where they are monitored. Therefore, it is recommended that a pressure-tap nozzle and thermowell be added within 100 ft. of the orifice installation. This will allow periodic checks of temperature and pressure changes caused by process-load variations or seasonal weather conditions. Correction factors can then be supplied monthly or quarterly to the flow readings. If the pressure or temperature varies enough in a 24-hr. period to cause a flow error of 3 to 4%, it is recommended that it be recorded continuously and compensated for daily.

Selection of Flanges and Gaskets

Steam flow to be measured with orifice plates may be done by the use of orifice flanges, or pipe taps and vena-contracta taps with regular flanges, or ring-joint orifice flanges. When mechanics are installing ½-in. taps on a 10-in.-dia. line located in a pipeline 20 ft. aboveground, it is very easy to be in error on some of the dimensions or the elevation. A better installation can be made by using weldneck orifice flanges. In this case, the holes for the taps are precision drilled, and both holes can be exactly aligned by bolting the flanges together before welding is started. This method will allow both legs to be on the same elevation when the impulse lines are extended in the horizontal plane.

Ring-joint orifice flanges were often used on high-pressure steam (650 psig., and up). Fig. 1 shows a ring-joint holder and flange. The plate fits inside the holder, and the holder fits between two ring-joint flanges. The oval-shaped outer ring of the plate holder is squeezed into a concave groove in the flange surface. Thus, no gasket is required.

However, experience has shown that a ring-joint installation guarantees a good seal the first time only. The area of contact between the oval ring and the flange is quite small. It is difficult to get the flanges to seal off again once the line is broken into. Sometimes, it is necessary to remove the flanges from the pipe, and remachine the sealing groove.

For these reasons, ring-joint orifice flanges are no longer recommended for high-pressure steam service. Instead, the raised-face weldneck orifice flange is now recommended when used with an asbestos-filled Type 304 stainless-steel spiral-wound gasket.

The following is a typical set of ordering specifications and prices for ring-joint and raised-faced flanges:

Ring-joint flange — Price

1. 8-in., 600-psi. ring-joint weldneck orifice flanges, carbon steel, for use in Schedule 20 pipe $296.74
2. Handle-type ring-joint orifice-plateholder, 8-in., for use in 600-psi. flanges, with hold-down screws for 8-in. orifice plate....... 39.60
3. 8-in. orifice plate, 1/8-in. thick, material Monel, plate O.D. 8.437 in., for use in 600-psi. ring-joint holder, bore 4.141 in.... 65.00

Total ... $401.34

Raised-faced flange — Price

1. 8-in., 600-psi. weldneck raised-face orifice flanges with bolts, nuts and gaskets. Gasket material to be asbestos-filled, Type 304 stainless-steel spiral wound. Flange material to be carbon steel for use in Schedule 20 pipe.......... $263.43
2. 8-in. orifice paddle, 1/4-in. thick, 600 psi., material Monel, plate O.D. 12 5/8 in., bore 4.141 in. 130.00

Total.... $393.43

The price of the materials is virtually the same for either type. However, the weldneck raised-face flanges with the asbestos-filled spiral-wound gaskets are, by far, the best installations. If it is necessary to remove the orifice plate for any reason, the flanges may be easily resealed by installing a new set of $8 gaskets.

When planning steam-flow installations for pressures above 300 psig., it is necessary to check closely the temperature and pressure specifications of all materials. When considering thermowells, ask the vendor for a curve of temperature vs. pressure ratings. Some thin-wall thermowells will only withstand 300 to 400 psi. at 750 F. Standard differential-pressure-cell bodies are generally limited by the gasketing to a temperature rating of 400 F.

If the meter impulse-lines are steamed out and the steam is left on too long, the cell body could reach temperatures approaching the line temperatures. High-temperature gasketing is available from the cell manufacturer.

Pressure ratings for the transmitter's body should be reviewed before a brand is selected. Most cell manufacturers can furnish meters with body ratings of 1,500 psig. for measuring a differential of 20 to 800-in. water column. But very few transmitters designed for measuring less than 20-in. water differential are rated for more than 500 psig.

Installation Details

With the exception of flow-switches (shown later in Fig. 3), the steam-flow measurements discussed here all use two impulse lines from the measuring element in the pipeline to a differential-pressure meter located below the pipeline. The steam in the impulse line condenses, and the meter actually measures the difference in head pressure on the two liquid legs. Fig. 2 shows the details of a 640-psi., 750-F. steam-flow installation. If the impulse lines are 10 ft. or more in length, no condensing pots are required. Note that the liquid legs are traced with 25-psi. steam and insulated to prevent freezing.

The insulation kit shown in Fig. 2 is a snug-fitting enclosure that is commercially available for standard differential-pressure meters. The kit is built in two halves that fit around the cell body, and it is retained by a strap. No liquid caulking is required, and the kit's appearance is always neat. It is easily removed for meter calibration or inspection, and can be reused. Initial material cost ($50) is higher than asbestos wrapping and caulking ($15), but savings are derived from initial labor costs, and the reusable aspect. The insulation kit is considered an integral part of the differential-pressure cell and, therefore, the instrument mechanic does not require the assistance of an insulator.

If ambient temperature will permit, the differential-pressure cell should be put into operation and closely inspected for leaks before installing any insulation. A small leak in the low-pressure leg will make the meter read high. A leak in the high-pressure leg will make the meter read low, or even make it go all the way to zero.

Flow Switches Measure Excess Steam

Some steam-driven turbines require 175-psi. steam input, and exhaust 25-psi. or some other low-pressure steam into a steam header. Occasionally, the turbines require higher than normal amounts of 175-psi. steam, which results in an excess of 25-psi. steam. In such cases, it may be necessary to vent the 25-psi. steam to atmosphere. Fig. 3 illustrates how inexpensive flow-switches can be used to measure steam being vented to the atmosphere. The flapper-type flow-switch can be added to the steam vent line by welding in a 2-in. Thredolet coupling. Also, a second pen can be added to an existing recorder that is used to measure the 175-psi. steam flow to the turbine. The flow-switch is set up as an on-off device to give a fixed pneumatic output to the second recorder pen if any flow goes through the vent line.

Personnel who read the charts should be instructed that whenever the second recorder pen is reading above zero, steam is being vented. The actual value of 25-psi. steam being vented will always be the same as the mass flow (lb./hr.) of 175-psi. steam input. Flow-switches can be installed for approximately one-third the cost of orifices.

Deluxe Pitot Tubes

Another practical method of measuring steam is to install a deluxe pitot tube. This tube, sometimes referred to as an Annubar, is quite different from the standard pitot tube. It covers a larger area of the pipe, contains more measurement holes, and therefore produces greater accuracy. While a standard pitot tube yields accuracies within ±2 to 5% of flow, an Annubar can produce accuracies within ±0.55 to 1.55%.

Fig. 4 shows how an Annubar goes across the pipe to measure each cross-section of flow. The inner tube produces an average of the flowrates seen across the pipe by each measuring hole. A high pressure is created at the upstream holes, which aids in preventing them from being plugged by pipe scale or other particles.

The averaging-type pitot tube should be considered:

1. When pressure drop is critical. The measuring element is small in diameter and creates a negligible amount of pressure drop.

2. On large existing pipelines. Rigging, scaffolding and cranes are generally required to hold up a pipe 8-in. and larger in diameter while the pipe is cut and orifice flanges are welded into place. Installation of the deluxe pitot tube only requires that a ½ to 1¼-in. coupling be added to the side of the pipe. On large pipes, this device can be installed for less than one-half the labor cost of orifice flanges.

3. On existing lines that cannot be taken out of service. Another version of the deluxe pitot tube can be installed through a 1¼-in. valve and coupling that can be added to the pipe with a "hot tap."

Fig. 5 shows one commercially available hot-tap Annubar installation for 10-in.-dia. pipe. The special flanges, long bolts and double nuts

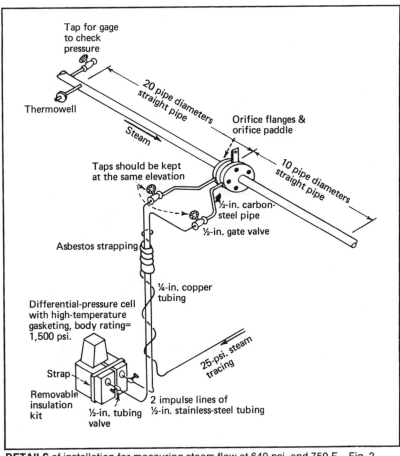

DETAILS of installation for measuring steam flow at 640 psi. and 750 F.—Fig. 2

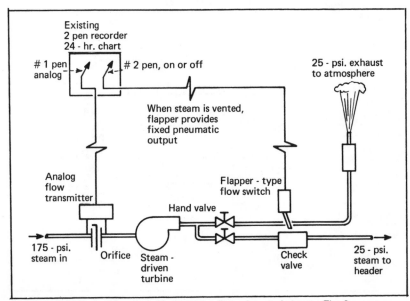

FLOW SWITCH provides signal to measure flow of vented steam—Fig. 3

are supplied with the unit. These are necessary to safely pull the unit through the valve while the steam line is still in service. The final installation leaves hardware 44 in. above the pipe. To install and remove the measuring element requires an additional 24 in. Therefore, there must be approximately 6 ft. of free space available at the side of the pipe to be hot tapped. The hot-tap machine works best when used on the top side of the pipe.

Because the pressure drop created by Annubars is low, there is generally a low differential pressure available for measurement. If the pipe size is large and the steam flow is low, the use of an Annubar may not be practical. For example, an 8-in. pipe carrying 450-psi. steam at 18,000 lb./hr. produces only about 1.5 in. water differential. For any full-range steam flow that creates an available differential pressure of less than 5 in. water column, this unit should not be used. An alternate solution to such a situation might be to install a smaller run of pipe at the point of flow measurement. This would increase the velocity of the steam and the differential available, and allow the use of an Annubar.

Problems and Solutions

Installation of Annubars on 175-psi., and lower pressure steam lines will present very few problems. Minor leaks may occur at the packing gland for the measuring element or the screwed fittings a few weeks after installation. These leaks can easily be prevented by using wire-core asbestos packing in the packing gland, and applying a high-temperature sealer to the screwed fittings.

Installing Annubars in 640-psi. and higher pressure steam lines is not so easy. The Annubars may be rated for 1,000-psi service but can only be bought with screwed fittings at the impulse lines. Screwed fittings are never sufficient to hold high-pressure steam and will develop leaks. These leaks can be prevented by welding the im-

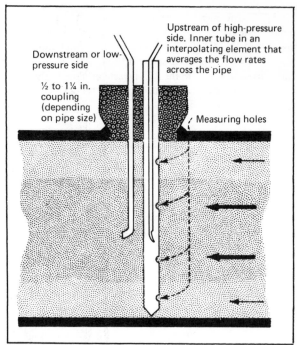

PITOT tube yields average flowrate across pipe—Fig. 4

pulse lines to the unit, but such action makes the instrument difficult to remove for repairs.

Fig. 6 shows the modifications recommended so that the device can be used on high-pressure steam. These include welding socket-weld valves, short nipples, and flanges to each impulse line. These modifications will prevent leaks at the screwed fittings but still provide a method of removing the unit. Should a leak occur at the flanged fitting, it can be repaired by temporarily shutting off the socket-weld valve. On hot-tapped installations, these socket-weld valves are necessary to shut off the

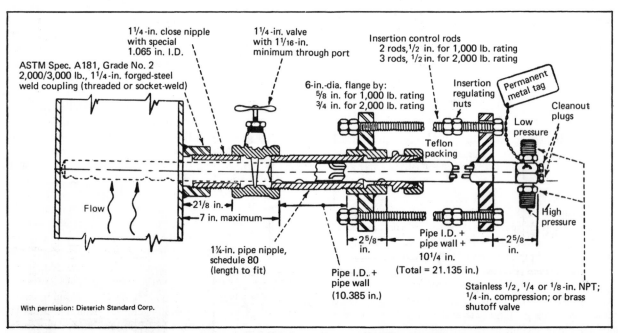

HOT-TAP installation of Annubar for measuring steam flow (details are for 10-in. dia. Schedule 40 pipe)—Fig. 5

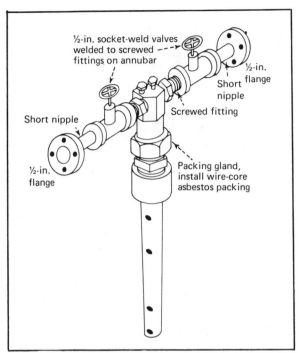

MODIFICATIONS to Annubar for 640 psi. steam—Fig. 6

steam while the measuring element is jacked in or out of the larger gate-valve.

The hot-tapped Annubar (Fig. 5) is somewhat unique since the measuring element fits through a valve. Therefore, mechanical, instrument and operating personnel installing or working around Annubars should be instructed on how the element must be clear before the valve can be closed. If any attempt is made to close the valve while the element is still in the pipe, the hollow tube can be distorted or even flattened.

Potential Hazards With Steam

Steam leaks at any pressure are hot and can cause burns. Steam leaking under high pressure is invisible until it condenses in the atmosphere. A 1,200-psi. steam leak can cut in two a 2-in. × 6-in. piece of lumber. Personnel assigned to install steam-flow instrumentation should be fully trained on the hazards of steam.

Several crews will often be involved during the installation of orifices. The instrument mechanics may install the orifice plate between cold flanges on the ground. Pipe fitters may weld the flanges into the steam line. Then plant operators may put the steam line into service. In such cases, each crew's work should be closely coordinated and individual responsibilities clearly defined to prevent mishaps. For example, the heat from steam causes expansion of metals. Therefore, tubing connections and bolting, made when cold, should be re-tightened after the pipe reaches operating temperature.

Should an orifice for steam measurement be placed into service with the orifice-flange bolting not properly tightened, one or both of the following is possible: (1) the line may develop a large leak at the flanges, or (2) the orifice plate may become loose between the flanges.

A loose orifice plate can be vibrated, in place, by the

force of the steam. These vibrations can cause a Type 316 stainless-steel plate, ⅛ in. thick, to crack, warp, or even break into pieces. Any of these mishaps would affect the accuracy of the steam-flow measurement at that orifice. But more important is that leaking steam or fragments of metal leaving the flanges might cause injury.

Final Corrections for Steam-Flow Accounting

If total plant-steam-flow accounting is desired, distribution meters should be installed on steam headers leaving the boilers or powerhouse, to act as standards for the accounting system. This will allow a monthly comparison of the totals for all user and distribution meters.

Following installation of several meters, the percent error between the total of all user meters and the total of distribution meters may not be acceptable. By using a portable temperature instrument and pressure gage at each flow installation and reviewing all existing meter calculations, most of the errors can probably be found.

For example, the hole size in an orifice plate may have been calculated for 175-psig. saturated steam at 377 F. (for which the specific volume, v_d, is 2.40 cu. ft./lb.), while the steam may be discovered to be still superheated at 175 psig. and 425 F. (for which v_a is 2.59 cu. ft./lb.). Using Eq. (1) and the appropriate values for this condition, the correction factor would become:

$$w_a = 0.96w_r$$

As another example, for an orifice plate calculated for 377 F. and 175 psig. (v_d is 2.40 cu. ft./lb.) but found to be 425 F. and 160 psig. (v_a is 2.84 cu. ft./lb.), the correction factor would be:

$$w_a = 0.92w_r$$

Where possible, the orifice-plate hole sizes should be recalculated for the new operating conditions, and new orifice plates installed. Of course, this requires shutting down those steam lines. If this is not practical, the differential-pressure head calibrations required on the flowmeters should be recalculated and changed. If the meters have a limited fixed head (100 to 200-in. water column, etc.), or are already at the limit of their range of calibration, a tag bearing the correction factor can be attached to the flow-recording chart case.

With the applications of these corrections, the total steam flow shown by the distribution meters can be set up to agree with the total of all the meters showing steam used, with an accuracy of ±4 to 5% each month.

Meet the Author

Donald L. May is a senior instrument engineer with the Textiles Div. of Monsanto Co., P.O. Box 1507, Pensacola, FL 32575. His duties are systems engineering and instrument applications for improving process control in the utilities, chemical and textile plants. He has a B.S. in electrical engineering from Auburn University. He is a registered professional engineer, a member of the Institute of Electrical and Electronic Engineers, and a senior member of the Instrument Soc. of America.

Are your steam traps wasting energy?

Stafford J. Vallery, E. I. du Pont de Nemours & Co.

It would be a rare chemical process plant that has no steam traps, but most chemical engineers know little about these devices. (The latest edition of Perry's "Chemical Engineers' Handbook" has no mention of them.) So they are generally ignored, not maintained, and, as a result, waste enormous amounts of energy. The three articles of this report will tell you how to stop this waste in your plant.

Steam traps—the quiet thief in our plants ....p. 171
A steam-trap checking program ....p. 173
Setting up a steam-trap standard ....p. 178

Originally published February 9, 1981

170

Steam traps— the quiet thief in our plants

In the first year of setting up a steam-trap program, one dollar spent on upgrading traps can return more than three dollars in energy savings.

☐ A malfunctioning steam trap is a great waster of energy. Almost any large plant in the chemical process industries has some (often many) energy-wasting traps.

The cost of energy is usually a large percentage of the cost of manufacturing. In most petrochemical plants, for example, it is the second most costly item (the greatest expenditure being for hydrocarbon raw materials). This energy portion of the cost of manufacturing ranges from 8 to 25%. Fuel used for producing steam is usually the major part of this. And most plants can save 10 to 20% of fuel costs simply by having a formal, active steam-trap program.

To carry this out requires someone on the plant's staff who is interested in energy, has a technical background (preferably an engineer), and can devote sufficient time to develop a simple, yet effective, program.

For field work, this engineer will need one mechanic or pipefitter for each 3,000 operating steam traps. For the first year, a return of $1 million in energy saving for each $300,000 spent in upgrading the system is the rule rather than the exception.

(For estimating purposes, you can assume that a modern petrochemical plant will have about one steam trap for each 100 lb/h of steam consumed.)

How to get started

The road to an effective steam-trap energy-saving program can be made much easier by following these six steps.

1. Select a steam-trap energy coordinator.
2. Develop a plant standard of four or fewer approved steam-trap types. (See the third article of this report, on p. 92.)
3. Monitor steam consumption for your different products (pounds of steam consumed per pound of product produced).
4. Go into selected plant areas, change out *all* steam traps and monitor steam savings.
5. Develop and implement a steam-trap checking procedure to detect bad traps (see the article on p. 87) and change them out using your plant trap standard.
6. Standardize on a set trap-checking frequency, say

every six months, with plant-staff followup to make sure the program is working.

These six steps can be divided into two sections—Steps 1-4 are designed to solve *misapplication*, and Steps 5 and 6 are designed to eliminate *neglect*. Recognizing the possibility of trap misapplication and neglect goes a long way toward solving trap problems.

Energy coordinator

Selecting the right person as the steam-trap energy coordinator is the key to the success or failure of an effective steam-trap program. This person should be a senior (at least five years of plant service) employee who has a keen interest in energy saving. A desire to learn is more important than previous steam-trap experience.

Steam-trap manufacturers provide schools or seminars on steam traps—types, failures, applications, etc—and all are anxious to tell their stories. The local steam-trap representative can arrange for attendance at such classes. Also, various state, local and engineering society groups have knowledgeable people who will usually be glad to assist you by sharing their steam-trap experience. Such experience can be more valuable than vendor information because of the sometimes conflicting claims from the various trap manufacturers. An important thing to remember is that the person selected for this assignment should be free for at least a year to devote *full* time to steam traps and be willing and technically able to learn and understand steam-trap types, applications, failures, testing, and the like.

Establish a cost base

One of the best ways to show the effectiveness of your steam-trap program is to choose an area of 150 or more traps that does not now receive routine steam-trap maintenance. Instrument the steam flow to this area so that steam consumption can be totaled. Operate the area at "average" production or steam consumption for a reasonable time—say a month. Then inspect each trap and record the percentages of cold, properly working, and bad (blowing steam) traps.

Having done this, change out *all* traps, good as well as

bad ones, to the newly established plant steam-trap standard. Then measure the steam consumption for the same time, again say a month, and show the "average" saving per steam trap. This will normally convince the most skeptical manager of the effectiveness of a steam-trap program. You will also probably notice that the steam plumes from venting steam traps disappear in this area—a more visible sign of the program's effectiveness.

It is not uncommon for the average saving to be at least 10 lb/h per steam trap for systems whose pressure is under 500 psig. This is an approximate energy saving of $300 per year for each trap.

Total plant changeout

Armed with the success of the test area, you can now expand to the entire plant. Select a crew of plant pipe-fitters (maintenance mechanics) to be the steam-trap repair and replacement crew. These again should be energetic and knowledgeable field people. One field person for each 3,000 steam traps that are easily accessible will be sufficient. Your ratio may change to one field person for each 1,000 steam traps if the traps are scattered, not accessible, or in very poor condition initially. A crew of six workers can replace about twelve traps per day. So for a plant containing 3,000 traps, it will take about a year for a complete changeout. Newly installed steam traps should conform to your plant standard, which should address itself to easily accessible and maintainable steam-trap "banks."

Steam-trap checking procedure

During this changeout time, a comprehensive steam-trap checking program should be formulated by the plant energy coordinator. Each trap should be numbered and listed on both a steam-trap map and a steam-trap list. These will be the guides used to find each trap during the routine six-month test. The "trap map" should show a plot plan of each process area, with the traps numbered in letter groups of 50 or fewer traps. Example:

        Area A . . . . . . 1A through 39A
        Area B . . . . . . 1B through 54B
        Area C . . . . . . 1C through 47C

        Total traps in area   140

The "trap list" should have each letter trap-area on a single sheet (using both front and back). This makes expanding the listing of traps in each area easy, by either adding to an existing letter group or creating a new one. Information on the "trap list" should be:
- Trap number (1A, 5B, etc.).
- Service description (instrument tracing, column heater, etc.).
- Trap condition (cold, good, etc.).
- Operating pressure (10 psi, 50 psi, etc.).
- Type trap from the plant standard.

Each trap should then be permanently marked with a stainless steel tag on its condensate piping with the trap's respective number (1A, 5B, 47C, etc.). This allows easy trap identification when using the map and list. Note that the tag is attached to the piping—not the trap itself. Thus, if a defective trap is replaced, the number remains in place and does not have to be changed from the old trap to the new one.

Routine checking frequency

With all traps changed out to good ones and a plant standard established for specifying replacement traps, the system is ready for the first semiannual checkup. After the first year, a normal trap failure rate of less than 10% can be expected. For 600-psig service, you may have a 70% failure; but for a 60-psig service, a 2% failure is more likely with properly selected traps.

A normal six-month check frequency is recommended initially for all traps. This should be done in January (winter conditions) and June (summer conditions). The six-month routine check of every trap in the plant will not require large manpower expenditures. A skilled trap checker and a helper can check about 500 traps a day. So for a crew of two with a plant of 3,000 traps, only six working days (twelve man-days) are used. At an average of $100/man-day, the cost per check for each trap is only 40¢. One trap found and corrected from blowing only 40 lb/h will pay for all 3,000 traps being checked.

It is very important to use the same people to check the traps each time. Their knowledge from repeating checks will find and solve some of the difficult trap problems. It will also guarantee uniformity in checking. Most plants have found that a set trap-checking crew can check at least twice as many traps effectively in a given time as a crew whose permanent assignment is not traps.

Here is an example of how the checking would work. For a plant with 3,000 traps, all are checked in January by the permanent trap mechanic and a helper. This can be accomplished in less than two weeks. Then, during the next two months, the lone mechanic will replace all "bad" traps. For the remaining three and a half months, the trap assemblies that were changed out are disassembled, and new assemblies made up for use during the second six-month changeout.

A typical timetable for a 3,000-trap, single-mechanic system may look like this:

Task	Date
Winter steam-trap check	Jan. 8-Jan. 16
Replace bad 600-psig traps	Jan. 17-Feb. 14
Replace bad 425-psig traps	Feb. 15-Feb. 29
Replace bad 310-psig traps	Mar. 1-Mar. 7
Replace bad 190-psig traps	Mar. 8-Mar. 23
Replace bad 130-psig traps	Mar. 24-Apr. 6
Replace bad 60-psig traps	Apr. 7-Apr. 19
Replace bad 25-psig traps	Apr. 20-May 7
Repair trap assemblies	May 8-July 7
Summer steam-trap check	July 8-July 16

This completes the cycle and starts it over again.

Conclusion

There are four essentials to an effective trap program:
1. Select the right person to head the program.
2. Set up a plant trap standard.
3. Establish a routine trap-checking procedure.
4. Set up a permanent trap-maintenance crew.

These few people will save a minimum of $100,000/year/person in energy cost. Your plant will also be quieter, have fewer steam plumes, and have greater utility—all from a simple yet effective steam-trap program.

In-plant steam-trap testing

A planned check of all the plant's steam traps—at regular intervals—will pay off in energy savings.

☐ In many plants, most of the steam traps are actually steam wasters. The higher the price of energy, the more important it becomes to bring these losses under control. Following is the procedure used in our plant.

AREA CFA	BLDG. 11	FLOOR 1st	DATE		
NO.	DESCRIPTION		CONDITION	STM. PRESS.	TRAP
1A	Cond. tk. vent inst.			35	I
2A	Regeneration steam heater			500	II
3A	" " "			"	"
4A	H₂O recorder trap			35	I
5A	Process line tracers			35	I
6A	" " "			"	"
7A	" " "			"	"
8A	" " "			"	"
9A	" " "			"	"
10A	" " "			"	"

Fig. 1—In order to systematize steam-trap testing, you must first know what steam traps there are in your plant, and where each is to be found. Survey the plant and group the traps at each location into lots of 50 or fewer. Give each trap an identification number, and include it on a list. This list also shows trap service (tracing, drip leg, etc.), pressure (in psig) and a code for the type and size of trap.

Fig. 2—If your plant has standardized on a few types of traps, you can save space on the trap list by giving each type a code number. These are the codes for Du Pont's Victoria plant.

CODE	DESCRIPTION	SERVICE
I.	Bucket trap [Model____]	Light loads Low pressures
II.	Disk trap [Model____]	Light loads High pressures
III.	Bellows trap [Model____]	Moderate loads All pressures
IV.	Level chamber and control automatic	Heavy loads All pressures

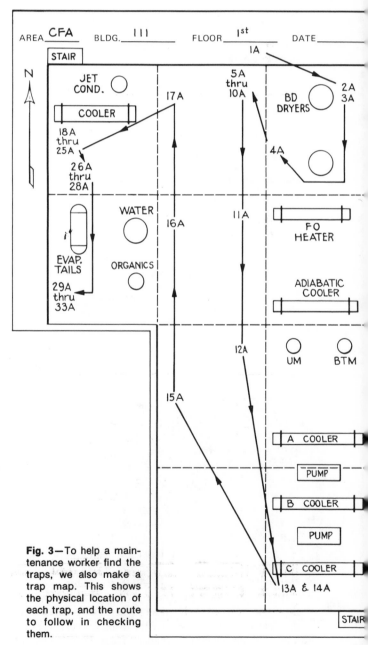

Fig. 3—To help a maintenance worker find the traps, we also make a trap map. This shows the physical location of each trap, and the route to follow in checking them.

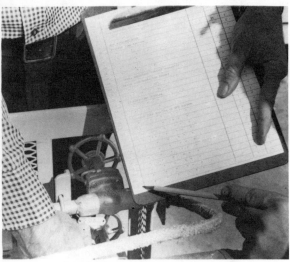

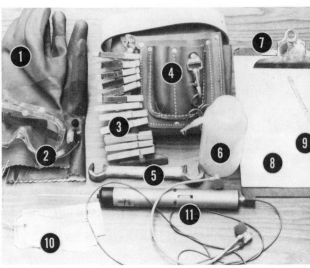

Fig. 5—Here are the tools and equipment used in checking the traps:

1. Gloves
2. Goggles
3. Spring-type clothespins (colored and plain)
4. Carrying pouch and belt
5. Valve wrench
6. Water-filled squeeze bottle
7. Clip board
8. Trap list and trap map
9. Pencil
10. "Maintenance required" tags
11. Ultrasonic sound detector.

Fig. 4—Use the trap list and trap map to make up stainless steel marker tags, and attach them as shown in the photo. The tags are attached on the trap's condensate side next to the guard valve. This location minimizes burn hazards and prevents loss during trap changes.

Fig. 6—Everything except the clipboard and lists is stored in the leather pouch. When not in use, the ultrasonic tester goes into the pouch forward of the water bottle.

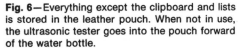

Fig. 7—Wet testing is done first. A few drops of water are squirted on the trap. The water should start to vaporize immediately. If it does not, this indicates a cold trap.

Fig. 8—A cold trap is marked by placing a colored clothespin on the trap's drain-valve handle. Wet-test each trap in the immediate area and mark those that are cold.

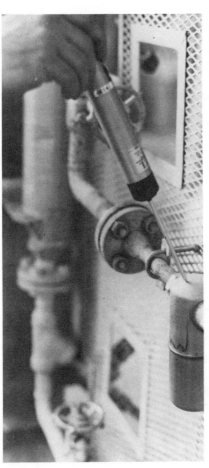

Fig. 9—Go back to the first trap and give it a sound test. When flow passes through a trap, an inaudible ultrasonic vibration is generated—the ultrasonic detector changes this to a frequency we can hear. A *bucket trap* (as shown here) should be relatively quiet, or cycle on and off at regular intervals (determined by condensate load and steam pressure). A ringing sound made by the bucket hitting against the trap wall tells that the trap is blowing. This sound is audible, but the ultrasonic detector will help pinpoint the noisy one in a bank of traps.

Fig. 10—*Disk traps* are checked by touching the detector directly to the top of the trap. A good trap cycles every 6 to 10 seconds. Cycling faster than once per second is called "machine-gunning," and indicates sufficient steam loss to merit replacement. A hot disk trap that does not cycle at all may have failed in the open position.

Fig. 11—On *bellows traps*, listen to the outlet piping. These traps should cycle or throttle depending on the condensate load. Ones that never close are either under a very heavy load or are blowing steam.

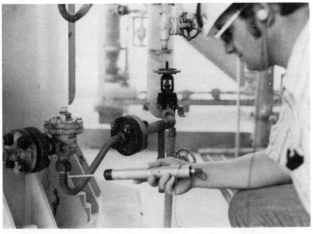

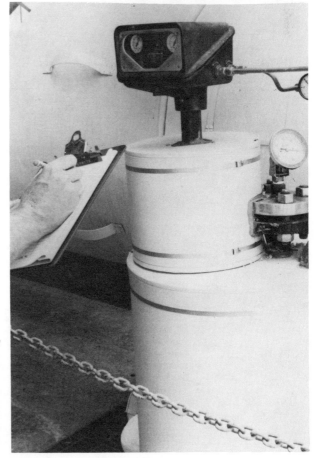

Fig. 12—For *level chambers and control automatics*, there is no way to use the ultrasonic detector, because a properly functioning system is always open, throttling the condensate flow. Such traps are usually maintained by the instrument repair group, separate from the normal trap-maintenance program. Record the percentage level as shown by the level indicator.

Fig. 13—Mark any "noisy" trap with an uncolored clothespin. Continue the ultrasonic testing of all traps in the immediate area.

Fig. 14—Go back to the first trap in the group, and check it by observation to determine what is actually passing through it. To do this, close the condensate valve and open the trap drain-valve. This bucket trap in 35-psig steam-tracing service is functioning properly.

Fig. 15—This bucket trap, also in tracing service with 35-psig steam, is blowing. Note the clothespin that shows it had been found to be defective with the ultrasound test. Often, small bucket traps that are blowing steam can be corrected by repriming. To do this, close the drain valve and leave the condensate valve closed for about 10 minutes. Condensate collecting during this time will sometimes reprime the trap.

Fig. 16—This is normal discharge for a 550-psig drip-leg disk trap. A large number of high-velocity water droplets are surrounded by flashing steam. With 550-psi condensate, about one third will flash into steam—do not confuse this with live steam blowing through.

Fig. 17—Here is a worn disk trap that is leaking steam during the off-cycle. Note the clear vapor next to the outlet pipe. *Caution:* With high-pressure steam traps, the condensate and steam exit at very high velocities when the trap is opened ("decked"). Maintenance workers should not deck traps larger than ½ in. without special permission, because of the large volumes of condensate that might be present.

Fig. 18—After all traps in a group have been given the water test and ultrasonic test and the noisy traps have been tested by observation, they are ready to be marked for maintenance. For each trap marked with a colored clothespin (cold traps), make out a "Maintenance required" tag marked "C" (for "cold") on the back. The production department should check to see why the trap is cold (supply cut off, plugged or crimped tracer, plugged trap, check valve stuck closed, etc.) and add this information to the tag, to guide the Maintenance Dept. in making repairs. Remove the clothespin after the tag is in place.

Fig. 19—Plain clothespins mark traps that are either blowing or wasting steam. The observation (decking) test determines whether the high sound level is from blowing steam or from a very heavy load. (Normally, traps in tracer or drip-leg service will not be subjected to such loads.) When the trap has been confirmed as bad, attach a "maintenance required" tag marked "B" (for "bad"). The Maintenance Dept. will replace these with new assemblies. Remove the clothespin after tagging the trap.

Traps may also be marked "D" for external trap damage (cracks, corrosion, etc.) or "L" for leaks at welds, fittings and the like.

Fig. 20—Now enter the appropriate code letter for each trap on the trap listing sheet:

B **Bad** trap, blowing live steam.
C **Cold** trap.
D Externally **damaged** trap.
G **Good**, properly working trap.
L Externally **leaking** trap.
W Trap is **wasting** steam, but not yet blowing.

Now recheck to see that all valves are back in the "as found" condition.

When the trap listing is completed, copies should be sent to a predetermined group. We circulate them to the area production supervisor, the appropriate mechanical foreman, the plant energy coordinator and the appropriate division energy coordinator.

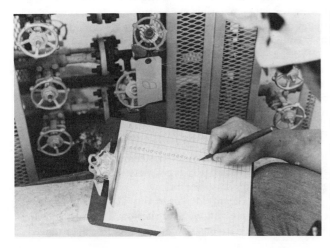

Setting up a steam-trap standard

Your plant will save money if you standardize on a type and size of trap for each of several pressures and condensate loads.

☐ The previous two articles have discussed how to set up an energy conservation program, to limit energy losses in steam traps by getting people involved at all levels. This article is directed at the hardware involved—"the right trap for the job." This basically means establishing a trap standard for all replacements and new installations.

Standardization is a vital part of any steam-trap program, to prevent premature failure caused by misapplication. Unfortunately, steam-trap misapplication is very common. Oversizing is probably the most usual malady. Far too often, steam traps are designed for heavy ($>$1,000 pph) condensate loads and then placed in services where the condensate load is very light ($>$50 pph). This keeps steam adjacent to a loosely fitting internal-trap-valve, resulting in live-steam loss.

To prevent such loss, a simple, effective steam-trap standard should be adopted and closely followed. The best way to gain acceptance and use of this standard is to: (1) keep it simple, (2) make it accessible, and (3) put it in the form of a drawing. This standard must be clear enough to be used by the field mechanic or foreman, the production supervisor, and the plant technical engineer.

A clear drawing is ideal because it can contain the most information in the simplest form.

Table I breaks down the various-sized steam traps by condensate loads for each pressure service: 94% light, 5% moderate, and 1% heavy. From this chart, you can see that the ideal trap standard would be to have three standard traps based on condensate loads alone. But the number of standard traps would not exceed nine if both pressure and condensate-load required separate traps.

Selecting traps for a standard

The problem here is which type of steam trap will give the best performance for each of the three condensate-load services—light, moderate and heavy. Trap testing and analysis was the method chosen for determining the best trap. Both existing process loads and trap test-stands were used in the evaluation process.

Fig. 1 shows one of the test stands used for evaluating various light-load steam traps. This test stand consisted of seven natural-convection steam condensers that provided a 6-lb/h condensate flow to the individual test steam-traps. Three-way valves were located downstream of each trap so that individual steam losses could be diverted for measurement. (This particular test installation was for 600-psig steam service.) A separate, smaller, stand was used for pressures under 100 psi. All testing of moderate- and heavy-load traps was done in the plant because of the large volumes of condensate involved.

The three main causes of steam-trap failures during testing were: (1) materials of construction, (2) weakness in steam-trap design, and (3) trash or solids buildup in the trap body.

The test work done by the author sifted the 38 different traps being used down to a standard of only four. These selected four trap systems are not ideal by any means, but are much superior to the multiplicity of types and sizes commonly used before the standard was adopted.

Selecting trap systems

Here is the basic way the four steam-trap systems were selected. All nine basic trap types were tested using the six general categories discussed below:

Steam Loss—Steam loss over the life of a trap was considered the most important criterion. Here, basic trap design, materials of construction, and size orifices played

Distribution of steam traps in a petrochemical plant		Table I
Total traps, by pressure, %	Total traps by pressure/condensate-load, %
9% 400 psig and up (High-pressure)	<1% Large condensate loads (>3,000 lb/h)
	1% Medium condensate loads (500-3,000 lb/h)
	8% Small condensate loads (<500 lb/h)
13% 100-400 psig (Moderate-pressure)	~1% Large condensate loads (>2,000 lb/h)
	2% Medium condensate loads (200-2,000 lb/h)
	10% Small condensate loads (<200 lb/h)
78% 5-100 psig (Low-pressure)	<1% Large condensate loads (>1,000 lb/h)
	2% Medium condensate loads (100-1,000 lb/h)
	76% Small condensate loads (<100 lb/h)

very important roles in steam loss. A trap valve that is erosion-resistant, just large enough to carry the required condensate load, designed to be tolerant of steam-line trash, and that has a steam loss, when new, of <1 pph was rated highest in this category. (See Tables II, III and IV for ratings.)

Materials of construction in steam traps have a dramatic effect on steam loss. The material around the flashing-condensate part of a trap should be stainless steel. It is common knowledge that flashing condensate is very erosive. If it is also slightly corrosive (and most condensate is), attack on carbon steel is certain. The inspection of several hundred steam traps that were removed from plant service (and those from the trap test stands), vividly points out the problems of carbon steel in steam traps. Fig. 2 shows how steam and condensate have bypassed the steam-trap valve by eroding the threaded portion of the carbon-steel trap cap. Fig. 3 shows examples of a safety problem—the carbon-steel threads were eroded away sufficiently to weaken the top of the trap to the extent that it would blow off while in service. In each case, the stainless-steel parts of the trap showed no signs of wear.

One carbon-steel trap on a 600-psig steam-header drip-leg was passing 4,200 lb/h of steam into a low-pressure collection header, which was vented to the atmosphere. This high volume of steam also contributed to additional steam loss by overpressuring the low-pressure header and caused traps that are sensitive to backpressure (disk and impulse types) to open prematurely and waste steam. Although a carbon-steel trap is the least expensive to buy, it normally is the most expensive to own!

Life—The life of a trap is governed by how long the valve system lasts, how long the body stays leakfree (both internally and externally), and how tolerant the trap is to steam-line trash. Tests show that all steam traps wear because they do not have a positive on/off action. Example: A disk trap appears to have an on/off valve actuation, but it actually does not. During condensate discharge, the disk does not fully retract in the valve cap, but flutters or wobbles very rapidly. All condensate has small suspended particles of scale and rust that the disk and seat constantly close on; this accounts for the faster wear rate on the moderate-load traps compared with the very-light-load traps.

Reliability—Trap reliability was determined by how well the steam trap responded when the condensate load or pressure changed. Occasionally, a trap would stay closed and back up condensate, or lock open. But most traps passed this test satisfactorily.

Size—This was a measure of the physical size and weight of the trap. Orientation of the trap was considered, but to a lesser degree.

Noncondensable Venting—All steam traps must have some way of discharging noncondensables, such as air. Most traps use a small leak somewhere in the valve design to do this (which results in some steam loss). The liquid-expansion design uses no direct gas venting but subcools the condensate enough so that most noncondensables will go back into solution. This will cause corrosion problems if the noncondensables include oxygen or carbon dioxide.

Cost—This was the last evaluation parameter, and is *least* important. When a ½-in. steam trap costs $25 but

Test stand used to evaluate light-load traps   Fig. 1

How steam eroded a carbon-steel trap cap   Fig. 2

Eroded carbon-steel threads pose a danger   Fig. 3

Criterion	Importance, range	Disk	Bucket	Impulse (piston)	Bellows	Float	Bimetallic	Orifice	Expansion	Instrumented
Steam loss	(0-10)	6	8	4	5	7	3	3	9	NA
Life	(0-8)	6	7	5	4	5	3	5	4	NA
Reliability	(0-6)	5	4	3	3	3	2	3	2	NA
Size	(0-3)	3	2	3	2	1	1	3	2	NA
Noncondensible venting	(0-2)	1	1	1	2	2	1	1	0	NA
Cost	(0-1)	1	1	1	0	0	0	1	0	NA
Small-trap types, total	**(30)**	**22**	**23**	**17**	**16**	**18**	**10**	**16**	**17**	**NA**

Steam traps for small condensate loads — **Table II**

wastes an average of 100 lb/h of steam ($3,000/yr), the initial cost of the trap becomes insignificant.

The ratings for each trap type are shown in Tables II, III and IV for the light, moderate and heavy condensate-load trap types. There was not a perfect "30" among any of the tested types. All steam-trap systems have their weaknesses, with the ones selected being judged best by the author. There are traps in each category that rated lower that undoubtedly would also make a workable choice. The important thing to remember here is to take a position and make a selection of the trap that appears to do the job well. Standardize on this trap for all applicable services and see that it gets used.

Adoption of the steam-trap standard will improve overall plant performance. It will:

■ Reduce energy loss through steam traps by minimizing misapplication.

■ Provide higher plant utility and product quality by yielding more-temperature-stable processes.

■ Lower trap installation costs by providing standardized trap stations.

■ Lower company stores cost through the stocking of 80% fewer steam traps.

■ Provide a nicer looking and safer plant by reducing steam plumes and wet spots.

Arranging traps in "banks" makes maintenance easier **Fig. 4**

Specifics of the steam-trap standard

As stated before, a drawing is the most useful form for a steam-trap standard. This single document can contain most of the necessary information for sizing, purchasing and installing steam traps. It is also easy to update and—since it is easily reproducible—it is simple and inexpensive for all plant areas to have copies.

Items that should be covered in the standard drawing:

■ Sizing-chart and instructions for use.
■ Drip- and tracer-trap-station layout.
■ Small-process-load trap layout.
■ Medium-process-load trap layout.
■ Large-process-load trap layout.
■ Steam-header drip-leg general design.
■ Condensate-header general design.
■ General notes.

The easiest way to describe a recommended "steam-trap standard" drawing is to show the one adopted for our local plant. The steam-pressure range of this standard covers 5 psig to 600 psig, and steam condensate flows from 2 to 80,000 lb/h. The piping schemes illustrated are a demonstrated minimum requirement for a safe, cost-effective installation. Arranging the traps in "banks" as shown in Fig. 4 is also recommended. The steam-trap models and sizes shown as standards for our petrochemicals plant may not necessarily be the best ones for the services specified, but they have provided satisfactory performance if monitored regularly on a 6-mo basis. Other trap types rating above "17" in the evaluation charts should also provide satisfactory service. The key is standardization and routine trap checks.

Trap-standard drawing

Steam-trap sizing chart—The sizing chart shown on the standard drawing should be made as simple and workable as possible. The sizing and selection chart is laid out so that only two pieces of information are needed: maximum operating pressure (psig), and maximum condensate load (lb/h). With only these two numbers, the chart tells you the trap size and type, and directs you to a recommended layout detail. This simple chart goes a long way toward solving the second greatest cause of steam loss—misapplication from oversizing. When setting up the sizing curve, be careful not to oversize the trap by more than 25%. It is very common for the field engineer to overestimate maximum condensate load for a given service. If, then, the

Steam traps for medium condensate loads   Table III

Criterion	Importance, range	Trap type								
		Disk	Bucket	Impulse (piston)	Bellows	Float	Bimetallic	Orifice	Expansion	Instrumented
Steam loss	(0-10)	6	7	5	8	7	4	2	NA	NA
Life	(0-8)	3	7	5	5	4	2	6	NA	NA
Reliability	(0-6)	4	4	4	4	4	3	3	NA	NA
Size	(0-3)	3	2	3	3	3	2	3	NA	NA
Noncondensible venting	(0-2)	1	1	1	2	2	1	1	NA	NA
Cost	(0-1)	1	1	1	1	0	0	1	NA	NA
Medium-trap types, total	**(30)**	**18**	**22**	**19**	**23**	**20**	**12**	**16**	**NA**	**NA**

sizing chart is also significantly oversized, a trap could be selected that may cause excessive steam loss.

Specifying each condensate-load group to use a different pipe size will make installing the wrong trap less likely. Example:

- Light loads—1/2-in. piping.
- Moderate loads—3/4-in. piping.
- Heavy loads—1-in.-and-larger piping.

Tracing and steam-main drip-pocket trap design details (Detail No. 1)—In Detail No. 1, the standard for a typical trap installation starts to take shape. A 15-in. standard trap assembly is established for easy removal and installation. The test valve is located on the removable part of the assembly, on the downstream side of the trap, because it is the valve most likely to leak and require replacing.

Normal trap maintenance is accomplished by removing the old trap assembly (piping, trap, drain valve, and flanges) and installing a complete replacement assembly. Then, back in the repair shop, the reusable parts of the removed assembly (piping, drain valve, and flanges) are used to make up a new assembly that is stored for later use. Use of standard piping sizes and dimensions makes trap changeout a 15-min, one-man assignment.

For many petrochemicals plants, the light-load traps—which includes tracing and drip-pockets—account for over 80% of all steam traps. These small devices in pressures above 100 psig should be very carefully monitored, with very close attention given to proper trap selection. It is strongly recommended that stainless-steel construction be used. Any leakage past the trap valve in these lightly loaded traps results in steam loss. Because steam conden-

sate can be both erosive and corrosive, a leaking trap nearly always gets worse.

Small-load-trap design details (Detail No. 2)—This system is very similar to the tracing and drip-pocket trap design above. Note that the design includes a second trap, in the form of an installed spare, for critical services. It should be emphasized that this is a spare trap and that only *one* trap should be "valved in" at a time. Running with both traps online invites early trap failure and steam loss.

Moderate-load-trap design details (Detail No. 3)—This covers about 15% of a petrochemical plant's traps. These are found on small unit-heaters, reboilers, and jacketed vessels where the steam condensate load is above 200 lb/h. Fast response, high reliability and dual traps are common installations. Because of the higher condensate loads, it is sometimes difficult to detect a steam loss of 100 lb/h when the condensate load is 500 lb/h. So, a high-reliability trap is also recommended for this service.

It is good engineering design to have a spare trap permanently installed on moderate condensate loads. This will greatly reduce equipment shutdown from steam-trap failures. Again, this is a *spare* trap assembly and should never be operated or sized as a dual-operation trap. If extra capacity is needed above the capacity of the standard single trap, the instrumented trap system as shown in Detail No. 4 should be used. Note that in all trap details the check valves are optional and should only be used when backflow of condensate is intolerable. An example of such a case would be a process that uses both steam and water alternately for heating and cooling. Check valves in

Steam traps for large condensate loads   Table IV

Criterion	Importance, range	Trap type								
		Disk	Bucket	Impulse (piston)	Bellows	Float	Bimetallic	Orifice	Expansion	Instrumented
Steam loss	(0-10)	4	5	6	NA	7	4	2	NA	9
Life	(0-8)	3	5	5	NA	4	3	4	NA	8
Reliability	(0-6)	3	4	4	NA	5	4	5	NA	5
Size	(0-3)	3	1	3	NA	2	2	3	NA	0
Noncondensible venting	(0-2)	1	1	1	NA	1	1	1	NA	1
Cost	(0-1)	1	1	1	NA	0	1	1	NA	0
Large-trap types, total	**(30)**	**15**	**17**	**20**	**NA**	**19**	**15**	**16**	**NA**	**23**

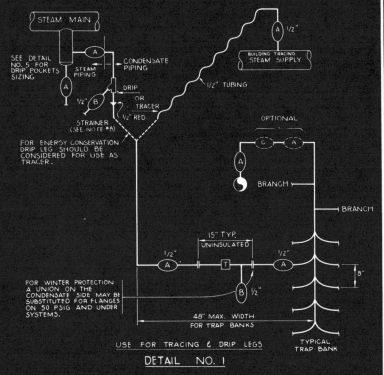

STEAM MAIN

SEE DETAIL NO. 5 FOR DRIP POCKETS SIZING

STEAM PIPING

CONDENSATE PIPING

½" B

DRIP OR TRACER

½" RED.

STRAINER (SEE NOTE #8)

FOR ENERGY CONSERVATION DRIP LEG SHOULD BE CONSIDERED FOR USE AS TRACER.

BUILDING TRACING STEAM SUPPLY

½"

½" TUBING

OPTIONAL

BRANCH

BRANCH

FOR WINTER PROTECTION A UNION ON THE CONDENSATE SIDE MAY BE SUBSTITUTED FOR FLANGES ON 50 PSIG AND UNDER SYSTEMS.

15" TYP. UNINSULATED

½" A

½" A

B ½"

8"

48" MAX. WIDTH FOR TRAP BANKS

TYPICAL TRAP BANK

USE FOR TRACING & DRIP LEGS

DETAIL NO. 1

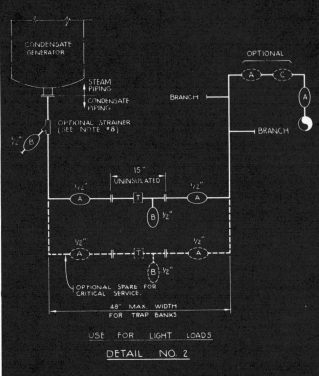

CONDENSATE GENERATOR

STEAM PIPING

CONDENSATE PIPING

OPTIONAL STRAINER (SEE NOTE #8)

½" B

OPTIONAL

BRANCH

BRANCH

15" UNINSULATED

½" A

½" A

B ½"

½" A

½" A

B ½"

OPTIONAL SPARE FOR CRITICAL SERVICE.

48" MAX. WIDTH FOR TRAP BANKS

USE FOR LIGHT LOADS

DETAIL NO. 2

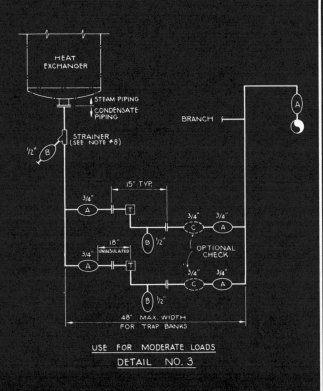

HEAT EXCHANGER

STEAM PIPING

CONDENSATE PIPING

STRAINER (SEE NOTE #8)

½" B

BRANCH

A

15" TYP.

¾" A

T

¾" ¾" C A

B ½"

OPTIONAL CHECK

18" UNINSULATED

¾" A

T

¾" ¾" C A

B ½"

48" MAX. WIDTH FOR TRAP BANKS

USE FOR MODERATE LOADS

DETAIL NO. 3

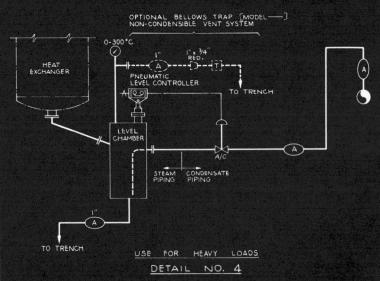

OPTIONAL BELLOWS TRAP [MODEL——] NON-CONDENSIBLE VENT SYSTEM

HEAT EXCHANGER

0-300°C

1"

1" x ¾" RED.

A

T

TO TRENCH

A

PNEUMATIC LEVEL CONTROLLER

LEVEL CHAMBER

A/C

STEAM PIPING

CONDENSATE PIPING

A

1"

A

TO TRENCH

USE FOR HEAVY LOADS

DETAIL NO. 4

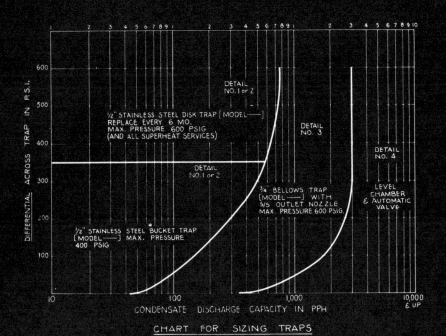

CHART FOR SIZING TRAPS

(Vertical axis: DIFFERENTIAL ACROSS TRAP IN P.S.I. — 100, 200, 300, 400, 500, 600)
(Horizontal axis: CONDENSATE DISCHARGE CAPACITY IN PPH — 10, 100, 1,000, 10,000 & UP)

DETAIL NO. 1 or 2

½" STAINLESS STEEL DISK TRAP [MODEL ——] REPLACE EVERY 6 MO. MAX. PRESSURE 600 PSIG (AND ALL SUPERHEAT SERVICES)

DETAIL NO. 3

DETAIL NO. 4

DETAIL NO. 1 or 2

¾" BELLOWS TRAP (MODEL ——) WITH S/S OUTLET NOZZLE MAX. PRESSURE 600 PSIG.

LEVEL CHAMBER & AUTOMATIC VALVE

½" STAINLESS STEEL BUCKET TRAP [MODEL ——] MAX. PRESSURE 400 PSIG

TRAP SIZING:

1.) DETERMINE MAXIMUM QUANTITY OF CONDENSATE TO BE HANDLED (FROM ENGR'G FLOW SHEETS) AND MULTIPLY BY A FACTOR OF 1.25.

2.) SELECT TRAP SYSTEM WITH FLOW EQUAL TO, OR SLIGHTLY GREATER THAN REQUIRED FLOW.

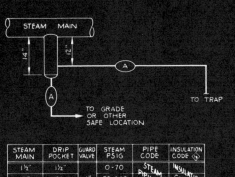

STEAM MAIN

14"   12"

A

A

TO TRADE

TO GRADE OR OTHER SAFE LOCATION

STEAM MAIN	DRIP POCKET	GUARD VALVE	STEAM PSIG	PIPE CODE	INSULATION CODE ◇
1½"	1½"	1"	0-70	STEAM PIPING CODES	INSULATION CODES
2"	2"		70-240		
3" & 4"	3"		240-600		
6" & 8"	4"				
10"	6"				
12" & 14"	8"	1½"			
12" & 18"	10"				
20"-24"	12"				
30"	16"				

STEAM

DETAIL NO. 5

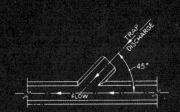

TRAP DISCHARGE

45°

FLOW

CONDENSATE HEADER INLET

TYPICAL DETAIL

VALVES	½" TO 2"		3" & LARGER		
STM. PSIG	ALL PRESSURES TO 600 PSIG		0-70 PSIG	70-300 PSIG	300-600 PSIG
PIPE CODES	SMALL STEAM PIPING CODES		LARGER STEAM PIPING CODES		
A	GATE VALVE CODES				
B					
C	CHECK VALVE CODES				

STEAM PSIG	INSULATION CODE ◇
0-70	INSULATION CODES
70-600	

CONDENSATE

DETAIL NO. 6

GENERAL NOTES:

1.) SELECTION OF MATERIALS, VALVES, FABRICATION AND INSTALLATION OF ALL PIPING TO BE IN ACCORDANCE WITH COMPANY CODE INDICATED ON DIAGRAMS.

2.) FOR DESCRIPTION OF VALVES SEE CO. VALVE CODE. ALL PIPING ASSEMBLIES SHOWN ON THIS DRAWING ARE DIAGRAMMATIC ONLY AND ARE NOT TO BE USED AS PIPING ARRANGEMENTS.

3.) WHERE MORE THAN ONE TRAP SERVES A SINGLE PIECE OF EQUIPMENT, TRAPS SHOULD BE INSTALLED IN SAME HORIZONTAL PLANE AND NOT STACKED ONE ABOVE ANOTHER.

4.) WHERE "STANDARD" DRAWINGS DIFFER FROM PROJECT SERVICE DIAGRAMS, PROJECT SERVICE DIAGRAMS ARE TO TAKE PRECEDENCE.

5.) ⊳——⊤——⊲—— ON SERVICE DIAGRAMS INDICATES ASSEMBLY SHOWN IN DETAILS 1 THRU 4. SERVICE DIAGRAMS TO STATE LINE SIZE (L.S.) AND PIPE CODE TO AND FROM TRAP.

6.) CONDENSATE PIPE CODE IS TO COMMENCE AT VESSEL AND NOT DOWNSTREAM OF TRAP. IN CASE OF TRACING IT IS TO COMMENCE WHERE THE TUBING ENDS AND PIPING STARTS.

7.) OPEN END OF 304 SST. SCH. 40 ½" BLOWDOWN VALVE NIPPLE TO BE THREADED.

8.) INLINE STRAINER — SOCKET WELD ENDS.

SIZE	TYPE	MAT.	DESIGN PRESSURE	DESIGN TEMP.	BLOW-OFF CONN.	STD. SCREEN
½" & 1"	PLANT STD.	C.S.	600	850	½" NPT.	.033 PERF. S/S
1½"	PLANT STD.	C.S.	600	850	½" NPT.	.033 PERF. S/S

9.) USE ½" STAINLESS STEEL DISK TRAP [MODEL ——] FOR ALL SUPERHEATED STEAM SERVICES.

SAFETY NOTES:

1.) EXCEPT WHERE PROHIBITED IN ASSEMBLY ABOVE, TRAPS AND PIPING ARE TO BE INSULATED FOR PROTECTION OF PERSONNEL. WHERE INSULATION IS NOT PERMITTED, FIELD IS TO PROVIDE GUARDS AROUND TRAPS AND PIPING.

GENERAL (PLANT) STEAM TRAPPING STANDARD INSTALLATION & SIZING DETAILS

DWG. **8993**

this instance may be required to prevent water or steam from backflowing into the wrong header.

Also note that all steam traps return their condensate to a collection header. It is good energy-conservation practice to collect this condensate in a storage tank and pump it to the powerhouse or a hot-water process user. Not to reuse the condensate represents a sizable cost penalty in the loss of the condensate's heat content and the original boiler-feedwater treatment expense. Steam condensate is normally good-quality hot water and should not be wasted.

Heavy-load-trap design details (Detail No. 4) — The very-large-quantity steam users require steam traps of very high capacities. Determining steam wasting in these large traps is difficult, if not impossible. These traps, with capacities in the thousands of pounds of condensate per hour, are normally constructed of carbon steel. As earlier mentioned, failures at the internal sealing or gasketing surfaces can result in large steam losses that are almost impossible to detect if the steam loss is less than the condensate flow.

As a result, the instrumented condensate-level chamber was designed to overcome the two major problems with commercially available large steam-traps—frequent failure, and no way to test for steam loss. These two problems can be solved by installing a water seal between the steam and the condensate systems. The design shown in Detail No. 4 and Fig. 12 of the "steam trap testing" article (p. 89) does just this by using a level chamber (or catch pot), a level-sensing controller, and a control valve. The condensate chamber has a workable 30-in. level range and is sized for a condensate velocity through the vessel of about 1 in./s or less.

Our standard now consists of only two basic vessel sizes: 18 and 24-in. dia. The level instrumentation, which has proven very successful, is a compact pneumatic controller. This self-contained pneumatic instrument provides a positive level-readout, a level-sensor, a control point, and a pneumatic-output control signal all in one small housing. The level-output signal goes directly to the stainless-steel control valve (where the actual level-control takes place). For most flashing-condensate services, a standard Type 316 stainless-steel control valve will provide long reliable life. Special trims and seats that have been coated with hard-facing materials are available for very-high-pressure hot-condensate services.

Always size the control valve for flashing service. In many instances, this will require using a large valve-body with reduced trim to maintain valve-position control, because the flash steam causes high backpressure in the small valve-bodies. Using the level chamber to replace large steel steam-traps will result in a steam consumption reduction of 2-17%, depending on the steam pressure and condition of the trap being replaced.

General steam-header drip-pocket design details (Detail

No. 5) — Most companies have a standard established for steam-main drip pockets. It should be reproduced—for clarity and easy access—on the steam-trap-standard drawing. Other company specifications are incorporated into this drawing in the form of codes for piping, insulation, and guard valves. Only the code numbers need be duplicated on the steam-trap standard. Details of these codes are generally found in company documents that are available elsewhere.

General condensate-header design details (Detail No. 6) — This detail incorporates the same general information as the steam-header details of No. 5, above. The general company codes and specifications are again duplicated here for the piping, valves and insulation.

General notes—These notes reflect good engineering practices. They call attention to special piping codes and other specifics not usually covered on the individual trap-layout detail.

Summary

Generally, a good steam trap is one that does its job by passing condensate and noncondensables, but without passing more than 1 lb/h of steam. Long trap life, high reliability, ease of installation, and low steam loss all should take precedence over initial trap cost.

In conclusion, here are the three main things to do for establishing an effective steam-trap standard:

1) Divide the plant steam traps into each of three condensate-load ranges—small, medium and large.

2) Select a steam trap or traps (if the steam pressure requirements dictate) for each of the three condensate ranges, using your own experience, plant tests, or the recommendations of a reliable source.

3) Combine items 1 and 2 above with applicable company standards for valves, piping insulation, etc., into a single plant-standard drawing.

If the above three steps are followed, and then put into practice, you will find that the second of the two greatest causes of steam loss from steam traps—misapplication—will be eliminated.

The author

Stafford J. Vallery is a Senior Engineer at the E. I. du Pont de Nemours & Co. Victoria Plant, P. O. Box 2626, Victoria, TX 77901. During the past 15 years, he has held various positions with Du Pont—development engineer at the Du Pont operated AEC facility at Aiken, S.C., and at the company's engineering department near Wilmington, Del. As part of his assignment at the Victoria plant, he has over the last five years worked to improve the plant's energy systems, especially in selection, testing and maintaining of steam traps. He holds a B.S.M.E. from Louisiana Tech University and is a member of the American Soc. of Mechanical Engineers.

Boiler-water control for efficient steam production

The mathematical relationships developed in this article enable one to adjust the blowdown rate and check the unit and total costs of steam generation through control of the concentration of boiler-water components.

Robert H. L. Howe, Eli Lilly and Co.

☐ Boiler water is never pure. Control of boiler-water solids concentration is critical both to the economic production of steam and to the maintenance of the boiler and its related equipment and piping.

One can analyze the components of boiler water to determine solids concentration in feedwater, condensate return, and the like. However, control of the concentrations is governed by the type of boiler, temperature, pressure, steam-production capacity, as well as by the characteristics of makeup water, availability of condensate return, rate of blowdown, and so on.

Based on these factors, it is possible to predetermine the most desirable boiler-water solids concentration by some simple mathematical expressions when the required conditions are known (from laboratory test and recorded data).

Mathematical analyses

The following are a few cases of mathematical analyses of the relationships between the makeup-water concentration and the feedwater, boiler water and steam-condensate return, as well as the rate of steam generation, etc. (See the Nomenclature box for definitions of symbols.)

Case I—When a boiler is newly placed online for operation, generally no blowdown is attempted immediately, and no steam condensate is returned for reuse. Then:

1st day: $C_{B1} = C_{F1}$
2nd day: $C_{B2} = C_{F1} + C_{F2}$

. . .
. . .
. . .
. . .

Nth day: $C_{BN} = C_{F1} + C_{F2} + \cdots + C_{FN}$

$$= \sum_{N=1}^{N} C_{FN} \qquad (1)$$

If the feedwater concentration is maintained constant:

$$C_{BN} = NC_F \qquad (2)$$

Case II—When the condensate is returned to the boiler for reuse on the $(N + 1)$th day, and still no blowdown is attempted, we have (neglecting the steam condensate concentration):

$$C_{B(N+1)} = \frac{YC_W + X\sum_{1}^{N} C_{FN}}{X} \qquad (3)$$

If the steam-condensate concentration is considered, then:

$$C_{B(N+1)} = \frac{YC_W + (X - Y)C_k + X\sum_{1}^{N} C_{FN}}{X} \qquad (4)$$

or

$$C_{B(N+1)} = \frac{YC_W + (X - Y)C_K + XC_{BN}}{X} \qquad (5)$$

Case III—When the steam condensate is reused, and blowdown is practiced on the Nth day after the boiler has been placed online, the boiler-water concentration at the end of the Nth day is:

$$C_{BN} = \frac{YC_W + (X - Y)C_K + XC_{BN} - BXC_{BN}}{X} \qquad (6)$$

But:

$$YC_W + (X - Y)C_K = XC_F$$

Therefore:

$$XC_{BN} = XC_F + XC_{BN} - BXC_{BN}$$
$$BXC_{BN} = XC_F$$

Or:

$$B = \frac{C_F}{C_{BN}} \qquad (7)$$

This indicates that the percent of blowdown for the Nth day is the ratio of the feedwater concentration and the boiler-water concentration. With a constant rate of blowdown maintained at all times, it is thus possible to determine the boiler-water concentration from the feedwater concentration.

Case IV—When it is desired to maintain a constant concentration in the boiler water for a particular type of

Originally published February 26, 1979

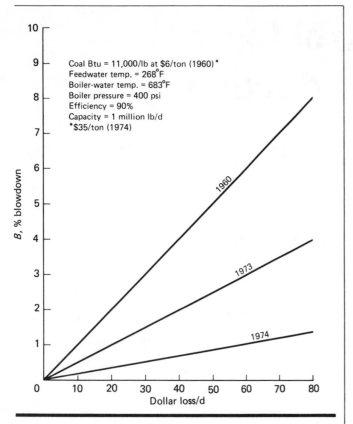

Coal Btu = 11,000/lb at $6/ton (1960) *
Feedwater temp. = 268°F
Boiler-water temp. = 683°F
Boiler pressure = 400 psi
Efficiency = 90%
Capacity = 1 million lb/d
*$35/ton (1974)

Operation loss due to blowdown   Fig. 1

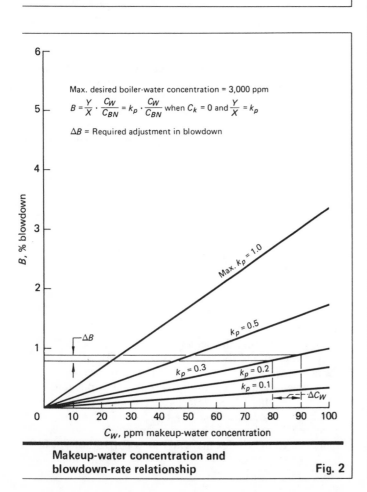

Max. desired boiler-water concentration = 3,000 ppm

$B = \dfrac{Y}{X} \cdot \dfrac{C_W}{C_{BN}} = k_p \cdot \dfrac{C_W}{C_{BN}}$ when $C_k = 0$ and $\dfrac{Y}{X} = k_p$

ΔB = Required adjustment in blowdown

**Makeup-water concentration and
blowdown-rate relationship   Fig. 2**

<table>
<tr><td>Nomenclature</td><td></td></tr>
</table>

B	Rate of continuous blowdown, %/d
C_W	Concentration of solids in makeup water, ppm
C_F	Concentration of solids in feedwater, ppm
C_B	Concentration of solids in boiler water, ppm
C_K	Concentration of solids in steam condensate return, ppm
N	Number of days boiler has been in operation
X	Steam-production capacity, lb/d
Y	Quantity of makeup water, lb/d
$X/Y = k_p$	Fraction of makeup water in boiler water

boiler and for the economical production of steam, we have:

$$C_{BN} = C_{B(N+1)} = C_{B(N+2)} \qquad (8)$$

The boiler-water concentration on any day should be

$$C_{B(N+1)} = C_{B(N+2)} = \cdots\cdots$$
$$= \frac{XC_F + XC_{BN} - BXC_{BN}}{X}$$

Then

$$0 = XC_F - BXC_{BN}$$
$$B = \frac{C_F}{C_{BN}} \qquad (9)$$

which indicates that if the makeup water and the steam-condensate return remain unchanged, a constant blowdown rate is desirable for maintaining a constant boiler-water concentration.

Component concentration in steam generation

Boiler-water concentration vs. makeup-water concentration— Rewrite Eq. (6) in the form:

$$XC_{BN} = YC_W + (X - Y)C_K + XC_{BN} - BXC_{BN}$$

or

$$0 = YC_W + (X - Y)C_K - BXC_{BN}$$
$$C_{BN} = \frac{YC_W + (X - Y)C_K}{BX}$$
$$= \frac{Y}{BX}C_W + \frac{X - Y}{BX}C_K$$

If all are constant except C_{BN} and C_W, then

$$\Delta C_{BN} = \frac{Y}{BX}\Delta C_W$$
$$= \text{Constant} \cdot \Delta C_W \qquad (10)$$

which denotes that, with a constant blowdown rate, a constant makeup rate and a constant condensate con-

Mathematical checking of boiler-water concentration — Table I

Day	X, steam production, lb	ΔX, change	B, blowdown	ΔB change	Y/X, makeup ratio	$\Delta (Y/X)$	C_{BN}, tested	ΔC_{BN}	C_{BN}', computed*	C_W
1	742,500		2.0/100=0.02		2/7=0.28		1,850	30	1,820	130
2	772,500	+ 30,000	2.7/108=0.025	+0.005	2.7/8=0.34	+0.06	1,700	40	1,740	128
3	690,000	− 82,500	2.7/108=0.025	0	2.7/8=0.34	0	1,700	70	1,770	130
4	720,000	+ 30,000	2.7/108=0.025	0	2.7/8=0.34	0	1,700	40	1,740	128
5	727,500	+ 75,000	3.3/106=0.031	+0.006	3.3/8=0.41	+0.07	1,800	50	1,850	140
6	757,500	+ 30,000	3.3/110=0.030	−0.001	3.3/10=0.33	−0.08	1,550	10	1,540	140
7	660,000	− 95,000	2.7/94=0.029	−0.001	2.7/9=0.30	−0.03	1,520	20	1,540	150
8	660,000	0	3.0/90=0.033	+0.004	3.0/8=0.37	+0.07	1,690	30	1,720	154
9	742,500	+ 82,500	3.3/94=0.032	−0.001	3.3/9=0.37	0	1,750	40	1,710	150
10	637,500	−105,000	3.0/100=0.030	−0.002	4.0/9=0.500	+0.13	2,250	50	2,300	140

*$C_{BN}' = \dfrac{Y}{BX} C_W$

centration, the change in boiler-water concentration is directly proportional to the change of makeup-water concentration.

Boiler-water concentration vs. change in makeup-water concentration, rate, etc.—When Y and C_W change, C_{BN} changes. It can be shown that:

$$\Delta C_{BN} = \frac{Y}{BX}\Delta C_W + \left(\frac{C_W}{BX} - \frac{C_K}{BX}\right)\Delta Y$$

$$= \frac{1}{BX}[Y \cdot \Delta C_W + (C_W - C_K)\Delta Y] \quad (11)$$

If the makeup-water concentration, the makeup rate, the daily steam production of the boiler, and the blowdown rate all change, then the boiler-water concentration will change as follows:

$$\Delta C_{BN} = \frac{Y}{BX}\Delta C_W + \frac{1}{BX}(C_W - C_K)\Delta Y -$$
$$\frac{1}{X}[YC_W + (X - Y)C_K]\frac{\Delta B}{B^2} -$$
$$\left[\frac{YC_W}{BX^2} + \frac{C_K}{BX} - \frac{(X - Y)C_K}{BX}\right]\Delta X \quad (12)$$

Blowdown rate vs. other variables—When there are changes in the daily makeup-water concentration and rate, in the daily steam production, and also in the boiler-water concentration, the blowdown needs to be changed in order to maintain a desired concentration in the boiler water. The change in the blowdown rate can be determined by:

From $B = \dfrac{YC_W + (X - Y)C_K}{C_{BN}X}$ and C_K = constant,

$$\pm\Delta B = \frac{Y}{XC_{BN}}(\pm\Delta C_W) + \left(\frac{C_W}{XC_{BN}} - \frac{C_K}{XC_{BN}}\right)(\pm\Delta Y) -$$
$$[YC_W + (X - Y)C_K]\left(\pm\frac{\Delta C_{BN}}{XC_{BN}}\right) -$$
$$\frac{1}{C_{BN}}\left[\frac{YC_K}{X^2} - \frac{YC_K}{X^2}\right](\pm\Delta X) \quad (13)$$

Eq. (13) is explained by Table I. If there is only a

Relationship between makeup water, feedwater and boiler-water concentration — Table II

Makeup	SiO_2 in makeup	SiO_2 in feedwater	SiO_2 in boiler	Blowdown
39.5%	2	0.8	30	2.7%
40.0%	1.8	0.7	26	2.7%

Makeup	Fe in makeup	Fe in feedwater	Fe in boiler	Blowdown
38%	0.80	0.30	10	3.0%
40%	0.95	0.38	12	3.15%

change in the makeup-water concentration, with Y and C_{BN} being maintained constant, the increase or decrease of blowdown is governed directly by the ratio of makeup water to the total concentration of the boiler water. Or:

$$\pm\Delta B = \frac{\% \text{ makeup}}{\text{boiler-water concentration}} \times (\pm\Delta C_W) \quad (14)$$

Fig. 1 explains Eq. (14).

Economic considerations

Blowdown is necessary for maintaining a constant desired boiler-water concentration, because solids are added to the boiler every day. However, excessive blowdown means a loss of heat. Fig. 2 shows the relationship between the blowdown rate and heat loss in terms of the cost of coal. Together, Fig. 2 and 3 point up how important proper water treatment is to the economical production of steam.

Makeup rate vs. heat loss and treatment cost—From Eq. (5), we have:

$$YC_W + (X - Y)C_K = XC_F$$
$$Y(C_W - C_K) = X(C_F - C_K)$$
$$\frac{Y}{X} = \frac{C_F - C_K}{C_W - C_K} \quad (15)$$

If

$$C_K \rightarrow 0, \quad \frac{Y}{X} = \frac{C_F}{C_W} \qquad (16)$$

Which indicates that the makeup ratio is equal to the ratio of feedwater concentration and the makeup-water concentration (see Table II). The ratio of condensate reused to the total feedwater is:

$$\frac{X - Y}{X} = \frac{C_W - C_F}{C_W - C_K} \qquad (17)$$

or, when $C_K \rightarrow 0$,

$$\frac{X - Y}{X} = \frac{C_W - C_F}{C_W} \qquad (18)$$

To render C_F a minimum, either Y (the quantity of makeup) must be a minimum, or C_W (makeup-water concentration) must be. If either Y or C_W is large, there will be either a heat loss or additional water-treatment costs.

Further, using α for unit steam cost, β for unit water cost, $A = \alpha X$ for total daily steam cost and $B = \beta Y$ for daily water cost. Eq. (16) can then be written:

$$\frac{(B/\beta)}{(A/\alpha)} = \frac{C_F}{C_W}$$

or

$$A = \frac{C_W}{C_F} \cdot \frac{\alpha}{\beta} B \qquad (19)$$

which illustrates the relationship between steam cost and water cost with respect to feedwater concentration and makeup-water concentration.

Summary

■ It is possible to predict and control the boiler-water solids concentration on the $(N + 1)$th day with the Nth day's data of related components that may affect the boiler-water concentration.

■ With data available, it is reasonable to predetermine, by mathematical means, the necessary adjustment of the blowdown rate in order to maintain a constant desired concentration in the boiler water.

■ It is possible to check the blowdown rate, the makeup ratio and the accuracy of laboratory tests on various boiler-water components by some simple mathematical expressions.

■ It is possible to check the unit and total costs of steam generation through component-concentration control.

The author

Robert H. L. Howe, 106 Drury Lane, West Lafayette, IN 47906, works in Environmental Technical Services, Eli Lilly and Co., as an Environmental Technical Services Consultant and Research Scientist. He also serves as an Honorary Science Advisor, National Science Council, Taiwan. He has taught at many universities—this article is based on lectures that have been delivered to graduate students in Turkey, Italy, China and (most recently) at the University of Notre Dame, where he delivered a yearly series of lectures based on his original research and work. Dr. Howe holds a B.S. in chemistry from Methodist University, a B.S. in engineering from St. John, an M.S. in physics and an M. Eng. from Cornell University, a Ph.D. in environmental engineering and microbiology from Purdue University, and a D.Sc. in environmental and chemical engineering from World Open University (California). He is a Fellow of the Royal Soc. of Health; a Fellow of the American Public Health Assn.; a Senior Fellow of the Institute for Advanced Sanitation Research, International; a Certified Specialist, American Environmental Engineering Intersociety Board, and a Registered Professional Engineer.

Energy-saving schemes in distillation

Distillation systems can add up to a significant amount of the total energy requirements of a processing scheme, and there are many techniques that can be used to reduce such energy consumption. But before applying any technique, it is necessary to recognize the working relationships between capital, operating costs, and plant operability.

William C. Petterson and **Thomas A. Wells**, *Pullman Kellogg Div., Pullman Inc.*

☐ Designing strictly for energy conservation is as ill-advised as ignoring trends in energy prices. Capital is equally critical; it must be effectively used to purchase reliable systems. Gross margins in the chemical process industries (CPI) are such that most potential energy savings cannot offset product loss resulting from an unreliable system.

Here, some basic chemical engineering techniques are discussed, which are being applied to distillation sequences in light-hydrocarbon (i.e., olefins) recovery plants to reduce energy consumption, while maintaining high levels of flexibility and operability.

Utility values

Specification of utility values for a particular evaluation has a controlling effect on the conclusions that are drawn, because one must consider the use of the utility as it relates to the entire pattern of the unit. For the purpose of following our discussion, the stipulations listed below will be used to develop utility values:

■ The plant is highly "work-oriented," with extensive use of high-pressure steam turbines. Low-pressure steam is supplied by turbine exhaust to the low-pressure steam header. The relationship between low-pressure and high-pressure steam costs must account for the work extracted in the turbine in reaching the low-pressure header. It will be assumed that the startup steam requirements for the unit prescribe a high-pressure boiler in excess of any normal steam-combination capacity. The utility-cost factor for steam levels will contain no capital charges for incremental production.

■ The cooling-water system capacity is a direct function of anticipated normal operation. Therefore, incremental water-cooling consumption must be included in capital charges. Table I summarizes utility costs estimated for a plant in the 1979–1980 period.

Plant utility costs in the year 1979-1980	Table I
Material	**Cost**
Fuel	$3.00/$10^6$ Btu
High-pressure steam, 1,500 psig	$4.15/$10^3$ lb* $3.33/$10^6$ Btu
Low-pressure steam, 50 psig	$1.75/$10^3$ lb† $1.60/$10^6$ Btu
Cooling water, 30°F rise	12 ¢/10^3 gal** 48 ¢/10^6 Btu
Electrical power	3 ¢/kWh

*Based on 90%-efficiency boiler.
†Based on 75%-efficiency turbines.
**Includes capital charges.

Note: Incremental high-pressure steam used for the production of work is for condensing water at a rate of 6.75, which results in $230/[(hp)(yr)].

Originally published September 26, 1977

Interreboiler application   Table II

Equipment duty	System with interreboiler	System without interreboiler
Rectifying trays	90	90
No. of stripping trays	45	28
Condenser duty, 10^6 Btu/h	135	135
Interreboiler duty, 10^6 Btu/h	50	0
Reboiler duty, 10^6 Btu/h	50	100
Top section:		
Height, ft	155	155
Diameter, ft	18	18
Thickness, in.	2.2	2.2
Bottom section:		
Height, ft	89	64
Diameter, ft	16	18
Thickness, in.	1.8	2.2
Tower cost, $\$10^6$	5.1	4.8
Exchanger cost, $\$10^6$	3.8	3.7
Refrigerating system cost, $\$10^6$	−0.1	0.0
Total capital cost, $\$10^6$	8.8	8.5
Operating cost, boiler hp, $\$10^6$/yr	14,850 4.15	17,050 4.77

Interreboilers and intercondensers

Interreboilers and intercondensers—when applied in accordance with typical economic and operating criteria—can produce significant reductions in the operation cost of a distillation system.

The generally accepted approach of applying heat only at the bottom of the tower, and withdrawing heat only at the top, is most often directed by the economic and operability requirements imposed on the design. In situations where energy costs are low, the thermodynamic inefficiencies inherent with this approach are usually not worth reducing.

However, in multistage distillation, it is possible to add and remove heat at numerous locations in the distillation column. It is theoretically possible—but seldom practical—to apply this concept to each equilibrium stage in the column by adding finite quantities of heat to every stripping stage, and removing finite quantities of heat from every rectification stage.

For a specific case where the feed and product rates and purities are constant, and for a particular condenser duty, the total heat applied to the stripping section has a unique value, regardless of the number of places where it is put into the stripping section. The economics of multiple reboiling lie in the ability of the system to utilize multiple levels of heat. In a single reboiler system, the entire heat load must be applied to the base of the column and, therefore, must have a high temperature. Since the cost of heat energy is usually a function of departure from ambient temperature, this single input of high-temperature energy is the most expensive method of reboiling a distillation system.

When the same amount of energy is divided up and added to several intermediate points—between the feed tray and the bottom tray—the temperature levels of the energy can be progressively lower as the feed tray is approached. The temperature of the energy source at a particular location must be higher than the tower

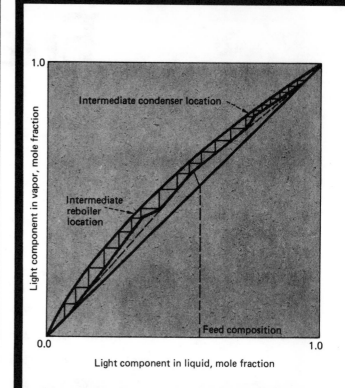

McCabe-Thiele diagram for system using intermediate condenser and intermediate reboiler   Fig. 1

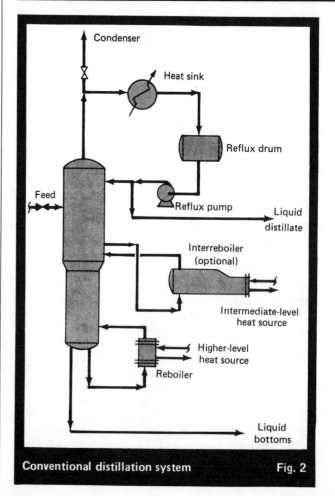

Conventional distillation system   Fig. 2

Comparison of heat pump and alternative systems — Table III

Equipment costs and dimensions	Cooling-water steam-system	Heat-pump system	Cooling-water/ waste-heat system
Tower overhead temperature, °F	100	40	100
Optimum reflux ratio	6.9	5.6	7.2
Above minimum reflux, %	5	7	13
Tower diameter, ft	15.9	13.9	16.3
Tower height, ft	224	157	190
Shell thickness, in.	1 5/16	9/16	1 3/8
Condenser duty, 10^6 Btu/h	121	–	130
Condenser area, ft^2	61,000	–	65,900
Reboiler duty, 10^6 Btu/h	124	123	134
Reboiler area, ft^2	3,600	123,000	32,000
Compressor requirement, boiler hp	–	5,550	–
Total system capital cost, 10^6	4.2	4.9	4.3
Condensing medium	Cooling water	Heat pump	Cooling water
Medium unit cost, $/10^3$ gal	0.12	–	0.12
Total condensing cost, 10^6	1.9	–	2.0
Reboiling medium	Low-pressure steam	–	Circulating hot water
Medium unit cost, 10^6 Btu	1.60	–	(0.07)
Total reboiling medium, 10^6	7.5	–	(0.30)
Total four-year operating cost, 10^6	9.4	5.0	1.7
Total capital + 4-yr operating cost, 10^6	13.6	9.9	6.0
Total annual cost, 10^6	3.4	2.5	1.5

liquid at that point, but only by the amount that results in an economic quantity of heat-transfer area in the intermediate reboiler. The same concepts apply to the rectification section, where instead of using a single low-temperature energy sink, heat can be removed at several locations, which use successively warmer heat sinks as the feed tray is approached.

Using multiple condensers and multiple reboilers can have significant effects on the design of a distillation column itself. The number of distillation stages required to achieve a particular separation is determined by the ratio of liquid-to-vapor flowrates in the column. The column diameter is set by the magnitude of these flows.

Fig. 1 shows a McCabe-Thiele diagram for a binary system, which employs an intermediate condenser and an intermediate reboiler. This diagram shows the significant changes in the liquid-to-vapor ratio that occur above the intercondenser and below the interreboiler. These changes cause an increase in the number of distillation stages required in both the stripping and rectification sections. These increases are a function of the location of the intermediate level systems, and the percentage of the total duty requirements satisfied by these systems.

There is also a potential opportunity to reduce the column diameter above the intercondenser and below the interreboiler, because of the reduced vapor and liquid traffic in these sections. The application of these systems to a distillation column should be studied and optimized considering the following points:

1. The applicable levels of heat or cooling sources determine the point of application of intercondensers and interreboilers. The magnitude of the potential en-

ergy savings dictates their applicability to a particular system.

2. The increased loading of the intermediate systems causes changes in the overall tower height and diameter, as well as in the heat-transfer area.

3. The use of less-expensive energy levels reduces the overall operating cost of a particular distillation system. This changes the capital and energy cost relationship, which shifts the optimum operating point to greater percentages above the minimum reflux.

4. The reliability of the heat sources, and the accu-

Economic summary for split-tower operations — Table IV

Equipment cost and dimensions	High-pressure tower	Low-pressure tower
Tower height, ft	225	205
Tower diameter, ft	10.0	9.3
Shell thickness, in.	1 1/16	3/4
Number of trays	123	110
Compressor, boiler hp	5,240	
Capital for total, $ millions (System including exchangers, pumps compressor, etc.)	5.84	
Operating cost (4 yr), $ millions	5.13	
Total capital + 4-yr operating cost, $ millions	10.97	
Total annual cost, $ millions	2.74	

Note: This system is comparable to the examples treated in Table III.

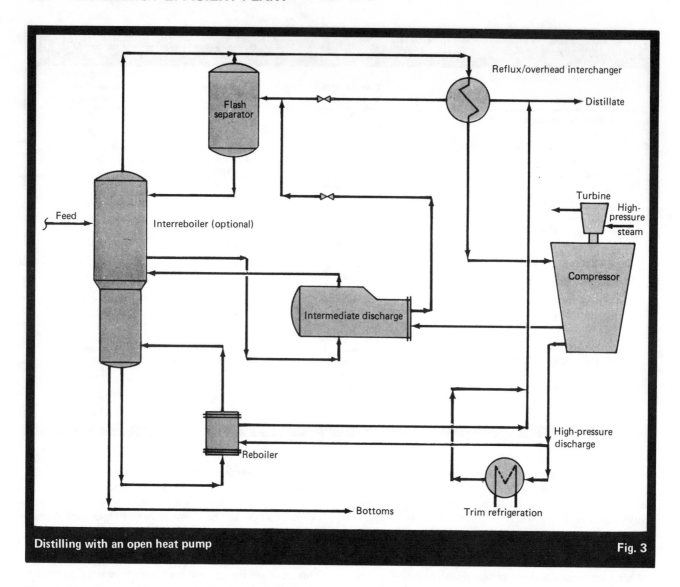

Distilling with an open heat pump   **Fig. 3**

rate determination to the column internal loading and conditions, must be considered when designing for different levels of flexibility and reliability. The effects of alternate feed composition must also be carefully evaluated, before committing the system to a particular configuration.

As an example of the attractiveness of interreboilers, Table II shows a summary of key factors of two distillation systems. One is equipped with an interreboiler (Fig. 2), while a similar one is not. Each system produces identical products from a given feed. The use of an interreboiler system for this example shows that it takes half a year to recover the capital investment.

It should be emphasized that not every system can economically utilize intermediate reboilers and intermediate condensers. Even less common are systems that can justify more than one intermediate system. The greatest potential exists with systems that use medium- or high-pressure steam, or low levels of refrigeration in the distillation process.

Heat pumps

Distillation systems that use heat pumps have long been used in chemical processing. The recent upsurge in

popularity is attributable to potential savings in operating expenses. In a conventional distillation system, energy is used on a once-through basis, entering from a high-temperature heat source and exiting to a low-temperature heat sink. The cost of separation is very high, because the gross energy is totally degraded from a combustion of over 3,500°F in the steam-generation system to ambient temperature in the cooling-water air exhaust. This system owes its existence to simplicity, low investment and cheap energy.

The heat pump, on the other hand, takes the energy from the condenser and uses work to elevate it to a level high enough to be transferred to the column reboiler. This energy—that is totally degraded—is the net work that is required to transfer the energy from the overhead to the reboiler (work of separation), plus the lost work resulting from converting fuel to steam, and then to the work of separation.

The heat pump represents a significant reduction in energy consumption, but at the sacrifice of capital and simplicity. It is important to note that the heat pump configuration is most advantageous when the fractionation system has a low-temperature difference across the column. When the temperature difference expands, the cost of recycling energy increases, and the heat pump loses some of its attractiveness.

Fig. 2 and 3 are a comparison between a conventional distillation system—which utilizes steam in the reboiler and cooling water as a condensing medium—and the same separation carried out in an "open" heat pump. In the open heat-pump system, the gross tower overhead is compressed to a pressure at which its dew-point temperature is high enough to provide a satisfactory temperature approach in the reboiler, where the overhead is condensed while supplying heat to the reboiler.

When there is concern with possible product contamination by compressor oils, the system can be closed by simply providing a conventional overhead condenser to isolate the tower system from the refrigerant system. The closed system imposes the additional thermodynamic inefficiency of a temperature difference between the condensing-tower overhead vapors and the vaporizing refrigerant liquid. The condensed liquid leaving the reboiler—which serves as reflux for the tower—is subcooled against tower overhead vapors. This reduces flash vapors when the reflux is introduced into the column. In most applications, the saving in the reduction of flash vapor recycle is greater than the expense of a superheated compressor suction, but the system should be evaluated for each application.

The heat-pump system requires a certain balance of thermal loads to minimize the use of external heat or refrigeration sources. The thermal balance of the system is particularly sensitive to the thermal condition of the feed and products. For instance, the requirement for a vapor feed and liquid overhead product may impose a case where insufficient reboiler duty is available to condense the necessary reflux plus the product. An external refrigeration source would be required to make up the difference.

Table III shows an analysis of a distillation system that is first developed as a tower—condensed by cooling water and reboiled by steam—and then treated as a heat-pump system. The economic attractiveness of a heat pump is readily apparent when this comparison is made. However, the existence of an alternative free-energy source is not an uncommon occurrence when critically examining an integrated process for energy conservation when a temperature of 175°F is available. This temperature might be too low to provide an attractive heat source in other applications and would, therefore, require cooling by air or cooling water to 110°F before reintroduction to the processing scheme.

As is usually the case, waste-heat recovery represents a very significant improvement in any system when it can be used as a direct replacement for steam. In addition to the utility values given in Table I, the circulating hot water that serves as a heat source is assigned a 7¢/10⁶-Btu credit, because its reuse in the process eliminates the operating cost of cooling water and air cooling that must be used to otherwise cool the circulating water to its required temperature. However, these capital items in the hot-water cooling loop must remain in place to provide flexibility.

The results of this brief analysis indicate the advisability of using waste heat to its fullest advantage. How-

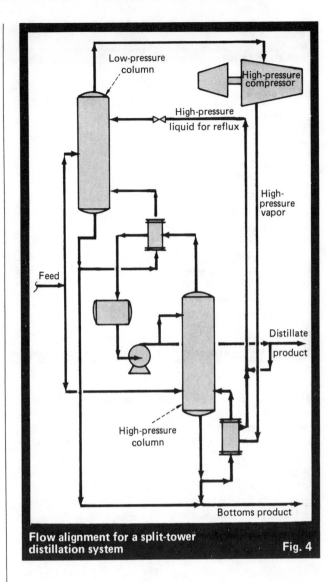

Flow alignment for a split-tower distillation system   **Fig. 4**

ever, in situations when this option is not feasible, then the heat pump can be an attractive substitution for steam heat, in spite of the higher capital cost and increased complexity. There are several variations of heat-pump and tower configurations; a brief literature survey will yield several practical techniques and guidelines for their applications.

Split-tower arrangements

There are several possible arrangements that utilize two separate towers that are "thermally linked." Basically, they consist of a high-pressure tower and a low-pressure tower operating in parallel—such as in Fig. 4—so that the high-pressure tower condenser is used as a source of heat for the low-pressure tower reboiler. For the example considered below, the products from both towers have the same composition, which result from parallel operation on the same feed. The feed-split between the towers depends on the thermal condition of the feed (Table IV).

The linking of the towers allows the same fractionation as the heat-pump case (Table II) with about half the reboiler and condenser duty. However, the much greater temperature difference across the overall system

Interreboiler versus feed heater   Table V

Equipment and duty	Feed heater		Interreboiler
	Top section	Bottom section	
Tower			
Number of trays	20	16	45
Diameter	6 ft 6 in.	5 ft 0 in.	5 ft 6 in.
Condenser			
Duty, million Btu/h	26.0		22.3
Cooling-water cost, $/yr	104,100		89,500
Reboiler			
Duty, million Btu/h	10.6		10.0
Steam cost, $/yr	141,900		134,300
Interreboiler or feed-heater			
Duty, million Btu/h	15.3		12.3
Total capital cost, $	536,500		524,400
Annualized capital cost, $/yr (distributed over 4 years)	134,100		151,400
Annual utility cost, $/yr	246,000		223,800
Total annual cost, $/yr	380,100		375,200

Note: Assuming a cooling-water condenser and free heat for feed heater or interreboiler.

results in increased energy consumption. Table V summarizes the operation of this system with the same heat-pump equipment discussed previously (Table II). The results show near equivalency with the single-tower heat pump. There are several versions of split-tower designs. They can provide significant savings when properly designed and optimized. However, there are sacrifices in flexibility and operability that accompany the increase in complexity.

Feed optimization

A distillation column feed can vary from subcooled liquid to superheated vapor. The thermal condition of the feed is an important parameter in the design of a distillation column because changes in the condition can affect both the capital and operating costs for a given system. The following examples show how operating costs can change significantly with changes in the feed condition indicating that, in the optimization of a distillation system design, the feed condition cannot be ignored.

A distillation column feed may come from one of many types of processing equipment, such as another distillation column, a heat exchanger or a reactor. An example of how the feed condition can be easily modified occurs when the feed stream comes from the overhead condenser of a preceding distillation column. It can be taken either as all vapor (if it can be pressured into the next column) or as all liquid. In other instances, an additional heat exchanger may be needed to either heat or cool the feed stream. Sometimes temperature limitations for exchanger-fouling considerations limit the heat input to a feed stream. These limitations cannot be ignored in the design of a distillation system.

Cooling water, steam, and various other heat sources may be used to modify the feed condition. If steam is used, it must be at a lower pressure, and thus at lower cost than that being used in the column reboiler. The

ideal source of heating or cooling is either direct or indirect exchange with another process stream. These sources are usually free.

Consider a typical distillation column that produces 80% of the feed as the overhead product. The changes in condenser and reboiler duties for changes in feed condition are shown in Fig. 5. The condenser duty changes only moderately, while the reboiler duty is more than halved as the feed is shifted from saturated liquid to saturated vapor. For a tower condensed with cooling water and reboiled with 50-psig steam, the utility costs are 37% lower for the all-vapor feed. Capital costs change very little over the range of feed conditions, so the optimum is an all-vapor feed (see Fig. 5). Therefore, if the column feed comes from the overhead of a preceeding tower, it should definitely be taken as all vapor, if possible.

Under other circumstances, such as when the feed is available as a liquid, the appropriate economic analysis must be performed to justify additional capital expenditures. When an essentially free source of heat is readily available—such as from a process stream which must be cooled—it is usually economically justifiable to vaporize, at least partially, an all-liquid feed in systems such as considered in the paragraph above.

A much different situation exists if the fractionation column separates only 20% of the feed into the overhead product. Now, the reboiler duty is almost constant, while the condenser duty almost doubles as the feed changes from all liquid to all vapor (see Fig. 6). For a tower condensed with cooling water, the net annual cost is minimized when the feed is about 40% vapor. If the stream is available as a liquid, the cost of heating it is not justified even if the heat is free. The savings between zero and 40% vaporization are not enough to justify the cost of the feed heater. If the stream is available as a vapor and can be condensed with cooling water, heating should, again, not be considered. The

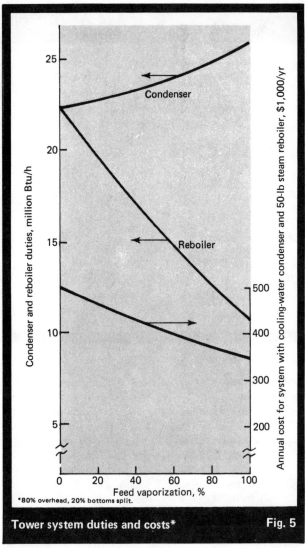

Tower system duties and costs*　Fig. 5

*80% overhead, 20% bottoms split.

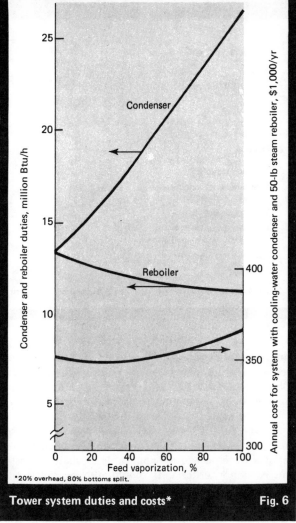

Tower system duties and costs*　Fig. 6

*20% overhead, 80% bottoms split.

cost of exchanger surface, as well as cooling water, exceeds any saving that might be incurred.

However, if the tower has a refrigerated condenser, the high cost of refrigeration, relative to cooling water, greatly effects the economic comparison. Increasing condensing duty significantly affects operating cost to such an extent that it is justifiable to condense the feed in an added cooling-water exchanger, if this is at all possible.

There are many variations that can be considered; it is difficult to make any generalizations as to the optimization of feed condition. The preceding examples provide an indication of the effects of the overhead- and bottoms-product splits, and shows variations in condensing costs on the optimization of the feed condition.

Feed heater versus interreboiler

Both feed heaters and interreboilers provide the use of a lower-level, and thus lower-value, heat source to reduce the total utility cost for a fractionation system. However, both also affect the capital cost of the system. The interreboiler causes an increase in capital costs because of an increased tray requirement. This increase can be partially compensated for by taking advantage

of the lower vapor rates below the interreboiler, and reducing the tower diameter in that section. However, reducing the tower diameter does restrict operating flexibility. The feed heater tends to reduce the capital cost of the system.

For a cooling-water, condensed-distillation system producing 80% of the feed as the overhead product, the optimum design for the tower corresponds to a totally vaporized feed, if the heat source is free. This system is compared to a tower with an all-liquid feed and an interreboiler in Table V. The net annual cost for the system, which utilizes a feed heater, is $375,200/yr, compared to $380,000/yr for the system with an interreboiler. In this example, the feed-heater system would probably be selected because of operating advantages over the interreboiler system. The most important of these advantages is the larger temperature approach in the feed heater, as compared to the interreboiler, if the same heat source is used. This results from the lower temperature of the tower feed as compared to any interreboiler feed.

If this tower required a refrigerated condenser, the economics and conclusions would be different. Because of the high cost of refrigeration, the optimum design is

Equipment and duty	Feed heater		Interreboiler
	Top section	Bottom section	
Tower			
Number of trays	20	16	45
Diameter	6 ft, 6 in.	5 ft, 0 in.	5 ft, 6 in.
Condenser			
Duty, million Btu/h	24.8		22.3
Refrigerating cost, $/yr	714,000		643,400
Reboiler			
Duty, million Btu/h	12.5		10.0
Steam cost, $/yr	167,000		134,300
Feed heater or interreboiler			
Duty, million Btu/h	12.4		12.3
Total capital, $	576,900		654,400
Annual capital cost, $ (distributed over 4 years)	144,200		163,600
Utility cost, $/yr	881,000		777,700
Total annual cost, $/yr	1,025,200		941,300

Interreboiler versus feed heater — Table VI

Note: Assuming a refrigerated condenser and free heat for feed heater or interreboiler.

compared to an interreboiler system with an all-liquid feed in Table VI. This economic comparison indicates a $38,900/yr (about 8%) advantage for the interreboiler, making it the preferred alternative for use of the free waste-heat source.

For the final selection between design alternatives, the actual systems must be evaluated in much greater detail with respect to the applicable mechanical design and economic criteria. It should also be noted that many of the optimum design points determined in this article exist on very shallow cost curves. Therefore, the penalty for missing the optimum design point is often small.

Because of this, it is desirable to consider both the potential operating costs over the expected life of a process plant and the plant's operating stability, in addition to the criteria for the early life of a plant, which are often the bases for a design. By considering the circumstances that might exist over the life of a process plant, a better design may be accomplished.

References

1. Null, H. R., Heat Pumps in Distillation, *Chem. Eng. Progr.,* July 1976, pp. 58–64.
2. Wolf, W., et al., Energy Costs Prompt Improved Distillation, *Oil and Gas J.,* Sept. 1975 (Reprint).
3. Reus, T. V., and Luyben, W. L., Two Towers Cheaper Than One?, *Hydrocarbon Process.,* July 1975, pp. 93–96.
4. M. W. Kellog Co., U.S. Patent No. 3,000,188, Sept. 9, 1961, Gas Separation (Heat Pump).

The authors

William C. Petterson is a senior process engineer with the Pullman Kellog Div., Pullman Inc., Three Greenway Plaza East, Houston, TX 77046. His responsibilities include process design of petrochemical and specialty chemical processes that have extensive application in fractionation equipment. He holds a B.S.Ch.E from the University of Cincinnati and a M.S.Ch.E from Rice University.

Thomas A. Wells is a senior process engineer with the Pullman Kellogg Div., Pullman Inc., Three Greenway Plaza East, Houston, TX 77046. His activities include process design and development in all areas of the olefins flowscheme. He holds a B.A. in chemistry from Drury College and a B.S. and M.S. in chemical engineering from the University of Missouri-Rolla.

Section III
EQUIPMENT AND MATERIALS

Refractories
   Saving heat energy in refractory-lined equipment
Insulation
   Cost-effective thermal insulation
   The cost of missing pipe insulation
   Insulation saves energy
   Economic pipe insulation for cold systems
Boilers and Furnaces
   Calculating boiler efficiency and economics
   Improving boiler efficiency
   CO control heightens furnace efficiency
   Cutting boiler fuel costs with combustion controls
   Burner makers stay hot in a volatile market
Saving Fuel Oil
   Fired heaters—How to reduce your fuel bill
   Fuel-efficiency thrust ups sales of additives
Waste-Heat Boilers
   How to avoid problems of waste-heat boilers
   Boiler heat recovery
Heat Exchangers
   New heat-exchange units rely on enhanced transfer
   Continuous tube cleaning improves performance of condensers
      and heat exchangers
Immersion Heaters
   Heat more efficiently—with electric immersion heaters
   Energy-efficient motors spark an old controversy
   Energy-efficient motors gain wider interest
Variable-Speed Drives and Motor Controllers
   Variable-speed drives can cut pumping costs
   Motor controllers spell savings for CPI
Pumps
   Saving energy and costs in pumping systems
   Select pumps to cut energy cost
Other Equipment
   Practical process design of particulate scrubbers
   Low-cost evaporation method saves energy by reusing heat

Saving heat energy in refractory-lined equipment

High-temperature equipment operating under severe conditions must use the least amount of energy in order to economically process many materials.

James E. Neal and *Roger S. Clark,* *Johns-Manville Corp.*

☐ Refractories are vital to processes using great heat, such as those found in the chemical, hydrocarbon, metallurgical and ceramic industries. Originally, refractories were seen merely as a means for containing heat within a given space, but they are now considered critical to the success of an efficient energy-conservation program and a cost-efficient manufacturing process. The proper use of refractories also promotes safety and health for the workers and creates a more comfortable environment overall.

In our discussion, we will provide information on the types and properties of refractories, refractory selection, installation and/or construction, heat-flow theory, and economics.

A basic approach

The traditional approach to insulation is to only consider the amount of heat escaping through the insulation when we should instead be thinking of the amount of heat contained or restricted from flowing.

If we look at insulation in a resistive sense, we cannot help but be impressed by its performance. For example, let us examine:

1. The radiative heat transfer between two radiantly black surfaces, each 1 ft². One surface is at 100°F, and the other at the temperature, T, as indicated by the abscissa in Fig. 1. This heat transfer does not include convective and conductive heat flow of the air, both of which are small by comparison.

2. The heat transfer resulting from placing 1 in. of insulation in the form of a refractory fiber having a density of 10 lb/ft³ between the two plates.

There will be quite a difference in heat transfer for each condition. For instance, suppose we have a hot-face temperature of 2,000°F. In this situation, and with no insulation, we find a radiant heat flow of over 60,000 Btu/(h)(ft²) from Fig. 1. With insulation, the total heat flow from the hot face to the cold face is about 1,500 Btu/(h)(ft²). Over 58,000 Btu/(h)(ft²) has been retained. This is a ratio of 40 to 1.

Obviously, it would be economically as well as practically impossible to operate this system without the insulation. The capacity of the power source that would be necessary to maintain the temperature of 2,000°F would have to be increased 40 to 50 times if insulation were not used.

Insulation-selection criteria

An insulation allowing no heat flow would be ideal, of course. Several insulation systems approach this goal, but they must operate in a vacuum. In most applications, this condition is impractical to realize. Insulation selection is not limited to resistance to heat flow. Other criteria that may be involved are: temperature limitations, thermal-shock resistance, coefficient of expansion, strength, hardness, compressibility, specific heat, erosion resistance, chemical resistance, environmental resistance, space limitations, ease of application, and economics for the various refractory materials.

The choice is generally a compromise governed to a great extent by the cost of insulation and its installation. However, the increasing cost of energy introduces a new parameter, life-cycle cost. This is based on the initial cost plus maintenance and operating costs (in-

Originally published May 4, 1981

cluding energy cost) for a reasonable life expectancy of the system, rather than the lowest initial cost.

Compromise begins with the initial conception of the refractory or insulating material. The components of the refractory/insulating system are critical parts in this selection.

Types of refractories

The essential properties of the available refractories are shown in Table I. The basic advantages and disadvantages for each material are:

Dense castables, for use in temperatures to 3,300°F, are modern monolithic versions of the age-old fired-clay brick. Their insulating qualities are marginal. However, they have lower thermal conductivity than fired brick, and provide greater strength and lower first cost than true insulating firebrick (IFB).

True insulating firebrick is available to 3,300°F as are insulating castables. In general, these materials weigh substantially less than dense refractories, have much better insulating qualities (being two to four times more resistant to heat transmission), but are not as strong. In most applications, the thicknesses called for by non-structural considerations provide ample strength.

Refractory fiber products such as felts and blankets are available up to 3,000°F. They have three major advantages: excellent insulating values, ease of application, and low heat storage. Low heat storage is of great advantage in cyclical heat applications, such as for a periodic kiln. Heat absorbed by the walls of such a kiln, and then lost when the kiln cools, can exceed the amount of heat actually used for production.

Calcium silicate, mineral-wool, fiberglass boards and blankets, all at the lower fringe of refractory-temperature ranges, are often used because of their combination of qualities. For example, calcium silicate usually offers structural-strength and high thermal-insulating values, and is not affected by moisture or humidity. Mineral wool and fiberglass generally combine the properties of light weight, heat resistance, low conductivity and high sound absorption.

Castables

Castables, also called refractory concretes, are available in dense and insulating versions. Castables are supplied dry, and are mixed with water before installation. They are installed by pouring, trowelling, and pneumatic gunning, or occasionally, by ramming. As a result, castables give a smooth, practically jointless construction to monolithic hearths, walls and roofs of furnaces and kilns.

Applications of castables in industries or processes using heat vary from aluminum furnaces or reformers for ammonia plants, to zinc furnaces. The growing use of castables stems directly from their advantages. Castables save money during installation because they go into place faster and easier. Mixing, conveying and placement of the unfired castable mixture are often mechanized. Furthermore, castables save time by eliminating the cutting of brick to fit a furnace or kiln, and eliminate the need for inventories of special brick shapes.

Gunning castables into place requires more skill than pouring them, and uses more material, but gunning

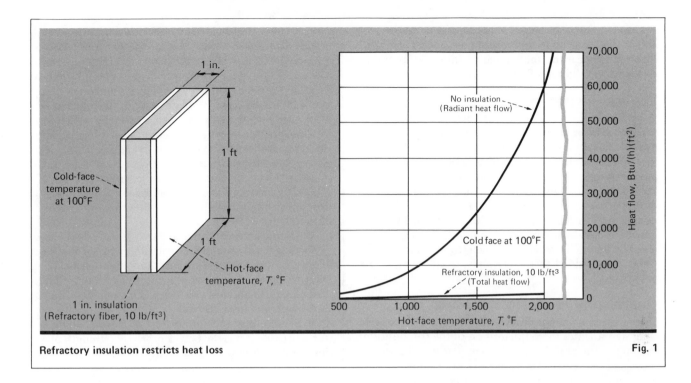

Refractory insulation restricts heat loss   Fig. 1

applies more material in less time than any other method. It also makes possible the installation of castables to horizontal, vertical or overhead structures without using forms.

Castables also have a number of performance advantages. First, dense castables have low permeability, partly because of the nature of the material and partly because the structure has few joints. As a result, castables are usually the best lining for vessels or chambers operating at other than atmospheric pressures. Second, most castable refractories have good volume stability within the specified temperature range, and many have good resistance to impact and mechanical abuse.

Installation practices for castables may cause prema-

ture failure and uneconomic life of the refractory. The reason is that the user or installer of such castables performs most of the ceramic manufacturing processes. The manufacturer of castables simply grinds and mixes raw materials together, then bags and ships them. The user adds water (whose amount and quality are critical to complete the mixing operation), and fires the emplaced castable to the necessary refractory quality.

The amount of water required for casting is probably the most commonly neglected variable associated with castable installation. Some castables are more sensitive to water requirements than others. Too much **or** too little water can produce poor physical characteristics, such as poor strength or poor abrasion resistance.

Refractories for use in fired heaters   Table I

	Mineral-wool block	Calcium-silicate block	Haydite, vermiculite	Insulating castables	Dense castables	Insulating firebrick	Dense firebrick	Refractory-fiber boards and shapes	Refractory-fiber blankets	Refractory-fiber modules
Maximum operating temperature, °F	1,500	2,000	1,800	2,800	3,300	3,200	3,300	2,600	2,600	2,600
Thermal insulation	1	2	3	3	4	3	4	2	1	1
Heat storage	1	1	2	3	4	2	4	1	1	1
Thermal-shock resistance	4	4	2	2	1	2	1	1	1	1
Chemical resistance	4	4	4	2	1	2	1	3	3	3
Erosion resistance	4	3	3	2	1	2	1	3	4	3
Strength	4	3	2	2	1	2	1	3	4	2
Resiliency	2	3	4	4	4	4	4	2	1	1
Acoustical insulation	2	3	3	3	4	3	4	2	1	2
Installed cost	Low	Medium	Low	Medium	Medium	Medium	Medium	High	Medium	High

Code: 1 - Excellent, 2 - Good, 3 - Fair, 4 - Poor.

Unfortunately, there are more failures of dense-castable installations than there should be. In practically all cases, failure can be pinpointed to improper preparation and installation of the material.

Physical properties of castables

To choose the right castable refractory for any application, we must balance the requirements of the process against the cost and capability of the castable. Capabilities and other characteristics of castables, such as maximum service temperature and modulus of rupture, are obtainable from manufacturers' data sheets. Test procedures for these physical properties are established by the American Soc. for Testing and Materials (ASTM), and are conducted by the refractory-manufacturers' laboratories. Other characteristics, such as the amount of water to be added to the castable, are determined by actual tests.

Chemical properties of castables

In refractory compositions, the alumina-silica ratio roughly indicates the refractoriness of the composition: the higher the alumina (Al_2O_3) content, the higher the refractoriness. Lime (CaO) in the composition appears as a result of the binder, and also contributes to the ultimate refractoriness.

Iron oxide (Fe_2O_3) becomes important in refractory compositions when the process being carried out in the refractory-lined vessel involves a reducing atmosphere. For example, carbon monoxide in the vessel's atmosphere at certain temperatures can react with the iron oxide to deposit carbon within the lining. If the reaction continues, the carbon will cause the lining to crack. Likewise, a hydrogen atmosphere will reduce Fe_2O_3.

Thermal expansion of castables

Unlike brick, castables are not fired before installation. Only after prolonged exposure to heat do castables acquire the reversible expansion and contraction characteristics found in fixed ceramics. Because of this, castables show the results of complex forces working on them during the initial heatup. The fired aggregate expands according to its chemical composition, the cement phase shrinks as it loses the water of hydration, and the castable itself expands or contracts as it sinters and mineralogical reactions occur.

Most of what we know about the thermal expansion of refractories is based on brick. This knowledge is difficult to transfer to castables. From an application standpoint, two factors are important:

- Fireclay castables do not normally require an expansion allowance.
- Certain dense, highly refractory castables have substantial reversible thermal expansion, up to 2,000°F or more. Castables of this type may require an expansion allowance in massive installations.

Insulating firebrick

The objective of any insulation manufacturing process is to create a product of great porosity. After all, only dead-air spaces insulate. The surrounding material is useless in the insulating sense, and it is there only to provide structural strength.

Insulating firebrick (IFB), made to this objective, reduces heat loss by two thermal mechanisms. As compared with dense refractories, IFB has high resistance to heat flow and, being lightweight, stores very little heat. Initially, IFB was applied to cyclically operated equipment, allowing energy conservation from reduced heat flow and low heat storage. Later, IFB was used in continuous heat-processing equipment such as tunnel kilns. In such equipment, IFB's high insulating ability conserves fuel and lowers the cost of construction by reducing the thickness of the refractory system.

Let us compare the capabilities and limitations of insulating firebrick with dense firebrick. We will find that IFB is:

- Highly resistant to heat flow.
- Very light in weight, and stores much less heat. The light weight expedites installation and bricklaying. Moreover, the supporting structure and foundations can be lighter and less expensive. This weight reduction can be as much as 80%.
- Structurally self-supporting at elevated temperatures because it has the best compressive strength of any insulating refractory. IFB is compatible with dense firebrick to firm up the whole construction.
- Machined to finished shape. It has tighter tolerance than dense fireclay brick. This provides a tighter refractory construction, less heat loss through the joints, and faster installation.
- Low in impurities that can adversely affect refractory performance—for example, a furnace atmosphere having hydrogen as one of its constituents.
- Available in a large variety of shapes, both as supplied by the manufacturer and as made from slabs without mortared joints.
- Readily cut or sawed on the job.

On the other hand, we find IFB to be:

- Typically lower in strength than dense fireclay brick. This will influence the design of large structures.
- Prone to hot-load deformation failure from heavy loads at high temperatures.
- Generally not good for abrasion resistance. For example, it should not be used in a flue that would have high velocities and/or abrasive matter present.
- More prone to thermal and mechanical spalling than most dense refractories. This prohibits its use in some very abusive applications. For example, the slab-and-billet reheat furnace found in rolling mills.
- Incompatible with a chemical environment that will flux an alumina-silica refractory body.

Shapes and sizes

IFB is available in almost all the standard refractory shapes in addition to special shapes such as arch, key-wedge, and circle brick, and in nonstandard thicknesses (obtained by mortaring two or more pieces).

The classification by ASTM (C 155-70), as shown in Table II, groups the brick in accordance with bulk density (a good indication of thermal conductivity) and the behavior of the reheat change test.

The reheat change test for each group is conducted at a temperature 50°F below the normal maximum recommended service temperature. For example, Group 23 brick (ordinarily referred to as 2,300°F brick, and recommended for this temperature) is tested at 2,250°F. A maximum of 2% linear shrinkage is allowed.

If precise operating conditions are known, fairly exact factors of safety can be determined. This is seldom the case. Hence, a rule of thumb has been developed by refractory users that a 200°F safety factor will be allowed between the furnace design temperature and published temperature-use limits.

For example, a Group 23 brick would not be used for a furnace having a design temperature above 2,100°F. This rule has evolved through long experience. Its rationale is the difficulty of operating heating equipment at a precise design temperature for long periods of time. During that period, we can expect temperature variations within the furnace, and it is almost inevitable that an upset condition will occur where the design temperature will be exceeded.

Excellent information on construction methods is published by refractory manufacturers for their products, and is equally applicable to IFB. These handbooks include design shortcuts, tables of brick shapes, calculation methods, and many details for building and supporting refractory structures.

Brick equivalents

Because refractory brick differs in size from products such as building brick, the refractories industry has developed a term to describe what is meant by a "brick." This term is a "brick equivalent" (BE), and the brick having the dimensions of 9 in. by $4\frac{1}{2}$ in. by $2\frac{1}{2}$ in. is defined as one brick equivalent. Thus, a brick having dimensions of 9 in. $\times$ $4\frac{1}{2}$ in. $\times$ 3 in. is 1.20 BEs.

To illustrate the concept of BE, let us consider a furnace wall that might require 1,200 BEs. The wall could be built from twelve hundred $9 \times 4\frac{1}{2} \times 2\frac{1}{2}$ bricks, or one thousand $9 \times 4\frac{1}{2} \times 3$ bricks. Or, it could be constructed from four hundred $11\frac{1}{4} \times 9 \times 3$ bricks because this brick contains 3.00 BEs.

The idea of brick equivalents is useful in estimating the brick requirements. Calculations can be done first in brick equivalents, and then converted into the exact number of pieces of each size desired. The information in Table III is helpful in estimating quantities. To calculate the number of BEs needed for walls of other thicknesses, use multiples of the values given in Table

III. These same values can be applied to curved walls by using the outside wall dimensions.

Mortar and mortaring methods

An incorrect selection of mortar not only slows down bricklaying, but can compromise the strength and life of the resultant structure. The mortar must have good water-retention properties, and be a strong air-setting one. Good air-setting properties give the refractory wall proper strength and stability. Brick not strongly bonded will loosen and bulge, making it necessary to rebuild the lining. Furthermore, IFB is most frequently used in furnaces where the temperatures are not high enough to develop a good bond in a heat-setting mortar. Even in furnaces operated as high as 2,600°F, the temperature drop through the IFB is so great that sufficient temperature for a ceramic bond may exist only in the first $\frac{1}{2}$ to 1 in. of the hot face of the brick. The remainder of the wall will not be bonded.

The most common and best mortaring practice is to use "dipped joints." The mortar is thinned to a creamy consistency, just pourable, and the brick dipped into this slurry before being placed in the wall. Mortar supplied wet to trowel consistency will commonly require additional water to reach this creamy consistency. One precaution in dipping is to minimize the amount of mortar deposited on the surface of the brick that will become the hot face of the furnace lining. This coated face is prone to spalling.

Troweled joints are often used but should be discouraged because they require a degree of skill mastered by only a few brickmasons. However, troweling is often used on crowns.

Insulating aggregates

Insulating aggregates are a comparatively unfinished form of refractory. There are two common types: expanded and burn-out:

Expanded aggregate is made by expanding a clay in such a manner that aggregate particles are formed having internal porosity and light weight. Such aggregates have very fine insulating properties. However, expanded insulating aggregates are not readily available. Most producers use their production in-house. An exception is haydite, a clay that bloats upon being heated. It is a low-temperature refractory of limited use. A common application involves mixing it with calcium aluminate cement and vermiculite, and gunning the resulting castable into low-temperature petrochemical heaters.

Burn-out aggregate is made by the burn-out IFB processes. Both reject-brick and dobies [molded blocks of ground clay or refractory materials, crudely formed, and fired] are crushed into aggregate. Various temperature grades are available. The crushed aggregate will often have excessive fines that can detract from its insulating value.

Insulating aggregates are used as fills in places such as tunnel-kiln roofs, flat suspended arches, and car-bottom furnace floors. The major use of the insulating aggregates is in the form of castables in which the aggregate is combined with calcium aluminate cements. Refractory manufacturers formulate these castables

Classifications for insulating firebrick		Table II
Group	Reheat* change, °F	Bulk density,† lb/ft^3
16	1,550	34
20	1,950	40
23	2,250	48
26	2,550	54
28	2,750	60
30	2,950	68
32	3,150	95
33	3,250	95

Source: ASTM C 155-70.
*Not more than 2% when tested at indicated temperature.
†Not greater than indicated density.

into numerous combinations of aggregate sizing, aggregate temperature grades, and types of calcium aluminate cement.

Refractory fibers

Many insulating fibers will block radiation, stop convective currents, and still have a minimum amount of thermal conduction. For high-temperature applications (1,000 to 3,000°F), fibers classified as ceramic or refractory are used. In general, these refractory fibers are considered in four broad categories:

- High-silica-leached and fired-glass fibers.
- Flame-attenuated pure silica fibers.
- Alumina-silica fibers.
- Pure metal-oxide fibers.

The silica fibers have their major application in the aerospace industry. They are very expensive and currently have limited use in the industrial field. Chemically produced metal-oxide fibers are also quite expensive, but their high-temperature capabilities are meeting some very critical needs. For example, pure alumina fibers can be used for insulations up to 3,000°F.

The bulk of the refractory fibers in use are the alumina-silica ones because of their relatively low cost and good thermal properties. Our discussion will primarily deal with alumina-silica fibers, but the principles generally apply to the other fiber types. The easiest way to understand the differences between refractory fibers is to examine their temperature limits (see Table IV).

Characteristics of refractory fibers

The features most important for standard applications are fiber diameter and thermal stability. Fiber diameter is influenced by complex manufacturing considerations. The alumina-silica fibers have average diameters ranging from 2.2 to 3.5 microns. The finer fiber produces a lower thermal conductivity at light density and high mean temperature, but has a marginal effect at higher densities.

The thermal stability of refractory fibers is, in most cases, the more important characteristic. The maximum temperature limit of any fiber is set by the manufacturer to indicate the point at which the fiber becomes thermally unstable. A phenomenon known as devitrification begins to set in at approximately 1,800°F. This crystal-forming process is responsible for shrinkage, loss of strength, and general physical degradation that occurs in such fibers at high temperature. It is a time/temperature phenomenon that rapidly accelerates at 2,300 to 2,400°F.

Let us consider the amount of shrinkage at 2,300°F. Products from different manufacturers, all rated for 2,300°F, have linear shrinkages ranging from 3.5 to 8.8%. Obviously, the thermal stability of the product depends on the amount of shrinkage. Hence, a user of refractory fibers should be aware of the difference in shrinkage of the various refractory fibers and should determine the appropriate design criteria for acceptability. The most efficient insulating product is of no value if it does not have structural integrity.

No matter which fiber is chosen, there are several characteristics inherent to all. Thermal conductivity of refractory fiber is very low compared with brick and castable refractories at the same temperature range. Low density allows a lighter steel-support structure in furnace design. Low heat storage means faster heating and cooling time in periodic kilns. Resiliency of the fiber allows it to be compressed and be packed in various areas, and also makes it resistant to mechanical shock and vibration.

Refractory fibers are completely resistant to thermal shock. They are used in severe applications such as the furnace doors of heat-processing equipment. The fibers are chemically stable except for certain acids and strong alkalies that attack them. After wetting with oil or water, the properties will return when the product is cleaned and dried.

Most products made from refractory fibers use a high percentage of fiber. Hence, the characteristics for the fibers generally apply to finished products.

Health and safety

The airborne particulates generated by handling a refractory fiber are classified as a nuisance dust that does not produce a significant toxic effect. The fiber can cause throat irritation if inhaled, and mechanical skin irritation on contact. After working with the fibers for a period of time, most people no longer react to the skin-irritating effects. It is recommended that workers exposed to the dust wear long sleeves, gloves and safety goggles. Respirators are also recommended where airborne concentrations exceed the limit for a nuisance dust.

Refractory-fiber products

Refractory fibers are processed into products such as felts, blankets, vacuum-formed shapes, sprays and paper. Almost all of these products are available for use from 1,600 to 2,600°F.

The basic material is the refractory fiber in bulk form. It is inorganic and can operate continuously at

Brick equivalents for different wall thicknesses	Table III
Wall thicknesses, in.	**Brick equivalents, BE/ft^2**
2½	3.6
4½	6.4
9	12.8

1 ft^3 = 17.07 BE
BE = 101.25 in.^{3}

Temperature limits for refractory fibers	Table IV
Refractory fiber	**Temperature range, °F**
Fluxed alumina-silica	1,600 to 1,800
Alumina-silica	2,300 to 2,400
Modified alumina-silica	2,600 to 2,700
Pure metal oxides	3,000 to 3,100

the maximum recommended service temperature. In single-use applications, the fiber can be taken well above this temperature limit.

There are three varieties of bulk fiber: cleaned or uncleaned, lubricated or unlubricated, and long or short length, or a combination of these.

Uncleaned fiber still has the "shot" or unfiberized material in it, and normally has the longest fiber length. It is not reprocessed before being used as packing or loose fill in expansion joints, furnace-crown cavities, or wall cavities. The long fiber length tends to tie the bulk material together. Lubrication added to bulk fiber makes handling easier. This uncleaned fiber is the most inexpensive type.

Cleaned fiber has been processed in order to separate the shot from it. Certain applications such as aerospace and paper making cannot tolerate a large amount of unfiberized material. The process of shot separation also tends to break the fiber into shorter lengths.

Lubricated fibers contain an organic lubricant that is added in the fiberizing process to give the fibers a thin coating. This allows them to be more easily worked when further handling is required, such as in packing joints. The lubrication allows the fibers to slip on one another, and also significantly reduces dusting. However, a lubricant must not be used on fiber intended for wet mixing since such fiber will not properly disperse in water.

All bulk forms of refractory fiber are highly resilient and are often used to reduce brick cutting in the field, to make emergency repairs, and to pack void spaces in refractory construction. Bulk-fiber density will range from a loose fill of 6 lb/ft^3 to a maximum for packed material of about 12 lb/ft^3. Packing to higher densities will cause the fibers to break up and lose resiliency.

Refractory-fiber felts and blankets

Refractory fibers in the form of felts and blankets are used as linings for furnaces, kilns, and other high-temperature equipment.

Felts are organically bonded mats in roll or sheet form. The product has good cold strength and is used when a semirigid insulating sheet is required. Upon firing, however, the phenolic binder burns out. This causes two problems. First, the binder emits an objectionable odor and irritating fumes. Second, the fired product without the binder is quite weak, which can be serious if strength is required for the material to stay in place.

High-density felts (10 lb/ft^3 or higher) are formed in the same manner as regular-density felts, but the binder is cured in a pressure press to achieve felt densities of up to 24 lb/ft^3. Such high-density felts are one answer where a lower-thermal-conductivity felt of exceptional strength is needed or where space (i.e., thickness) for a felt is limited.

Blankets are completely inorganic (i.e., no binder is used with the refractory fiber). Originally, the blankets were not "needled." In one process, the refractory-fiber mat was sent directly from the collection chamber through a set of heated compression rolls in order to heat-treat the mat into a blanket. In another process, the fiber was water-washed to remove the shot, and then refelted into a blanket. With either process, the resulting product was weak. The product from the first process was subject to laminar failure when flexed, while the weakness in the product from the second process was due to the short fibers.

The un-needled blanket is being replaced by an inorganic needled product. Here, the mat is run from the collection chamber through a needling machine. The needler has thousands of barbed needles that move up and down through the blanket. As the needles penetrate, the barbs hook some of the fibers and pull them through the blanket, binding the blanket together. Needling laces the blanket together with its own fibers. As these fibers become interlocked, the product becomes an integral unit having excellent flexibility. After needling, the blanket is heat-treated to relax the fibers and bring them into intimate contact with each other.

After firing, the strength of the blanket is as great as before because the interlocking mechanism is not affected by heat. This is very important when lining furnaces and kilns where the anchor spacing for the blanket may be as wide as 18 in.

Needled blankets have many applications such as (a) removable insulating blankets for turbines where the blankets must be flexible and resistant to physical abuse, (b) insulation wraps for investment-casting molds and for stress-relieving field welds, and (c) high-temperature sound-absorption systems. Since blanket properties such as strength vary with the manufacturer, detailed product performance data should be studied.

Another material made predominantly of refractory fiber is a board for expansion joints in brick constructions. This board is a refractory felt having a high phenolic-binder content. As brick structures are laid up, a space must be provided for the brick to expand, but this space should also be packed so heat will not escape.

The standard practice is to carefully gauge the joint spacing, and build a free-standing wall away from the first wall. This is tedious and time-consuming, and requires the stuffing of the space at a later time if insulation is desired. An expansion-joint board provides the necessary spacing in this type of construction, as well as the fiber to fill the joint. It is rigid enough to build against, yet resilient enough to spring back and refill the joint when the structure cools. Much time and installation cost can be saved by using such board.

A thin and flexible paper is made from refractory fibers in thicknesses ranging from $\frac{1}{32}$ to $\frac{1}{8}$ in. The fiber is held in a latex-binder matrix. This paper is expensive but its qualities are desired in high-temperature gaskets, linings for combustion chambers, and expansion-joint materials for molten-metal troughs. It can also be used for thermal insulation wherever small clearances are encountered.

Applications for refractory fibers

Perhaps the largest single application of refractory fibers is for the lining of furnaces and kilns. Table V lists the several categories of furnaces that use these fibers as the primary insulation in the roof and walls. Since service conditions for each application vary, different types of fibers will be required. For lining heating equipment with refractory-fiber materials, there are two basic

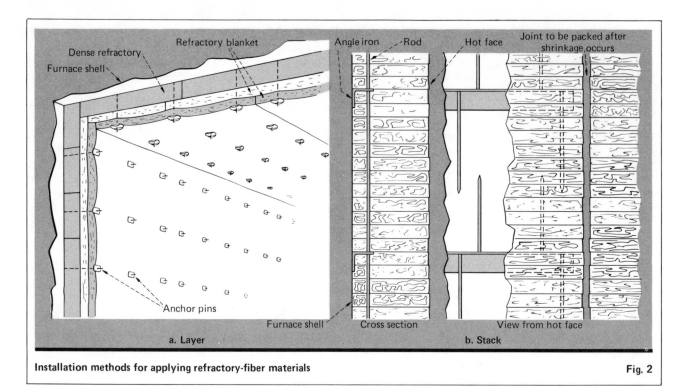

Dense refractory
Furnace shell
Refractory blanket
Anchor pins
Furnace shell
a. Layer

Angle iron — Rod
Hot face
Joint to be packed after shrinkage occurs
Cross section
View from hot face
b. Stack

Installation methods for applying refractory-fiber materials — Fig. 2

methods—layer or stack construction. Typical construction features for each method are shown in Fig. 2.

Layer construction involves impaling one thickness of insulation on top of another on anchors attached to the furnace shell, until the required thickness is achieved. However, major differences exist in layer construction for furnaces operating up to 2,250°F and those up to 2,600°F in the anchor system and in the type of hot-face fibers. The layer method also enables the use of various types of refractory felts or blankets in combination. For example, a needled blanket (useful to 1,600°F) becomes the backup insulation for the hot-face refractory-fiber material. The needled blanket is less expensive than the hot-face alumina-silica lining, so cost savings are possible.

Stack construction uses refractory fibers or blankets cut into strips that extend from the hot surface to the cold surface of the furnace. These pieces are stacked upon each other in the field or formed into a module that is subsequently mounted as a single unit.

Anchoring systems

In layer construction, a variety of anchors are used to meet varying temperature and operating conditions.

Applications for refractory fibers	Table V
Type of furnace	**Temperature, °F**
Brick firing	1,800 to 2,200
Ceramic firing	1,800 to 3,250
Forging	2,250 to 2,450
Heat treating (metal)	1,500 to 2,100
Petrochemical heaters	1,600 to 2,100

All forms of alumina-silica fiber require anchors that will match, or exceed, these conditions.

Metallic anchors are available in a variety of lengths and alloys for temperatures to 2,250°F. These anchors are welded to the furnace shell, and refractory-fiber insulations and blankets are impaled on them. An anchor washer is then installed to hold the insulation in place.

Ceramic anchors are used where temperature limits exceed 2,250°F. One system uses a Type 304 or 310 alloy stud, welded to the furnace shell. A ceramic pin is threaded over the metal stud. A ceramic washer locks onto the head of the ceramic pin. A major advantage for this system is that a less expensive alloy can be used for the stud as long as the ceramic pin extends into the refractory lining far enough.

Another method uses a ceramic locking cup. The cup extends into the insulation 2 in., and locks onto the standard metal anchor. The cup is then filled with either bulk fiber or insulating cement. The ceramic cup does not penetrate the lining very far, and requires the use of a high-alloy stud. Even then, the temperature on this stud could easily exceed 2,250°F in high-temperature furnaces.

Ceramic anchors are required in carbon-rich reducing atmospheres. The nickel in alloy-metal anchors catalyzes the reduction of carbon monoxide, and carbon accumulates around stud locations where the temperature is 1,000 to 1,200°F, to cause fiber deterioration.

Modular construction

The modular system is a method of prefabricating the refractory blanket into 12-in.-square modules that can be rapidly installed. The modules are attached to the furnace steel and eliminate the layer-by-layer buildup.

In one patented modular system, the strips of blanket

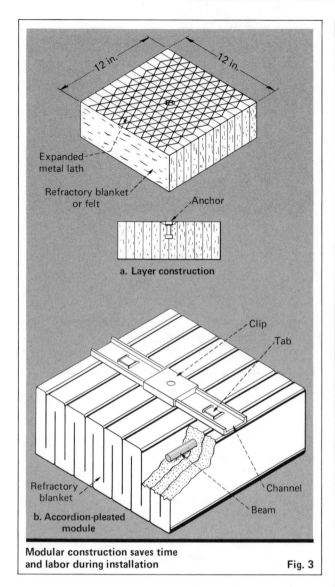

Expanded
metal lath

Refractory blanket
or felt

Anchor

a. Layer construction

Clip

Tab

Refractory
blanket

Channel

Beam

**b. Accordion-pleated
module**

Modular construction saves time
and labor during installation

Fig. 3

restraints are removed and the refractory fiber expands. Since shrinkage occurs at high temperature, the accordion pleats expand and offset any shrinkage.

Economic considerations

There are several tradeoffs to consider when comparing layer methods to stack or module ones. In layer construction, less-expensive backup insulation is easily used. Stack and module construction employs one material that extends from the hot surface to the furnace shell. Material costs, of course, will be higher for the stack method.

The opening of shrinkage joints at high temperatures is another factor. Layer methods offset the joints from layer to layer, and provide a small amount of compression between the blanket edges. In the modular method, the joints fill during fiber expansion. The success of this depends on how well the job is installed. In some cases, the joints may need to be packed after the initial firing.

The undisputed advantages of modular construction are the large decreases in installation time and labor costs. The buildup of layers and the intricate pin-layout required in layer construction are eliminated. Installation time is saved because the modules are light and more easily handled than large rolls of blanket or felt. Essentially, two-thirds of the onsite labor time is replaced by shop prefabrication. Thus, installation costs and furnace downtime are reduced.

Installation advantages also come from the hidden anchoring system of the modules. Since each module has the hardware located close to the insulation's cold face, lower-cost alloys may be safely used at furnace temperatures that would normally require ceramic anchors. Atmospheric deterioration of the anchors is eliminated because the hardware is located in a relatively cool zone. When repair or replacement is necessary, only the damaged modules are removed. With blankets or felt rolls, such spot repairs are difficult, and large areas must often be replaced.

Layer construction has a definite advantage over modular in relation to thermal efficiency. Fiber orientation in layer construction is perpendicular to the direction of heat flow; whereas in the stack module (Fig. 3a), the fibers are parallel to the flow. This orientation can increase thermal conductivity 20 to 40% as indicated by manufacturers' tests. Hence, more insulation is needed in stack-module construction than in layer construction to achieve the same thermal resistance. The accordion-pleated module (Fig. 3b) will have a thermal efficiency that is greater than the stack module but not as good as the layered construction, due to the orientation of the blanket. Here, the fiber orientation is both parallel and perpendicular to the heat flow.

Retrofit applications

Rising fuel costs are causing a closer look at kiln and furnace operations in relation to heat losses. If the unit is inefficient but the refractory lining is still serviceable, its efficiency can be increased without replacing the entire refractory lining. This can be done by installing a refractory-fiber blanket directly to the hot face of the existing surface. Since refractory fibers do not soak up a lot of heat, they do not require large amounts of heat to

or felt are stack-constructed. The stacks are attached to a 12-in.-square expanded-metal lath by using a refractory cement. The module is then fastened to the furnace shell with a one-step blind-welding and threading operation. In another version, the lath is welded directly to the furnace shell. The strips are compressed when mounted to the metal lath. The module tends to expand to fill the shrinkage joints that occur at high temperatures. However, the joints in this system do open, and must be stuffed with bulk fiber or blanket strips during the first cooldown cycle.

Another patented module is made with a single piece of inorganic, needled, refractory-fiber blanket 12 in. wide. This blanket is accordion-pleated so each section is as deep as the final thickness of the lining (between 5 and 12 in.). Stainless-steel support beams are buried in the fiber pleats next to the cold face of the block. A stainless-steel suspension channel attached to the beams fits into a stainless-steel clip that is fastened to the furnace wall.

Banding and kraft-covers keep the pleated blanket in compression prior to and during application. After the module is attached to the furnace shell, the compression

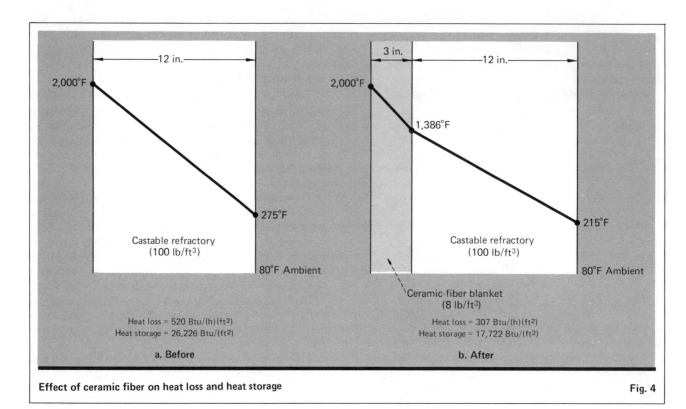

Effect of ceramic fiber on heat loss and heat storage   Fig. 4

reach equilibrium conditions. In this application, the overall heat loss will be reduced because of additional insulation and the elimination of cracks on the hot face that allow heat to escape. Generally, fiber insulation can be attached to the existing refractory surface by using the wallpaper or modular-veneering technique.

Wallpaper construction—Major cracks or spalled areas should be repaired. A desired stud pattern is laid out. Layers of blanket are then impaled on the studs to the desired thickness. The reduction in overall heat loss for a typical installation is shown in Fig. 4.

Veneer construction—A module is made from strips of refractory fiber 12 in. long, turned edge-grain up and held together with an open-mesh organic cloth. Using an air-setting refractory mortar, the modules are installed over refractories such as dense brick, insulating firebrick, bubble alumina, semisilica brick, castables of any density, and plastic and ramming mixes. The module's thickness can vary from 1½ to 5 in. The organic cloth burns off when the furnace is first fired.

Performance

Insulating firebrick (IFB) and refractory-fiber materials are used in heating equipment for many process operations. Such heating equipment does not normally undergo aggravated chemical attack such as that found in a basic-oxygen steel furnace, or the aggravated physical abuse found in a rotary cement kiln.

Insulating firebrick and refractory fibers have very long service lives. It is not uncommon to find IFB linings in steel-annealing furnaces last more than 20 years. Refractory fibers are completely resistant to thermal shocking—so cyclic heating equipment can be elevated and reduced in temperature as quickly as the burner system will allow. On the other hand, an IFB-lined fur-

nace must be heated and cooled with reasonable caution to prevent thermal spalling.

Neither material resists mechanical abuse such as a direct blow by a steel ingot. IFB will withstand higher air velocities and will better withstand minor abuse such as gouging and small blows.

Compensating for the thermal movement of IFB makes the design and construction of an IFB-lined furnace more difficult. Refractory fibers experience none of this thermal movement.

Refractory fibers are not considered a good backup insulation to dense refractories, because they are resilient and compressible. On the other hand, IFB is thoroughly compatible with dense-refractory construction.

Refractory fibers are the optimum material for periodic heating equipment up to 2,250°F. IFB will be favored above 2,250°F. For cyclic heating equipment operating at any temperature, it is doubtful that dense refractories would be specified unless abuse of the lining or chemical resistance were a major factor.

In a very strong thermal-shocking environment (to 2,600°F), refractory fiber would be selected. Its low heat storage favors periodically operating heating equipment, and the lining will not deteriorate from thermal spalling. In such an environment, the choice is not between IFB and refractory fibers, but between refractory fibers and dense refractories. IFB is too prone to thermal spalling in viscous situations.

Until 1973, dense refractories were used almost exclusively for continuous-operating heating equipment such as tunnel kilns. Now, IFB in these applications saves energy, and reduces foundation requirements and plant space. Refractory fibers are seldom used because low heat storage is not sufficient in such equipment. Only at low mean temperatures (below 1,800°F) is a refractory fiber more cost-effective than IFB. Above 1,800°F, IFB is the optimum material for most applications.

There is an exception as regards continuous-heating equipment such as the fired heater common to refineries and petrochemical plants. In recent years, some furnace builders have made portions of these units of refractory fibers. The reasons are twofold:

■ Downtime for these units is expensive. With refractory-fiber construction, the unit can be shut down and cooled rapidly if repair is needed; the repair is done quickly, and the unit returned to service with minimum downtime.

■ The amount of and cost for the supporting steel is lower. Typically, these units are tall, and the savings in structural steel make refractory fibers cost-effective for portions of the equipment.

Almost any heating equipment that operates above 2,500 to 2,600°F is not practical for refractory fibers. The higher the temperature, the less cost-effective they are. This applies to fibers that are useful to 2,800 to 3,000°F.

Refractories in furnace design

A basic knowledge of refractories is not enough to enable one to specify a refractory for a given installation. Furnace design is a highly specialized field, and few aspects of it are more specialized than refractories. Hence, refractory manufacturers can be invaluable for selecting the optimum material. They do not design heat-processing equipment but do furnish technical data and suggestions on applications.

Specific information is required regarding a kiln or furnace and its operation. Before an intelligent recommendation can be made for a refractory, one should have a complete understanding of the unit, including operating data such as function, cycle time and temperature range; and type of fuel, including its characteristics and impurities. After analyzing the need for a refractory, choosing the optimum product, and engineering and building the refractory system, the next step is performance evaluation.

Refractory application skills are largely developed through on-the-job experience. Moreover, many decisions are based on judgment—carefully made after evaluating a great deal of information. Although innumerable factors influence refractory performance, most come under one of the following subjects: heating-equipment design, refractory-system design and type of refractory, operating practice, refractory quality, and workmanship.

Heating-equipment design

Furnaces are designed to accomplish consistently and economically certain results, such as production of high-pressure steam, reduction of an ore, refining of a metal, or heat treating. The efficiency of heat-processing equipment is normally reported as the ratio of production to fuel consumption.

The effect of furnace design on refractory life too often receives only secondary consideration until after the furnace is in operation, and high refractory costs force a redesign. There are cases where an otherwise efficient and successful furnace design was abandoned because no satisfactory lining material could be found or developed. Furnace design is necessarily the first requirement. Ideally, the refractory-system design is then done, and the furnace design altered as needed and if possible. Thus, refractory and furnace people must work as a team to optimize performance.

The refractory members of the team are not expected to be furnace designers, but should be able to recognize design features that are inherently dangerous from the standpoint of refractory service. Such features are most frequently found in home-made furnaces and in well-designed furnaces that have been altered to reduce fuel consumption or increase production.

Insufficient combustion space (frequently referred to as "a bottled-up condition") is a common cause of refractory failure. More heat is released within the combustion space than is absorbed by the charge, dissipated through the furnace walls by heat flow, or carried out with the flue gases. As a result, refractory-wall temperatures can approach flame temperatures (above 3,000°F). Coupled with impurities, such temperatures will quickly destroy even high-quality refractories.

It would be convenient to say that a heat release of X Btu/(h)(ft³) of combustion space were the upper limit for the satisfactory use of a given refractory. Unfortunately, it is not this simple. Many other factors—including refractory-wall thickness and height, effect of the amount of insulation, fuel type, amount of excess air, and especially the amount of heat absorbed by the charge or the water tubes—influence the amount of heat that can be released without disastrous effects. Judgment based on experience must be the guide, rather than a formula.

Theoretically, a single combustion space operates at a uniform temperature. This rarely occurs in practice. Only one thermocouple can control temperature in a firing zone, and it is usually placed to control the temperature of the ware or load.

Furnace drawings should be carefully studied to determine the location of control and shutdown thermocouples, and of furnace areas that could be expected to exceed the thermocouple settings. One must study the intended thermocouple settings to determine the maximum controlled-upset condition in the furnace. From this, factors of safety for refractory selection can be established, including possible zones where different refractories can be used.

When portions of the furnace lining (division walls, bridge walls, piers, door jambs, or the nose of an arch) are exposed to high temperatures on more than one side, they frequently fail before the rest of the lining. Failure may be due to load, shrinkage, excessive vitrification and spalling, slagging, or a combination of these. Sometimes, a change in the design of the refractory lining or in the refractory material will eliminate, or at least improve, conditions in the vulnerable spots.

Many burner designs are in use, and some can affect refractory selection. For example, a long-flame front burner can cause flames to lick the opposite wall. A bag wall or muffle in front of a burner can reradiate heat onto the refractories around the burner. A top-mounted flat-flame burner can direct excessive heat to a roof or onto a skew.

210 EQUIPMENT AND MATERIALS

Refractory-system design

The study of a refractory system should involve three specific areas, whether one is designing a system or evaluating a refractory. These are: (1) thermal—the refractory system must resist the attack of the heat that is expected; (2) structural—the refractory structure must be mechanically sound; and (3) chemical—the system must resist the attack of whatever chemical materials are present.

Thermal considerations

Temperature—Obviously, the refractory system must withstand the expected temperature of the heat-processing equipment in which it is to be used. However, this is much too simple. The hot-load strength of the refractory, while it might appear to be a mechanical consideration, must be studied as a thermal consideration because it is the thermal design of the system that will determine the refractories to be used.

Load-bearing strength—The refractory's load-bearing strength, measured as hot-load deformation, is not usually of major importance where the refractory is exposed to heat on only one face. In most instances, there is a fairly steep temperature gradient through the lining, and the load is largely carried by that portion of the lining cool enough to be below the temperature at which any softening may occur.

Load deformation—In other refractory constructions, hot-load deformation is the primary factor for refractory selection. Potential trouble spots would arise when the refractory was exposed to high temperature on more than one face, or where the lining would be heavily insulated, thus decreasing the temperature drop through the wall. The ceramic process of why deformation takes place is covered by Norton,[*] and others.

There is no ironclad rule or formula to predict the amount of deformation that occurs in actual service. A fireclay refractory may experience hot-load deformation when the refractory is exposed to soaking heat, or when the temperature gradient through the refractory is very slight. Although it is risky to make a general rule, it is safe to say that hot-load deformation may occur in a fireclay brick, including IFB, when the refractory will experience anywhere within it a temperature of 2,250°F, or above, with a load of more than 10 psi. At higher temperatures, say above 2,400°F, even lower temperatures within the refractory, and lower loads, may cause trouble via hot-load deformation over an extended period of time.

Backup insulation—Since a load-bearing refractory system might use dense refractory, it is necessary to consider the effect of heavy backup insulation. As an example, let us consider a wall built of $13\frac{1}{2}$ in. of firebrick, and backed with 1 in. of insulating block. With a hot-face temperature of 2,600°F, about 4 in. of the inner brick wall will be exposed to a temperature above 2,250°F. With a concentrated heavy load on the inner portion, this part of the wall could deform. However, with a uniformly distributed load over the entire $13\frac{1}{2}$ in. of brick, there would be little chance of trouble because the relatively cool $9\frac{1}{2}$ in. of brick in the back

*Norton, F. H., "Refractories," 4th ed., McGraw-Hill, New York, 1968.

would carry the entire load, and the wall would remain structurally stable.

This example illustrates the occasional necessity for trial-and-error approaches in refractory-system design. Hot-load deformation properties are used to compare one manufacturer's refractory brick to another's. Experience must be added to hot-load properties to determine which refractories tend to work in given situations, and which do not.

Expansion—Expansion can cause a wall to bow, cause pinch spalling, and create other problems. Many failures have been identified as due to not providing for expansion. The correct procedures, allowing for expansion spaces, are available in refractory manufacturers' installation and construction manuals.

Thermal spalling—IFB has been used in some thermal-spalling situations but is not considered as resistant to spalling as dense castables, fireclay brick, or fireclay-ramming mixes.

Structural considerations

Roof construction—Heat-processing equipment generally requires a roof. The options include a sprung arch or dome, a flat suspended arch, and variations of these. A sprung arch is the least expensive approach, but is limited to small spans. Otherwise, one must use a flat suspended arch, or division walls so that several arches can cover the required roof area. Arches and domes are more successful with low-rise (down to 2 in./ft or even $1\frac{1}{2}$ in./ft of span) roofs, particularly at higher temperatures. This helps to eliminate pinch spalling.

Most insulating refractories require some sort of anchoring. This is one of the most critical aspects of castable and IFB constructions. Therefore, each application requires careful study in order to get the optimum anchoring system.

Abrasion/erosion resistance—Usually, this is a straightforward problem because the materials' limitations are known. What is not appreciated, however, is that erosion occurs more rapidly in turbulent-flow zones such as a corner than in nonturbulent zones such as a straight flue. Thus, the design must recognize that if erosion can occur, it will certainly occur in the turbulent areas. These areas might benefit from a more abrasion/erosion-resistant construction.

Permeability—An important factor with IFB is its very high permeability. Insulating-refractory constructions commonly use lower-temperature grades of refractory as the cold face is approached. If a flow of hot gases can take place through the lining, then damage to the backup insulation can be expected. Hence, we should be very cautious in furnaces where a high positive pressure exists within the furnace, and flow to the outside is possible.

Mechanical spalling—This occurs from high differential stresses placed on the refractories. Refractory fibers, being resilient, do not encounter this problem.

Chemical considerations

Refractories are affected by the action of the atmosphere in the heating equipment, and by chemical attack on permeable materials.

In heat-treating equipment, nonoxidizing atmo-

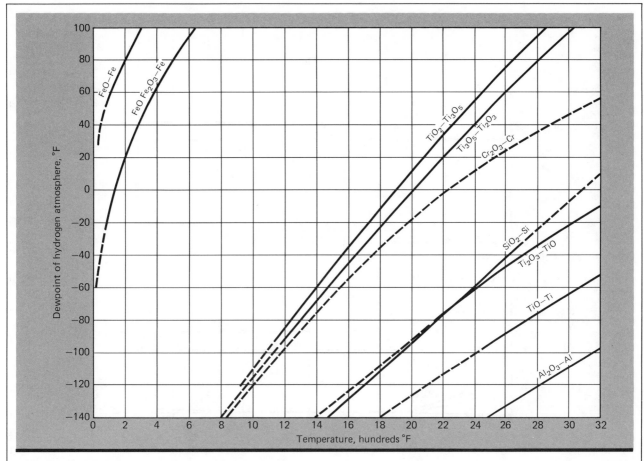

Metal/metal-oxide equilibria in a hydrogen atmosphere   Fig. 5

spheres create some problems for insulating refractories. Low oxygen pressure (i.e., concentration) reduces Fe_2O_3 to FeO. Since FeO is less refractory than Fe_2O_3, this condition promotes hot-load deformation of the refractory at temperatures above 1,800°F. This effect will also reduce SiO_2 to SiO (a gas), which also causes refractory failure. Upon cooling, the SiO tends to form deposits on surfaces in heat exchangers, boilers and reformers.

Equilibria conditions for a number of metal/metal-oxide combinations in a hydrogen atmosphere are shown in Fig. 5 for the relationship between dewpoint and temperature. As long as operating conditions are maintained to the left and above a particular curve, the metal's oxide will be stable. If the conditions are to the right and below a curve, the oxide will be reduced. This chart is true only for IFB and high-density brick. Refractory fibers should not be used above 900°F, or at a dewpoint lower than −20°F.

As the dewpoint is lowered, the service temperature must also be lowered to prevent a particular oxide from being reduced. For example, if the dewpoint for SiO_2 is −60°F, the temperature must be maintained below 2,400°F to prevent reduction of SiO_2.

A disintegration triggered by the catalytic decomposition of carbon monoxide or hydrocarbons such as methane also occurs in prepared atmospheres. Let us review the mechanism for failure due to this reaction. Ferric oxide (Fe_2O_3) is present in the refractory in local-

ized concentrations. This is converted to iron carbide (Fe_3C) that catalyzes the decomposition of carbon monoxide at 750 to 1,300°F to carbon dioxide and carbon. The carbon is deposited on the Fe_3C. Carbon builds up on the catalytic surfaces that are under stress, and ultimately causes disintegration of the refractory. In many cases, such stresses are severe enough to burst the steel shell of the furnace.

A similar condition exists in hydrocarbon atmospheres, which persists up to 1,700°F.

The risk of carbon disintegration can be minimized by using refractories above the carbon-deposition temperature, and by using products having low iron content. This effect is more noticeable in dense-brick or castable refractories.

Thermal conductivity

Furnace atmospheres also have an effect on the thermal conductivity of insulating refractories. Let us review how the thermal conductivity of the furnace gases affects these refractories. Fig. 6 shows the thermal conductivities for air and hydrogen. The component of the thermal conductivity affected by changing the gas constituents in the furnace atmosphere is the gas conduction. It is quite apparent that the gas (normally air) in the pores of an insulating refractory can be readily replaced by other gas constituents found in the furnace atmosphere.

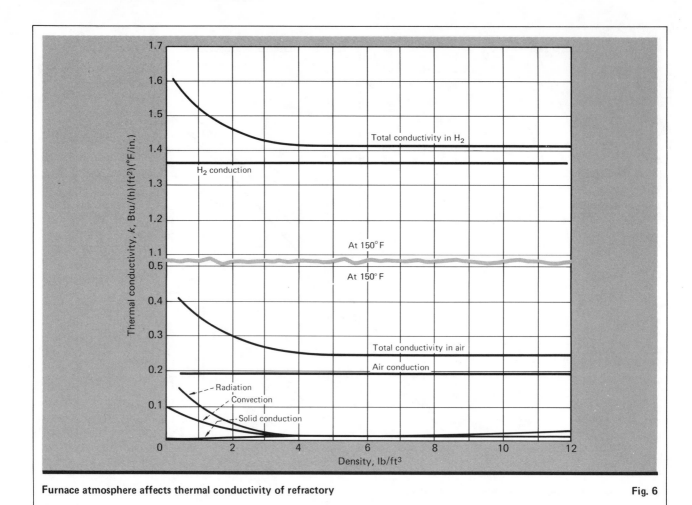

Furnace atmosphere affects thermal conductivity of refractory

Fig. 6

Replacement or dilution of this air constituent by other gases will change the insulation's k value. The amount of change is dependent on the k value of the replacement gas, and the porosity of the insulation. Most gases involved in a heating atmosphere have essentially the same k value as air (Fig. 7). However, hydrogen has a very high k value, causing a significant change in insulation effectiveness.

Other compounds can be found in small quantities in many furnace atmospheres. They can originate in the fuel, the refractories, or even the charge in the kiln or furnace. One of these is vanadium pentoxide (V_2O_5) from low-grade fuel oils such as Bunker C. Another vanadium compound, sodium vanadate, may be found in oil flames as droplets. It appears to decompose and cause alkali attack at about 1,240°F.

Sulfur occurs in fuels and in some clays. Depending on its content and chemical form, it can become part of a furnace atmosphere as sulfur oxides.

Alumina-silica (45 to 54% Al_2O_3 range) maintains the highest hot strength of the several refractory materials cycled at 1,400 to 1,800°F in the presence of an SO_3 atmosphere.

Test data show that disintegration of the refractory may occur if Na_2SO_4, $MgSO_4$, $Al_2(SO_4)_3$, or $CaSO_4$ are formed in a sulfur dioxide atmosphere.

Insulating refractories are seldom used in installations where chemical attack is expected. Unfortunately,

it often arises unexpectedly. Primarily, such attack is due to the permeability of insulating refractories. Many times, dense refractories resist chemical attack, not because of chemical resistance but because their very high density and low permeability prevent damaging materials from entering.

Slagging is defined by the American Soc. of Thermal Manufacturers as "the destructive chemical reaction between refractories and external agencies at high temperatures, resulting in the formation of a liquid with the refractory."

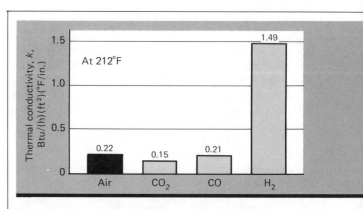

Hydrogen conductivity affects insulation effectiveness Fig. 7

Attack from mill scale occasionally causes slagging. As ferrous-metal objects are heated, their surfaces oxidize, and the resulting iron oxides flake. These oxides fall onto the hearth, and will readily attack an alumina-silica refractory because the iron oxides act as a fluxing agent. Should the oxides become airborne, they can contaminate refractories on the furnace wall.

Other materials heated in a furnace can throw off other oxides, considered to be fluxes for an alumina-silica refractory. Incinerators are the worst of all; they burn or oxidize everything put into them.

Refractory fibers are troublesome in fluxing situations. Experience dictates extreme caution when applying such fibers above 1,800°F. However, these fibers have worked well in incinerator afterburners above 1,800°F, perhaps because the fluxes by then have become so oxidized that they are no longer able to attack the refractory.

Chemical attack of refractories has become much easier to diagnose. Chemical analysis of refractory specimens can be made quickly and inexpensively by most refractory manufacturers. The methods used include spectrographic analysis, x-ray diffraction, and wet chemistry. The best approach is to analyze specimens of the damaged refractory, and of the original refractory before firing. On the basis of these data, a decision is made. One comparison may be between the damaged refractory and its original composition. If a fluxing element is found that should not be present, its source is searched for. If the source cannot be eliminated, an alternative refractory must be used.

Direct reactions such as a flux from a ceramic glaze being fired and an alumina-silica IFB are relatively easy to evaluate. However, there are some secondary reactions in which a material in the furnace atmosphere will act as a catalyst, reacting with a second element in the atmosphere and a component in the refractory. These are considerably more difficult to evaluate.

Operating practices

Premature refractory failures may arise because of unintentional changes in furnace conditions. These changes, while unobserved by the operator, may be of long enough duration to destroy a refractory lining but not long enough to be considered a change in practice. The changes may arise from: burners being out of adjustment, furnace atmospheres that are highly reducing for a period of time, furnace temperatures becoming abnormally high, or variations in composition of the raw materials or fuels.

Such changes in practice are usually difficult to discover. They leave no record other than their effects on the refractory lining. Failure of the refractory may not become apparent until weeks or months after the actual damage has been done. This makes investigation difficult, and often raises credibility problems.

Sometimes, the premeditated abuse of heat-processing equipment arises in order to get increased furnace throughput and/or greater return on assets. In almost every case, operating a furnace and refractory system beyond its capacity will reduce refractory life. This abuse is common, and there are really only two alternatives. One, stop doing it; or two, expect a shorter life from the refractories.

Workmanship

It is easy to blame brickmasons for every failure of refractory brick. By the time failure occurs, the evidence is practically destroyed. While good brickwork is extremely important, there are other factors involved. These probably occur more frequently than either poor brickmasonry, or refractory products that are not up to standard quality.

These other factors include the proper size and location of expansion joints, the rise of a furnace arch, and the thickness of a wall. These are not factors of masonry workmanship, and should be taken care of in the design of the refractory system. In practice, many design details are left to the brickmason doing the work. In some cases, this is the best way to handle such details.

Sometimes, however, brickmasons called on to do refractory work have only had experience in the building trades. Such masons make mortared joints up to ½ in. thick. This practice is not compatible with IFB where commonly the mortar may not even be visible. The problem seldom happens in large companies employing their own masons. Smaller companies should keep this in mind, and make sure that the person in charge of the masons understands high-temperature-furnace work.

Steven Danatos, Editor

The authors

James E. Neal is manager of refractory engineering and technical services for Johns-Manville Corp., Ken-Caryl Ranch, Denver, CO 80217. He has had more than 35 years of experience with refractories. A graduate of Rutgers University, he has a B.S. in ceramic engineering. Mr. Neal is a member of the American Ceramic Soc., Tau Beta Pi and Keramos. He is chairman of the refractory-fiber committee for the American Soc. for Testing and Materials.

Roger S. Clark is merchandising manager of refractories for Johns-Manville Corp., Ken-Caryl Ranch, Denver, CO 80217. He has been associated with the refractory industry for many years. Prior to joining Johns-Manville in 1972, he was sales manager of Babcock & Wilcox's ceramic fiber division. He has a B.S. in mechanical engineering from Texas A&M University, and is a member of ASME and the American Ceramic Soc.

Cost-effective thermal insulation

With increasing energy costs, increased insulation can give a greater return on investment than increased plant capacity. Here is what you need to know about insulation selection, thickness calculations and economics.

Michael R. Harrison and *Charles M. Pelanne*
Johns-Manville

☐ It is estimated that the U.S. chemical process industries could, by the early 1980s, save 252 trillion Btu/year by increasing its use of insulation and improving system maintenance. About 200 trillion Btu of this would come from improved insulation of steam lines.

For the U.S. petrochemical sector, energy costs are some $4.75 billion, or 23% of the total operating cost.

Under today's energy costs, increased insulation frequently pays a better return on investment than increased capacity. Hence, insulation is moving up from its traditional 2% of total plant construction costs to figures ranging from 4 to 8%.

Consequently, economic calculations are more critical and insulation system design and installation are given more engineering thought. The process industries are also upgrading the types of insulation being used, to make sure that the energy savings will continue through the life of the plant.

Among the recent trends in the use of insulation in chemical process plants have been:

■ Creation and use of computerized methods for the calculation of insulation thickness.

■ Greater use of contract companies that are specialists at installing insulation.

■ An upgrading of the insulation materials used for process plants.

■ More attention to jacketing. Almost all insulation for process plants is now jacketed, to provide protection against weather, mechanical damage, and fire. Most jacketing is aluminum or, where acid-corrosion problems exist, stainless steel. Some plastics, such as polyvinylacetate, are also used because of their resistance to a wide range of chemicals.

■ A long-term effort to get better values for the heat transfer properties of insulation as used. The laboratory goal now is a systems approach, giving accurate numbers for the total heat-loss of a whole piping system.

■ Greater use of engineered insulations, designed to give specific combinations of insulating value, size, weight and damage resistance. Combinations—layers of different insulating materials and jackets—are also used this way.

■ Newly fabricated forms of insulation designed to make easier the most difficult insulation installation jobs: around valves, flanges, and other fittings.

Originally published December 19, 1977

Particular CPI insulation requirements

Most insulations employed in plants of the chemical process industries are in the lower intermediate range, from ambient to about 500°F. Most applications are on steam lines, process vessels and process piping.

In these applications, the insulation generally should meet five criteria:

1. The material must be compatible with the chemicals being processed. It should be essentially inert regarding flash-point depression and similar properties, when chemicals may contact the insulation.

2. The material's properties, particularly thermal-transport properties, should be reproducible so that they do not vary significantly from batch to batch. This is particularly important when process materials must be kept within a specified temperature range.

3. Both the insulation and its barrier system should resist mechanical abuse over the life of the plant, since both piping and vessels in CPI plants are generally accessible to workers and are walked on, have ladders leaned against them, and tools dropped on them.

4. The barrier system has to be effective, allowing little or no water leakage. Penetration of the insulation by water seriously diminishes its insulating properties until it dries out.

5. Installation of the insulating material on valves and pipe fittings should be as easy as possible. Normally, the cost of insulating a steam valve is about three times that of insulating a similar length of straight pipe.

The material most often used in insulating process vessels and piping for intermediate temperatures is calcium silicate, because of the need for resistance to mechanical abuse. For further protection against mechanical abuse, as well as protection against weather and other damage, jacketing is almost always employed. The corrosion resistance of the jacket is also an important consideration.

For low-temperature applications (minus 150°F to chilled-water temperatures), expanded polyurethane and polystyrene, or foamed glass are most often used. Permeability to water vapor is the major problem faced in these applications. Water vapor penetrates and deposits as ice at the cold surface, progressively destroying the insulation properties of urethane or styrene, even though the warm surface appears intact. Foamed glass, which both resists moisture penetration and better withstands the periodic heating and drying necessary to counteract this moisture penetration, is growing in popularity over the plastic foams.

Fiberglass is the insulation most used for applications ranging from the temperature of chilled water up to the boiling point (212°F). Fiberglass would work as well as or even better than other insulations at colder temperatures (it is an extremely efficient insulator even in the cryogenic range) if the vapor-barrier could be maintained intact. But at below freezing temperatures, a vapor-barrier breakdown can produce a massive failure with fiberglass rather than a localized failure as foam might undergo. With fiberglass, the vapor could freeze and destroy the insulation for a whole tank, whereas with foam, because it is a series of air cells, the freezing vapor would not penetrate as far.

Lower-temperature cryogenic systems generally employ either massive solid insulations or vacuum types. In the latter, highly polished metal supporting-walls

Estimated short-term energy-conservation potential in the six largest energy-consuming industries				Table I
	Estimated energy savings by insulation and maintenance*		Estimated total industry process energy usage†	
Industry	10^{12} Btu/yr	10^{15} joules/yr	10^{12} Btu/yr	10^{15} joules/yr
Chemical	252	266	2,662	2,810
Primary metals	500	528	5,900	6,230
Petroleum	343	362	3,044	3,210
Paper	208	219	2,563	2,700
Glass, cement, structural clay products, others	190	201	2,000	2,110
Food	47	50	1,283	1,350

*Estimated precision, ±30%.

†Process energy only; does not include feedstock energy usage. Estimated precision, ±10%.

Industry (or process) and operation	Radiation, convection, conduction, and unaccounted losses[†]	Rejected heat, (10^{12}Btu/yr*)				Total, all ranges
		Below 100°C	100–250°C	250–800°C	800–1,800°C	
Chemical						
Chlorine/caustic soda	40	254	51			
Ethylene/propylene	20	151	79	40		
Ammonia	32	115	139	40		
Ethylbenzene/styrene	1	24	28			
Carbon black	2		32			
Sodium carbonate	2	36	16			
Oxygen/nitrogen	1	171	147			
Cumene		4	4			
Phenol/acetone		8	8			
Other	150	628	290	79		
Total	**248**	**1,391**	**794**		**159**	**2,345**
Primary metals						
Steel	298	79	397	596	795	
Aluminum	119	378	79	20	79	
Other	119	397	119	40	79	
Total	**536**	**854**	**595**	**656**	**953**	**3,058**
Petroleum	318	994	497	695		2,186
Paper	199	1,391	755	119		2,265
Glass, cement, other						
Cement	119	79	40	219		
Glass	99	60	12	87	36	
Other	99	79	40	159	20	
Total	**317**	**218**	**92**	**465**	**56**	**1,049**
Food	60	795	159	119		1,073
Grand totals	**1,679**	**5,643**	**3,110**	**2,213**	**1,009**	**11,975**
Ratio to grand total Σ_T		0.47	0.26	0.18	0.09	

Estimated heat-rejection in the six largest fuel-consuming industries (1972–1973)   Table II

*To convert to 10^{15} J/yr, multiply by 1.0543. The estimated accuracy of the values is ±20%.
[†]Radiation, etc., losses are distributed within other losses shown in the table.

have a vacuum space, reflective foils, and various inorganic or organic materials coated with metals between them.

The high-temperature range above 1,000°F calls for more-heat-resistant materials. Eight principal types of thermal insulation are employed: 1. calcium silicate based material (on piping and vessels); 2. mineral fibers; 3. ceramic fibers; 4. oxide fibers; 5. carbon fibers; 6. rigid ceramic insulating-brick; 7. castable ceramic insulating-refractories; and 8. multiple metal-foil systems for vacuum applications (similar to the cryogenic insulation).

Insulations available

Insulation materials are normally classified according to the service-temperature ranges they are suitable for. These tend to be arbitrary. When operating temperatures reach a certain upper limit, the materials may become structurally unstable, or become uncompetitive because they have a relatively high thermal conductivity. Whenever lower-temperature limits are given for most insulations, it usually simply means that they are not cost effective in these applications, not that they are technically unable to do the job. Within each temperature range, choice among the materials available is generally based on other properties and cost.

Cryogenic range (455°F to −150°F)

Insulations fall within two types: vacuum and massive, the latter consisting of one or more solid phases distributed with a gas, such as dry air, to produce a very low thermal conductivity.

Vacuum insulation systems, consisting typically of highly polished metal supporting walls with a vacuum space between them (sometimes with multiple metal reflective foils or opacified powders inside) are usually custom-designed and installed by the insulation vendors. Performance is generally guaranteed.

Representative equipment insulations				Table III
Insulation type	Temperature range, °F	Conductivity, k Btu/(h)(ft²)(°F/in.)	Density, lb/ft³	Applications
Urethane foam	−270 to 225	0.11 to 0.14	2.0	Tanks and vessels
Fiberglass blankets	−270 to 450	0.17 to 0.60	0.60 to 3.0	Chillers, tanks (hot and cold), process equipment
Elastomeric sheets	− 40 to 220	0.25 to 0.27	4.5 to 6.0	Tanks and chillers
Fiber glass boards	Ambient to 850	0.23 to 0.36	1.6 to 6.0	Boilers, tanks, and heat exchangers
Calcium silicate boards, blocks	450 to 1,200	0.22 to 0.59	6.0 to 10.0 to	Boilers, breechings, and chimney liners
Mineral fiber blocks	to 1,900	0.36 to 0.90	13.0	Boilers and tanks

The theory behind vacuum cryogenic insulation is that of the double-walled dewar flask, which relies in part on vacuum between the walls, and in part on reflection of radiant heat. Without gas inside, heat transfer will be mainly by radiation. Coating the inside hot surface facing the evacuated area reduces heat transfer to a level proportional to the emissivity of the coating (0.01 for silver, for example).

The floating-shield approach takes advantage of the fact that thermal radiation can often be cut in half by floating a radiation shield between the cold and warm surfaces. Powders such as expanded perlite, silica aerogel, carbon black, calcium silicate, diatomaceous earth and fiber are used; the material is packed in before the air is pumped out.

The multilayer type, a series of reflective foil shields of aluminum separated by low-conductivity fillers such as fiberglass, comes in blanket form. A related type—a number of aluminum-coated layers of polyester film, crinkled to reduce heat transfer by conduction from layer to layer—withstands high acceleration loads without loss of insulating effectiveness.

Less costly than the evacuated forms described above are foam types. Foamed polyurethane and polystyrene, either foamed into flexible sheets, foamed in place or foamed in rigid insulation sections, initially showed great promise, but deterioration from permeation by water vapor and air have caused dissatisfaction. Foamed glass stands up better to drying-out processes and is less permeable. Foam types work through their microstructure of closed cells containing air or other gas. The cells, although not as small as in some newly developed insulations, are small enough to greatly reduce heat transfer by gas-phase convection, assuming they are filled with a gas other than air.

Low-temperature range (−150°F to 212°F)

Some of the evacuated types of insulation for cryogenic service are used in the lower end of this range, but foams are employed more. The main problems are moisture permeation, and fire hazards—and, at lower temperatures, heat-transfer rates that are too great to be acceptable.

Since the cost of refrigeration is greater than the cost of heating, more insulation is often justified in low-temperature applications.

Moisture problems—In the low-temperature range, moisture is by far the major insulation problem. Water permeability of four major materials is as follows:

Material	Water permeability, perm-in.
Foam glass	0.00
Urethane foam	0.3 to 0.6
Polystyrene foam	1 to 4
Fiberglass	100 to 200

Permeation by water or water vapor can dramatically increase the conductivity of the insulation. Water-vapor permeation is not a serious problem until the moisture condenses—except with foam, where the vapor replaces air in the cells. Ideally, the insulation should either absorb no moisture or readily give up any that enters, and it should resist water deterioration. Moisture barriers are extensively used, but in practice it is almost impossible to achieve the perfect barrier needed. Extra thicknesses of insulation, even beyond what would be economically dictated for cold-line applications, are sometimes employed to keep the warm surface temperature above the dewpoint of the water vapor in the air.

Moisture interferes with the insulating properties, at least temporarily, in all temperature ranges but is not a serious problem except at lower temperatures. Above ambient temperatures, the vapor pressure of water is higher at the insulation's inner surface than in the surrounding area, and this pressure tends to push moisture out. Protection from weather and spillage is what is needed.

But when the temperature of the inner surface is less than ambient, this flow reverses—the vapor is pushed in, to an extent proportional to the difference in temperatures between the insulated body and the outside air.

Even with water-impervious insulation, where water vapor may enter through unsealed joints or cracks, localized freezing can cause minor damage. When ice melts and runs along the piping to collect at low points

Secondary properties of representative insulations†						Table IV
Property	Calcium silicate	Mineral fiber	Fiberglass	Elastomeric, closed cell	Cellular glass	Urethane foam
Alkalinity, pH	8 to 12.5	7 to 9	7.5 to 9	—	7 to 8	6.5 to 7.5
Capillarity	Will wick	Will wick	Neg	None	Neg	None
Flame-spred index	—	Less than 25 *	Less than 25	25 to 75	—	75 or less*
Smoke-density index	—	Less than 50 *	Less than 50	150 to 490*	—	450 or less *
Non-combustible	N.C.	Some N.C.	—	—	N.C.	—
Compressive strength lb/in^2 @% deformation	100* to 250 @ 5%	1 to 18 @ 10%	0.02 to 3.4 @10%	2 to 6 @ 25%	100 @ 5%	16 to 100 @ 5%
Specific heat, btu/(lb)(°F)	0.20 to 0.28	0.22	0.20	0.19 to 0.27	0.20	0.23 to 0.25

† These data extracted from "Thermal Insulation," by John F. Malloy, Van Nostrand Reinhold, 1969.
* Indicates change from referenced table.

during a shutdown, then refreezes, major damage can result.

Vapor barriers

Vapor barriers take three forms:

Structural barriers—Often prefabricated to exact dimensions needed and ready to install, these are rigid sheets of reinforced plastic, aluminum or stainless steel jacketing—flat, corrugated or embossed.

Membrane barriers—Metal foils, laminated foils and treated papers, plastic films and sheets, and coated felts and paper—these are either part of the insulation as supplied, or can be supplied separately.

Coating barriers—In fluid form as a paint (or semi-fluid of the hot-melt variety), the material may be asphaltic, resinous or polymeric. These provide a seamless coating but require time to dry.

For low-temperature insulation, the vapor barrier should perform considerably better than 1 perm. The colder the equipment, the better the resistance should be, ranging from 0.2 perm asphalt-saturated shiny glazed papers for chilled water, down to 0.01 perm laminated paper and aluminum products for below freezing.

When the virtually inevitable happens in low-temperature applications, and water vapor gets in, the insulation must be heated and dried. With polyurethane and polystyrene foam, which suffer deterioration, (polystyrene softens at about 90°C, for example), care must be exercised in this procedure.

In the higher part of this temperature range—from about −20°F to the upper limit, 212°F—a variety of organic and inorganic massive insulations are used, sometimes in loose fill forms. These include:

- Compressed and granulated cork.
- Sandwiched cellular glass and felt board.
- Glass fibers bonded with organic resins.
- Expanded and cellular forms of polystyrene.
- Polyurethane foams.

- Rubber and resin combinations.
- Vinyl chloride cellular foams.
- Wood fibers with suitable binders.
- Polyvinyl acetate.
- Cork-filled mastic.
- Expanded vermiculite and perlite.
- Aluminum foil on paper.

Selection among these insulations is based on a number of factors, generally considered in this order:

1. Form available, such as rigid board, flexible board, loose fill, etc., and how it fits the insulation design.
2. Price.
3. Availability.
4. Thermal conductivity.
5. Moisture permeability.

Intermediate range (212°F to 1,000°F)

This temperature range includes conditions encountered in most chemical and petrochemical processes and in steam systems, so it is by far the most important to the chemical process industries.

The most important type of insulation in this range is fiber-reinforced hydrous calcium silicate.

Insulation types used in this service range include:

1. Calcium silicate.
2. Diatomaceous silica.
3. Cellular glass (to about 850°F).
4. Glass fiber bonded with high-temperature binders.
5. Magnesium carbonate with asbestos or other fibers and binders (to about 600°F).
6. Rock wool or mineral-derived fibers.
7. Expanded perlite with binders.
8. Metal-sheet reflective systems.

Selection of material in this temperature range is much more strongly dictated by the value of thermal conductivity than it is in the lower temperature ranges. But other factors, such as mechanical properties, forms available and cost of installation are also strong influ-

Insulation type	Temperature range, °F	Conductivity, k Btu/(h)(ft^2)(°F/in.)	Density, lb/ft^3	Applications
Urethane foam	−300 to 300	0.11 to 0.14	1.6 to 3.0	Hot and cold piping
Cellular glass blocks	−350 to 500	0.20 to 0.75	7.0 to 9.5	Tanks and piping
Fiberglass blanket for wrapping	−120 to 550	0.15 to 0.54	0.60 to 3.0	Piping and pipe fittings
Fiberglass pre-formed shapes	− 60 to 450	0.22 to 0.38	0.60 to 3.0	Hot and cold piping
Fiberglass mats	150 to 700	0.21 to 0.38	0.60 to 3.0	Piping and pipe fittings
Elastomeric pre-formed shapes and tape	− 40 to 220	0.25 to 0.27	4.5 to 6.0	Piping and pipe fittings
Fiberglass with vapor barrier jacket	20 to 150	0.20 to 0.31	0.65 to 2.0	Refrigerant lines, dual--temperature lines, chilled-water lines, fuel-oil piping
Fiberglass without vapor barrier jacket	to 500	0.20 to 0.31	1.5 to 3.0	Hot piping
Cellular glass blocks and boards	70 to 900	0.20 to 0.75	7.0 to 9.5	Hot piping
Urethane foam blocks and boards	200 to 300	0.11 to 0.14	1.5 to 4.0	Hot piping
Mineral-fiber preformed shpaes	to 1,200	0.24 to 0.63	8.0 to 10.0	Hot piping
Mineral-fiber blankets	to 1,400	0.26 to 5.6	8.0	Hot piping
Fiberglass field applied jacket for exposed lines	500 to 800	0.21 to 0.55	2.4 to 6.0	Hot piping
Mineral-wool blocks	850 to 1,800	0.36 to 0.90	11.0 to 18.0	Hot piping
Calcium silicate blocks	1,200 to 1,800	0.33 to 0.72	10.0 to 14.0	Hot piping

ences. Various products from different manufacturers within a given generic insulation type have different upper temperature limits, depending on the manufacturing process and raw materials. Specific data should be obtained for each specific product.

Principal intermediate-range materials

Calcium silicate—This is a mixture of lime and silica reinforced with organic and inorganic fibers and molded into the shape desired. Temperature range covered is 100 to 1,500°F. Flexural strength is very good. Compressive strength is high. These are the reasons this material is so widely used in the chemical process industries. Calcium silicate meets specifications for prevention of stress-corrosion cracking of austenitic stainless steel. It is available in both half-round and quarter-round segments for pipes, as well as in the form of blocks.

Mineral fiber—For this material, rock and slag fibers are bonded together with a heat-resistant binder. Upper temperature limit is up to 1,200°F (generally, about as high as for calcium silicate). The material has a practically neutral pH. It can be made to meet requirements involving stress-corrosion cracking of austenitic stainless steel. Its compressive strength is much less than for calcium silicate. It is available in both rigid molded

form for piping or vessel use, and as a flexible blanket for irregular surfaces.

Expanded silica, or perlite—This insulation is an inert siliceous volcanic rock with some combined water. The rock is expanded by heating to above 1,600°F—the water vaporizes expanding the rock volume and creating a structure of minute air cells surrounded by vitrified product. Added binders hold it all together and resist moisture penetration, and inorganic fibers reinforce the structure. The material has low shrinkage. Water absorption is low. The chloride content is low enough to eliminate its causing stress-corrosion cracking in austenitic stainless steel.

Glass—In various forms, glass products serve from the cryogenic level to around 1,200°F, with some specialty, high-purity products good up to 1,800°F. Among the forms available are flexible fiber blankets without binder, glass fiber with an organic binder, (generally a thermosetting resin good to 1,000°F or higher); semirigid fiberglass boards (for up to 850°F); molded sections for pipe; and foam or cellular form, used for low-temperature insulation applications. Glass has a low moisture absorption and can be easily dried. Upon drying, strength and thermal qualities are recovered. Its low chloride content keeps it from promoting stainless-steel stress-corrosion cracking.

Loose fill—A number of insulation materials are available in the form of loose fill: granulated insulation materials or loose fibers, diatomaceous silica, etc.

High-temperatures (1,000°F to 1,600°F)

This approaches the refractory range of materials, which begins at 1,600°F. The higher the temperature, the fewer the materials available, with major dropoff points at about 1,200°F, 1,600°F and 2,400°F.

Major materials in the high-temperature range are (listed in ascending order of temperature resistance):
1. Mineral fiber: 1,000 to 1,900°F
2. Calcium silicate: 1,200 to 2,000°F
3. Multiple-metal-foil systems for vacuum applications: to 2,500°F
4. Ceramic fibers based on the Al_2O_3-SiO_2 systems: 1,600°F to 2,600°F.
5. Castable-ceramic insulating refractories: 2,000 to 3,000°F.
6. Oxide fibers, primarily Al_2O_3 or ZrO_2: 2,800 to 3,000°F.
7. Rigid ceramic insulating brick: 2,000 to 3,200°F.
8. Carbon fibers: to 3,600°F (above this temperature, irreversible conductivity increases take place).

In high-temperature applications, service conditions can be severe. Among the abuses the insulating refractory must withstand are abrasion and erosion by molten materials, direct flame impingement, corrosive atmospheres and severe thermal shock. Although heat-transfer principles and energy economics are similar for both high-temperature and intermediate-temperature applications, use and design of refractories is an engineering art in its own right and will not be treated more fully here.

Slowing heat-transfer

Rather than thinking of insulation as a material, the chemical engineer should consider it as a strategy for slowing the transfer of heat. This point of view will guide the engineer through problems of material incompatibility, space limitations, and so on. The strategy starts with keeping in mind the three modes through which heat is transferred: radiation, conduction and convection.

Electromagnetic (infrared) radiation—This is the primary mode of heat transfer. It can operate even in a vacuum, where neither other mode can. The amount of heat transferred by radiation is governed by both the emissivity and the temperature of the surfaces involved in the emission and absorption of the heat energy, and by the relative area of these surfaces. Because radiant-energy-flow is proportional to the difference of the fourth power of the absolute temperatures of the hotter and cooler surfaces, the rate of heat-transfer increases dramatically as the temperature increases.

Control of radiation heat-transfer is achieved by changing the emittance level of the surfaces involved, or by the insertion of absorbing or reflecting surfaces (sheets, fibers, particles, etc.) between the two temperature boundaries.

Conduction—In this mode, atomic or molecular motion conducts heat through any gaseous, liquid or solid material. The tighter the molecular network, the more it will conduct. Thus, steel is more conductive than air. The motion, and therefore the heat transfer, will continue until the whole network is terminally balanced.

Control of conduction heat-transfer is achieved by use of less-conductive elements (e.g., glass instead of steel) and by breaking the continuous structure (fibers instead of solid mass).

Convection or mass motion—This is a function of density and happens only in a fluid. Here, a whole mass of the material moves, carrying heat energy with it (as opposed to simply having the individual molecules move). The lighter, heated fluid will rise and be replaced by colder, and thus heavier, gas or liquid. This exchange is governed by the fluid's freedom of movement and by its density; the denser the fluid, the more effective the heat transfer. The size, shape and orientation of the cell containing the fluid are factors affecting the convective heat transfer. Control of convection is achieved by the creation of small cells within which the temperature gradients are small.

Radiation is the primary mode of heat transfer; the other two, conduction and convection, come into play only as a result of radiation. If the molecules did not absorb the radiant energy, there would be none of the difference in temperature necessary to create molecular and mass motion.

Thus, the insulation strategy is to minimize radiation transfer, minimize convective transfer, and introduce a minimum of solid conduction media (insulation has, in general, a negligible effect on air conduction). A compromise game is played between these modes to create an insulation that will work well for a given set of conditions.

As large a number of radiation absorbers as possible are placed between the temperature boundaries, while minimizing the solid conduction. Since a gas is often present, the strategy calls for creating as many small cells as possible to inhibit convection.

All of these mechanisms work together in the insulation to form a very complex system. To simplify examination of the heat-transfer mechanisms and their combined effect, the familiar term "thermal conductivity" or simply "k" is used.

Theoretically, there are too many variables for "k" to be a precise description of an intrinsic property of a homogeneous material, as it purports to be. But, generally, it is the only practical way to define the characteristics of an insulation, so it is widely used.

An important variable is the specific temperature at which the insulation is functioning. The relative importance of the various modes of heat transfer will vary considerably as the temperature conditions vary.

Even more important in insulation design is that the engineer quickly gets moved off any theoretical base into hard problems involving criteria other than optimum resistance to heat flow. Some of these problems are: temperature limitations, thermal-shock resistance, coefficient of expansion, strength, hardness, compressibility, specific heat, erosion resistance, chemical resistance, environmental resistance, space limitations and—certainly not least—cost.

The choice is generally a compromise governed by an attempt to balance the cost of the installed insulation and the amount of money that is expected to be saved by lowered energy requirements. The increasing cost of energy, of course, is bringing more and more portions of CPI plants under scrutiny for possible insulation. The uncertainty about how these energy costs will rise over future decades makes the cost/benefit engineering decisions tougher.

Insulation design theory

In making these compromises when designing the insulation system for a process plant, it is useful to know the kinds of compromises made by the designer of the insulation materials available.

The apparent thermal conductivity of mass insulations is a combination of the following heat-transfer mechanisms: thermal conductivity of air, thermal conductivity of the solid components of the insulating structure, the convective heat transfer within the pore structure, the radiation heat transfer within the structure, and the interactions of these mechanisms.

The thermal conductivity of air in low-density materials is nearly a constant, since the proportion of air to solids in most insulations ranges from 99% air in 1.6-lb/ft^3-density mineral fiber, to no less than 75%.

The thermal conductivity of the solid is related to the number of discontinuities in the network (e.g., contact between the fibers or other solid particles) and to the gradient along the fiber or structural component as influenced by the orientation of the individual fiber.

When the solid portion of the insulation is small, as it should be, convective transfer appears. With a low portion of solids, cells are formed large enough to allow natural convective air turbulence. For the phenomenon to become apparent, the cells usually must be several millimeters across. At room temperature, for example, in a 0.50-lb/ft^3-density fiberglass insulation with a uniform 4-μm-fiber-diameter distribution, the average pore size is about 1 μm. A small amount of convection will occur in slightly higher densities because of the nonuniformity of the fiber distribution.

Radiation is a significant portion of the total heat transfer at room temperature in fibrous insulation of densities below 2 lb/ft^3; as temperature increases, radiation is significant at considerably higher densities. The radiation component can be divided into two major parts: (1) that which is not absorbed but is scattered and reflected by the fibrous structure, and (2) the fiber-to-fiber reradiation, dependent on the absorptance and emittance of the fiber and the difference in temperature between two fibers within view of each other. Interaction between the air and the solid portion of the insulation is the final heat-transfer mode. Air conduction acts as a short-circuiting process, since the air contacts all radiation-absorbing surfaces.

A typical fiberglass insulation in an ambient-temperature situation illustrates how this all works. First, take the heat flux without the presence of the insulation—that is, extend the density curve to zero. The total apparent "k_a" equals 1.667, of which the air contributes 0.178, convection 0.435 and radiation 1.054. Then add fiberglass insulation into the heat-flow path, giving a rapid reduction in "k_a." The convection is eliminated at about 0.6 lb/ft^3 bulk density. The radiation transmission is being continually reduced and partially transformed into radiation conductivity as the fibers are heated and start reradiating between themselves.

Solid conduction contributes very little at densities below 2 to 2.5 lb/ft^3. The curve shows that there is no advantage in increasing the density beyond 2.5 lb/ft^3 for this set of conditions. Economically, it may be advantageous to limit the density to 0.5 lb/ft^3, which gives a "k_a" of about 18% of the original apparent "k", considering that a further increase in density to 2 lb/ft^3 would only bring it down to 12.5% of the original apparent "k" with no insulation.

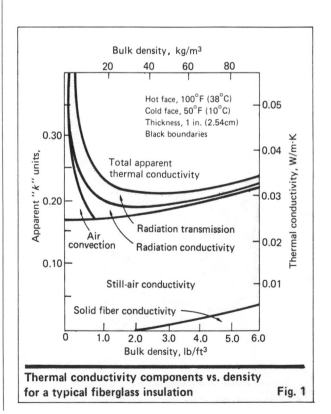

Thermal conductivity components vs. density for a typical fiberglass insulation   Fig. 1

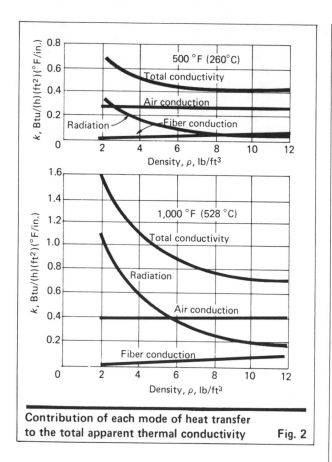

Contribution of each mode of heat transfer
to the total apparent thermal conductivity   Fig. 2

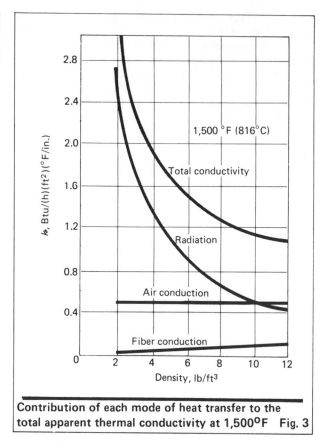

Contribution of each mode of heat transfer to the
total apparent thermal conductivity at 1,500°F   Fig. 3

Density requirements increase as the temperatures increase. With this insulation system, for example, optimum densities would be 6 lb/ft³ at 500°F; above 12 lb/ft³ at 1,000°F; and so on. At each temperature, a minimum apparent thermal-conductivity value occurs at a density range at which the sum of the variable components is at a minimum. The density for minimum apparent-conductivity increases with increasing temperature, because considerably greater density is required to reduce the radiation to the point where the

sum of the variable components attains a minimum value. The minimum apparent thermal conductivity also increases with temperature because of the increase of air conduction, as well as the higher total of the other variable components. No definite generalization can be made concerning the superiority of higher- or lower-density material without carefully considering all of the factors in any specific application.

It must be realized that all of the above is calculated in simple terms. Heat flow in actual material is much more complicated and virtually impossible to reduce to formula. Instead, theory has been correlated with experimental data by the insulation manufacturers in specific instances, giving limited density ranges.

Engineered insulations

Insulation manufacturers, in an effort to strike the best bargain among desired qualities of mechanical strength, chemical compatibility, weight, compactness, and so forth, with the least heat conductivity at the best possible cost, engineer their insulations toward particular applications.

The engineering of insulations in practice is less simple. The influences of the variables—radiation and conduction among solids and gases, convection in gases—must be considered closely.

Introduction of thermally opaque solids in the path of a radiant heat flux restricts its flow. The more solids introduced, the greater the reduction in radiant flow. But, of course, this results in an increase in solid conduction, which is influenced by the structural bonds in the network, which connect in series the solid fiber or

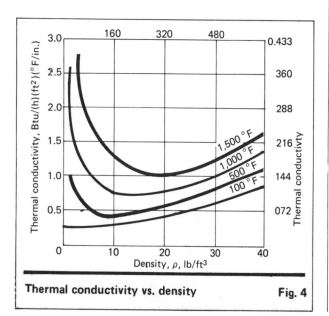

Thermal conductivity vs. density   **Fig. 4**

other particles in the direction of heat flow. The practice in fiber insulation is to lay the fibers in place perpendicular to the heat flow to minimize this conduction, which tends to move along the length of the fiber.

Gas conduction within the air or other medium between the solid particles can be eliminated or minimized by evacuation or reduction of pressure, or by replacement of the air by a high-density gas such as a fluorocarbon (as in polyurethane foam). The gas conduction is approximately proportional to the square root of its molecular weight. Another method is to try to reduce the spacing between solid particles without at the same time increasing the parallel-solid conduction component.

Gas conduction is an important constant. Thus, if the air or gas is removed or made less effective, the overall "k_a" will be improved. A gradual reduction of the gas conduction can be obtained by reducing the cell spacing in the insulation to the range of the free mean path of free gas molecules. This in effect blocks off the molecules from each other, reducing molecule-to-molecule collisions. This reduces thermal energy transfer from gas molecule to gas molecule.

This principle was followed in engineering Min-K* insulation for the space program, where a highly efficient insulation with low weight and low space requirements was needed. The developers reduced the process of gas conduction by creating very small cells in which the gas does not operate as a free gas. This minimized the gas conduction, stopped convection and created a multitude of discontinuous, poorly conductive solid barriers to absorb radiation.

Other insulation systems demonstrating extraordinary thermal transmission values are: "Super" insula-

*Min-K, a registered trademark of Johns-Manville.

tions, which can only function effectively in a vacuum, as in the multi-layer reflective insulation used for cryogenic purposes, and in polyurethane foams in which a low-conductivity gas such as fluorocarbon is substituted for air. This is only a temporary situation since air will ultimately replace the gas in the cells unless the foam is contained within gas-impermeable barriers.

Even with the reduction of the mean free path, radiation at high temperature is still a significant heat-transfer factor. This can be reduced by the use of opacifying agents.

The more-dense insulations perform better at higher temperatures. At lower temperatures, all densities converge in their "k" values, so little benefit is derived from an increase in density. For ambient-temperature applications, lower densities are the most economical.

Another important factor is the size of the fiber or particle. The smaller the fiber diameter, the greater the surface area to obstruct radiation. But, again, this is expensive. At very high temperatures, fine 1-μm fibers perform more effectively because of the shorter wavelengths of radiant energy generated as the temperatures increase.

In summary, heat flow can be effectively controlled by insulation within which:

1. Radiation has been reduced to a minimum level. It remains as the controlling effect on the lower part of the "k"-versus-density curve.

2. Solid conduction can be held to a minimum.

3. Convection is minimal, particularly at the structural density required to obtain reasonable radiation absorption.

4. Gas conduction, though present in most common insulating materials, in some instances can be reduced or eliminated.

Economic thickness calculations

There are many design criteria involved in specifying insulation, particularly in the process industries. But one, economic considerations, which until the past three years was comparatively low on the list, is steadily becoming more important.

By economic considerations, we mean the calculation of the amount of insulation that would save enough energy over a set time period to justify the investment. Economic thickness calculations, based on equations first published in 1926, have been refined and standardized over the years. New input, particularly concerning methods for developing investment return analyses, have made these calculations more useful just when they are most needed.

The method has stood up under 50 years of close scrutiny by engineers and has proven accurate. First performed by L.B. McMillan, a former Johns-Manville engineer, the calculations have been noted for their flexibility in the number of variables they have been able to accommodate.

Interest in economic thickness has increased considerably with the rise in energy costs. Previously the cookbook approach was the standard way to specify thermal-insulation thickness in chemical plants. The lower cost of energy did not justify an engineer's taking great pains in custom-designing insulation thicknesses for each pipe size and temperature range, unless the process itself demanded it. As the last item installed, it was often a matter of putting in whatever amount of insulation fit the space left in pipe racks—or it was governed by the amount of money left in the budget.

The concept of economic thickness has now been refined by the Federal Energy Administration into the ETI (Economic Thickness for Industrial Insulation) method. This was computerized by insulation manufacturers and others so plant data can be fed into the computer to get insulation-thickness specifications offering the best economics for a particular pipe in a particular plant.

These computer programs determine both the optimum thickness of insulation for piping in new plants, and the economics of retrofitting more insulation in old plants, according to energy costs and company accounting procedures.

One chemical plant was subjected to the economic thickness calculations and the company found that bringing old insulation, which had deteriorated to 75% of its original effectiveness, back to 95% effectiveness gave a return on its investment of 103%. Without such hard facts, however, the engineers would have had a hard time convincing management that this kind of investment was justified. Traditionally, investment capital just was not spent on maintaining insulation.

Such decisions are made by management after comparing the financial return on insulation to other investment opportunities, such as plant expansion, new and more efficient equipment, and so forth.

Below are four common investment analysis methods, as applied to upgrading insulation.

Simple Payback

Essentially, this is the time required to repay the capital investment with the operating savings.

Example:

	Thickness current std.	ETI	Difference
Insulation investment, $	200,000	250,000	50,000
Annual fuel cost, $	40,000	30,000	10,000

$$\text{Simple payback} = \frac{\text{Investment difference}}{\text{Annual fuel savings}} = \frac{50,000}{10,000} = 5.0 \text{ Years}$$

(While easy to calculate, and hence often referred to, this is generally used only to measure the degree of financial risk of various investment opportunities).

Discounted payback period

This indicator is defined as the length of time an investment requires to generate discounted, after-tax cash flows equal to the initial insulation cost.

Minimum annual cost

This measure annualizes all cost of insulation and the costs of heat lost through the insulation, discounting them at an assigned interest rate.

	Thickness, current standard	ETI	Comments
(1) Annual energy cost, $	40,000	30,000	
(2) After-tax energy cost, $	20,800	15,600	(1) × (1.0 − 0.48)
(3) Insulation depreciation, $	10,000		$200,000/20 Years
		12,500	$250,000/20 Years
(4) Total annual cash costs, $	10,800	3,100	(2) minus (3)
(5) Capital recovery factor* for 20 years at 10% = 1/8.514 = 0.1175.			
(6) Equivalent annual insulation cost	23,491		(200,000 × 0.1175)
		29,363	(250,000 × 0.1175)
(7) Total annual cost	34,291	32,463	(4) plus (6)

Conclusion: The lowest annual cost ($32,463) is the better of the two investment options.

*Capital recovery factor—Converts a one-time cost to an annual cost. The summation of the discount factor over a 20-year period = 8.93.

Maximum net present value

This takes the sum of the annual expenses, reduces this sum by the tax rate, and subtracts annual depreciation charges to produce an annual after-tax cash expense. The present value of the annual after-tax cash expense is found and to this is added the capital cost of the insulation.

	Thickness, current standard	ETI	Comments
(1) Annual energy cost, $	40,000	30,000	
(2) After-tax energy cost, $	20,800	15,600	(1) × (1.0 − .48)
(3) Insulation depreciation, $	10,000		$200,000/20 Years
		12,500	$250,000/20 Years
(4) Total annual cash costs, $	10,800	3,100	(2) minus (3)
(5) Present-value factor for 20 years at 10% = 8.514			
(6) Present value of annual cash flows*	91,951	26,393	(4) times (5)
(7) Present value of cash flow for insulation purchase	200,000	250,000	
(8) Net present value	291,951	276,393	(6) plus (7)

Conclusion: The ETI thickness provides the lowest net present value—$276,393 (which is the better of the two investment options.)

*Assumes no fuel-price escalation.

In performing economic thickness calculations, the plant's engineer and its insulation contractor must supply the computer with the following data (shown with sample answers, for a typical chemical process plant).

1. Type and cost of fuel (oil, 40¢/gal).
2. Heating value of fuel (140,000 Btu/gal).
3. Expected annual fuel-price increase (10%).
4. Efficiency of conversion of fuel to heat (80%).
5. Capital investment for heat plant ($2 million).
6. Cost of money to finance plant (10%).
7. Heating-equipment depreciation period (20 years).
8. Expected total annual heat production (600,000 million Btu).
9. Basis for economic thickness calculation (minimum annual cost).
10. Income tax rate (48%).
11. Insulation depreciation period (20 years).
12. Cost of money, or return on investment requirement, for insulation (12%).
13. Insulation material (fiberglass, pipe).
14. Jacket emissivity (0.2).
15. Surface resistance (0.7).
16. Installed insulation prices ($3.00 per linear ft for 1 in. through $21.86 per linear ft for 6 in.).
17. Pipe size (6 in.).
18. Piping complexity factor (1.15—for average on a scale ranging from simple to complex, as provided on table).
19. Ambient temperature (60°F).
20. Process temperature (350°F).

Industrial energy prices projected to 1990 — **Table VI**

FEA Region I: New England
1977 dollars/10^6 Btu

	Industrial electricity	Industrial natural gas	Industrial residual oil	Industrial coal
1977	10.75	2.01	2.58	1.47
1978	10.95	3.34	2.70	1.57
1979	11.15	3.66	2.82	1.67
1980	11.35	3.98	2.94	1.77
1985	12.17	2.99	2.96	1.85
1990	13.71	3.87	3.06	1.90

FEA Region II: New York/New Jersey
1977 dollars/10^6 Btu

1977	8.92	1.76	2.60	1.30
1978	8.93	1.80	2.76	1.42
1979	8.95	1.85	2.92	1.54
1980	8.97	1.90	3.08	1.66
1985	8.92	2.45	3.10	1.74
1990	9.69	3.15	3.19	1.79

FEA Region III: Mid-Atlantic states
1977 dollars/10^6 Btu

1977	8.21	1.45	2.62	1.11
1978	8.30	1.58	2.81	1.26
1979	8.40	1.72	3.00	1.40
1980	8.50	1.85	3.19	1.55
1985	9.73	2.41	3.21	1.64
1990	10.90	2.87	3.31	1.68

FEA Region IV: South Atlantic states
1977 dollars/10^6 Btu

1977	6.73	1.11	2.38	1.28
1978	7.19	1.20	2.55	1.43
1979	7.65	1.30	2.72	1.58
1980	8.10	1.40	2.89	1.74
1985	9.02	2.13	2.91	1.83
1990	10.04	2.71	3.01	1.87

FEA Region V: Midwest
1977 dollars/10^6 Btu

1977	6.97	1.34	2.70	1.19
1978	7.23	1.42	2.83	1.30
1979	7.49	1.50	2.96	1.41
1980	7.75	1.58	3.09	1.52
1985	9.09	2.13	3.12	1.58
1990	10.40	3.11	3.21	1.64

FEA Region VI: Southwest
1977 dollars/10^6 Btu

	Industrial electricity	Industrial natural gas	Industrial residual oil	Industrial coal
1977	7.06	1.21	2.48	1.67
1978	8.25	1.43	2.63	1.58
1979	9.45	1.64	2.79	1.49
1980	10.65	1.86	2.94	1.40
1985	10.74	1.88	2.96	1.50
1990	12.41	3.04	3.06	1.59

FEA Region VII: Central states
1977 dollars/10^6 Btu

1977	6.96	1.13	2.80	1.34
1978	7.32	1.33	2.90	1.38
1979	7.68	1.52	3.00	1.41
1980	8.04	1.72	3.10	1.45
1985	9.23	3.04	3.12	1.52
1990	10.27	3.33	3.22	1.56

FEA Region VIII: North Central states
1977 dollars/10^6 Btu

1977	4.75	1.16	2.88	1.03
1978	5.08	1.41	2.95	1.08
1979	5.41	1.66	3.02	1.14
1980	5.74	1.91	3.08	1.20
1985	6.84	1.94	3.18	1.29
1990	8.80	3.14	3.34	1.36

FEA Region IX: Western states
1977 dollars/10^6 Btu

1977	8.49	1.74	2.92	1.71
1978	9.26	2.07	2.93	1.65
1979	10.02	2.40	2.94	1.58
1980	10.79	2.72	2.96	1.52
1985	10.74	2.48	2.98	1.63
1990	11.49	3.21	3.07	1.74

FEA Region X: Northwestern states
1977 dollars/10^6 Btu

1977	2.78	1.54	2.91	1.12
1978	3.26	1.75	2.95	1.14
1979	3.75	1.95	2.98	1.16
1980	4.24	2.16	3.01	1.19
1985	4.76	2.17	3.03	1.59
1990	5.78	3.80	3.12	1.77

21. Annual hours of operation (8,760).
22. Length of pipe (600 ft).
23. Previous thickness specified (2 in.).

Given these parameters, our Energy Reduction Analysis computer program concluded that in this case, for the 600 ft of 6 in. pipe, an insulation thickness of 5 in. would provide a surface temperature of 73°F, and would save 206 million Btu's per year over the previously specified 2 in., for a dollar saving over the life of the plant of $15,762 (present value).

Planning insulation

The thermal efficiency of the insulation chosen, and the economic thickness, are not the whole story.

Chemical engineers designing a plant's insulation must also be concerned with such things as chemical attack (from both spillage and fumes), the possibilities of physical abuse, fire safety, personnel protection, difficulties in cleaning, future requirements of gaining access to insulated pipes and equipment, and so on.

Also, in chemical processes, frequently more is involved in controlling heat losses than saving energy. Many processes have rigid temperature parameters, and a product cannot cool too much in being transferred from one part of the plant to another.

In the design phase, it must be decided whether added insulation can sufficiently control this heat loss, or if a heated tracer must be added to the pipeline. (Where tracers are to be added, a specially shaped pipe insulation, allowing space for the tracers, can be ordered.)

Similarly, insulation beyond what conventional economic analysis indicates might pay for itself by saving pumping costs and avoiding other problems in handling viscous liquids.

Design factors

In dealing with high temperatures, it is important to look at the whole system and make sure that the insulations chosen can withstand the full range of temperatures that will be encountered.

With the trend toward thicker insulations, gaps in the covering, caused by the expansion of metal pipe as it heats up, become more of a problem. Expert insulation contractors prepare joints to minimize these future gaps, but some loss of insulating value must be counted upon.

Form of insulation

Having picked the insulating material and thickness, the form of the material is the next most important selection. Getting as high a ratio of factory fabrication to jobsite fabrication is a key factor in keeping insulating costs to a minimum. Other important insulation-cost factors that should be given attention are ease of installation, and avoidance of damage to the installed insulation, particularly while other construction work continues.

Manufacturers of insulation have devised a number of ways to increase factory fabrication of their products, cutting back on field work. Reflective insulation is virtually all prefabricated to exact dimensions provided by drawings. Solid insulation—mass types such as calcium silicate—is more factory fabricated than heretofore.

Pipe insulation is generally formed in semicircles that are clamped or strapped together over the pipe. Insulation for larger pipes, over about 25 in., are available in quad segments. Calcium silicate pipe insulation is available with either aluminum jacketing or—when superior corrosion- and fire-resistance or strength are required—stainless-steel jacketing already attached.

For lower-temperature applications, fiberglass pipe insulation is available with either a water-vapor barrier or metal jacketing.

Pipe insulation is generally provided in either 2 or 3-ft lengths, which have been found to be the most advantageous sizes for field use. Such lengths are easier to handle without damage and require less cutting to adjust either for changes in pipe directions or for fittings.

For larger-diameter pipe, and other curved surfaces, with diameters in excess of about 34 in., scored blocks of calcium silicate are available for wrapping around the curves. These scored blocks or sheets, as well as flexible blankets (with or without vapor barriers), or metal mesh, or corrugated metal jacketings, are also useful in insulating irregular shapes.

A number of flexible insulation fittings for piping are available. One-piece elbows made of polyvinyl chloride over fiberglass are used on lines insulated with fiberglass or calcium silicate. Aluminum, molded plastic and other preformed fittings for elbows are also available and are widely used.

Tank insulation

There is a trend now toward insulating large tanks, even in places with comparatively mild climates, such as Texas. The energy that would otherwise be needed to reheat materials justifies it.

Generally, either polyurethane foam or fiberglass is used. Urethane works well up to about 250°F; the fiberglass up to 450°F. In some installations, urethane is avoided because of the potential fire hazard. On the other hand, it is less costly to apply.

Insulating an old, dented tank requires an expert, using proper scaffolding and welding equipment. Particularly where the insulation is to be jacketed, as it usually is, tank owners tend to bring in contract insulation units, rather than try to tackle the job with their own maintenance men.

Generally, metal-jacketed fiberglass is either attached by means of studs welded to the tank wall, or held in place by banding. Banding is used on old tanks that are corroded and have thin spots that often make welding difficult.

Jacketing

Jacketing serves several functions in process-pipe and -equipment insulation:
- Protects the insulation against mechanical damage from workmen and equipment throughout the life of the plant.
- Protects exposed insulation from weather.
- Serves as a water-vapor barrier.
- Protects against spills and leaks.
- Protects against fire.

Metal jacketing provides the best protection against mechanical damage. But the designer must be certain that the jacketing specified is adequate for the service conditions encountered. For example, we recently subjected four jacketed insulations to three common service conditions. Tested were calcium silicate with 0.016-in.-

thick jackets, and mineral wool with 0.016-in., 0.032-in., and 0.050-in. preformed aluminum jackets.

First a ladder was rested on the insulation supporting the full weight of a 190-lb man. All but the 0.016-in. jacketed calcium silicate and the 0.050-in. jacketed mineral wool deformed. Then the 190-lb man walked on the insulation. Again, only calcium silicate and the 0.050-in. jacketed mineral wool were unaffected. The 0.016-in. jacketed mineral wool was badly dented; the 0.032-in. jacketed sample was deflected, but sprung back. Similarly, a hammer was dropped from about 4 ft., which dented all three specimens, creasing the two lighter thicknesses on mineral wool.

Corrosion resistance is a particular concern in pipe jacketing and equipment insulation in chemical plants. Where there are corrosive atmospheres, spills or leaks, the jacketing material must resist such conditions, which generally means using stainless steel rather than aluminum. Some types of plastic jacketing, such as the polyfluorides, are excellent against both corrosives and weather, but others, such as PVC, which also offers fine protection against most corrosives, can run into problems in the presence of sunlight, ozone or weather. Ability of a plastic to withstand fungal or bacterial growth can also be a consideration, as in food plants.

In chemical plants, fire safety is always a prime concern. In determining the flame-spread and smoke-production ratings of an insulation system, the insulation and its jacketing should be considered as a composite unit, with the system tested as a whole, rather than by relying on tests of the components separately. Experiment and experience have shown that components do not always yield the same results when put together, as did the insulation, jacketing and cement individually.

Basic data on insulation fire performance derives from the basic ASTM E-84 tunnel test (National Fire Protection Assn. test NFPA 255 or Underwriters Labs test UL 723). Plain calcium silicate insulation normally has flame-spread, fuel-contributed and smoke-developed ratings all of zero. Mineral wool is virtually as good. Up to 25 for flame-spread and 50 for smoke-developed are generally acceptable. Urethane foams have the greatest problems in this area, with flame-spread ratings up to 75, and smoke-developed ratings up to 400, although newer varieties, such as isocyanurate foam, are rated as low as 20 for flame-spread and 30 for smoke-developed.

An insulation with poor fire ratings adds to the danger when the product handled already poses a fire hazard—a problem that is holding down use of urethane foams for insulating tanks in the petroleum and petrochemical industries. But even an insulation with excellent fire ratings can become part of the hazard if it soaks up spilled or leaked flammable materials. Good jacketing that has stayed intact and leakproof is the best answer for this on straight pipe. Valves, pumps, fittings pose special problems because of the greater likelihood of leaks. Such leaks surrounded by molded insulation can result in the soaking up of considerable flammable material without anyone realizing it.

Valves and fittings

Valves, flanges and other fittings are the main stumbling blocks for anyone insulating a piping system. Not only must potential leaks be guarded against, and provisions for access for inspection and maintenance made, but wrapping the troublesome shapes with a heat barrier equaling that on the straight pipe is the most time-consuming and expensive part of the insulation job. Insulating a valve can be three times as expensive as doing a similar length of straight pipe. Because of this, when energy was comparatively cheap, valves were frequently left uninsulated.

Under present conditions, however, it pays to insulate. A 6-in. valve under 60 psi at 70°F ambient can lose 70 Btu/ft^2/h, or 40 million Btu/yr, at a cost of $140–150 a year. An uninsulated flange can lose $216 a year worth of energy. This can mean, at a conservative figure of $120 per fitting each year, and some 250,000 such fittings in a plant, annual losses of $30 million.

Current valve and flange insulating methods have brought the cost down to under $70 for a 6-in. valve. In most cases, insulating a valve or flange will pay for itself in six months to a year. Molded insulations, shaped to fit the valve, flange or other fittings are available. More important in CPI plants—particularly those handling flammable products—are special fitting-insulation systems using refractory fiber that is bonded to corrugated stainless steel. Insulation time and cost are saved; a leak detection system can be incorporated; and the insulation can be removed for maintenance without being destroyed.

Older methods of insulating such fittings involved covering the valve or fitting with insulating cement, smoothing the surface and finishing it with a canvas jacket indoors or a wire-mesh-reinforced emulsion outdoors. In other cases, sleeves of a larger-diameter pipe insulation (for flanges), or structures built up out of block insulation, might be employed.

Reflective and encapsulated types

Reflective-type insulation, which blocks heat transfer by reflection rather than by resistance to conduction and convection through solid mass, is essential in nuclear power plants and is finding growing use in other applications where the qualities needed within a nuclear reactor containment are desirable enough to justify their much higher cost.

These qualities are:

■ Resistance to contamination, moisture damage, vibration, mechanical damage.

■ Removability and replaceability without destruction of the insulation. (Nuclear insulation, if it had to be destroyed, would pose the same radioactive-materials disposal problem as other nuclear wastes.)

■ Protection from fire.

Reflective insulation is virtually all shop-fabricated according to specifications and drawings that are meticulously worked out. Although, in some construction-speedup situations, reflective insulation had been ordered and fabricated from plant drawings before piping was completed, this had led to generally unacceptable amounts of both wasted insulation materials and field work. Shops set up on the plantsite can alter and even build new reflective insulation to adjust for piping changes, but this work is slow and expensive. Reflective insulation installed too soon also leads to waste, both via piping and other structural changes, and through damage by following work crews.

Encapsulated insulation is used where qualities of reflective insulation (such as precision form, physical strength, removability, high degree of protection from the environment) are desired, but space or flooding conditions preclude use of reflective insulation. Any of a number of standard mass-insulation materials, including refractory- and ceramic-fiber blankets and felts, calcium silicate, Min-K, can be encapsulated in a variety of metals, including stainless steel, aluminum, titanium or Inconel, depending on the temperature and other service characteristics involved. Where less protection is required, semiencapsulated insulation can be employed.

Scheduling and installation

As the trend continues toward greater shop fabrication of insulation, more preplanning is also required for the mass types. In simplest terms, if you want 7 in. of insulation on a pipe, you have to be sure that it is 7 in. away from the next pipe, wall, piece of equipment or other obstruction (and, of course, if you want 7 in. on both pipes, they must be more than 14 in. apart). More space must be allowed at curves, bends, elbows, flanges, valves, etc. Also, some extra space must be left for the insulation installers to work.

Future developments

While research continues on devising special insulations to fit special requirements, such as Min-K, the most important work now going on is development of better data on existing insulations, such as calcium silicate, to allow the engineer to better plan the insulation of his plant. Criticism has been leveled, often justifiably, that the kind of exact performance data needed for today's economic analysis and planning of process plant insulation are not available. The "k" factor itself is an approximation, and other manufacturing, installation and service conditions throw the theoretical numbers off even further.

The research centers of leading insulation manufacturers are working to narrow the gap between theoretical data and actual performance. Therefore, insulation designers have begun planning a whole new way of calculating heat transfer of insulations.

Traditionally, such calculations have involved taking the "k" factor of an insulation and multiplying it by the inches of thickness, throwing in the temperature of the product, the size and length of the pipe, and coming up with a theoretical heat-loss number for the plant or system.

But this is not quite good enough now. Considerable work is being done with a new systems approach, based on experimentally derived data. The problem has been that variations in installation; gaps developing at insulation joints (as the pipe expands under the heat of startup); the special heat-transfer calculation problems of insulated curves, valves, pumps and other piping—all add up to a different set of numbers than the theoretical calculations do.

In the systems approach, the insulation laboratory would put together small sections of piping, including insulation gaps, curves and fittings, and then closely monitor heat loss as a measurement of the heat difference between the beginning and the end of the piping. In extending this number to the actual plant, the engineer would be able to consider not only the length of the piping system but also its relative complexity; that is, how many bends and valves it has. When this method is fully developed, far-more-reliable numbers will be available.

The rationale behind this research emphasis is that the greatest energy and money savings will result from better use of the insulation materials we now have available. Both engineers and corporate financial management, given this kind of reliable data, will be assured that more insulation than previously specified, both in new and old plants, will pay. Short of basic process changes, which means building new plants, this is how the chemical process industries will save energy. The amounts saved will be considerable.

The authors

Michael R. Harrison is a technical services engineer for the Industrial Products Div. of Johns-Manville, Greenwood Plaza, Denver, CO 80217, where he provides engineering and technical support for his division's market managers and sales force. He holds a B.Sc. in mechanical engineering from the University of Colorado, Boulder, Col.

Charles M. Pelanne is a senior research specialist in the Johns-Manville Applied Research Section of the Central Research Dept., Research & Development Center, Ken-Caryl Ranch, Denver, CO 80217. He is an expert in the development of thermal conductivity measurement equipment, particularly heat-flow meter techniques, and is also expert in the field of heat transfer through insulations.

The cost of missing pipe insulation

Here's a graphical method for finding the annual expense of heat loss from uninsulated piping—at any efficiency of energy conversion, and for any cost of fuel and wind condition.

Rene Cordero, Allied Chemical Corp.

☐ For piping whose surface has been left uninsulated or whose insulation has been so severely damaged as to be useless, the yearly (8,760 h) cost of the heat loss per square foot can be found from Fig. 1. Costs so determined will generally be within the accuracy of data that can be obtained in the field, and are suitable for most applications without further refinement.

Fig. 1 costs are based on a net efficiency of energy conversion of 100%, a fuel price of $1.00/million Btu, and heat transfer to still air. Corrections are necessary to arrive at annual costs for other conditions:

■ *Net efficiency of less than 100%*—To correct for a different net efficiency, divide the cost derived from Fig. 1 by the actual net efficiency of energy conversion. (For intermediate-size boilers most common in Allied plants, the net efficiency is typically 72%.)

■ *Fuel cost other than $1.00/million Btu*—To correct for a different cost, multiply the cost derived from Fig. 1 by the actual cost of the fuel.

■ *Outdoor installation exposed to moving air*—To correct for variant wind conditions, multiply the cost obtained from Fig. 1 by a correction factor from Fig. 2 for the actual wind velocity. For most of the continental U.S., 10 mi/h represents a reasonable annual average.

Assumptions made in developing the graphs

The unit costs provided by Fig. 1 are average values for a range of pipe sizes in still air. Although there is some variation for different pipe sizes and ambient temperatures, the variations are small compared to the cost of heat loss caused by low air velocities.

The heat loss due to missing insulation is the difference between the heat loss from bare pipe and that through normal insulation. Insulation standards have changed over recent years because of varying fuel and insulation costs and other economic factors. And although the loss through insulation will vary with the standard adopted for a project, such variations will be negligible compared to the heat loss from bare pipe.

In general, corrections for precise ambient conditions, pipe size and insulation thickness will not be jus-

Originally published February 14, 1977

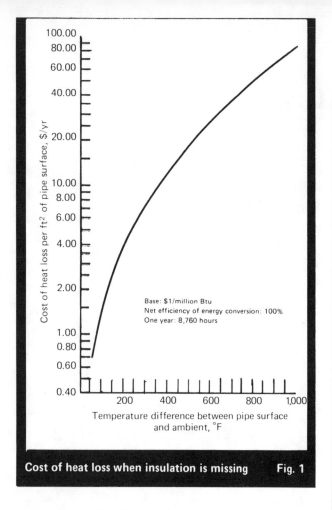

Base: $1/million Btu
Net efficiency of energy conversion: 100%
One year: 8,760 hours

Cost of heat loss when insulation is missing　　**Fig. 1**

tified because they will go beyond the reasonable accuracy of field data.

Sample problems illustrate the procedure

The first problem requires estimating the annual cost for the extra energy lost from a pipe in still air without insulation:

Pipe dia., in.	6
Pipe temperature, °F	375
Average ambient temperature, °F	75
Length of uninsulated pipe, ft	10
Fuel cost, $/million Btu	1.44
Net energy conversion efficiency, %	72

For the ΔT of 300°F, Fig. 1 gives a unit economic loss of $7.60/yr/ft² of pipe surface. This figure is converted into the cost per linear foot by multiplying it by 1.734 ft²/ft, the surface area per foot of a nominal 6-in. pipe. The answer is $13.20/ft of pipe.

The annual cost for the entire 10 ft of pipe is, of course, $132.00. The corrections for net efficiency and fuel cost (132.00 × 1.44/0.72) give an annual cost for missing insulation of $264.00.

The second problem is similar to the first, except that the pipe is exposed to an average wind velocity of 10 mi/h during the year.

From Fig. 2, the wind correction factor is 2.14. Correcting the previously calculated annual cost for the uninsulated pipe in still air ($264.00 × 2.14), gives an annual cost for missing insulation of $565.00.

References

1. Crocker, S., and King, R. C., "Piping Handbook," 5th ed., McGraw-Hill, New York, 1967.
2. "Engineering Manual on Building Materials and Power Products," Johns-Manville.

The author

Rene Cordero is a piping and process mechanical equipment design engineer for Allied Chemical Corp. (Corporate Engineering Dept., P.O. Box 2105R, Morristown, NJ 07960). Holder of a B.M.E. degree from Catholic University and an M.M.A.E degree from the University of Delaware, he is a registered engineer in the State of New Jersey and a member of ASME. He is a co-author of "How to Select Insulation Thickness for Hot Pipes" (*Chem. Eng.*, July 21, 1975).

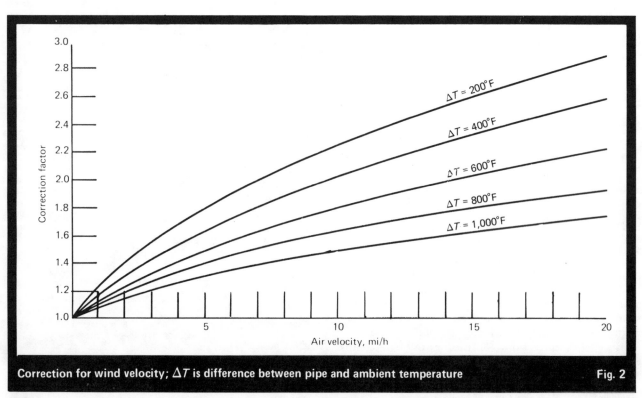

Correction for wind velocity; ΔT is difference between pipe and ambient temperature　　**Fig. 2**

Insulation Saves Energy

Shortcut graphical methods allow rapid evaluation of heat losses from uninsulated and insulated vessels. Depending on the temperature difference between vessel and ambient air, the method also sets the required thickness of outside insulation.

RICHARD HUGHES and VICTOR DEUMAGA, The Badger Co.

Accurate calculations for determining the heat flow between the contents of a vessel and its surroundings are often complex. Multiple resistances in series are usually present in the heat-flow path. In addition, conduction, convection and radiation flows may be simultaneously involved. And there may be heat flow by evaporation and condensation within the confines of the vessel.

Therefore, the particular assumptions and simplifications involved in the method used to calculate heat losses must be understood, and the accuracy of the method must be consistent with a particular requirement.

For example, a simple shortcut approximation will usually indicate whether outside insulation of a given thickness is required. On the other hand, it is often desirable to accurately check a final design (sometimes at varying ambient or internal tank conditions) to determine the average annual heat losses.

Accordingly, we will present two methods of heat-loss calculation:

1. A rapid graphical method that can be used for exploratory evaluation of heat losses.

2. A more accurate algebraic method that is suitable for a computerized solution if multiple calculations are required.

Graphical Method Is Rapid

We can obtain an approximate value for the unit heat loss from Fig. 1 or 2. The unit heat loss when multiplied by the surface area under review yields the overall hourly

Guidelines for Heat-Loss Calculations From Storage Tanks — Table I

Condition	Neglect:	Graphical Solution. Use:	Algebraic Solution. Calculate:
Vessel is uninsulated. Contained liquid is agitated.	Film resistance of contained liquid.	Fig. 1. Correct for wind velocity or outside surface emissivity, if required.	Heat loss through outside air film only, by using Eq. (7b) and (4b).
Vessel is uninsulated. Contained vapor condenses appreciably at wall temperature.	Film resistance of contained vapor.	Fig. 1. Correct for wind velocity or outside surface emissivity, if required. Temperature correction factor, W = 1.0	Heat loss through outside air film only, by using Eq. (7b) and (4b).
Vessel is unisulated. Contained liquid fouls or solidifies at wall temperature.	Film resistance of both contained liquid and vapor adjacent to fouled surface.	Fig. 1 with temperature correction factor W = 0.4; and wind velocity or emissivity correction as required.	Heat loss through fouling layer and adjacent air film by using Eq. (4a) or (8).
Vessel is uninsulated, and contains stagnant liquid or dry vapor.		Fig. 1 with temperature correction factor, W, obtained from table for type of contained material; and wind velocity or emissivity correction as required.	Heat loss through inside-fluid film and adjacent air film by using Eq. (6) or (6a), (7) and (4).
Vessel is insulated with calcium silicate or similar material.	Film resistance of both contained liquid and vapor adjacent to insulated surface.	Fig. 2 with wind-velocity correction factor obtained from table. Insulation thickness may be obtained from Fig. 2.	Heat loss through insulation and adjacent air film by using Eq. (7a) and (3).

Originally published May 27, 1974

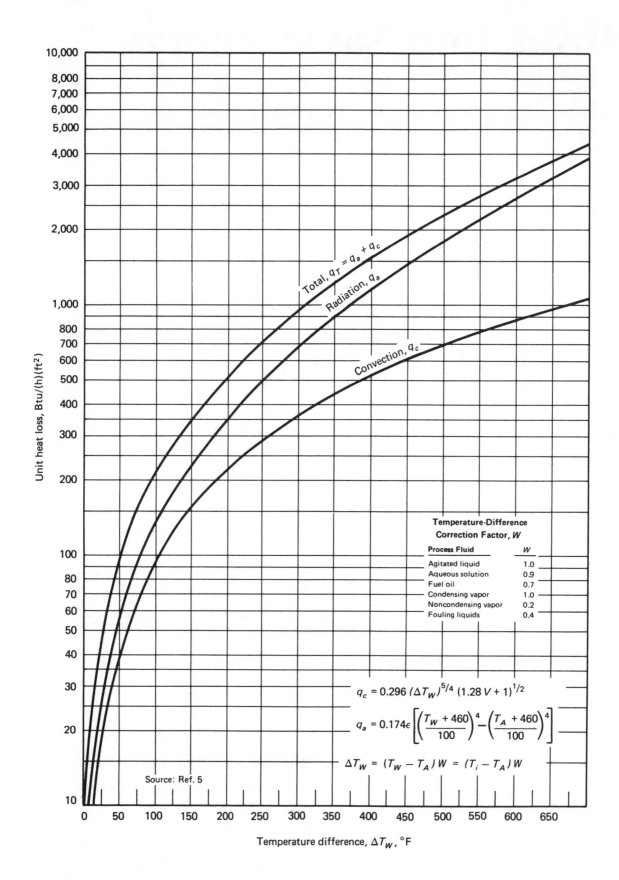

UNINSULATED tanks have heat losses that depend on nature of tank contents. Values in chart are for wind velocity of zero, surface emissivity of 0.9, and ambient air temperature of 70°F—Fig. 1

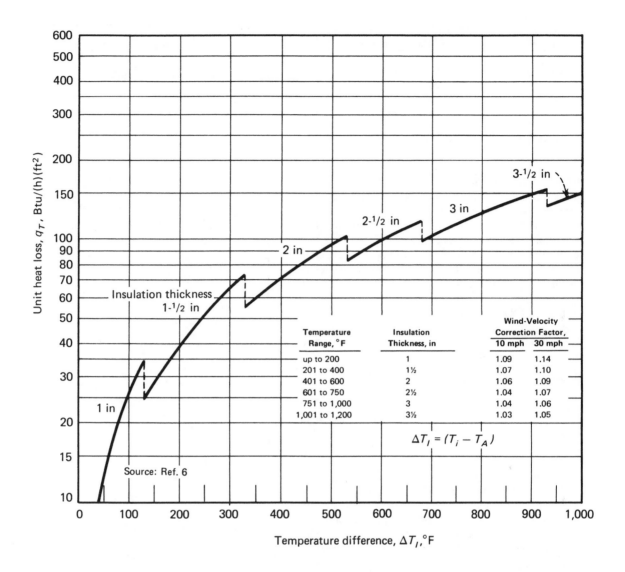

INSULATED tanks, covered with calcium silicate, have heat losses based on negligible resistance to heat flow on process side. Values in chart are for wind velocity of zero, emissivity of 0.8, ambient air temperature of 70°F—Fig. 2

heat loss. A summation of the discrete losses through the tank roof and tank side, exposed to both liquid and vapor, gives the total hourly heat loss to the atmosphere. In addition, heat loss from the bottom of the tank to the ground is easily calculated from Eq. (5) to complete the heat-loss approximation. Table I summarizes conditions for which Fig. 1 and 2 are recommended.

Fig. 1 is used to approximate heat losses from uninsulated tanks. A table of correction factors for temperature difference is included to compensate for common internal resistances to heat flow. In addition, the unit heat loss is divided into a convection and a radiation component. This allows rapid, independent correction for wind velocity and surface emissivity, if required. Fig. 1 also contains a total unit-heat-loss curve at zero wind velocity and 0.9 radiation emissivity.

Fig. 2 is used to approximate heat losses from insulated tanks. A table of recommended insulation thickness for

various temperature ranges is included, and wind-loss correction factors at the recommended insulation thickness are given.

Analysis of Algebraic Method

Heat losses from tanks are calculated by using the natural-convection heat-transfer relations for the sides and roof of the tank, and the conduction heat-transfer relation for the bottom where the tanks rest on the ground.

The total heat loss, Q_T, is divided into four components:

$$Q_T = Q_L + Q_V + Q_R + Q_G \qquad (1)$$

where Q_L is heat loss from liquid in tank to atmosphere through sidewall; Q_V is heat loss from vapor in tank to atmosphere through sidewall; Q_R is heat loss from vapor

in tank to atmosphere through roof; and Q_G is heat loss
from liquid in tank to ground through the bottom of the
tank.

Table I summarizes conditions for which specific equations are recommended to calculate the component heat
losses.

Heat Losses: Sidewall and Top

The method for calculating side and top heat losses,
Q_L, Q_V and Q_R, is essentially the same. In all cases, heat
loss is calculated through two series resistances to heat
transfer. These resistances are: (1) the tank inside-film or
the tank insulation, and (2) the air film (when insulation
is used, the inside-film resistance is neglected). Note that
in all cases the tank wall itself is neglected as a separate
resistance. If the tank is fabricated of heat-transfer-resist-
ant material, then the wall itself can be treated as an
insulation layer. Similarly, if appreciable fouling or so-
lidification occurs within the tank, the fouling layer becomes equivalent to an insulation layer.

The usual individual and overall relationships for heat
transfer are given by:

$$Q/A = q_a + q_c = q = U(T_i - T_A) = h_i(T_i - T_W) =$$
$$(k_I/X_I)(T_W - T_I) = h_A(T_I - T_A) \quad (2)$$

When the tank surface under consideration is insulated,
the unit heat loss, q, is given by:

$$q = (k_I/X_I)(T_i - T_I) = h_A(T_I - T_A) \quad (3)$$

Note that the inside-film resistance is neglected, and
$T_W = T_i$. The inside liquid-film resistance can also be
neglected, even in the absence of insulation, if the liquid
contents are agitated. In this case, a direct resolution is
possible by using Eq. (4b). Similarly, the inside vapor-film
resistance can be neglected if appreciable condensation
occurs in the portion of the tank's surface that is exposed
to vapors.

When the tank surface under consideration is not insulated, the unit heat loss, q, is given by:

$$q = h_i(T_i - T_W) = h_A(T_W - T_A) \quad (4)$$

For either insulated or uninsulated tanks, the method
of computation is the same. The air temperature, T_A, and
the tank-contents temperature, T_i, are known. Individual
film heat-transfer coefficients [as defined by Eq. (6) and
(7)] are used, together with insulation resistance, where
applicable, to calculate the remaining unknown intermediate temperature, T_I, in Eq. (3), or T_W in Eq. (4), and
hence to calculate the unit heat loss. The total heat loss
is then calculated as the product of the unit heat loss and
tank surface area under consideration.

When the tank surface is coated internally with a fouling or solidified layer, the unit heat loss is found from
a modification of Eq. (4), which is:

$$q = (1/R_i)(T_i - T_W) = h_A(T_W - T_A) \quad (4a)$$

When the inside-film resistance is small and the tank
surface is insulated, the unit heat loss becomes a particular
case of Eq. (4), which is:

$$q = h_A(T_i - T_A) \quad (4b)$$

Nomenclature

A	Area of heat transfer, ft^2
C	Specific heat, Btu/(lb)(°F)
D	Diameter of tank, ft
g	Gravity acceleration, 4.17×10^8 ft/h^2
H	Tank height, ft
k	Thermal conductivity, Btu/(h)(ft^2)(°F/ft)
P	Pressure, psia
Q	Heat transferred, Btu/h
q	Unit heat transferred, Btu/(h)(ft^2)
R	Fouling factor, (hr)(ft^2)(°F)/Btu
T	Temperature, °F
U	Overall heat-transfer coefficient, Btu/(h)(ft^2)(°F)
V	Wind velocity, mph
W	Temperature-difference correction factor
X	Thickness, ft
β	Coefficient of volumetric expansion, 1/°F
ϵ	Emissivity
μ	Viscosity, lb/(h)(ft)
ρ	Density, lb/ft^3

Subscripts

A	Environment
a	Radiation
c	Convection
F	Fluid film
G	Ground
i	Tank content
I	Insulation
L	Liquid
M	Insulation mean
R	Roof
r	Fouling layer
T	Total
V	Vapor
W	Tank wall

Heat Losses: From Bottom of Tank

When a tank rests on the ground, the heat loss through
the bottom, Q_G, is given by:

$$Q_G = 2Dk_G(T_L - T_G) \quad (5)$$

Eq. (5) was derived [1] from an equation given by
McAdams [3]. The inside-film resistance to heat transfer
is neglected.

If the tank is mounted aboveground, the bottom tank
surface, A_G, is then added to the side-surface, A_L, when
calculating Q_L by Eq. (3) or (4).

Finally, Q_T is found as the sum of the component heat
losses from Eq. (1).

Film Coefficients and Fouling Factor

Inside liquid-film coefficient for a liquid in contact with
the tank surface ($h_i = h_L$) is given by:

$$h_L = 0.45k_F\left[\frac{\rho_F^2 g\beta_F C_F(T_i - T_W)}{\mu_F k_F}\right]^{0.25} \quad (6)$$

Eq. (6) was derived [2] from an equation given by
McAdams [4].

Inside vapor-film coefficient for a vapor in contact with
the tank surface ($h_i = h_V$) is given by:

$$h_V = 9.686 \times 10^{-2}P^{0.5}(T_F)^{-0.1}(T_V - T_W)^{0.25} \quad (6a)$$

Eq. (6a) is a particular case of Eq. (6), and can be used for both nitrogen and air at moderate temperatures (50 to 400°F) and pressures (0 to 500 psig). When the vapor present is appreciably different from air, the vapor-film coefficient is calculated from Eq. (6) by using vapor-film physical properties. For a condensing vapor, the vapor-film resistance is neglected, and hence the coefficient is not calculated.

The outside air-film coefficient is calculated from one of the following equations.

When the tank surface is uninsulated, Eq. (7) is recommended for substitution into Eq. (4):

$$h_A = 0.296(T_W - T_A)^{1/4}(1.28V + 1)^{1/2} +$$
$$0.174\epsilon \left[\frac{\left(\frac{T_W + 460}{100}\right)^4 - \left(\frac{T_A + 460}{100}\right)^4}{T_W - T_A} \right] \quad (7)$$

Eq. (7) includes a wind-velocity correction factor and a radiation loss factor [5].

When the tank surface is insulated, Eq. (7a) is recommended for substitution into Eq. (3):

$$h_A = 0.296(T_I - T_A)^{1/4}(1.28V + 1)^{1/2} +$$
$$0.174\epsilon \left[\frac{\left(\frac{T_I + 460}{100}\right)^4 - \left(\frac{T_A + 460}{100}\right)^4}{T_I - T_A} \right] \quad (7a)$$

Eq. (7a) differs only from Eq. (7) in that the surface temperature of the insulation, T_I, is used instead of the tank-wall temperature, T_W.

In either case, Eq. (3) or (4) is then solved by trial and error for the intermediate temperature, T_I or T_W, which is then used to calculate the unit heat loss, q.

When the tank surface is uninsulated and the liquid-film or vapor-film resistance for the material contained in the tank is small, Eq. (7b) is recommended for substitution into Eq. (4b):

$$h_A = 0.296(T_i - T_A)^{1/4}(1.28V + 1)^{1/2} +$$
$$0.174\epsilon \left[\frac{\left(\frac{T_i + 460}{100}\right)^4 - \left(\frac{T_A + 460}{100}\right)^4}{T_i - T_A} \right] \quad (7b)$$

For these conditions, Eq. (4b) can be used directly to calculate the unit heat loss, q.

If the thickness of the fouling layer can be estimated, the inside fouling factor used in Eq. (4a) can be calculated from:

$$R_i = X_r/k_r \quad (8)$$

The thermal conductivity, k_r, is usually taken as the fluid thermal conductivity at the mean temperature across the fouling layer.

Temperatures and Physical Properties

Initially, we can assume that the intermediate temperature, T_W or T_I, is midway between T_i and T_A. The inside fluid properties are used at the average inside-film temperature of $(T_i + T_W)/2$. Similarly, outside properties at the average outside-film temperature of $(T_W + T_A)/2$ are

Emissivity of Surface — Table II

Surface	Fraction of Black-Body Radiation	
	at 50° - 100°F	at 1,000°F
Black nonmetallic surfaces such as asphalt, carbon, slate, paint, paper.	0.90 - 0.98	0.90 - 0.98
Red brick and tile, concrete and stone, rusty steel and iron, dark paints (red, brown, green)	0.85 - 0.95	0.75 - 0.90
White or light-cream brick, tile, paint or paper, plaster, whitewash.	0.85 - 0.95	0.60 - 0.75
Bright aluminum paint, gilt or bronze paint.	0.40 - 0.60	–
Dull brass, copper, or aluminum; galvanized steel; polished iron.	0.20 - 0.30	0.30 - 0.50
Polished brass, copper, Monel metal	0.02 - 0.05	0.05 - 0.15
Highly polished aluminum, tin plate, nickel, chromium.	0.02 - 0.04	0.05 - 0.10

Source: "ASHRAE Guide (Heating, Ventilating, Air-Conditioning Guide)," Chap. 5, p. 52, American Soc. of Heating, Refrigerating and Air-Conditioning Engineers, New York, 1959.

used to evaluate physical properties for the film coefficients in Eq. (4).

When the tank is insulated, the average insulation temperature, T_M, becomes $(T_i + T_I)/2$. Similarly, the average outside-film temperature becomes: $(T_I + T_A)/2$. These average temperatures are used to evaluate physical properties for the film coefficients in Eq. (3).

For greater accuracy, the average film temperatures can be recalculated by using T_I or T_W, as found from Eq. (3) and (4). Hence, a new heat loss can be obtained by using physical property data at the new average film temperature.

Problem Illustrates Procedures

A fuel oil at 12° API at 60°F with a viscosity of 50 SSF (Saybolt seconds furol) at 122°F is stored at 300°F in a 20-ft-dia. by 30-ft-high carbon-steel tank at atmospheric pressure. The oil level in the tank is 18 ft. Air temperature is 70°F. Let us calculate heat losses to the environment for the following two cases:

Case 1—Total surface of the tank is uninsulated and black. Wind velocity is 0 mph. Surface emissivity of tank is 0.9. Thermal conductivity of ground is 0.8 Btu/(h)(ft²)(°F/ft).

We will begin the computations by finding the heat loss from the wetted inside surface of the tank. The wetted area is:

$$A_L = \pi D H_L = \pi(20)(18) = 1,130 \text{ ft}^2$$
$$\Delta(T_i - T_A) = (300 - 70) = 230°F$$

From Fig. 1, we find the unit heat loss to be 664 Btu/(h)(ft²) for this temperature difference. Therefore:

$$Q_L = A_L q_L = 1,130(664) = 750,320 \text{ Btu/h}$$

Next, we calculate the heat loss from the dry inside surface and roof surface of the tank:

$$A_V = \pi D H_V + \pi D^2/4$$
$$A_V = \pi(20)(30 - 18) + \pi(20)^2/4$$
$$A_V = 754 + 314 = 1,068 \text{ ft}^2$$
$$\Delta T_W = (T_i - T_A)W = 230(0.20) = 46°F.$$

Note that W is the temperature-difference correction factor for a noncondensing vapor (Fig. 1).

For this temperature difference, we find the unit heat loss from Fig. 1 to be: 84.3 Btu/(h)(ft²). Hence:

$$Q_V = A_V q_V = 1,068(84.3) = 90,030 \text{ Btu/h}$$

Assuming that $T_G = T_A$, we substitute the appropriate quantities into Eq. (5) to find Q_G as:

$$Q_G = 2(20)(0.8)(300 - 70) = 7,360 \text{ Btu/h}$$

Therefore, the total heat loss is:

$$Q_T = Q_L + Q_V + Q_G$$
$$Q_T = 750,320 + 90,030 + 7,360 = 847,710 \text{ Btu/h}$$

Alternatively, we can find the exact solution by using the algebraic equations. After making the necessary calculations, we find that:

$$Q_T = Q_L + Q_V + Q_G$$
$$Q_T = 750,980 + 135,782 + 7,360 = 894,122 \text{ Btu/h}$$

Case 2—Roof of tank is uninsulated and is coated with aluminum paint. Side-wall is insulated with calcium silicate, or equivalent, and has a surface emissivity of 0.8. Wind velocity is 30 mph. Tank contents are not agitated. Thermal conductivity of ground is 0.8 Btu/(h)(ft²)(°F/ft).

As in Case 1, the wetted surface of the tank, A_L, is 1,130 ft.² This represents the interior circumferential area of the tank wetted by the fuel oil to a height of 18 ft.

Heat loss from the vessel wall to ambient air is a function of:

$$\Delta T_I = (T_W - T_A) = (300 - 70) = 230°F$$

From Fig. 2, we find the recommended insulation thickness to be $1\frac{1}{2}$ in, and the wind-velocity correction factor to be 1.10. Also, from Fig. 2, we find that the unit heat loss, corrected for wind velocity, becomes:

$$q_L = 46(1.10) = 50.6 \text{ Btu/(h)(ft}^2)$$
Hence: $Q_L = A_L q_L = 1,130(50.6) = 57,178 \text{ Btu/h}$

Now, we calculate the heat loss from the insulated dry-side surface:

$$A_V = \pi(20)(30 - 18) = 754 \text{ ft}^2$$
$$\Delta(T_W - T_A) = 230°F$$
$$q_V = 46(1.10) = 50.6 \text{ Btu/(h)(ft}^2)$$
$$Q_V = A_V q_V = 754(50.6) = 38,152 \text{ Btu/h}$$

In order to calculate the heat loss from the uninsulated roof, we must use Fig. 1. We find that $A_R = \pi(20)^2/4 = 314$ ft,² and that surface emissivity = 0.4 from Table II.

$$\Delta T_W = (T_i - T_A)W = 230(0.20) = 46°F$$

Using this corrected temperature difference and Fig. 1, we find that the unit heat loss for radiation and convection (correcting the radiation loss for an emissivity of 0.4 and the convection loss for the wind velocity) is:

$$q_a = 54.2(0.4) = 21.7 \text{ Btu/(h)(ft}^2)$$
$$q_c = 35.5[(1.28)(30) + 1]^{1/2} = 222.8 \text{ Btu/(h)(ft}^2)$$
$$q_R = q_a + q_c = 21.7 + 222.8 = 244.5 \text{ Btu/(h)(ft}^2)$$

Hence: $Q_R = A_R q_R = 314(244.5) = 76,773 \text{ Btu/h}$

Heat loss to the ground is the same as Case 1, i.e. $Q_G = 7,360$ Btu/h.

Therefore, the total heat loss is the sum of the component parts, or:

$$Q_T = 57,743 + 38,529 + 76,773 + 7,360$$
$$Q_T = 180,405 \text{ Btu/h}$$

Again, we can calculate a more-exact answer by using the algebraic equations. For this case:

$$Q_T = 54,419 + 36,279 + 49,887 + 7,360$$
$$Q_T = 147,945 \text{ Btu/h}$$

This problem illustrates the ease with which rapid approximations to heat loss can be made.

References

1. Stuhlburg, D., How to Design Tank Heating Coils, *Petrol. Refiner*, Apr. 1959, p. 143.
2. Ibid, p. 146.
3. McAdams, W. H., "Heat Transmission," 3rd ed., p. 54, McGraw-Hill, New York, 1954.
4. Ibid, p. 172.
5. Malloy, J., "Thermal Insulation," p. 35, Van Nostrand Reinhold, New York, 1969.
6. "Heat Insulation Manual," pp. 72–79, Pabco Div., Fibreboard Corp., Emeryville, Calif.

Meet the Authors

◀ **Richard E. Hughes** is a process supervisor with The Badger Co., One Broadway, Cambridge, MA 02142. He joined Badger in 1959 and has a wide range of experience in chemical and petroleum processes. He has a B.S. in chemical engineering from the University of Witwatersrand, South Africa, and an M.S. in chemical engineering from Yale University. He is a member of AIChE.

Victor Deumaga is a process design engineer with The Badger Co., Cambridge, ▶ MA 02142. He joined Badger in 1972, and has experience in vinyl chloride, acrylonitrile and phthalic anhydride processes. He has a B.S. in chemical engineering from Pratt Institute, and an M.S. in chemical engineering from Massachusetts Institute of Technology.

Economic pipe insulation for cold systems

Preventing cold-pipe sweating once usually outweighed other considerations in determining insulation thickness. Nomographs in this article are geared to selecting the thicker insulation now demanded by higher energy costs.

Jules L. Abramovitz, Coaltek Associates

☐ Cold-insulation design presents fundamentally different problems from those of hot systems.* One basic difference is that capital costs for refrigeration equipment add substantially to the cost of generating refrigeration. This "capital equipment fuel surcharge" can tack 25% onto prevailing electric rates.

A second major difference is that cold-insulation specifications must satisfy two requirements: an economic one, and the functional one of preventing sweating, which damages insulation. A third difference is that in selection nomographs one independent variable must be stated in ¢/kWh, rather than in the 10^6 Btu of hot systems.

Occasional and moderate condensation upon the surface of cold insulation can be tolerated (it can never be completely eliminated), but frequent and excessive condensation damages the insulation. To prevent this destruction, cold insulation must be designed to limit surface wetting. In the past, this necessity has overshadowed economic considerations. But under current conditions, and those likely in the future, energy economics will prevail in many instances. Proper design now requires that thickness be determined for both criteria, and that the more demanding criterion be applied.

Cold systems vary in temperature of operation and in capacity of refrigeration machines, factors that materially affect the cost of equipment and operation. On the basis of 10^6 Btu/h capability, refrigeration equipment is far more expensive than heating equipment. The investment in refrigeration equipment cannot be ignored; annualized, it adds substantially to operating costs.

The nomographs presented are based on costs for a recent installation. Because installations differ, methods of adjusting for these costs and operations have been incorporated in the solutions.

Differences between refrigeration-system and average-ambient temperatures are generally moderate, usually in the range of 100°F or less. Nevertheless, the nomographs take into account the difference between average annual ambient temperature and that of the fluid inside the pipe, rather than only the temperature of the fluid (as do nomographs for hot insulation).

Developing a refrigeration energy audit

The basic technical data of the reference installation upon which the nomographs are based are:

Capacity, tons	1,050
Operating temperature, °F	10
Brake horsepower, full load, hp	2,300
Motor efficiency, full load, %	95
Duty, continuous, high load, h/yr	8,450

The system requires auxiliary power as follows:

Cooling-water pump, hp	145
Cooling-tower fan, hp	166
Brine-circulation pump, hp	70

The total power needed, therefore, is 2,681 hp, or 2.55 hp/ton of refrigeration.

The energy liability is therefore:

$$(2.55 \text{ hp/ton})(0.745 \text{ kW/hp})(1/0.95) = 2.0 \text{ kW/ton}$$

Or:

$$(2.0 \text{ kW/ton})(10^6/12,000 \text{ Btu/h/ton}) = 167 \text{ kWh}/10^6 \text{ Btu}$$

The capital equipment cost, including a part of the auxiliaries, is \$645/ton of refrigeration. Paying 9% interest on the capital financed over a 13-year economic life adds a 1.73 cost factor to the foregoing base cost. Allowing 1% per year for maintenance results in an additional cost factor of 0.13.

The total equipment cost is therefore:

$$645 (1.73 + 0.13) = \$1,200/\text{ton}$$

Originally published October 25, 1976

(text continues on p. 241)

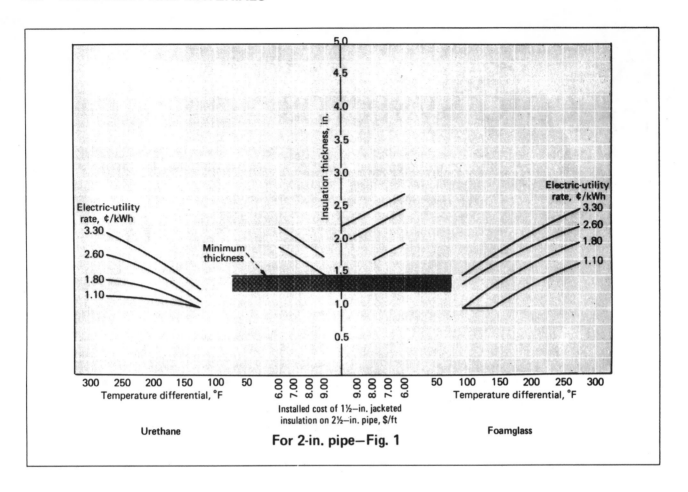

For 2-in. pipe—Fig. 1

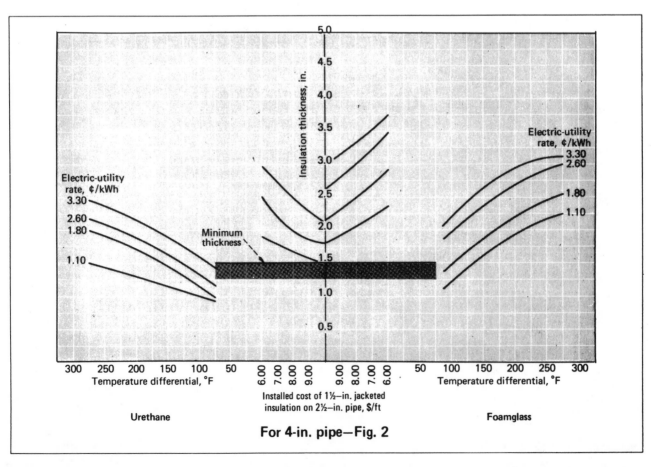

For 4-in. pipe—Fig. 2

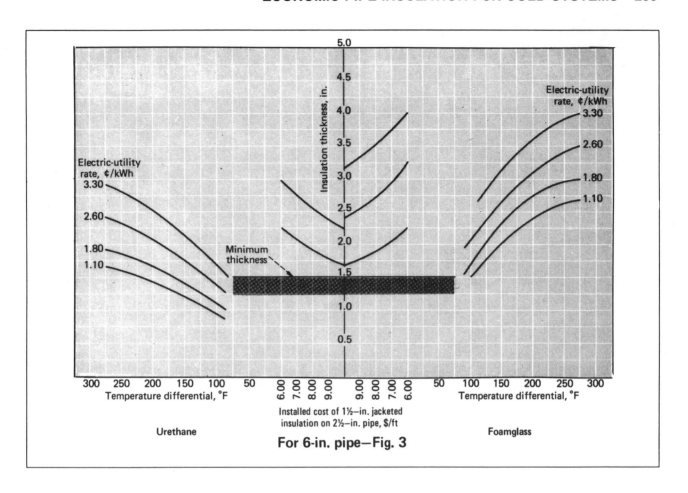

For 6-in. pipe—Fig. 3

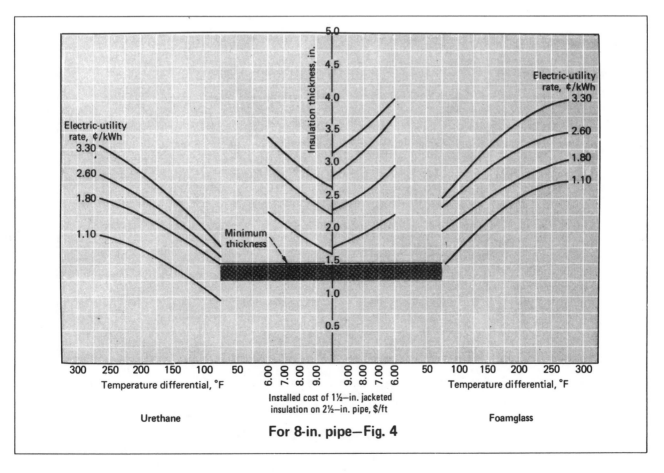

For 8-in. pipe—Fig. 4

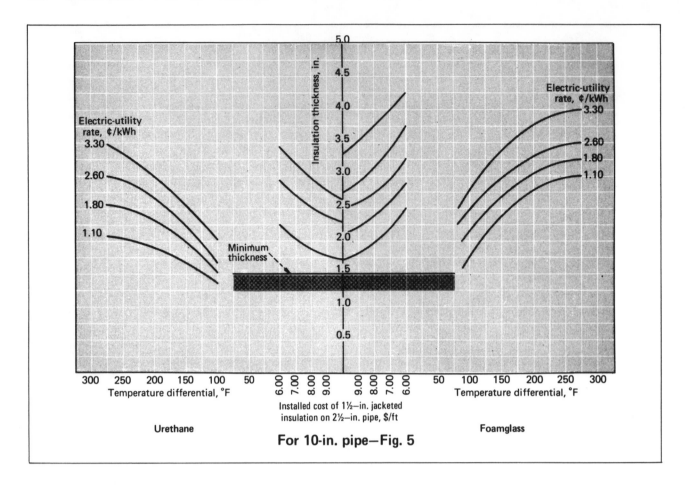

For 10-in. pipe—Fig. 5

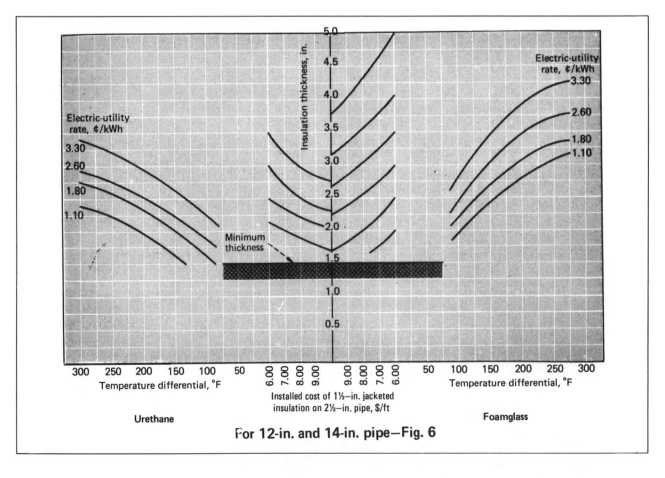

For 12-in. and 14-in. pipe—Fig. 6

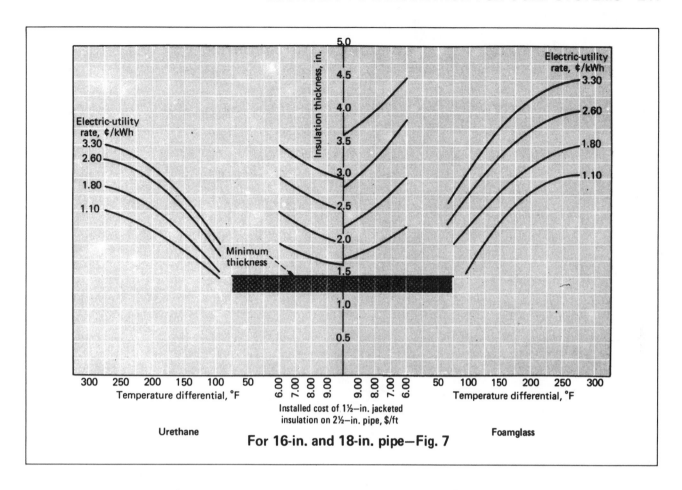

Installed cost of 1½-in. jacketed
insulation on 2½-in. pipe, $/ft

Urethane   **For 16-in. and 18-in. pipe—Fig. 7**   Foamglass

Spreading this cost over 13 years (8,450 hours of operation per year) results in a capital cost ratio of:

$$\left(\frac{\$1,200/\text{ton}}{8,450\ \text{h/yr} \times 13\ \text{yr}}\right)\left(\frac{10^6}{12,000\ \text{Btu/h/ton}}\right) =$$
$$\$0.91/10^6\ \text{Btu, or } 91\text{¢}/10^6\ \text{Btu}$$

The foregoing values are subject to frequent fluctuation, and adjustment procedures are presented later.

Basis and use of the nomographs

Fig. 1 through 7 represent graphic solutions of the economic thickness of insulation plotted against temperature differential. Two arrays of curves are included in each figure, one for urethane, the other for foamglass, insulation. For each material, parameters of electrical rates are provided. Capital equipment cost does not appear separately; instead, its equivalent of 91¢/10^6 Btu has been incorporated into the stated electric rates.

The following information must be known to use the nomographs: pipe size, in.; temperature difference, °F, between the pipe fluid and the average ambient; electric utility rate, ¢/kWh, projected to the midlife of the plant; insulation material; and installed cost.

For the sake of simplicity, installed cost is based on one thickness, 1½ in., of jacketed insulation applied to a single pipe size, 2½ in. Only this installed cost is needed, regardless of the actual pipe size or insulation thickness.

Common to all the nomographs is a minimum insulation thickness of 1½ in. (Several manufacturers have indicated they are discontinuing making insulation thinner than 1½ in. because of dwindling interest.)

Fig. 8 and 9 represent graphic solutions for the insulation thickness necessary to limit surface condensation. The temperature determinants here are fluid temperature (horizontal scale) and atmospheric dewpoint at design conditions. Two dewpoints are shown (75°F and 83°F); interpolation between them is linear. (In our work, "5% values"—the atmospheric conditions that are exceeded only 5% of the cooling-season hours—are used as design points.) A 90°F dry-bulb temperature is used for setting the heat-transfer rate from the pipe outward.

The thicker insulation required by the two criteria should be installed.

Adjusting for variations from basis

Adjustments are likely when (1) a derived capital equipment cost factor other than 91¢/10^6-Btu output is necessary, all other variables remaining constant; and (2) derived capital cost and energy-consumption rate differ from the stated base.

A capital cost ratio of 91¢/10^6-Btu net output of refrigeration was calculated for an equipment service life of 13 years. At a power consumption rate of 167 kWh/10^6 Btu, this cost can be expressed as an equivalent "energy surcharge" by:

$$\left(\frac{91\text{¢}}{10^6\ \text{Btu}}\right) \div \left(\frac{167\ \text{kWh}}{10^6\ \text{Btu}}\right) = 0.55\ \text{¢/kWh}$$

(By analogy, the capital expense for a jet costing $500,000, having a service life of 2 million miles, and consuming 2 gallons of fuel per mile would be converted into an equivalent fuel surcharge as follows:

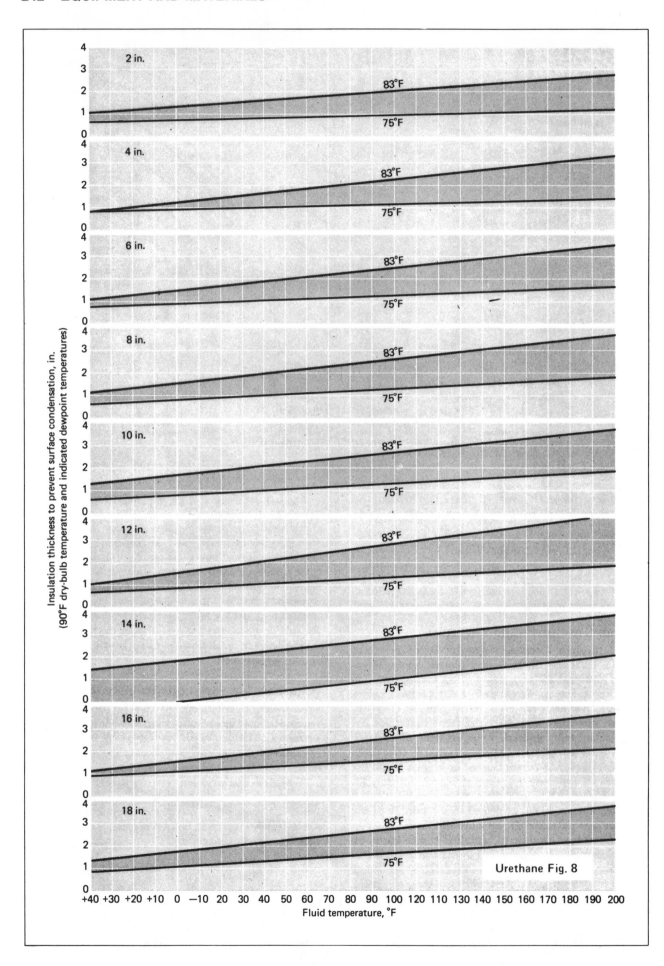

Insulation thickness to prevent surface condensation, in. (90°F dry-bulb temperature and indicated dewpoint temperatures)

Fluid temperature, °F

Urethane Fig. 8

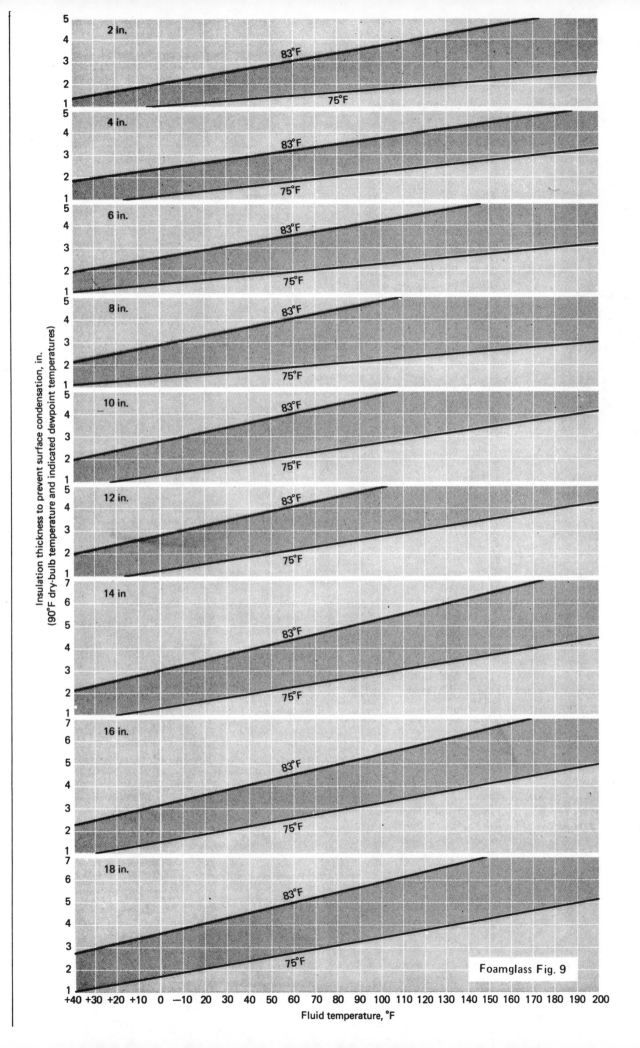

Foamglass Fig. 9

Amortization factors for Eq. (1) to solve cost of equipment

Amortization period	Interest rate, %												
	6.0	6.5	7.0	7.5	8.0	8.5	9.0	9.5	10.0	10.5	11.0	11.5	12.0
1	1.060	1.065	1.070	1.075	1.080	1.085	1.090	1.095	1.100	1.105	1.110	1.115	1.120
2	0.5454	0.5493	0.5531	0.5569	0.5608	0.5646	0.5685	0.5723	0.5762	0.5801	0.5839	0.5878	0.5917
3	0.3741	0.3776	0.3811	0.3845	0.3880	0.3915	0.3951	0.3986	0.4021	0.4057	0.4092	0.4128	0.4163
4	0.2886	0.2919	0.2952	0.2986	0.3019	0.3053	0.3087	0.3121	0.3155	0.3189	0.3223	0.3258	0.3292
5	0.2374	0.2406	0.2439	0.2472	0.2505	0.2538	0.2571	0.2604	0.2638	0.2672	0.2706	0.2740	0.2774
6	0.2034	0.2066	0.2098	0.2130	0.2163	0.2263	0.2229	0.2263	0.2296	0.2330	0.2364	0.2398	0.2432
7	0.1791	0.1823	0.1856	0.1888	0.1921	0.1954	0.1987	0.2020	0.2054	0.2088	0.2122	0.2157	0.2191
8	0.1610	0.1642	0.1675	0.1707	0.1740	0.1773	0.1807	0.1840	0.1874	0.1909	0.1943	0.1978	0.2013
9	0.1470	0.1502	0.1535	0.1568	0.1601	0.1634	0.1668	0.1702	0.1736	0.1771	0.1806	0.1841	0.1877
10	0.1359	0.1391	0.1424	0.1457	0.1490	0.1524	0.1558	0.1593	0.1627	0.1663	0.1698	0.1733	0.1770
11	0.1268	0.1301	0.1334	0.1367	0.1401	0.1435	0.1469	0.1504	0.1540	0.1575	0.1611	0.1648	0.1684
12	0.1193	0.1226	0.1259	0.1293	0.1327	0.1362	0.1397	0.1432	0.1468	0.1504	0.1540	0.1577	0.1614
13	0.1130	0.1163	0.1197	0.1231	0.1265	0.1300	0.1336	0.1372	0.1407	0.1444	0.1482	0.1519	0.1557
14	0.1076	0.1109	0.1143	0.1178	0.1213	0.1248	0.1284	0.1321	0.1357	0.1395	0.1432	0.1470	0.1509
15	0.1030	0.1064	0.1098	0.1133	0.1169	0.1204	0.1241	0.1277	0.1315	0.1352	0.1391	0.1429	0.1468
16	0.0990	0.1024	0.1059	0.1094	0.1130	0.1166	0.1203	0.1240	0.1278	0.1316	0.1355	0.1394	0.1434
17	0.0954	0.0989	0.1024	0.1060	0.1096	0.1133	0.1170	0.1208	0.1247	0.1285	0.1325	0.1364	0.1405
18	0.0924	0.0959	0.0994	0.1030	0.1067	0.1104	0.1142	0.1180	0.1219	0.1259	0.1298	0.1339	0.1379
19	0.0896	0.0932	0.0968	0.1004	0.1041	0.1079	0.1117	0.1156	0.1195	0.1235	0.1276	0.1316	0.1358
20	0.0872	0.0908	0.0944	0.0981	0.1019	0.1057	0.1095	0.1135	0.1175	0.1215	0.1256	0.1297	0.1339

$5 \times 10^5/2 \times 10^6$ miles ÷ 2 gal/mile = $0.125/gal.)

For example, if the equipment capital-cost factor were $1.50/10⁶ Btu instead of 91¢/10⁶ Btu, the equivalent "energy surcharge" would be:

$$\left(\frac{150¢}{10^6\,\text{Btu}}\right) \div \left(\frac{167\,\text{kWh}}{10^6\,\text{Btu}}\right) = 0.90¢/\text{kWh}$$

The 0.35¢/kWh *difference* between 0.90¢/kWh and 0.55¢/kWh is *added* to the applicable electric-utility rate when using the nomographs.

In Case 2, the adjustment mechanism is more involved: an adjusted electric utility rate is derived, and the nomographs are entered with this adjusted rate.

When the derived capital cost (including the financing) and energy consumption rate differ from the reference installation, the first step is to calculate a realistic capital-cost ratio, in $/10⁶ Btu:

$$S_b = (83.3\, S/H)(A + M) = \$/10^6\,\text{Btu} \qquad (1)$$

In Eq. (1), S = the cost of the equipment and a proportionate part of the cost of the auxiliaries, $/ton; H = projected hours of operation per year; A = the amortization factor, which may be calculated with the formula $A = i/1 - (1 + i)^{-n}$ (in which i is the interest rate, and n is the number of payment periods), or may be found in the table above; and M = maintenance as an annual percentage of S.

The second step is to determine the adjusted electric utility rate, in ¢/kWh, for entering the nomographs:

$$R_{adj} = \left[R + \left(\frac{S_b - 0.91}{1.67}\right)\right]\frac{E}{167} = ¢/\text{kWh} \qquad (2)$$

In Eq. (2), R = any projected electric-utility rate, ¢/kWh; S_b = the capital-cost ratio derived with Eq. (1),

$/10⁶ Btu; and E = any total-energy-consumption rate, kWh/10⁶ Btu.

Example illustrates second case

Assume the following data:

Energy consumption rate, kWh/10⁶ Btu	190
Projected life of system, yr	11
Projected annual service, h	7,800
Projected electric rate, ¢/kWh	1.95
Capital equipment cost, $/ton	750
Maintenance, %/yr	2
Cost of capital, %/yr	10.5

The adjusted capital-cost ratio, calculated with Eq (1), is $1.42/10⁶ Btu. The amortization factor, A, is 0.1575.

The adjusted electric-utility rate, determined with Eq (2), is 2.56¢/kWh. Enter this in nomographs.

Mechanically driven refrigeration units—The procedures discussed are not applicable to systems using mechanically driven units. Of course, the condensation criteria are the same whether the unit is electrically or mechanically driven, so Fig. 8 and 9 can be used for either.

The author

Jules L. Abramovitz is a senior engineer with Coaltek Associates, a partnership of Allied Chemical Coal Technology Corp. and an affiliate of Salem Corp. (P.O. Box 1013R, Morristown, NJ 07690). He holds a B.S. in mechanical engineering from Newark College of Engineering and an M.S. in mechanical engineering from Fairleigh Dickinson University.

Calculating boiler efficiency and economics

This calculator program enables you to figure excess air and combustion efficiency from Orsat-apparatus readings. You can also determine the economic advantages of adding economizers or excess-air-control instrumentation.

Terry A. Stoa, ADM Corn Sweeteners

Steam generation in direct-fired boilers accounts for about 50% of the total energy consumed in the U.S. Hence, boiler efficiencies have a significant impact on conservation [1].

During burning of a fuel, perfect combustion occurs when the fuel/oxygen ratio is such that all of the fuel is converted to carbon dioxide, water vapor and sulfur dioxide. If an insufficient amount of oxygen is present, not all of the fuel will be burned, and products such as carbon monoxide will be formed. Conversely, if excess oxygen is present, it serves no purpose and is in fact a major contributor to poor boiler efficiency. The oxygen and its associated nitrogen that passes through the boiler is heated to the same temperature as the combustion products. This heating uses energy that would otherwise be available to produce steam.

Boiler efficiency is the ratio of heat output (steam and losses) to the heat input (fuel, feedwater, combustion air). Flue-gas analysis and stack-temperature measurements can be used to monitor efficiency.

The percentage of excess combustion-air is determined by analyzing the boiler exit gases for oxygen or carbon dioxide (or both). Assuming that the gases consist solely of O_2, CO_2 and N_2, the following equation can be used:

$$A_x = \frac{\%O_2}{0.266(100 - \%O_2 - \%CO_2) - \%O_2} \times 100$$

where $\%O_2$ and $\%CO_2$ are found by an Orsat-type device [2].

Equations have been developed using just $\%O_2$, e.g.:

$$A_x = \frac{a \times \%O_2}{1 - 0.0476 \times \%O_2}$$

where Factor a is characteristic of the fuel being burned [3]. Based on curves presented in Ref. 4:

a, natural gas $= 4.5557 - (0.026942 \times \%O_2)$
a, No. 2 fuel oil $= 4.43562 + (0.010208 \times \%O_2)$

Boiler efficiency and net flue-gas temperature* follow a linear relationship:

$$E = 1/m \times (T - b)$$

where slope $1/m$ and intercept b/m vary with the type of fuel and the percentage of excess air. The following equations for finding m and b are based on curves presented in Ref. 4:

Natural gas: $\log(-m) = -0.0025767A_x + 1.66403$
$\quad\quad\quad\quad\ \log(b) \ = -0.0025225A_x + 3.6226$

No. 2 fuel oil: $\log(-m) = -0.0027746A_x + 1.66792$
$\quad\quad\quad\quad\quad \log(b) \ = -0.0027073A_x + 3.6432$

The calculation procedure becomes:

1. Determine flue-gas analysis of O_2 or CO_2, or both.
2. Determine the net stack exit temperature, T.
3. Calculate the percentage of excess air, A_x.
4. Calculate m and b.
5. Calculate boiler efficiency, E.

The calculated efficiency does not account for radiation or carbon losses. It is a measure of stack heat losses.

Once the efficiency is found, steam costs can be determined by the equation:

$$C_s = \frac{1,000 \times C_f \times H}{E}$$

This cost accounts only for the fuel portion. For a more accurate figure one must include chemical treatment costs, electric costs, labor costs, etc.

Efficiency calculations provide a sound basis for evaluating conservation projects such as installation of economizers and excess-air controls. Potential dollar savings can be based on either constant steam outputs or present fuel costs, as seen in Fig. 1 and 2.

Similarly, efficiency improvements at constant fuel input will result in capacity increases, as shown in Fig. 3.

A computer program was written to perform the described calculations, using a Texas Instruments TI-59 programmable pocket calculator. The calculator eliminates the need for charts, tables and nomographs, while providing the user with fast, dependable results. The

*Net flue-gas temperature is the difference between ambient temperature and the stack temperature measured after the last heat-transfer surface of the boiler.

Originally published July 16, 1979

245

Program for TI-59 calculator determines boiler efficiency and percentage of excess air. Inputs required are flue-gas temperature

Step	Key	Code	Step	Key	Code	Step	Key	Code	Step	Key	Code	Step	Key	Code	Step	Key	Code
000	76	LBL 1	062	42	STO	124	04	4	186	00	0	248	02	2	310	06	6
001	11	A	063	15	15	125	02	2	187	00	0	249	02	2	311	02	2
002	42	STO	064	00	0	126	07	7	188	00	0	250	01	1	312	02	2
003	00	00	065	91	R/S	127	01	1	189	00	0	251	03	3	313	06	6
004	91	R/S	066	76	LBL 6	128	07	7	190	00	0	252	69	OP	314	95	=
005	76	LBL 2	067	17	B'	129	69	OP	191	69	OP	253	02	02	315	22	INV
006	12	B	068	22	INV	130	01	01	192	02	02	254	03	3	316	28	LOG
007	42	STO	069	86	STF	131	03	3	193	69	OP	255	06	6	317	42	STO
008	02	02	070	00	00	132	05	5	194	05	05	256	00	0	318	08	08
009	91	R/S	071	02	2	133	06	6	195	25	CLR 10	257	00	0	319	76	LBL 14
010	76	LBL 3	072	02	2	134	02	2	196	69	OP	258	00	0	320	30	TAN
011	14	D	073	08	8	135	00	0	197	00	00	259	00	0	321	94	+/-
012	42	STO	074	08	8	136	00	0	198	53	(	260	00	0	322	85	+
013	03	03	075	02	2	137	00	0	199	43	RCL	261	00	0	323	43	RCL
014	91	R/S	076	42	STO	138	00	0	200	11	11	262	00	0	324	02	02
015	76	LBL 4	077	05	05	139	00	0	201	65	×	263	00	0	325	95	=
016	19	D'	078	93	.	140	00	0	202	43	RCL	264	69	OP	326	55	÷
017	42	STO	079	00	0	141	69	OP	203	00	00	265	03	03	327	43	RCL
018	01	01	080	02	2	142	02	02	204	85	+	266	69	OP	328	07	07
019	91	R/S	081	06	6	143	69	OP	205	43	RCL	267	05	05	329	95	=
020	76	LBL 5	082	09	9	144	05	05	206	12	12	268	98	ADV 12	330	42	STO
021	16	A'	083	04	4	145	25	CLR 8	207	54	)	269	93	.	331	09	09
022	86	STF	084	02	2	146	69	OP	208	65	×	270	00	0	332	06	6 15
023	00	00	085	94	+/-	147	00	00	209	43	RCL	271	00	0	333	01	1
024	01	1	086	42	STO	148	01	1	210	00	00	272	02	2	334	00	0
025	09	9	087	11	11	149	06	6	211	55	÷	273	05	5	335	00	0
026	04	4	088	04	4	150	01	1	212	53	(	274	07	7	336	03	3
027	05	5	089	93	.	151	03	3	213	01	1	275	06	6	337	02	2
028	08	8	090	05	5	152	03	3	214	75	-	276	07	7	338	00	0
029	42	STO	091	05	5	153	07	7	215	93	.	277	94	+/-	339	03	3
030	05	05	092	05	5	154	69	OP	216	00	0	278	65	×	340	69	OP
031	93	.	093	07	7	155	01	01	217	04	4	279	43	RCL	341	04	04
032	00	0	094	42	STO	156	01	1	218	07	7	280	06	06	342	43	RCL
033	01	1	095	12	12	157	07	7	219	06	6	281	85	+	343	00	00
034	00	0	096	93	.	158	06	6	220	65	×	282	01	1	344	69	OP 16
035	02	2	097	01	1	159	02	2	221	43	RCL	283	93	.	345	06	06
036	00	0	098	07	7	160	00	0	222	00	00	284	06	6	346	03	3
037	08	8	099	42	STO	161	00	0	223	54	)	285	06	6	347	07	7
038	42	STO	100	13	13	162	00	0	224	95	=	286	04	4	348	05	5
039	11	11	101	01	1	163	00	0	225	42	STO 11	287	00	0	349	07	7
040	04	4	102	93	.	164	00	0	226	06	06	288	03	3	350	06	6
041	93	.	103	09	9	165	00	0	227	66	PAU	289	95	=	351	05	5
042	04	4	104	42	STO	166	69	OP	228	66	PAU	290	22	INV	352	02	2
043	03	3	105	14	14	167	02	02	229	66	PAU	291	28	LOG	353	01	1
044	05	5	106	01	1	168	69	OP	230	87	IFF	292	94	+/-	354	69	OP
045	06	6	107	05	5	169	05	05	231	00	00	293	42	STO	355	04	04
046	02	2	108	93	.	170	25	CLR 9	232	39	COS	294	07	07	356	43	RCL
047	42	STO	109	05	5	171	69	OP	233	68	NOP	295	93	. 13	357	02	02
048	12	12	110	42	STO	172	00	00	234	02	2	296	00	0	358	69	OP 17
049	93	.	111	15	15	173	02	2	235	01	1	297	00	0	359	06	06
050	01	1	112	00	0	174	07	7	236	04	4	298	02	2	360	06	6
051	05	5	113	91	R/S	175	03	3	237	01	1	299	05	5	361	01	1
052	42	STO	114	76	LBL 7	176	02	2	238	01	1	300	02	2	362	00	0
053	13	13	115	15	E	177	01	1	239	07	7	301	02	2	363	00	0
054	93	.	116	25	CLR	178	03	3	240	69	OP	302	05	5	364	04	4
055	08	8	117	69	OP	179	69	OP	241	01	01	303	94	+/-	365	04	4
056	42	STO	118	00	00	180	01	01	242	02	2	304	65	×	366	03	3
057	14	14	119	01	1	181	01	1	243	07	7	305	43	RCL	367	06	6
058	01	1	120	04	4	182	06	6	244	06	6	306	06	06	368	69	OP
059	04	4	121	03	3	183	06	6	245	02	2	307	85	+	369	04	04
060	93	.	122	02	2	184	02	2	246	00	0	308	03	3	370	43	RCL
061	02	2	123	02	2	185	00	0	247	00	0	309	93	.	371	06	06

nd the percentage of oxygen in flue gases (as determined by Orsat analysis). Table I

Step	Key	Code	Step	Key	Code	Step	Key	Code	Step	Key	Code	Step	Key	Code	Step	Key	Code
372	69	OP 18	434	00	00	496	66	PAU	524	03	3	552	01	1	580	00	0
373	06	06	435	93	.	497	66	PAU	525	00	0	553	01	1	581	00	0
374	06	6	436	00	0	498	65	×	526	00	0	554	07	7	582	69	OP
375	01	1	437	00	0	499	93	.	527	00	0	555	02	2	583	03	03
376	01	1	438	02	2	500	01	1	528	01	1	556	07	7	584	02	2
377	07	7	439	07	7	501	55	÷	529	04	4	557	00	0	585	07	7
378	02	2	440	07	7	502	43	RCL	530	69	OP	558	00	0	586	01	1
379	01	1	441	04	4	503	09	09	531	03	03	559	69	OP	587	04	4
380	02	2	442	06	6	504	95	=	532	03	3	560	01	01	588	03	3
381	01	1	443	94	+/-	505	42	STO	533	07	7	561	01	1	589	06	6
382	69	OP	444	65	×	506	10	10	534	04	4	562	05	5	590	06	6
383	04	04	445	43	RCL	507	25	CLR 23	535	01	1	563	03	3	591	02	2
384	43	RCL	446	06	06	508	69	OP	536	03	3	564	02	2	592	00	0
385	09	09	447	85	+	509	00	00	537	06	6	565	03	3	593	00	0
386	69	OP	448	01	1	510	01	1	538	06	6	566	06	6	594	69	OP
387	06	06	449	93	.	511	05	5	539	02	2	567	03	3	595	04	04
388	25	CLR	450	06	6	512	03	3	540	00	0	568	07	7	596	69	OP
389	69	OP	451	06	6	513	02	2	541	00	0	569	06	6	597	05	05
390	00	00	452	07	7	514	03	3	542	69	OP	570	03	3	598	43	RCL
391	98	ADV	453	09	9	515	06	6	543	04	04	571	69	OP	599	10	10
392	43	RCL	454	02	2	516	03	3	544	69	OP	572	02	02	600	58	FIX
393	09	09	455	95	=	517	07	7	545	05	05	573	02	2	601	03	03
394	91	R/S	456	22	INV	518	69	OP	546	43	RCL	574	00	0	602	99	PRT
395	76	LBL 19	457	28	LOG	519	02	02	547	03	03	575	01	1	603	58	FIX
396	39	COS	458	94	+/-	520	06	6	548	99	PRT	576	00	0	604	09	09
397	02	2	459	42	STO	521	03	3	549	02	2 24	577	01	1	605	98	ADV
398	01	1	460	07	07	522	03	3	550	01	1	578	00	0	606	98	ADV
399	04	4	461	93	. 21	523	00	0	551	04	4	579	01	1	607	98	ADV
400	01	1	462	00	0										608	91	R/S
401	01	1	463	00	0
402	07	7	464	02	2
403	69	OP	465	07	7
404	01	01	466	00	0
405	02	2	467	07	7
406	07	7	468	03	3
407	06	6	469	94	+/-
408	02	2	470	65	×
409	00	0	471	43	RCL
410	00	0	472	06	06
411	05	5	473	85	+
412	01	1	474	03	3
413	00	0	475	93	.
414	03	3	476	06	6
415	69	OP	477	04	4
416	02	02	478	03	3
417	00	0	479	02	2
418	00	0	480	95	=
419	03	3	481	22	INV
420	02	2	482	28	LOG
421	02	2	483	42	STO
422	04	4	484	08	08
423	02	2	485	61	GTO 22
424	07	7	486	30	TAN
425	00	0	487	68	NOP
426	00	0	488	76	LBL
427	69	OP	489	10	E'
428	03	03	490	43	RCL
429	69	OP	491	03	03
430	05	05	492	65	×
431	98	ADV 20	493	43	RCL
432	25	CLR	494	01	01
433	69	OP	495	66	PAU

Comments

1. % O_2
2. T
3. C_f
4. H
5. No. 2 fuel oil
6. Natural gas
7. Print "BOILER:"
8. Print "DATE:"
9. Print "LOAD:"
10. Calculate A_x
11. Print "FUEL: GAS"
12. Calculate m
13. Calculate b
14. Calculate E
15. Print " ____ % O_2"
16. Print " ____ T, °F"
17. Print " ____ % XS"
18. Print " ____ % EFF"
19. Print "FUEL: #2 OIL"
20. Calculate m
21. Calculate b
22. Calculate C_s
23. Print "COST/MM BTU:"
24. Print "FUEL COST/1,000 LB"

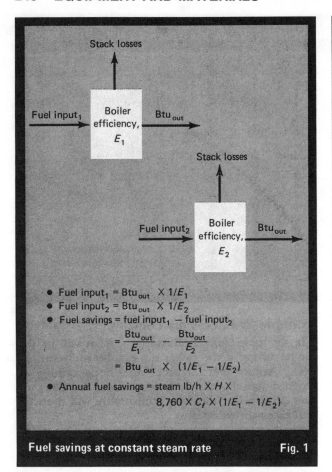

- Fuel input$_1$ = Btu$_{out}$ × 1/E_1
- Fuel input$_2$ = Btu$_{out}$ × 1/E_2
- Fuel savings = fuel input$_1$ − fuel input$_2$

$$= \frac{Btu_{out}}{E_1} - \frac{Btu_{out}}{E_2}$$

$$= Btu_{out} \times (1/E_1 - 1/E_2)$$

- Annual fuel savings = steam lb/h × H ×
  8,760 × C_f × (1/E_1 − 1/E_2)

Fuel savings at constant steam rate   **Fig. 1**

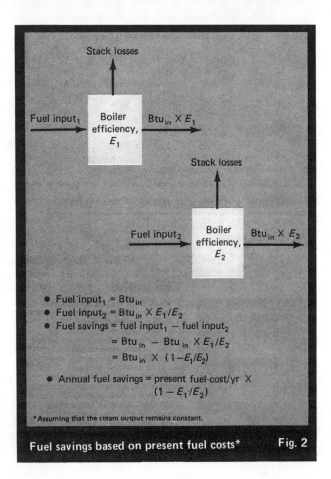

- Fuel input$_1$ = Btu$_{in}$
- Fuel input$_2$ = Btu$_{in}$ × E_1/E_2
- Fuel savings = fuel input$_1$ − fuel input$_2$

$$= Btu_{in} - Btu_{in} \times E_1/E_2$$

$$= Btu_{in} \times (1 - E_1/E_2)$$

- Annual fuel savings = present fuel-cost/yr ×
  (1 − E_1/E_2)

*Assuming that the steam output remains constant.

Fuel savings based on present fuel costs*   **Fig. 2**

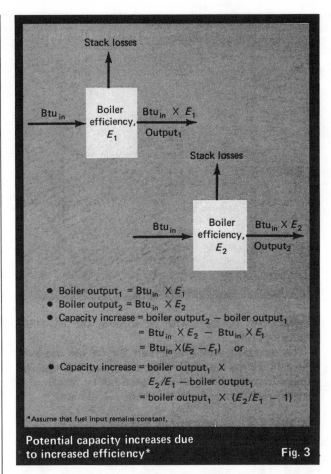

- Boiler output$_1$ = Btu$_{in}$ × E_1
- Boiler output$_2$ = Btu$_{in}$ × E_2
- Capacity increase = boiler output$_2$ − boiler output$_1$

$$= Btu_{in} \times E_2 - Btu_{in} \times E_1$$

$$= Btu_{in} \times (E_2 - E_1) \quad \text{or}$$

- Capacity increase = boiler output$_1$ ×
  E_2/E_1 − boiler output$_1$

$$= \text{boiler output}_1 \times (E_2/E_1 - 1)$$

*Assume that fuel input remains constant.

Potential capacity increases due to increased efficiency*   **Fig. 3**

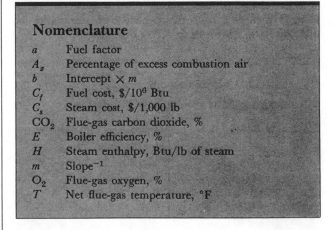

Nomenclature

a	Fuel factor
A_x	Percentage of excess combustion air
b	Intercept × m
C_f	Fuel cost, \$/10^6 Btu
C_s	Steam cost, \$/1,000 lb
CO_2	Flue-gas carbon dioxide, %
E	Boiler efficiency, %
H	Steam enthalpy, Btu/lb of steam
m	Slope^{-1}
O_2	Flue-gas oxygen, %
T	Net flue-gas temperature, °F

program can be used with or without the PC-100A printer option.

The program and instructions for its use are presented in Tables I and II. Note that variables may be entered in any order, and they need not be reentered if they do not change for subsequent calculations.

Example 1: Economizer saving

Boiler analysis without an economizer is found to be: 5%O_2 in the exit gases, where net flue temperature is 550°F. The economizer vendor claims that a 200°F reduction can be realized.

What is the annual saving if C_f = \$2.79/10^6 Btu

Step	Procedure	Enter	Press			Display
	User instruction for loading and running program that calculates boiler efficiency and % excess air.					**Table II**
1	Partition calculator	2	2nd	OP	17	799.19
2	Enter program, sides 1–4	1				1
		2				2
		3				3
		4				4
3	Enter variables in any order:					
	a. %O_2	%O_2			A	O_2
	b. T, net flue-gas temperature	T			B	T
	c. C_f, fuel cost \$/10^6 Btu*	C_f			D	C_f
	d. H, steam enthalpy Btu/lb steam*	H		2nd	D'	H
4	Enter fuel type:					
	a. No. 2 fuel oil			2nd	A'	0
	b. Natural gas			2nd	B'	0
5	Calculate efficiency				E	A_x flashes 3 times, stops showing E
6	Calculate C_s, steam cost/1,000 lb			2nd	E'	H flashes 3 times, stops showing C_s

*These variables are required only for steam cost calculations.

(No. 2 fuel oil), and the capacity of the boiler is 100,000 lb/h steam ($H = 1,160$ Btu/lb)?

Procedure: 1. Find E_1.
2. Find E_2.
3. Calculate saving.

Find E_1	Find E_2
BOILER: Existing	BOILER: With economizer
DATE:	DATE:
LOAD:	LOAD:
FUEL: *2 OIL	FUEL: *2 OIL
5. % O2	5. % O2
550. T, °F	350. T, °F
29.44002625 % XS	29.44002625 % XS
80.6386846 %EFF	85.82422822 %EFF

Solution: Annual fuel saving

$= \text{steam lb/h} \times H \times 8{,}760 \times C_f \times (1/E_1 - 1/E_2)$

$= \dfrac{100{,}000 \times 1{,}160 \times 8{,}760 \times 2.79 \times 0.0749}{10^6}$

$= \$212{,}348$

Equations for coal and #6 fuel oil

The above program can be modified to accommodate two more fuels by changing the appropriate steps in sections 5, 6 and 10 of the program. For low-sulfur coal, use the following:

$$a = 4.4477 + 0.025446 \times \%O_2$$
$$m = -(0.0226225 + 0.00015719A_x)^{-1}$$
$$b = 96.8609\,(0.0226225 + 0.00015719A_x)^{-1}$$

For #6 fuel oil, use the following:

$$a = 4.3957 + 0.7078/\%O_2$$

$$\log(-m) = -0.0027A_x + 1.6544$$

$$\log(b) = -0.0027A_x + 3.6345$$

For HP-67/97 users

The HP version of the program operates in a very similar manner to the TI version, except that there is no alphabetic printout with the former. Table III provides a listing of the HP program, and Table IV gives user instructions. Table V lists the constants that must be stored prior to running the program. Note: At step 4, the HP-97 prints %O_2, T, A_x, and % efficiency. With the HP-67, %O_2, T, and A_x will appear briefly, followed by % efficiency, which will remain in the display. At step 5, the HP-97 prints C_f and C_s. The HP-67 flashes C_f, followed by C_s.

Step	Procedure	Enter	Press	Display
	User instructions for HP version			**Table IV**
1	Load program			
2	Enter variables in any order:			
	a) % O_2	%O_2	A	%O_2
	b) T, Net flue-gas temperature, °F	T	B	T
	c) C_f, Fuel cost, \$/10^6 Btu*	C_f	D	C_f
	d) H, Steam enthalpy Btu/lb steam*	H	d	H
3	Enter fuel type (select one only):			
	a) No. 2 fuel oil, or		a	0
	b) Natural gas		b	1
4	Calculate efficiency, E		E	See text
5	Calculate C_s, steam cost/1,000 lb		e	See text

*These variables are required only for steam cost calculations.

Listing of boiler efficiency program—HP version   Table III

Step	Key	Code	Step	Key	Code	Step	Key	Code	Step	Key	Code	Step	Key	Code
001	*LBL0	21 00	046	CLX	-51	091	1	01	136	PRTX	-14	181	.	-62
002	P⇄S	16-51	047	RCL1	36 01	092	.	-62	137	RCL9	36 09	182	6	06
003	F1?	16 23 01	048	X=Y?	16-33	093	6	06	138	PRTX	-14	183	4	04
004	GTO1	22 01	049	GTO0	22 00	094	6	06	139	CLX	-51	184	3	03
005	0	00	050	1	01	095	4	04	140	SPC	16-11	185	2	02
006	R/S	51	051	R/S	51	096	0	00	141	RCL9	36 09	186	+	-55
007	*LBL1	21 01	052	*LBLE	21 15	097	3	03	142	R/S	51	187	10^x	16 33
008	1	01	053	RCLA	36 11	098	+	-55	143	*LBLC	21 13	188	STO8	35 08
009	R/S	51	054	ENT↑	-21	099	10^x	16 33	144	CLX	-51	189	GTOc	22 16 13
010	*LBLA	21 11	055	.	-62	100	CHS	-22	145	.	-62	190	*LBLe	21 16 15
011	STOA	35 11	056	0	00	101	STO7	35 07	146	0	00	191	RCLC	36 13
012	R/S	51	057	4	04	102	.	-62	147	0	00	192	ENT↑	-21
013	*LBLB	21 12	058	7	07	103	0	00	148	2	02	193	RCLD	36 14
014	STOB	35 12	059	6	06	104	0	00	149	7	07	194	x	-35
015	R/S	51	060	x	-35	105	2	02	150	7	07	195	SPC	16-11
016	*LBLD	21 14	061	CHS	-22	106	5	05	151	4	04	196	SPC	16-11
017	STOC	35 13	062	1	01	107	2	02	152	6	06	197	SPC	16-11
018	R/S	51	063	+	-55	108	2	02	153	CHS	-22	198	.	-62
019	*LBLd	21 16 14	064	STO6	35 06	109	5	05	154	ENT↑	-21	199	1	01
020	STOD	35 14	065	RCL2	36 02	110	CHS	-22	155	RCL6	36 06	200	x	-35
021	R/S	51	066	ENT↑	-21	111	ENT↑	-21	156	x	-35	201	RCL9	36 09
022	*LBLa	21 16 11	067	RCLA	36 11	112	RCL6	36 06	157	1	01	202	÷	-24
023	SF0	16 21 00	068	x	-35	113	x	-35	158	.	-62	203	STO0	35 00
024	CF1	16 22 01	069	RCL3	36 03	114	3	03	159	6	06	204	CLX	-51
025	2	02	070	+	-55	115	.	-62	160	6	06	205	RCLC	36 13
026	2	02	071	RCLA	36 11	116	6	06	161	7	07	206	PRTX	-14
027	8	08	072	x	-35	117	2	02	162	9	09	207	F1?	16 23 01
028	8	08	073	RCL6	36 06	118	2	02	163	2	02	208	P⇄S	16-51
029	2	02	074	÷	-24	119	6	06	164	+	-55	209	RCL0	36 00
030	X⇄Y	-41	075	STO6	35 06	120	+	-55	165	10^x	16 33	210	DSP3	-63 03
031	CLX	-51	076	F0?	16 23 00	121	10^x	16 33	166	CHS	-22	211	PRTX	-14
032	RCL1	36 01	077	GTOC	22 13	122	STO8	35 08	167	STO7	35 07	212	DSP9	-63 09
033	X=Y?	16-33	078	SPC	16-11	123	*LBLc	21 16 13	168	.	-62	213	SPC	16-11
034	GTO0	22 00	079	.	-62	124	CHS	-22	169	0	00	214	SPC	16-11
035	0	00	080	0	00	125	ENT↑	-21	170	0	00	215	SPC	16-11
036	R/S	51	081	0	00	126	RCLB	36 12	171	2	02	216	RTN	24
037	*LBLb	21 16 12	082	2	02	127	+	-55	172	7	07	217	R/S	51
038	CF0	16 22 00	083	5	05	128	RCL7	36 07	173	0	00			
039	SF1	16 21 01	084	7	07	129	÷	-24	174	7	07			
040	1	01	085	6	06	130	STO9	35 09	175	3	03			
041	9	09	086	7	07	131	RCLA	36 11	176	CHS	-22			
042	4	04	087	CHS	-22	132	PRTX	-14	177	ENT↑	-21			
043	5	05	088	ENT↑	-21	133	RCLB	36 12	178	RCL6	36 06			
044	8	08	089	RCL6	36 06	134	PRTX	-14	179	x	-35			
045	X⇄Y	-41	090	x	-35	135	RCL6	36 06	180	3	03			

Data storage locations—HP version   Table V

Storage area	Value
Primary 1	19458
2	0.010208
3	4.43562
6	A_x*
9	% Eff*
A	%O_2
B	T, (°F)
Secondary 1	22882
2	−0.026942
3	4.5557
6	A_x*
9	% Eff*

*Indicates values calculated by the program.

References

1. Fundamentals of Boiler Efficiency, Lubetext D250, The Exxon Corp., 1976.
2. Sisson, Bill, Combustion Calculations for Operators, *Chem. Eng.*, June 10, 1974, p. 106.
3. Shinskey, F. G., "Energy Conservation Through Control," Academic Press, New York, 1978.
4. Schmidt, Charles M., Finding Efficiencies of Stoker-Fired Boilers, in "The 1977 Energy Management Guidebook," by the editors of *Power* magazine. McGraw-Hill, Inc., p. 85.

The author

Terry A. Stoa is Plant Engineer for ADM Corn Sweeteners, Inc., a division of Archer Daniels Midland Co. P.O. Box 1470, Decatur, IL 62526. His responsibilities at the wet-corn-milling facility include plant expansion, energy conservation, process control, and providing technical assistance to the production group. He also has had experience in soybean solvent-extraction and synthetic-fiber spinning. He received a bachelor's degree in chemical engineering from the University of North Dakota and is a member of AIChE.

Improving boiler efficiency

Boilers can produce more steam from less fuel if heat losses are minimized and combustion is optimum. Downtime will be reduced with proper water quality.

J. C. Wilcox, Jr., Babcock & Wilcox

☐ Improving efficiency in the chemical process industries' boilers—boilers that produce steam for both process and power—can reduce consumption of expensive fuel by as much as 6%. More than 90% of the approximately 400 boilers sold to the U.S. petroleum industry and 95% of the approximately 850 boilers sold to the U.S. chemical industry in the past decade are burning gas or oil—fuels that are becoming more expensive and less available.

Controlling heat losses is one way to improve boilers. These losses, especially in older units, reduce the overall efficiency of many steam generating systems to below 70%. Another effective measure that is recommended for limiting fuel consumption is to install modern, effective combustion controls.

Minimizing heat losses

In any boiler system, heat lost up the stack represents a considerable waste of energy. Combustion products having temperatures above ambient, and moisture that leaves with the flue gas, constitute major heat losses.

Efficiency can be increased by reducing stack-gas temperatures through the addition of heat traps—either economizers or air heaters—that use the heat from the flue gas in other areas of the boiler.

Economizers heat feedwater with the captured energy, increasing boiler efficiency by 1% for every 10 to 11°F increase in feedwater temperature.

Air heaters transfer flue-gas heat to the furnace's incoming combustion air. For each 100°F increase in air temperature, efficiency is improved about 1.7%.

Generally, boiler efficiency increases about 2.5% for each 100°F drop in exit-gas temperature. And economizers and air heaters can improve combustion process efficiency by as much as 6%.

In deciding whether to install heat traps, a company must balance capital costs, maintenance expenses and possible added fan-power requirements against the substantial fuel savings. Most boilers in the chemical and petroleum industries are large enough to justify such equipment.

The typical boiler in these industries has a steam

Operator monitoring, even on automated boilers, can anticipate problems and increase efficiency

capacity of 100,000 lb/h. Current economic tradeoffs favor adding heat traps to boilers with steam capacities greater than about 30,000 lb/h.

Another factor in heat losses is radiation. Proper boiler insulation will minimize these losses. Burn spots and corrosion marks on outer casings indicate flue-gas leakage, which, as well as adversely affecting efficiency, can be irritating to operators.

Small leaks can occur, too, in welded inner casings.

Originally published October 9, 1978

These are often difficult to locate. One method of finding them is to discharge a smoke bomb in the furnace when the unit is down for maintenance. Leaking smoke streams will pinpoint locations where repair is needed. In these cases, it is strongly advised that the equipment manufacturer be contacted for recommendations.

Optimum combustion

In the furnace itself, combustion occurs most efficiently when the correct fuel-to-air ratio exists.

To maximize the efficiency in the boiler's furnace, three factors are important:

■ Excess air in the combustion chamber should be held to the minimum consistent with complete combustion.

■ The ash should contain no unburned combustible matter.

■ The exit flue gas should contain no unburned combustible gases, such as carbon monoxide.

Efficient combustion-control systems can regulate fuel and air firing-rates in response to load changes, and they are extremely important when boilers are regularly subjected to wide load-swings. Oxygen monitoring equipment, specifically, can control the fuel-to-air ratio most precisely because it continuously fine-tunes the combustion control system.

This equipment analyzes the flue gas and provides continuous readings of oxygen, carbon dioxide and carbon monoxide. Oxygen monitors of this kind can be incorporated into the automatic control system or used as a constant visual indicator for manual adjustment.

Almost all chemical and petroleum plant boilers are equipped with fully automatic control systems. To make sure the controls work effectively operators must watch for changes that might affect reliability.

Burner flame scanners must be kept clean; they will fail to "see" the flame if soot builds up on the instruments' tubes. This could result in false trips.

Scanners are placed in the burner wall "behind" the flame or located in the rear wall of the furnace. This rear-wall location has proved effective for oil-fired package boilers. In such units, front-wall positions may give a false, bright reading when, in fact, the furnace may be smoky and dangerous.

Pneumatic combustion-control equipment should be watched for changes in relative control pressures. Deviations from constant pressure, relative to load, can mean potential problems.

Because a boiler normally burns four times its original cost in fuel every year, the importance of combustion control in maintaining high efficiency becomes very clear. Control changes should be noted and brought to the attention of the manufacturer's service engineer.

It is wise to keep spare control-system components on hand. For example, an extra program-controller module for the startup and shutdown interlock systems, along with spare scanner cells, will eliminate lead time and unnecessary boiler downtime.

The potential of furnace explosion is reduced by safety interlock flame-failure systems and startup programming. Operators should never bypass or jumper them out, and should be able to recognize danger signals that spell potential problems.

Fuel should never be allowed to enter the furnace without proper ignition, and the fuel- and air-flow should always be controlled within the proper specifications. Failure in either area could produce explosive conditions. Charts on the control panel often warn of impending danger.

Operators should be wary if changes occur in the normal relative air- and steam-flow pen position on the 24-hour chart. Abnormal changes in steam and flue-gas temperatures could also indicate trouble.

Wide fluctuations in furnace pressure also show unstable ignition at the burner. Pulsating or erratic burner operation, abnormally dark fire, unusual impingement on furnace walls or an unusually dark stack-gas should all alert the operator that the boiler is not operating most efficiently. Operators can also check and replace electron tubes in the program controller. If further work is required, however, the manufacturer's service engineer should be called.

Burner and furnace conditions

Burners should be inspected regularly to make sure they are free of fouling, which would detract from operating dependability. "Dual atomizer" burners are available. These allow fuel inputs for full boiler load while the main gun is being cleaned.

Improper burner assembly after cleaning also can cause poor operation, loss of efficiency and, in extreme cases, failure to start.

Normally, operators should not attempt to adjust burners and registers. If adjustment is required because of flame impingement, for example, it is advisable to request help from the manufacturer's service engineers. Changes in burner setting can often mean that the combustion-control equipment also must be reset. Because of the extensive and critical interrelationships between burners and controls, trial-and-error methods can prove costly.

A thorough purge before every lightoff is the only effective means of preventing puffs. The program controller's interlock system maintains this purge cycle and prevents the introduction of any fuel until the required number of furnace air changes are achieved and verified. However, to further ensure safe startups, operators should periodically clock the purge cycle independently. This will make sure the controller is operating reliably.

Operators also should make certain that the burner

Plant lab technician checks boiler water

immediately lights and continues to burn. Properly operating flame-scanners will shut off fuel feed immediately upon improper combustion. However, operators also should check the flame visually on a scheduled basis.

The more attention the equipment receives, the less likelihood that serious problems will occur. If any danger signal appears and the reason is obvious and immediately correctable, operators can take action.

Water quality increases reliability

Equipment availability and dependability are critical to economical operation. Both are enhanced by good boiler-water quality. In maintaining tight boiler-water standards there are three major concerns: boiler-feedwater makeup, condensate flowing into the boiler feedwater, and the boiler water itself.

Water leaving raw-water treatment for makeup should have essentially no hardness. This can be assured by any of several types of commercially available softeners. Condensate returned to the boiler can range from a very small to very high percentage of the boiler's total water requirements, depending on plant operations and design. Makeup water is added to the recovered condensate to provide the remainder of the boiler's feedwater.

Taking advantage of the heat contained in the condensate can reduce total fuel consumption. If contamination in the condensate can be treated economically, the condensate should be returned to the boiler rather than dumped. This has the additional benefit of reducing makeup requirements.

The most common of the many contaminants that can be picked up by the condensate stream are suspended iron, hardness salts, and organics. The most common organic substance is oil, which can incite foaming in the boiler water.

Operators should watch the condensate returns to be sure that contaminants are removed before they can enter the feedwater system.

Most of the common contaminants can be removed by filtration. In some plants, however, a sodium zeolite filter will be needed to further purify the condensate. Oxygen should always be removed from the feedwater stream by a deaerating heater, and neither condensate nor makeup should bypass it. The deaerating heater should be kept under pressure at all times. To further control oxygen, sulfite may be added as an oxygen scavenger.

To control hardness in the system's water, either phosphate treatment of the boiler water or chelation of the feedwater may be employed. Hardness, if left untreated, can cause scaling inside of boiler tubes and lead to overheating and tube rupture. Higher-pressure boilers require much more stringent water quality than low-pressure units; what is acceptable at 150 psi is well beyond the operating limits of a 900-psi system. Power-boiler water requirements, therefore, are much tighter than those for process steam units.

Plants that cannot maintain a completely equipped and staffed water laboratory should retain a competent water consultant to set up treatment procedures and perform periodic inspections to make sure quality remains at acceptable levels. In comparison to the costs of maintenance and loss of steam production caused by tube failure, the cost of the service is worthwhile.

Analyze alternative fuels

Most boilers equipped to burn either oil or gas can be retrofitted to fire both fuels. However, many of these units cannot be modified to burn waste fuels. Some alternative fuels require specially designed boilers.

After being ignored for many years, alternative energy sources have been pushed to the forefront of public attention by the recent shortages and escalating costs of oil and gas. Many forms of solid, liquid and gaseous hydrocarbon waste fuels are now being burned in boilers.

A hydrocarbon byproduct must be thoroughly analyzed to determine its suitability as a boiler fuel. Fuel-burning equipment and combustion-air intake methods should also be examined to make sure that the unit will operate with high efficiency and low carryover of solids in the stack gas. Special design considerations are required for liquid waste streams such as pitch, acid sludge or residual tars, and for waste gases such as carbon monoxide.

It is important to identify the waste types, their quantities and characteristics before a waste processing system is designed. The following chemical and physical properties of all byproduct fuels should be identified:

- Heating value.
- Ignition temperature.
- Volatile matter.
- Fixed carbon and hydrogen contents.
- Elemental analysis.
- Ash analysis.
- Ash and waste content.
- Density.
- Ash fusion temperature and fouling characteristics.
- Corrosiveness and erosiveness of fuel and combustion gases.
- Toxicity and odor of gases.
- Explosiveness.
- Viscosity characteristics (for liquid fuels).
- Variability of quantities and characteristics.
- Special handling problems (such as polymerization of liquid fuels).

The supplier of the boiler and burner should be called in to examine the potential waste fuel, and to determine whether the existing boiler can be modified to burn it. In many cases, these byproducts can readily supplement conventional fuel and cut energy costs.

The energy situation that the chemical and petroleum industries face today poses serious economic problems. More-efficient boiler operation is one way for these industries to meet these challenges.

The author

John C. Wilcox, Jr., is General Sales Manager, Industrial and Marine Div., Power Generation Group, The Babcox & Wilcox Co., P. O. Box 2423, North Canton, OH 44720. He received his degree in mechanical engineering from Stevens Institute of Technology, and is a member of the American Iron and Steel Institute, the American Institute of Mining, Metallurgical and Petroleum Engineers, the American Soc. of Mechanical Engineers, and currently represents Babcock & Wilcox as cochairman of the American Boiler Manufacturers Assn.'s Industrial Boiler Committee.

CO control heightens furnace efficiency

Control via carbon monoxide ensures furnace firing at low excess air —even when the heat content, temperature and pressure of the fuel fluctuate, and the temperature and distribution of the air vary.

Lyman F. Gilbert, Sr., Gilco Engineering Inc.

☐ Air infiltration into a furnace often renders oxygen measurement questionable, even useless. Yet a carbon monoxide analyzer measuring the same flue gas can extract useful data and provide correct control action.

The CO analyzer measures the reaction at the flame's tip. Therefore, dilution from excess air, or infiltration from leakage or maladjusted burner registers, will not cause the measured data to be meaningless.

Some advantages of CO control:
- 1 to 7% direct fuel saving.
- 10 to 40% auxiliary power saving.
- 10 to 20% reduction of flue-gas corrosion products.
- 15 to 35% reduction of NO_x pollutants.
- Greater safety of operation through better measurement and control.
- Reduced failures and downtime because of early detection of abnormal operation.

Causes of low combustion efficiency

Nitrogen, and the inert gases formed via combustion, all carry heat out the flue-gas stack. Therefore, any air in excess of that required to burn the fuel increases the loss of heat up the stack, and additional fuel must be burned to offset this loss. The heat in the flue gas minus that in the water vapor is called the dry-flue-gas loss. This loss is usually expressed as a percentage of the total heat of combustion of the fuel.

This dry-gas loss is partially controllable because it is made up of the combustion air required to burn the fuel stoichiometrically, plus any excess of air that did not enter the combustion reaction (Fig. 1). It is convenient to compute these values on the basis of pounds weight required per million Btu, because most fuels are purchased on a cost per million Btu.

In the absence of a continuous measurement of excess air, most combustion control systems measure fuel and air flow, when gas or oil (or both) is burned. Steam flow may be substituted for fuel flow in coal firing. Because of differences in the fuel's heating value and preparation, this flow-metering method of control requires considerable excess air to assure complete combustion

under all operating conditions. Considerable excess air usually means 20% to 35% in excess of stoichiometric.

A typical fuel-and-air-flow-metering combustion control system is shown in Fig. 2, and a typical steam-and-air-flow-metering combustion control system in Fig. 3. The steam flow (as Btu of heat leaving the boiler) divided by the boiler efficiency represents the heat input of the unmetered fuel (coal).

These metering combustion-control systems are subject to large errors from: changes in fuel heat content, flow-metering errors, changes in the metered fluid's pressure, temperature and viscosity. These fuel-and-air-flow-metering errors cannot be easily overcome by more-precise measurement of the air and fuel feed.

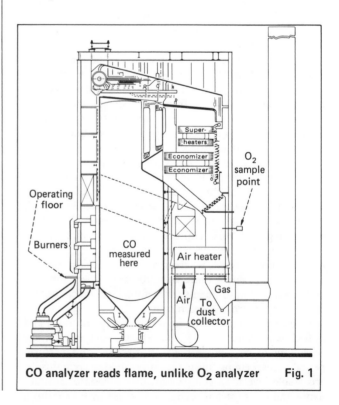

CO analyzer reads flame, unlike O$_2$ analyzer   **Fig. 1**

Originally published July 28, 1980

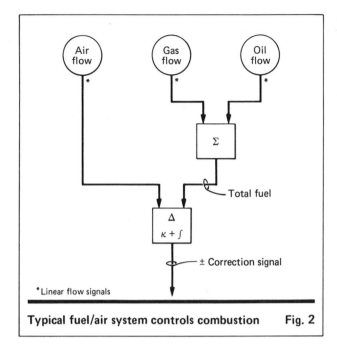

Typical fuel/air system controls combustion   Fig. 2

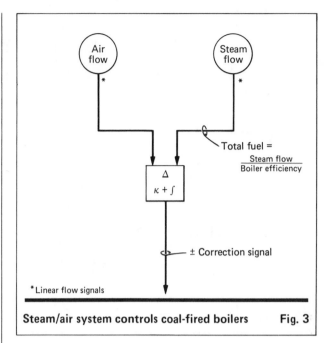

Steam/air system controls coal-fired boilers   Fig. 3

Oxygen analyzers have long been used to measure excess air in flue gases. However, there have been numerous problems with flue-gas analyzers, and oxygen analyzers in particular. These problems have frequently been associated with the flue-gas sampling system.

Special care must be exercised in designing such a system so that it runs as troublefree as possible. A well-engineered sample system will operate for several months on gas and light oils without maintenance. Coal-fired systems usually require weekly maintenance.

A second problem with the oxygen analyzer is its inability to analyze the flame's chemical reaction. Oxygen measured in the flue gases can come from idle burners, overfire air ports, infiltration, etc. Therefore, O_2 is usually carried at different excess levels in the flue gas, depending on the fuel and the furnace load.

Rules of thumb quickly sprang up for O_2 levels on different furnace and fuel types, and on furnace loading. Typical examples are: gas, 1.5% O_2 at full loading to 3.5% O_2 at 33% loading; and oil, 3% O_2 at full loading to 5% O_2 at 33% loading. The variable excess O_2 is used in the absence of an analyzer to measure the flame's chemical reaction. Therefore, an arbitrary increase in O_2 is made with a decrease in firing rate, on the assumption that it will promote better burning.

Measuring CO

Carbon monoxide has long been regarded as an indicator of improper combustion. Most methods of measuring CO have been chemical (Orsat) or via thermal conductivity. These methods have required high concentrations of CO before the measuring apparatus functions. Neither method measures concentrations below 0.2% CO.

Efficiency calculations for typical boilers show that the CO should be maintained between 0.012% and 0.025% (120 to 250 ppm). The CO analyzer should have a range of 0 to 1,000 ppm for most combustion-control applications. The analyzer will be at 10 to 25%

of its range (100 to 250 ppm) most of the time but can monitor transient concentrations of up to 800 ppm.

Oil-fired furnaces will smoke slightly at 600 to 800 ppm CO, if the burners are in good condition. Dirty burners will usually cause slight smoking at 300 ppm CO, or less. CO in a gas-fired furnace may exceed 2,000 ppm (0.2%) before it smokes perceptibly.

These CO levels are so dilute that the CO will not burn even when mixed with air. The lower limit of flammability of CO in air is 15.7% CO at 25°C.

Although a CO reading is affected by the same infiltrants as O_2 and CO_2, the results differ greatly. Fig. 4 illustrates a typical CO vs. O_2 curve. The CO control point is usually 120 ppm. If the operator or control system continuously maintains this point, even when there is leakage, a 100% dilution of the furnace CO represents only a 1.2% decrease in theoretical air (i.e., from 102.6% theoretical air at 120 ppm CO to 101.4% theoretical air at 240 ppm CO). Similarly, the CO analyzer can have a calibration drift of 10% or more without significantly affecting system performance.

Air leakage into a furnace increases the oxygen reading. A 5% infiltration of air into the flue gas where the O_2 analyzer takes its sample can cause a 100% error in the O_2 reading at 5% excess air. If the operator or control system maintains 1% O_2, the infiltration of 5% air through idle burners, overfire air, leakage, etc., will decrease the theoretical air from 105% to 100%.

Small excursions in O_2 control will cause substoichiometric firing without the operator's knowledge. With gas fuel, this could cause high CO, and efficiency losses equivalent to 15% excess air, or more. With oil fuel, this would cause some smoking, and result in violations and operation at higher O_2 levels. Because the CO analyzer measures trace levels, its control at a constant level (i.e., 120 ppm) provides continuous correction of the air/fuel ratio despite changing fuel heating value, air infiltration and other factors. The CO analyzer can control a constant end-point in the flame's reaction.

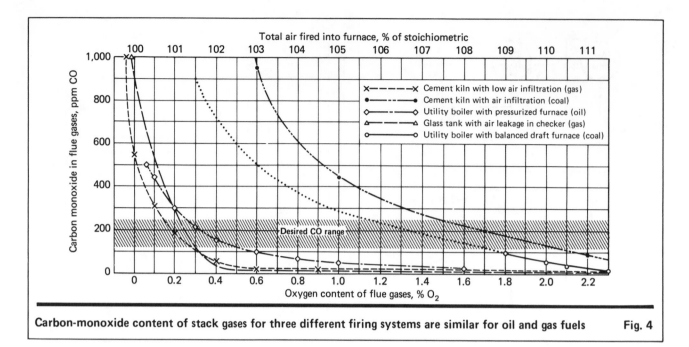

Total air fired into furnace, % of stoichiometric

Cement kiln with low air infiltration (gas)
Cement kiln with air infiltration (coal)
Utility boiler with pressurized furnace (oil)
Glass tank with air leakage in checker (gas)
Utility boiler with balanced draft furnace (coal)

Desired CO range

Oxygen content of flue gases, % O_2

Carbon-monoxide content of stack gases for three different firing systems are similar for oil and gas fuels   **Fig. 4**

The CO analyzer provides an excellent feedback-control measurement. In Fig. 5, the combustion control of Fig. 2 has been modified to accept this analyzer as a feedback controller. The instrument is a conventional proportional-integral controller with a signal-limiting feature. It should have a manual/automatic station for routine analyzer maintenance.

The system of Fig. 3 can be modified in the same way. In each case, the measured feedback variable, CO, only modifies a metered signal—the air flow. The air flow measurement is rapid, usually less than 2 s. Therefore, the CO only trims the air flow to compensate for changes in either the fuel's quality, pressure or temperature, or in the combustion air's pressure or temperature. The CO control will also compensate for varying flame patterns that arise from flue-gas recirculation and overfire air injection. The air flow meter will read less than the fuel flow meter by 10 to 20%, depending on the excess air criterion used to calibrate the air flow meter.

It is desirable but not essential to plot the curve of Fig. 5 for a particular firing system. With an O_2 analyzer, plot O_2 on the abscissa and CO on the ordinate. In the absence of an O_2 analyzer, a CO_2 analyzer may be used. The CO_2 analysis should be plotted with the higher CO_2 values to the left, the lower ones to the right. The CO_2 is diluted by the excess air, and O_2 increased by it.

The O_2 vs. CO, or CO_2 vs. CO, relationship can be used to troubleshoot the instrument and firing system. Fig. 6 shows a curve of excess air vs. CO for a particular furnace. The percent excess air can be computed from O_2 or CO_2 via the equations:

$$EA = \left[\frac{\%CO_2 \text{ theoretical}}{\%CO_2 \text{ measured}} - 1\right]100 \qquad (1)$$

$$EA = \left[\frac{\%O_2 \text{ measured}}{21\% - \%O_2 \text{ measured}}\right]100 \qquad (2)$$

If the fuel is oil (CO_2 theoretical = 16%), the excess

air 9% (1.73% O_2, 14.68% CO_2) and the CO 155 ppm, the CO will be in the desired operating band, but the excess air will be considerably greater than the curve indicates it should be (i.e., excess air 2%, or 0.41% O_2 and 15.68% CO_2). Two simple tests can be conducted to identify the problem:

1. Increase the excess air by a small amount, 5%, from 9% to 14% (2.58% O_2 and 14.04% CO_2), and the CO should drop to 28 ppm.

2. Decrease the excess air from 14% to slightly less than the original amount, or to 8% (1.56% O_2 and 14.81% CO_2), and the CO should rise to 290 or 300 ppm.

If the CO follows the curve fairly closely, there probably has been a shift in the O_2 analyzer, a large source

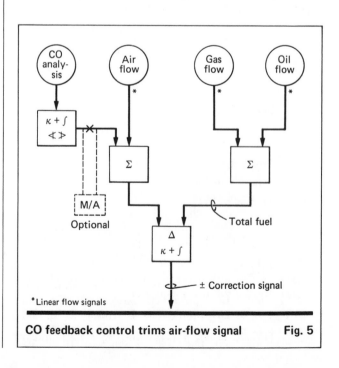

CO feedback control trims air-flow signal   **Fig. 5**

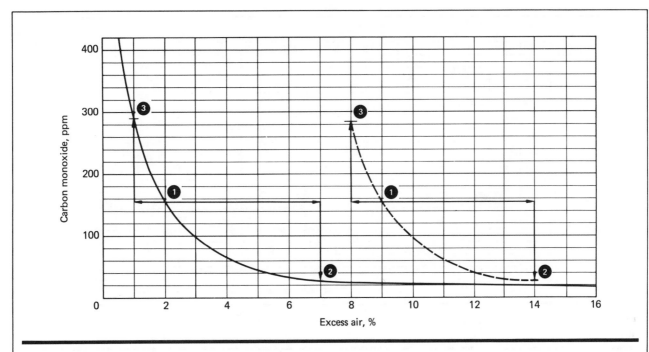

Test for identifying problems shows effect of air infiltration through an open ash-pit door   Fig. 6

of air infiltration, an air register open on a burner out of service, or some other fault (Fig. 6).

If the response to the 5% increment in air flow (9% to 14% excess air) is only a slight decrease in CO (from 155 to 148 ppm), and the decrease in air flow from 14% to 8% causes the CO to rise to only 158 ppm, the small change in CO makes it obvious that a lot of excess air is available (Fig. 7). The things to check include dirty burners, closed air registers, and a faulty CO recorder or analyzer.

Dilution of the CO, whether by infiltration, maladjusted air registers, air heater leaks, etc., will affect the output of the CO analyzer only slightly. The dilution effect (error) is linear on the CO and CO_2 analyzers, and exponential for O_2. The error in the O_2 reading caused by dilution (infiltration) is grossly affected by the true O_2 in the furnace and the amount of dilution.

If the infiltration is constant (i.e., 5% dilution of the flue gases at the point of the O_2 measurement), the O_2 analyzer error will read 210% high at 3% excess air in the undiluted flue gases, 105% high at 5% and 53% high at 10%. An 8% change in the furnace excess air produces a 4 to 1 change in the O_2 analyzer error. This variable O_2 measurement error makes low excess-air firing using only an O_2 analyzer impractical.

How fuel is saved

When the excess air is decreased, the largest saving results from the reduction of the volume of flue gas leaving the stack. The flue gas heat saved is proportional to the reduction of air flow through the furnace. Excess air only affects the dry-flue-gas loss, which is described by the following equations:

$$DG = P\,C(T_s - T_a)100 \qquad (3)$$

In Eq. (1), DG = dry-flue-gas loss, %; P = pounds of

dry flue gas ($O_2 + N_2 + CO_2$) per million Btu fired; C = specific heat of flue gas at stack temperature, Btu/(lb)(°F); T_s = stack temperature after the last heat exchanger (air heater, economizer, etc.), °F; and T_a = ambient air temperature, °F.

Although the specific heat of a gas varies with temperature, over a small temperature range its value is nearly constant. In most cases, the flue-gas temperature will remain constant enough to permit using a specific heat value of 0.24 Btu/(lb)(°F). Rewriting Eq. (3):

$$DG = (P/10^6\ \text{Btu})(0.24)(T_s - T_a)(100)$$
$$= (24 \times 10^{-6})P(T_s - T_a) \qquad (4)$$

Because the concern is to reduce the dry-gas loss, solve Eq. (4) twice: once for the typical excess-air conditions, and again for low excess-air conditions. The difference represents the reduction in dry-gas loss.

Calculate the pounds of dry gas per million Btu fired via Eq. (5):

$$P = (P_a)[(EA + 100)/100] + P_f - P_w \qquad (5)$$

In Eq. (5), P_a = lb air/10^6 Btu at 100% theoretical (stoichiometric) air; EA = excess air, %; P_f = lb fuel/10^6 Btu; and P_w = lb water vapor/10^6 Btu from burning H_2; and the 100s are in percents.

The percentage of excess air may be determined via Eq. (6) or (7):

$$EA = \left[\frac{(\%\ CO_2\ \text{theoretical})(100\%)}{\%\ CO_2\ \text{measured}}\right] - 100\% \qquad (6)$$

$$EA = \frac{(\%\ O_2\ \text{measured})(100\%)}{21\% - \%\ O_2\ \text{measured}} \qquad (7)$$

For estimating purposes, the following of P_f, P_w and P_a per 10^6 Btu values are provided:

P_f—coal = 100 lb (10,000 Btu/lb coal); gas = 45 lb

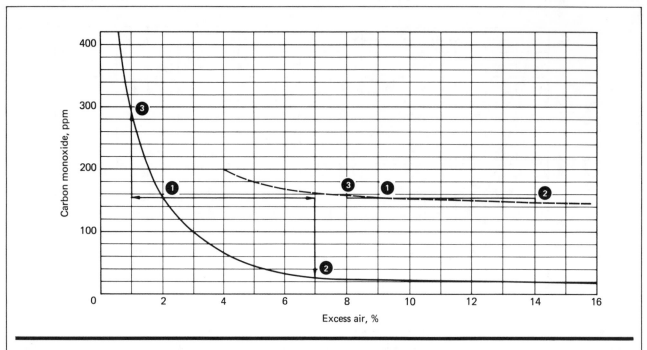

Partly closed air register is revealed as the cause of insufficient air to one burner   Fig. 7

(1,050 Btu/ft³ gas); and oil = 55 lb (150,000 Btu/gal oil).

P_w—coal = 48 lb; gas = 134 lb; and oil = 52 lb.
P_a—coal = 780 lb; gas = 745 lb; and oil = 755 lb.

This is sufficient information for computing the dry-gas loss for a boiler or process plant. First, calculate the percent dry-gas loss for high excess-air operation, then again for low excess-air operation. The difference will be the fuel efficiency improvement.

Example illustrates calculations

Assume the following boiler: size = 500,000 lb/h; fuel = No. 6 oil at \$3.17/10⁶ Btu; operating days/yr = 330; O_2 in flue gas = 6.0%; stack temperature = 500°F; ambient temperature = 70°F; average load factor = 90%; steam pressure = 600 psig; steam temperature = 700°F; feedwater temperature = 240°F; and boiler efficiency = 81%.

Using the equations, solve for: *EA*, % excess air; *P*, lb dry gas per million Btu fired; *DG*, % dry-gas loss at 6.0% O_2 (typical excess-air operation), and at 0.6% O_2 (low excess-air operation); and dollars saved from reduced excess-air operation.

Solve for the excess air via Eq. (7):

$$EA = (6.0\% \times 100\%)/(21\% - 6.0\%) = 40.0\%$$

Find lb dry flue gas per 10⁶ Btu via Eq. (5):

$$P = 755\left(\frac{40.0\% + 100\%}{100\%}\right) + 55 - 52 = 1,060$$

Determine % dry-gas loss via Eq. (4):

$$DG = (24 \times 10^{-6})1,060(500 - 70) = 10.94\%$$

Now reduce the excess air until the CO begins to increase from a very low value, usually less than 25 ppm

(0.0025% CO) to about 200 ppm (0.02% CO). The O_2 will decrease to about 0.6%. If the burners are in good mechanical condition, being fed the proper air/fuel ratio and not dirty, there will be no smoke (i.e., the opacity will be less than 3%).

At 0.6% O_2, the excess air via Eq. (7) is *EA* = 2.94%. Substituting this % excess air into Eq. (5), *P* = 780.20 lb dry gas. Inserting this *P* value into Eq. (4), *DG* = 8.05%. Therefore, the efficiency gain is: 10.94% dry-gas loss − 8.05% dry-gas loss = 2.89%.

If the oil flow is known, use it directly to calculate the heat input. Heat in the steam at 600 psig and 700°F = 1,350.19 Btu/lb. Heat in the feedwater at 240°F = 208.40 Btu/lb. The Btu input from the oil = (500,000 lb-stm/h)(1,350.19 − 208.40)/(81% efficiency/100%) = 704.81 × 10⁶ Btu/h. And the saving per hour at full loading is: (\$3.17/10⁶ Btu)(704.81 × 10⁶ Btu/h)(2.89%/100%) = \$64.57.

If the unit operates at an average loading of 90%, the annual saving amounts to: (\$64.57/h)(24 h/d)(330 d/yr)(90%/100%) = \$460,255.

The author

Lyman F. Gilbert, Sr., is president of Gilco Engineering Inc. (1850 So. Second Ave., Arcadia, CA 91006), which specializes in combustion controls and flame safeguard systems for the efficient operation of boilers and furnaces, as well as high-performance energy-control systems using onstream or sampling-probe analyzers.

Gilbert has been engaged in the research and development of fuel-burning systems and controls for 33 years, and holds 34 U.S. and foreign patents for such systems. A graduate of Washington University, with a B.S. in education (majoring in mathematics and physics), he is a member of IEEE and ISA.

Cutting boiler fuel costs with combustion controls

It is possible to achieve real savings by operating boilers with a flue-gas O_2 content of 1–2% on natural gas and 2½–3½% on No. 6 fuel oil. Here's how.

Donald L. May, Monsanto Co.

□ If perfect combustion were achieved in a boiler, there should be no CO, H_2 or O_2 in the flue gas, and the level of CO_2 would be at a maximum. [1] Among the many factors required for perfect combustion, some of the major ones are: Proper air-to-fuel ratio in the firebox, optimum atomization and distribution of fuel, and proper angle of adjustment for burners and air dampers.

Because any of these factors can have an interaction on the others, it is very seldom possible to obtain perfect combustion in a commercial boiler. However, combustion efficiency can be improved by monitoring the flue gas and making adjustments toward the optimum. Since it is possible to relate the flue-gas contents of CO, H_2 and CO_2 to each other stoichiometrically, it is not necessary to measure all four.

If only one of those gases is to be measured, it is best to measure O_2, because: new types of O_2 analyzers are extremely reliable and easily maintained; the O_2 indication can be used directly to determine the percentage of excess air, which is necessary to prevent smoking or unsafe mixtures; and consequently the O_2 measurement can be used as a feedback signal to actually control the amount of air being mixed with the fuel.

It can readily be seen that combustion efficiency is improved by lowering the amount of O_2 in the flue gas, once an O_2 analyzer has been installed. Assuming a constant stack-gas temperature, calculations show that for natural gas or No. 6 fuel oil each 2% reduction in

O_2 content raises the combustion efficiency at least 1%. Actually, when the O_2 content is lowered, the stack temperature will also go down, which indicates less heat loss and an additional efficiency improvement. Furthermore, lowering the incoming air requirements lowers the load on the air blowers, and if these are powered by steam turbines, additional fuel savings will be realized by the reduction in steam demand. Collectively, all these improvements mean that each 2% reduction in O_2 content in the flue gas can result in a fuel savings approaching 2%. On boilers utilizing continuous O_2 analyzers, it is quite easy to operate with an O_2 content of 1–2% on natural gas and 2½–3½% on No. 6 fuel oil. This compares to oxygen levels as high as 5–10% on some boilers.

The effect on the fuel bill

The savings this means for any one boiler can be determined in relation to its annual operating cost, by the following calculation:

Annual fuel cost = $PWU(8,760)/GL$

Where: P = price of oil, \$/bbl
G = gal of oil per bbl
L = lb/gal of oil
W = average fuel feed rate, lb/h
U = utility onstream time
8,760 = number of h/yr

Thus if a boiler produces 180,000 lb/hr of steam at

Originally published December 22, 1975

259

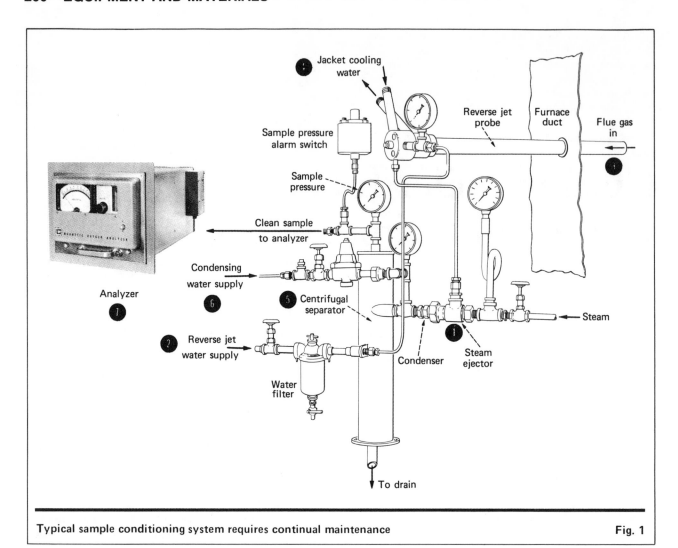

Typical sample conditioning system requires continual maintenance   Fig. 1

700°F and 650 psig, this boiler might consume 11,700 lb/hr of oil weighing 8 lb/gal and costing $10/bbl. If this boiler is operated 92% of the time, and its efficiency is increased 3% by continuously monitoring the O_2 in the flue gas, the savings would be:

$$(\$10)(11,700)(0.92)(8,760)(0.03)/$$
$$(42)(8.0) = \$84,015/yr$$

Such savings indicate that continuous monitoring of the flue-gas O_2 content should be a part of any economically operated boiler.

What type of analyzer

This focuses the problem on the type of O_2 analyzer to be used. The conventional types include: paramagnetic or thermomagnetic, catalytic combustion or rapid oxidation, electrochemical or galvanic cell, and inferential thermal conductivity. [2] All these analyzers are generally very reliable and do not require much maintenance.

However, these analyzers share one common drawback: They are set apart from the boiler duct, and therefore require a sampling system to obtain a representative portion of the flue gas, to condition it, and to

transfer it to the analyzer. A typical sampling system includes (Fig. 1): A steam ejector (1), and a water jet (2) to suck the gas in through a probe (3) that is inserted into the flue-gas duct. The sucked-in gas is cooled with water (4) and sent to a centrifugal separator (5) to remove solid particles and be contacted with cold water (6) to remove water vapors, and finally transferred to the analyzer (7).

This sampling system is subject to the following problems:

■ If the tubes between the analyzer and the probe in the flue duct are lengthy, they are likely to develop leaks and suck in air, thus altering the amount of O_2 in the sample.

■ The filters can plug and change the jet vacuum required to draw in the sample.

■ The intake probes or sample tubing can plug with fly ash, slag or sulfur deposits from the fuel.

■ The separator drain can plug up and allow water to run over into the measurement cell inside the analyzer, thus damaging it.

Because of these problems, the sampling and conditioning system of Fig. 1 typically requires minor attention weekly and major maintenance every four to six weeks. This serves as a major deterent to continuous monitoring of O_2 in boiler flue gases.

In addition to the conventional analyzers, there is also a newer analyzer that can be mounted directly in the flue-gas duct and thus not require a sample conditioning system. This new type of analyzer is of a dry-cell electrochemical probe type, in which the measurement cell is a small wafer of zirconia. A difference in O_2 partial pressures on the two sides of a zirconia wafer leads to ion migration, which causes a proportionate variation in an electric current conducted by the wafer.

In the O_2 analyzer, such a wafer cell is mounted (Fig. 2) in a pipe shield (1) inside the flue-gas duct. Plant or instrument air (3) is fed to the back side of the cell, while the front side is exposed to the O_2 in the flue gas. Since the zirconia is affected only by O_2, the back side sees the 20.9% O_2 in the plant or instrument air as the reference gas. The difference between the partial pressures on the two sides of the cell generates a millivolt output that is representative of the O_2 level present in the flue gas.

Furthermore, the resulting output signal is inversely proportional to the logarithm of the concentration of the O_2 being sampled, so that a small change at the low end of the operating range causes a relatively large variation in millivolt output. This means that there is far greater accuracy, resolution and readability at the lower end of the operating range (Fig. 3).

Since the cell is sensitive to temperature changes, it is temperature-controlled to 1,550°F. This temperature was chosen because it is well above the operating range of most stack gases and the millivolt output is high for the concentrations of 0.1–10.0% O_2 (0–80 mV).

These analyzers have been known to operate without maintenance on boilers using natural gas and No. 6 fuel oil for periods of four to six months. Such reliability brings the feasibility of continuous monitoring of O_2 to most boilers.

Where to locate the analyzer

Although the problems of the sampling system are resolved by mounting the analyzer directly within the duct, the ducts themselves on an induced-draft boiler continue to be a source of error. Generally constructed of thin sheet metal, they permit air to be sucked in by the negative draft pressures present. It is therefore important to locate the analyzer in those sections of duct that are least susceptible to errors caused by such leaks.

Since it is almost impossible to measure the leaks, a graphical method of leak analysis is recommended. A number of sample stations are established, each consisting of a piece of ¼-in tube extending 18 in into the duct and connected to a tubing-type valve on the outside.

When the sample stations are ready, the boiler is put under manual control at a constant load, and the draft pressure and stack temperature are monitored until they become constant, indicating a stabilized condition. Then O_2 samples are taken at all stations with a portable analyzer.

The O_2 content from this set of samples is averaged and the average made a base line above and below which the individual samples are plotted. The boiler is then stabilized at another load and a new set of sam-

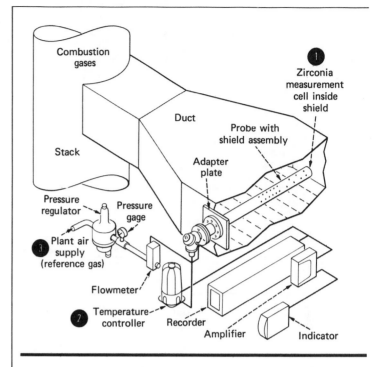

New type oxygen analyzer avoids sampling system   Fig. 2

ples taken, averaged, and individually plotted in the same fashion.

This procedure is repeated for a range of boiler loadings. When the plots are compared in a graph, they show which station will be nearest average throughout the range of boiler operations, as for example station 1 in Fig. 4.

Closed-loop control

After a probe-type analyzer has been installed in a suitable location, the continuous measurement of O_2 can be considered extremely reliable. The next logical step is to consider closed-loop control to allow the analyzer output to control the ratio of fuel to air being fed into the boiler.

Fig. 5 shows how this can be done: Air flow through a boiler is usually measured by taking the pressure drop across the entire boiler, and the ratio of air to fuel is established by selecting the proper full-range scales for the signals into an air-flow controller (1) and fuel-flow controllers (2 and 3). However, as the boiler becomes fouled internally, its pressure drop per unit air flow becomes greater and causes an error in the signal read by the air-flow transmitter. Consequently, it has been common practice to use a ratio station between the air flow controller and the fuel flow controller, and to adjust this station to provide more air flow than actually required for good combustion. Otherwise the fuel-to-air ratio might get too rich with fuel.

With the probe-type analyzer, the oxygen measurement is more accurate and reliable than the air-flow signal. Therefore the ratio station between the air and fuel flow controllers is no longer required. The O_2 analyzer signal (4) is fed to the O_2 controller (5), whose

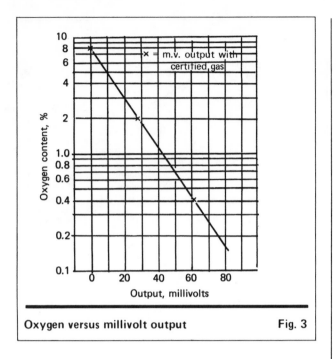

Oxygen versus millivolt output   Fig. 3

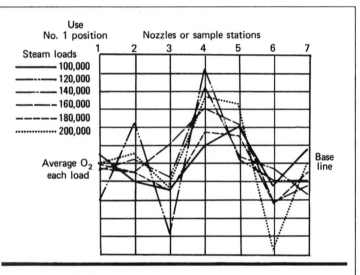

Sample stations versus oxygen content   Fig. 4

output biases the air-flow measurement through a divider or multiplier unit (6). This adjustment of the feedback signal causes the air flow controller to always provide the proper ratio of fuel to air. Because of the superior control thus achieved, the ratio of air to fuel can be lowered, increasing combustion efficiency and lowering the fuel required. In case the O_2 analyzer (4) should fail, the high-low limit relays (7 and 8) limit the effect of the bias, and prevent the O_2 analyzer from shutting down the boiler. Under these conditions, the boiler may be operated with the air-flow controller only.

When the steam load changes, the action of the valves controlling fuel flow is much faster than the action of the steam turbine that controls air flow. Con-

sequently, it is possible for a sudden increase in load to cause the fuel-to-air ratio to become too rich with fuel. This is prevented by a lead-lag system provided by high and low selector relays (9 and 10), respectively.

With this system, a call for more steam by the boiler master (11) causes the air to increase first, while a call for less steam causes the fuel to decrease first. The excess air provided by this system during a load change affords a safeguard that allows the combustion controls to be adjusted closer to the optimum fuel-to-air ratio, thus further reducing the fuel requirements.

On boilers that are sometimes fired by two fuels simultaneously, the fuel controllers should be hooked up in a primary and secondary fashion, with the prime-level-set station (12), the low selector relay (13) and the subtractor (14) working together. The prime-level-set station (12) is adjusted to the maximum amount of the cheapest fuel available. With this arrangement, all of the cheap fuel is used before any of the higher-priced fuel is called for. The Btu-computer portion of subtractor (14) is required if the two fuels do not have the same Btu content per unit of flow. The prime fuel must be converted to an equivalent quantity of secondary fuel before the subtraction is made.

Tuning techniques

Tuning a control scheme such as the one shown in Fig. 5 can be a real challenge. The controllers must function as a system rather than individually; and specific procedures should be developed and documented to enable plant personnel to maintain the tuning. These procedures should be designed for joint accomplishment by operating personnel and instrument mechanics. Steam load changes, the switching of controls from manual to automatic, and other normal control-room functions should be done by the operating technician, while all adjustments to the controller limits, proportional bands, and calibrations should be performed by instrument mechanics.

One successful technique requires the use of a four-pen recorder (about 6 in per hr). The recorder is hooked up to measure O_2 content, total fuel flow, air flow and furnace draft simultaneously. Then as each controller's gain and response time are adjusted, its effect on the other control signals can be quickly reviewed. Testing the control capabilities is accomplished by making a step change in the steam load required, and observing the patterns of the signals drawn on the recorder. Typically, after final controller adjustments:

a. The fuel and air flows should track together. This means that the percent movement on the chart, and the time required to move the recorder pens, should be about equal.

b. The induced-draft and air-flow signals should respond simultaneously. Because of the scale differences, the draft signal should move a larger amount on the chart.

When bringing a boiler on line, all controls should be on manual. The sequence of controller tuning would be: the induced-draft fan, forced-draft fan, fuel flow, and oxygen. It should be noted that, as each controller is placed on automatic, it may be necessary to go back and readjust the tuning of one or all controllers af-

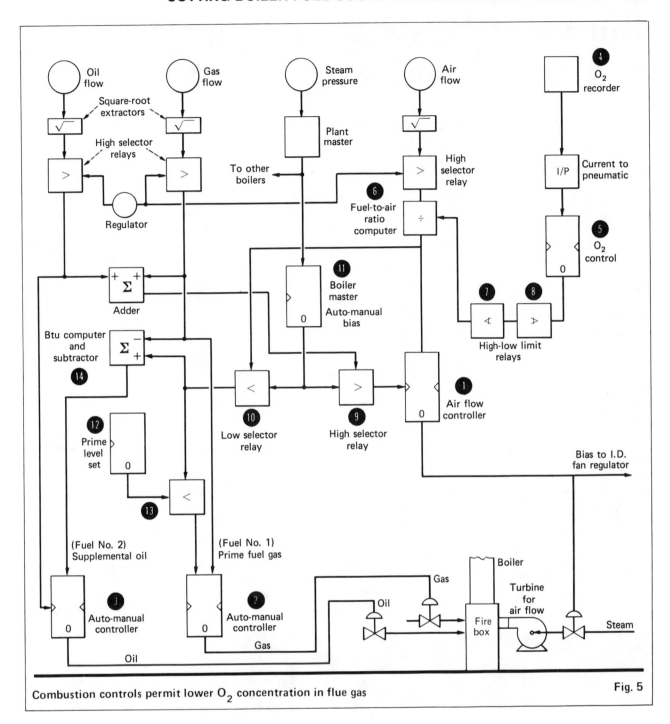

Combustion controls permit lower O_2 concentration in flue gas

Fig. 5

fected, until they perform as a system.

With such a control system functioning by means of an O_2 analyzer properly located inside the flue gas duct, it is feasible for the typical boiler now operating in the U.S. to reduce its O_2 level in the flue gas by 3–8%, for a 2–5% increase in efficiency.

References

1. "North American Combustion Handbook," North American Manufacturing Co., Cleveland, O. 44105
2. Considine, D. M., "Process Instrument and Control Handbook," McGraw-Hill Book Co., New York.

The Author

Donald Lynn May is a senior instrument engineer with the Textiles Div. of Monsanto Co., P.O. Box 12830, Pensacola, FL 32575. His duties are systems engineering and instrument applications for improving process control in utility, chemical and textile plants. He has a B.S. in electrical engineering from Auburn University. He is a registered professional engineer, a member of the institute of Electrical and Electronic Engineers, and a senior member of the Instrument Soc. of America.

Burner makers stay hot in a volatile market

Faced with high energy costs, growing government emphasis on cutting emissions of nitrogen oxides, and having to switch to dirtier fuels, U.S. industry is demanding—and getting—improved burners both for process heating and steam generation.

☐ The uncertain fuel-supply situation, stricter environmental controls, and increasing competition from novel fuel-burning methods are forcing U.S. makers of burners to develop better products.

To find out how the markets for burners are changing, and what the manufacturers are doing to adapt, CHEMICAL ENGINEERING has polled approximately 40 of them throughout the U.S. These companies have provided information on some of the products introduced for industrial steam-generation and process-heat applications* over the last few years, as well as on others ready to be commercialized within a year or so.

NEW DEMANDS—The burner makers cite a number of reasons why their business has changed. Among them:

*Numerous boiler/burner developments made by utilities cannot be applied in the CPI because of the great disparity in boiler sizes.

■ A looming U.S. Environmental Protection Agency (EPA) crackdown on nitrogen oxides (NO_x) emissions from stationary sources has created the need for units that pollute less.

■ Tightening supplies of nearly all types of fuels are fostering the use of more-versatile burners—i.e., those that can efficiently burn, say, oil and coal mixtures or liquid and solid wastes.

■ Natural-gas shortfalls are increasing the demand for supplementary units that convert more-plentiful liquid fuels into gas.

■ The growing reliance on coal has created a need for better pulverized-coal burners.

■ Stricter in-plant noise laws are forcing manufacturers to revamp burners to make them quieter.

■ Fluidized-bed combustion looms as a challenger in many conventional uses of burner-based boiler systems. Indeed, this type of technology is being tried out in several installations (*Chem. Eng.*, June 7, 1976, p. 69).

KNOCKING OUT NO_x—A number of manufacturers report efforts to build new burners that produce lower NO_x emissions.

TRW Inc.'s Energy Systems Group (Redondo Beach, Calif.), for one, is testing a new burner that is said to both slash NO_x by up to 40% and operate more efficiently. Two demonstration units are now onstream—one in a 100,000-lb/h industrial steam boiler used for a food-drying operation, the other in a small peak-power generator.

"We want to have two more demonstration units onstream in other types of boilers before we commercialize it," says R. R. Koppang, the company's manager of environmental product engineering. "If we get these installed by early next year, we will be able to go commercial by the end of 1979."

The new burners, which fire either gas or oil, are a second generation of low-NO_x burners that TRW introduced to the Japanese market about three years ago. (Japan has the world's strictest NO_x emission regulations—see *Chem. Eng.*, Feb. 14, 1977, p. 33. But the country's environmental agency is now readying new, more-relaxed standards).

HOW IT WORKS—TRW's technology is derived from its experience in making rocket engines. The new burner is similar to conventional ones. For instance, fuel is injected through a tube located inside a wider pipe through which air circulates. However, the fuel does not go out through orifices in the end of the tube, as in a conventional system, Koppang explains. Instead, the end of the tube is closed and the fuel is injected into the combustion-air stream through orifices in the tube wall. This makes for better air-fuel mixing, he says.

The mix is then directed radially to form a hollow, cone-shaped flame, in contrast with the long, usually bushy flame of a conventional burner. The cone shape is achieved through a combination of the geometry of the burner tip and the combustion-air venturis. Koppang notes that the cone configuration exposes more of the flame to its environment, which lowers the flame temperature by about 10%,

Section view of TRW's low-NO$_x$ burner shows annuli for air-flow control

Originally published August 14, 1978

Energy-Pak can be used to replace or supplement natural-gas supplies

which in turn cuts NO_x by up to 40%.

TRW's new burners have an additional feature—they can operate at as low as 20% of capacity and still reduce both fuel consumption and NO_x emissions in comparison with conventional burners working under the same conditions. Both reductions are achieved by cutting air flow to maintain a constant fuel/air ratio, so that fuel efficiency is approximately the same as at peak load.

"We have operated as low as 20% of capacity with heavy fuel oil and substantially reduced excess air," says Koppang. "We can save 0.5% fuel at full capacity, in comparison with a conventional burner, and up to 2% or more for low-capacity operation." He calculates that a burner installation on a large boiler could pay for itself in two or three years.

"If you reduce the load in conventional burners, you have to add excess air to maintain flame stability," says Koppang. "This reduces flame efficiency and increases NO_x production."

In TRW's system, air flow is controlled by two or three air annuli around the fuel pipe. When the boiler is operating at full load, air is injected through the annuli. When the load is reduced, air passes only through the inner annulus. This permits a reduction of air volume without a lowering of air velocity and mixing efficiency.

TRW is directing its marketing efforts to multiple-burner boilers of about 50,000 lb/h or more of steam. By contrast, the first-generation NO_x-control burner marketed in Japan has

been sold for single-burner package boilers.

In another low-NO_x development, John Zink Co. (Tulsa, Okla.) has recently installed a new burner at a California refinery. The device is said to release the same amount of heat as the conventional burner it replaced, but at a lower temperature. Although pressure drop is "slightly higher," the lower temperature guarantees NO_x won't exceed 100 ppm when the burner is fired with natural gas, or 150 ppm when fired with #6 fuel oil. (Field results are now running at about 55 ppm and 100 ppm for gas and oil, respectively.) Reduced temperature is achieved by adding a slit in the refractory that surrounds the burner tip; this allows recirculation of cooler gas across the flame.

National Air Oil Burner Co. (Philadelphia, Pa.) is developing a low-NO_x oil burner that it considers promising. The crucial feature is the burner block, which has hollow walls that contain several inlets. Hot combustion products recirculate to the flame, resulting in fuel-bound nitrogen being released as diatomic nitrogen instead of being converted to NO_x.

Aimed at natural-draft furnaces in process applications, the new unit will probably be commercialized within the next year, according to the developer. It has already been tested in three installations in chemical, petrochemical and steelmaking plants in Japan, with NO_x emissions running from 100 to 110 ppm—about half those of conventional oil burners, says National Air, and almost as low as those from gas burners.

NO_x SPINOFF—Where NO_x emission to the atmosphere isn't the main consideration, but NO_x introduction into the process stream is taboo, indirect heating or natural-gas firing has usually been the answer. This is the case in the food industry, for example, where nitrates/nitrites are undesirable contaminants.

However, indirect heating—with steam coils, for instance—is generally more wasteful of energy than direct heating. Therefore, some burner makers see a market for burners that emit negligible amounts of NO_x in such applications as spray drying of high-purity products.

Maxon Corp. (Muncie, Ind.), for one, has developed a new burner that will soon get its first field tryout in Europe in a spray-drying application. Maxon is initially aiming at Europe's dairy industry (where nitrates/nitrites restrictions in food processes are usually tighter than in the U.S.), and believes that regulations will ultimately be stiffened in the U.S. as well, providing a healthy market.

Key to the unit is a mixing tube—anywhere from 2 in. to 10 in. dia.—that premixes gas and air, producing a uniform flame that tends not to generate NO_2.

Laboratory tests indicate that the new unit can cut NO_2 levels in products to 0.5 ppm or less, versus typical levels of 1 to 2 ppm, says Maxon.

Process Combustion Corp. (Pittsburgh, Pa.), which custom-builds burners to solve specific problems, has also been developing new applications in the food industry.

PCC is interested in applying #2 fuel oil, instead of natural gas or indirect heating, for critically clean products. The firm says that the U.S. Dept. of Agriculture, for instance, is using the technology to spray-dry milk products, and a manufacturer is using it for direct drying of a table-salt additive. The unit will produce combustion products that are as clean as those emitted from forced-draft gas burners, PCC claims.

Over the past two years, PCC has also been working with a half-dozen PVC manufacturers to use fuel oil to directly fire processes that were formerly steam-heated or natural-gas-fired. One of these makers now has a commercial unit in operation that uses PCC's double-toroidal combustion chamber technology; two others will soon start up commercial units; the remaining three are conducting pilot tests.

PCC will say little about the burner's design, other than that it has "a basically unique method of combustion-air introduction in conjunction with an externally atomized oil gun." The units come in eight capacities, ranging from 2 million to 200 million Btu/h.

VERSATILE DEVICES—A number of burner makers are promoting units that can feed on a range of fuels.

One from Mechtron International Corp. (Orlando, Fla.) can fire any fuel or any combination of two or three fuels (in any percentages), according to Julian Downey, vice-president for corporate development.

In the works for five years, the new unit has recently won its first two commercial roles—one at a U.S. Navy base, for firing a steam generator with coal and/or coal-oil slurries; the other at a municipal sewage plant, for firing a fume incinerator with coal or waste rather than gas.

The new burner meets all pollution limitations in the U.S. and Canada, says Downey: "We can assure complete combustion, with burner turndowns of 10:1." Although an inventory has not yet been built up, standard units will eventually range from 3 million Btu/h to over 200 million Btu/h, the executive says.

In Addison, Tex., Forney Engineering Co. reports that "we are in the process of developing a burner that will handle everything from #2 oil to heavy oil. The goal is to improve combustion along a full operating range."

Forney will not reveal details yet, saying only that "the aim is to get complete stoichiometric combustion." The unit will be geared for applications in large processing plants and power plants, with capacities averaging about 100 million Btu/h.

North American Mfg. Co. (Cleveland, Ohio) is working on improving its line of dual-fuel (oil and gas), high-velocity process burners. According to the firm, the new line of Tempest burners will be available off-the-shelf within a year. The units, with capacities of up to 1.5 million Btu/h, are said to have greater fuel/air ratio flexibility, and to require less refractory-maintenance than previous models.

ON STANDBY—CPI plants faced with natural-gas curtailments often don't have the option to switch to liquid fuels, or they may find the switchover too costly and time-consuming. To get around this, several developers have recently come out with systems that produce a natural-gas substitute from liquid fuels.

For example, technology developed by Allied Chemical Co. (Morristown, N.J.)—and licensed to Heat Research Corp. (Philadelphia), a subsidiary of Pullman, Inc.—has just been commercially demonstrated in its initial tryout, according to HRC. And a second VFO (vaporized fuel oil) unit is now in startup, while four others will follow soon. Their capacities range from 72 million to 835 million Btu/h.

The system, which feeds on distillate fuels, can be applied to "almost any furnace which now fires gas," says HRC. For in-plant testing, the company has a 12-million-Btu/h test rig. (See *Chem. Eng.*, Aug. 15, 1977, pp. 111-112, for more details.)

Another manufacturer that has just come out with a standby system is Ransome Gas Industries, Inc. (San Leandro, Calif.). The device, dubbed the Energy-Pak, aims for smaller applications than the HRC/Allied unit—no more than 12.5 million Btu/h.

The compact (45 in. x 40 in. x 22 in.), self-contained unit blends liquid propane and air in a precise ratio, with the blend either completely replacing a natural-gas supply or supplementing it during peak loads. Essentially, it is a vaporizer that functions as a pressure regulator plugged into the gas-distribution system.

COAL FRONT—Increasingly, pulverized coal is becoming a more important fuel for the CPI (see *Chem. Eng.*, Feb. 14, 1977, pp. 40-44). Numerous burner and boiler manufacturers are trying to develop better coal-fired units, but most advances are still several years off.

Babcock & Wilcox Co. (Barberton, Ohio) is developing a new burner for huge industrial applications. The 250-million to 300-million-Btu/h unit will feed on pulverized coal, with hoped-for NO_x emissions of just 100 ppm, compared with typical outputs of about 450 ppm today. Commercialization of the device is still at least three to four years in the future, according to the company.

Another firm that hopes to commercialize an improved unit is Coen Co. (Burlingame, Calif.), a manufacturer of custom-made burners. The company is currently developing a 100-million-Btu/h, ceramic-lined coal scroll-burner for firing an air heater in the cement industry.

"We've made coal scroll-burners in the past," says R. Richard Vosper, chief engineer, "but they've had a high wear factor." Coen hopes that replacing traditional steel with abrasion-resistant ceramic will bring a solution to the problem.

Meanwhile, John Zink Co. expects to have a new pulverized-coal burner available "toward the end of the year," but will say no more.

NOISE CUTTERS—Several new burners have been designed with noise control in mind.

A John Zink Co. official notes, for instance, that "we have designed a burner that has lower combustion noise" to keep noise levels below OSHA requirements. "The first efforts were strictly mufflers," the official concedes, "but the new design involves modifications and construction changes to use the lowest ∆P possible. We knock high-frequency noise down by a combination of design features and materials—for example, proper refractories."

And Selas Corp. of America (Dresher, Pa.) has recently made noise-abatement improvements in its DN series of burners. The improved range—called the DNS—can handle both natural gas and #2 fuel oil for process applications, and readily meets OSHA noise limits, according to the firm.

"The most unusual part of the DNS burner is the nozzle," says David Gray, product manager. "It permits air and fuel streams to be mixed quickly and completely, and also imparts a swirling motion. Combustion takes place rapidly, and combustion gases burn very close to the refractory surface, resulting in a cloud of completely burned gases being uniformly discharged into the furnace." Uniform combustion, Gray explains, reduces burner noise.

Gas-only DNS units come in two sizes: 250,000 Btu/h and 2.5 million Btu/h. Dual-fuel units are available in two models: 500,000 Btu/h and 1 million Btu/h.

Larry J. Ricci

FIRED HEATERS—
How to reduce your fuel bill

This article takes a look at how energy conservation has affected the design and operation of fired heaters. Items of concern include excess-air reduction, enhanced heat recovery, combustion-air preheating, and the conversion of gas-fired heaters to liquid firing.

Herbert L. Berman, Caltex Petroleum Corp.

☐ Demands on the CPI for greater energy conservation have had direct impact on the design and operation of fired heaters. Fuel savings have been realized with a variety of measures, ranging from the inexpensive—such as the fine-tuning of operating procedures and the upgrading of maintenance techniques—to the capital-intensive—such as the installation of complex heat-recovery facilities at substantial initial outlay. The following discussion will focus on methods that have proved successful in raising the thermal efficiency of fired heaters.

Reduction of excess air

In existing installations, excess air is the most important combustion variable affecting the thermal efficiency of a fired heater. Although heater operation is easier to control at high excess-air levels, it is very costly. The higher the excess air, the greater the fuel consumption for a given heat absorption. The extra fuel is consumed in heating the excess air volume from ambient temperature to the temperature of the exiting flue gases.

In order to exercise greater control over excess air, the oxygen content of the flue gas should be monitored above the combustion zone. Often, excess air in the combustion section may run as low as 10 to 15%, but stack-gas analysis reveals an oxygen content equivalent to as much as 100% excess air. This differential results from air leakage into the heater that occurs between the combustion zone and the stack. Such leakage cannot be corrected by burner adjustments.

Air leakage into a heater can occur at many loca-

tions. One route of entry is through the seams of the steel casing, between adjacent plates and stiffening members. Air can also enter through distorted or poorly gasketed header boxes. The terminal tubes in the tube-coil, where they enter and leave the heater casing, can likewise be a source of air leakage. Only through a rigorous, continuing maintenance effort can air leakage into heater settings be controlled effectively.

There are various schemes for controlling excess air via the monitoring of flue-gas oxygen content. Four such monitoring schemes are listed here, in order of increasing sophistication and cost [1]:

1. The least costly scheme requires only periodic checking of oxygen content using a portable oxygen analyzer and a portable draft gage. With these readings as a guide, the operator can make the adjustments necessary to operate at minimum excess-air levels.

2. One of the most common monitoring systems employs a continuous oxygen analyzer equipped with a local readout device and a permanently mounted draft gage. Continuous readings enable the operator to adjust heater operation whenever necessary.

3. Bringing the oxygen and draft readouts into the control room and adding a remotely controlled, pneumatic damper positioner is the next step toward improving excess-air control.

4. Although generally justifiable only for large heaters, further sophistication in excess-air control can be achieved with automatic stack-damper control. In this control scheme, the damper is positioned automatically according to a draft signal received from a probe located below the convection section. The controller changes the damper position so as to hold a set-point

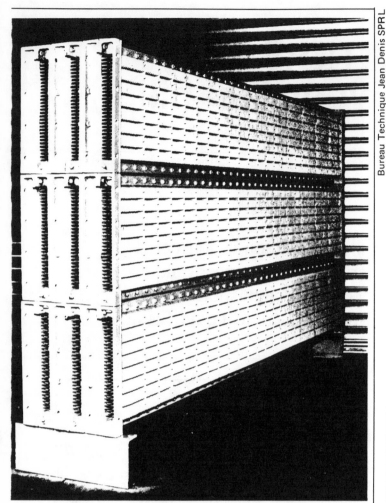

Recuperative preheater warms combustion air. Unit is equipped with finned, cast-iron tubes.   **Fig. 1**

Bureau Technique Jean Denis SPRL

draft corresponding to the targeted excess air. The draft set point is normally changed only to achieve variations in heater duty.

Further recovery of convection heat

The potential for additional heat recovery in the convection section exists when the heater operates at a relatively high flue-gas temperature. First consideration should be given to augmenting the convection-section surface area by adding several rows of convection tubes in the same heating service. Many current installations have anticipated such expanded service, and have been designed to accommodate the future installation of two rows of convection tubes.

The installation of supplementary heat-exchange units to reclaim additional convection heat often provides a quick economic return. Falling into this category are a number of units that recover process heat. Examples are feed preheaters for pyrolysis and steam hydrocarbon-reformer heaters, reboilers operated in conjunction with catalytic reformers, and superheaters providing process steam for petroleum refinery distillation units.

Very often the supplementary unit recovers convection heat for use in a steam-generating facility rather than in a process-stream heating application. Steam-generating units driven by convection heat are routinely coupled with catalytic-reformer heaters and steam hydrocarbon-reformer heaters. Fired heaters in such installations have convection sections equipped with several independent coils so that the same convection section furnishes heat for boiler-feedwater heating, steam generation, and steam superheating. Coils installed in the convection sections of steam-generating fired heaters are almost always of the forced-circulation type.

Occasionally, flue gases from several fired heaters are routed to a central waste-heat recovery facility. Typically, the recovered heat is utilized for steam generation. Waste-heat boilers of this type are normally designed with a flue-gas bypass around the boiler, which allows the heaters to operate even when the boiler is taken out of service. Where several heaters are involved, provisions for positive isolation at the flue-gas ducts from each heater will enable the user to take an individual heater out of service while the remaining heaters and the waste-heat boiler stay on line.

Before any add-on heat recovery device is retrofitted to a heater, the structural integrity of the existing steelwork and foundation must be examined to assess the effects of the increased loads.

Furthermore, it should be noted that the additional convection-section heat recovery will result in a lower stack-gas temperature, thereby reducing the stack draft. In addition, the installation of more surface area will increase the flue-gas pressure drop. Therefore, it is imperative that the effect of any alterations on the stack draft be analyzed beforehand in order to determine whether additional stack height will be required.

Replace bare tubes with extended surface

Many older generation fired heaters are equipped with bare-tube convection sections and operate with high flue-gas temperatures at 65 to 70% thermal efficiency. By replacing the bare tubes with extended surface tubes, efficiency improvement in the neighborhood of 10% may be realized in some installations. This corresponds to a stack-gas temperature reduction of about 300°F.

The most economical conversion from bare to extended-surface tubes can be made when the reduction in tube size is such that the tip-to-tip diameter across the extended surface tube is the same as the outside diameter of the original bare tube. On this basis, it is very likely that the existing convection section tube sheets can be retained. However, the reduction in tube I.D. will result in higher fluid pressure drops unless an increase in the number of convection-section parallel passes can be tolerated. Conversely, if the same tube I.D. is maintained, the conversion from bare to extended surface will necessitate the replacement of the convection-section tube sheets—with accompanying down-time and expense.

When conversion is contemplated for a heater firing liquid fuel, it is recommended that soot blowers be installed, in view of the greater fouling tendency of extended-surface devices. The cavities required to accommmodate the soot-blower lances can usually be created simply by omitting the installation of a row or two of tubes at selected locations in the convection section.

Again, it is mandatory that the stack draft be assessed to ascertain whether the existing stack is adequate.

Preheat the combustion air

Fuel consumption in a fired heater can be reduced markedly by preheating the combustion air. In the preheater, heat is transferred from the flue gas to the combustion air, reducing the exit temperature of the flue gas and raising the thermal efficiency. With air-preheat systems, exit flue-gas temperatures often range in the 300 to 350°F range and efficiency levels commonly reach 90 to 92% (LHV). With such systems, the attainable thermal efficiency is no longer controlled by the approach between the flue-gas and inlet-fluid temperatures.

The salient features of the more popular air-preheaters are noted here:

■ *Recuperative-type air preheaters.* These devices, typically of tubular construction, transfer heat by convection from flue gas to combustion air. Customarily, the air flows inside the tubes, whereas the flue gas flows across the tube bundle. The preheater can be installed in the fired heater above the process convection section or, as is more usually the case, at grade alongside the heater.

Materials selected for heat-transfer surfaces vary from designer to designer. If the flue gas is well above the acid dewpoint, a manufacturer may select cast iron tubes having internal and external fins. In the zone approaching the dewpoint, fins are provided on the gas side only, in order to keep tube-metal temperatures as high as possible. Below the dewpoint, plain tubes of borosilicate glass are used in order to minimize acid corrosion.

Tubular-type air preheaters are essentially tight between air and gas, with no leakage of air into the flue-gas side. Fig. 1 shows a bank of tubular cast-iron heating elements equipped with internal and external fins.

A schematic arrangement for a typical recuperative air preheater installation is shown in Fig. 2. (The arrangement shown is equally applicable to regenerative preheater systems, described below.) The system employs a forced-draft fan to supply combustion air, and an induced-draft fan to maintain a negative pressure and draw the flue gas through the system to the stack. A cold-air bypass enables the operator to route a portion of the incoming combustion air around the preheater when ambient temperature is very low. The preheater surface area can thus be maintained above the acid dewpoint, at a nominal sacrifice in thermal efficiency.

■ *Regenerative-type air preheaters.* This apparatus consists of heat-transfer elements housed in a subdivided cylinder, which rotates inside a casing. Hot flue gases pass through one side of the cylinder, cold air through the other side. As the cylinder slowly rotates, the elements continuously absorb heat from the flue gas and release it to the incoming air stream.

The cylinder is subdivided by baffles which, like the seals between the cylinder and the casing, help to mini-

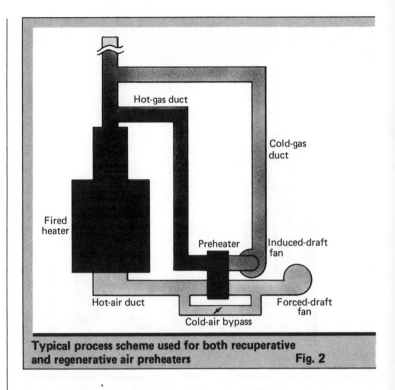

Typical process scheme used for both recuperative and regenerative air preheaters   **Fig. 2**

mize leakage of air into the flue-gas stream. This leakage, which results from the pressure differential that exists between the air side and the flue-gas side, is normally 10 to 15% of the total air flow. The preheater system, particularly the forced-draft and induced-draft

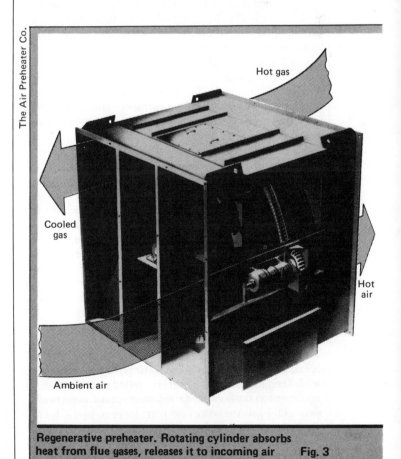

The Air Preheater Co.

Regenerative preheater. Rotating cylinder absorbs heat from flue gases, releases it to incoming air   Fig. 3

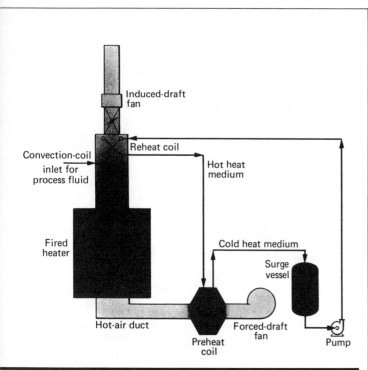

Induced-draft fan

Convection-coil inlet for process fluid

Reheat coil

Hot heat medium

Fired heater

Cold heat medium

Surge vessel

Hot-air duct

Preheat coil

Forced-draft fan

Pump

Heat-medium preheater transfers heat from flue gases to air via intermediate, circulating fluid   Fig. 4

fans, must be designed to accommodate this leakage.

The heating elements of the regenerative air preheater are installed in two or three layers. Corrosion of the elements is usually confined to the final portion of the cold-end layer, where cold air enters and cooled flue gas leaves. If such corrosion does occur, the elements of the cold-end layer can be removed and reversed to extend their service life. For those applications where the elements are exposed to very corrosive atmospheres, porcelain-enameled heating surfaces are available. Fig. 3 illustrates the basic construction of a regenerative air preheater. The main components of the overall process arrangement are shown in Fig. 2.

■ *Heat-medium air preheaters.* Instead of direct heat exchange between air and flue gas, these units employ an intermediate fluid to transfer heat from the flue gas to the incoming combustion air. The heat medium is contained in a closed loop that includes a reheat coil located in the flue-gas flow downstream of the process convection coil, and a preheat coil positioned in the air stream. The circulating fluid extracts heat from the flue gas, lowering the gas temperature and raising the fluid temperature. In turn, the hot fluid releases its heat to the incoming air.

The process scheme for a heat-medium preheat system (Fig. 4) includes, in addition to the preheat and reheat coils, a fluid-surge vessel and a circulating pump. Depending on the design of the heater and the available draft, the system may be provided with forced draft/induced draft, forced draft only, or induced draft only.

■ *Process-fluid air preheaters.* These systems [2] take off a portion of the process stream entering the convection section and send it to a preheat coil mounted in the air

stream, which warms the air. The subcooled process fluid is then returned to a reheat coil located in the flue-gas flow downstream of the main-process convection coil, which reheats the fluid to its original temperature level and returns it to the main process stream ahead of the convection coil.

Only a portion of the process fluid stream is drawn off for preheat service. The mass flows and specific heats of the combustion air and process fluid must be carefully balanced in order to achieve the desired temperature changes in each stream.

A schematic arrangement for a process-fluid air preheater is shown in Fig. 5. As with heat-medium preheaters, these systems can be supplied with forced draft/induced draft, forced draft only, or induced draft only.

Process effects of air preheating

Fired-heater operation with preheated combustion air results in higher adiabatic flame temperatures than with ambient air. From the combustion kinetics, then, one might expect heaters using preheated air to generate higher amounts of NO_x than those using ambient air.

However, nitrogen fixation from the atmosphere is usually a relatively minor factor in NO_x production. The nitrogen which most readily converts to NO_x is that which is chemically bound up in the fuel. The quantity of oxides of nitrogen formed from this source is not appreciably affected by elevated combustion temperatures. In fact, with an air preheater, the reduction in the amount of fuel consumed corresponds directly to a reduction in the amount of NO_x formed from fuel-bound nitrogen, and is quite likely to offset the additional atmospheric fixation due to the higher flame temperature.

Another process effect that manifests itself when an air-preheat system is retrofitted to an existing heater is the shift in duty split between the radiant and convection sections. The higher combustion temperature due to preheating, coupled with the lower convection transfer rate at reduced flue-gas flowrates, causes a greater proportion of the total heat absorption to take place in the radiant section.

Consequently, operation with preheated combustion air results in higher radiant duties than with ambient air. The magnitude of the shift in duty split is, of course, contingent upon the degree of preheat applied to the combustion air.

Gas-turbine exhaust used as combustion air

The exhaust gas from gas turbines, widely used as drivers for compressors, pumps, and electrical generators, usually ranges from 800 to 900°F and contains from 17 to 18% oxygen. These gases are well-suited as a source of preheated combustion air for fired heaters, and can be used to cut heater fuel consumption.

In most systems utilizing gas-turbine exhausts, an auxiliary source of combustion air permits independent operation of the heater and the gas turbine. Typically, the overall system is designed so that the gas-turbine exhaust can be vented to atmosphere whenever the heater is operated on ambient air.

Conversion from gas to liquid firing

In the face of curtailments of natural gas deliveries, many industries are seeking alternative fuel sources, primarily liquid fuels. The following areas of concern should be examined closely before any conversion from gas to liquid firing is undertaken.

Impact on radiant section. The major factor affecting the combustion chamber is the size of the oil flame compared to the size of the gas flame. At the same level of heat release, an oil flame is generally longer and wider than a gas flame. For this reason, the distance from burner to tube should be examined to assess the potential for flame impingement on the tube coil. Similarly, vertical clearance dimensions should be reviewed, since the longer oil flame may impinge on the shield tubes, the reradiating cone, or the baffle provided at the top of some older vertical-cylindrical heaters.

If the contemplated liquid fuel contains substantial vanadium and sodium concentrations, consideration should be given to protecting exposed tube supports from the corrosive combustion environment. Also, if the heater contains a reradiating cone or baffle, the heater manufacturer should be consulted regarding the possibility of removing such vulnerable equipment.

Impact on convection section. Convection-section extended surface in gas-fired heaters is very often of the high-density, finned-tube type. Under heavy liquid-fuel firing, such tubing is difficult to keep clean using conventional onstream cleaning techniques. Before conversion, therefore, the designer should consider replacing high-density finned tubes with either heavy low-density tubes, or studded tubes. Since replacement will reduce the total effective surface area, additional rows of convection tubes will be required to maintain thermal efficiency.

Facilities for onstream convection cleaning, such as soot blowers, should be installed as part of the conversion. The addition of such equipment, as well as the necessary ladder and platform access, requires that the structural integrity of the steelwork and foundation be assessed.

Adjustments to draft. Many heaters designed for gas firing operate virtually at their draft limit, usually because they are pushed well beyond their original design capacity. Because oil firing requires higher excess-air levels to achieve acceptable combustion, draft-limited gas-fired heaters will suffer a capacity decrease when converted to liquid firing, unless additional stack height is provided.

Alterations to burners. Burner replacement in a gas-to-liquid fuel conversion can, in most instances, be made on a one-for-one basis. However, for those conventional gas-fired heaters that operate with relatively small burner-heat releases in the range of 2 to 3 million Btu/h, substitution of oil burners on a one-for-one basis may result in unstable combustion. In these cases, a totally new firing arrangement should be specified as part of the conversion. The final burner arrangement should take into account such parameters as the combustion-chamber configuration, burner turndown requirements and fuel-oil combustion behavior.

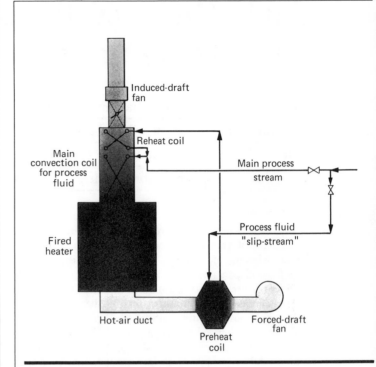

Typical arrangement for air preheater that uses process fluid as heat-exchange medium   **Fig. 5**

From a practical standpoint, an important consideration that must not be overlooked in any conversion study is the heater's mechanical condition and its remaining useful life. Investment costs for gas-to-oil conversions are expensive, particularly when convection sections must be revamped. Furthermore, the downtime necessary to effect the conversion is costly, due to lost revenues. Consequently, if conversion is contemplated for a heater that is in poor mechanical condition, an evaluation of maintenance costs, the conversion investment and revenue losses may well show replacement to be economically more attractive.

A good part of conversion cost lies in the fuel handling and distribution system. Conversion of only a few large units, particularly if they are located in the same general plot area, will reduce this cost.

References

1. Woodard, A. M., "Upgrading Process Heater Efficiency," *Chem. Eng. Progr.,* Vol. 71, No. 10, 1975, p. 53.
2. von Wiesenthal, P., "Furnace and Related Process Involving Combustion Air Preheating," *U.S. Patent 3,426,733,* 1969.

The author

Herbert L. Berman is a staff engineer with Caltex Petroleum Corp., 380 Madison Avenue, New York, NY 10017, specializing in fired-heater equipment. He has over 25 years of engineering and management experience, primarily in the fired-heater industry. He received a B.Ch.E. degree from the Polytechnic Institute of Brooklyn and did graduate work at New York University. A member of the AIChE, ASME and AACE, he holds several patents and is a licensed professional engineer in New York.

Fuel-efficiency thrust ups sales of additives

The market for additives that improve combustion has been gathering steam since the oil embargo of 1973.

High on the customer list are the CPI, which, intent on saving fuel, are buying more of these chemicals.

☐ In a way, boiler additives recall the "snake oils" of yore sold as cures for all ills. Makers of additives—which range from dispersants, stabilizers and ash modifiers to soot eliminators and combustion catalysts—claim that their wares can boost burning efficiency, eliminate corrosion, solve slagging and pollution problems, or prevent sludge buildup. All of which is luring the efficiency-conscious chemical process industries (CPI) into spending record amounts of money on these products.

The actual size of this outlay, however, remains a mystery; additive manufacturers do not like to mention specific sales figures. Still, they agree that demand for the products has risen dramatically. "We have seen a substantial increase in sales since the Arab oil embargo," says James Robinson, product manager for Betz Laboratories, Inc.'s (Philadelphia) boiler-products line. James Leopold, sales manager of fuel-oil additives for Mooney Chemicals, Inc. (Cleveland), concurs; he says that sales of the firm's additives line for the first eight months of 1977 were up 70% over the same period last year. "And 1976 was a good year too," he adds.

Many manufacturers maintain that even better days are ahead. For example, J. D. Martin, a product specialist for Nalco Chemical Co. (Chicago), views additives usage as "a big market that hasn't been fully tapped." And Patrick Lavery, sales and production manager for Lavery, Inc. (Bristol, Pa.), says that the overall market for the compounds could triple within the next five years.

TOP PERFORMERS—One of the biggest problems an engineer faces in selecting an additive is choosing from among the large number of products available. "Companies are offering all kinds of products," notes one additives expert, "and people don't understand what they do and how they work."

For practically every boiler problem, there is some kind of compound that reportedly provides an answer. In practice, however, although each additive has a specific function that may indirectly increase boiler efficiency by minimizing operating difficulties, it is combustion catalysts that are most often promoted by manufacturers as being capable of boosting fuel efficiency. And because fuel oil prices have surged in the last four years, it is not surprising that such catalysts are the top performers of the additive group—their sales having risen more sharply than those of other products.

Various kinds of materials have been billed as combustion improvers, including surfactants, organometallics, and low-molecular-weight polymers. At present, however, organometallic compounds are the most popular ones. Manganese is the preferred combined metal, although magnesium, cobalt, iron, barium and nickel are also considered effective. For added versatility, the metal is usually blended with dispersants and other ingredients to make a multipurpose product.

SHADOW OF DOUBT—Do the catalysts actually live up to their billing? Most users admit they are not sure. They point out that measuring small efficiency changes in boilers is very difficult, because the units' performance is given to fluctuations. In addition, it is difficult to accurately determine steam output and fuel consumption, so some operators maintain that what may seem to be an additive-related efficiency rise is actually due to an increased attention to a boiler's operating variables—that is, if there is any improvement.

Indeed, one technical superintendent of a chemical plant told CHEMICAL ENGINEERING that after using a combustion catalyst for more than a year, he still isn't sure the product is doing any good.

The flames of uncertainty are in this case being fanned by memories of the hard-sell tactics that additive manufacturers used to display in the leaner days of inexpensive fuel oil. "Even some reputable firms have made unrealistic, unsubstantiated claims," admits one producer. Adds Mooney Chemicals' Leopold, "I get very concerned when I see some of these claims of 5 to 10% more Btu output, because I think they are just sales pitches."

Many ineffective combustion-boosters have been driven out of the market by legitimate products, says Martin Stacey, managing director of Triple-E (U.K.) Ltd., a London-based manufacturer. Other additive makers claim that they are starting to overcome a "black-magic image" by operating more-sophisticated research and development programs, and better technical-service departments.

SEPARATING FACT FROM FANCY—By next summer, the question of how effective additives really are may have been partly resolved. Battelle Columbus Laboratories (Columbus, Ohio) is conducting a study of U.S. utilities' use of all types of boiler additives, under a $50,000 contract from the Electric Power Research Institute (Palto Alto, Calif.). The goal is to work up a manual that details just what can be expected from each type of compound, based on the operating experience of electric-utility firms (long-time consumers of these materials).

To be effective, a combustion catalyst must have the right weight and metal content, says Ira Kukin, president of Apollo Chemical Corp. (Whippany, N.J.)—an important supplier to utility companies. By this, Kukin means that formulations in liquid state should have a density of 8.5 to 14 lb/gal, and contain at least 5% beneficial metal. Those in slurry form should exhibit a minimum density of 10.5 lb/gal, with approximately 20% beneficial metal. (Other experts say that the boiler fuel, after being mixed with the additive, should have a metal content of at least 25 to 50 ppm.).

THE NEWER CROP—Some of the more prominent makers of combustion

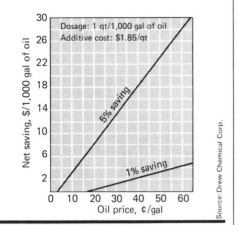

Dosage: 1 qt/1,000 gal of oil
Additive cost: $1.85/qt

5% saving

1% saving

Net saving, $/1,000 gal of oil

Oil price, ¢/gal

Source: Drew Chemical Corp.

Savings stem from use of Amergy 5000 combustion catalyst

catalysts are Drew Chemical Corp. (Parsippany, N.J.), Rolfite Co. (Stamford, Conn.), Ethyl Corp. (Richmond, Va.), Calgon Corp. (Pittsburgh), Apollo, Nalco, Betz, and Mooney Chemicals. Most of them introduced new products in the last few years of rocketing fuel-oil prices.

In 1974, for instance, Drew Chemical started commercial trials of Amergy 5000. Sales of this material are now growing rapidly, with this year's sales tripling the volume reached in 1976, according to Ronald C. Andrade, product supervisor for fuel treatment.

Amergy 5000, a blend of organometallic catalysts, solvents, dispersants, and water-emulsifying agents, typically cuts fuel costs by 1 to 5%, says Andrade, while lowering air pollution (see *Chem. Eng.*, Jan. 17, p. 89 for a detailed discussion of this product).

Although Drew Chemical recommends that boilers be run at different additive-to-fuel ratios to find the optimum conditions, it suggests a dosage of 1 qt/1,000 gal of fuel. At a price of $6 to $7/gal (depending on quantity purchased), Amergy 5000 is said to be cost-effective, even when fuel efficiency rises by 1% or so (see diagram).

In support of these claims, Andrade cites case histories of oil-fired boilers at two papermills. In one, the untreated boilers were being operated at "far below" rated capacity to avoid smoky stack emissions. At the recommended volumetric ratio of 1:4,000, the additive treatment improved combustion, allowing the steam load to be increased from 80,000 to 105,000 lb/h. At the other papermill, fuel consumption

reportedly dropped by about 6%—an annual saving of approximately $90,000.

THE WORSE, THE BETTER—Nalco has also introduced additives since the era of expensive fuel oil began, including a surfactant/combustion catalyst combination. The efficiency gains obtainable with the new compounds depend on how efficient combustion is prior to their addition, says Nalco's Martin. The worse a boiler's performance without additives is, the more the products will help, he adds.

"We've seen increases of between 2 and 6%, but this could also be more or less." But even with just a 1% efficiency boost, the saving can be sizable, Martin points out; it comes out to about $25,000/yr for a 100,000-lb/h boiler that uses fuel costing $2/million Btu.

Cost of the Nalco materials runs to about 2.5¢ to 10¢/bbl. The biggest clients are utilities, with 10 to 20% of total sales going to CPI firms. But utility industry purchases of additives should be in the order of 70 to 80% of total sales for these chemicals in five years, predicts Martin.

Betz has been producing a number of fuel-oil additives for more than ten years, including sludge dispersants, water emulsifiers and ash conditioners. In the past three years, says James Robinson, the company has launched several organometallic combustion catalysts. But while sales have been good so far, Robinson remains cautious about the future, maintaining that "it's hard to predict beyond one year what will happen to the fuel-oil additives market—especially with the uncertainties surrounding President Carter's proposed energy program."

OTHER PRODUCTS—Another firm that markets a wide range of compounds is Rolfite Co. In addition, the company sells the products of Basic Chemicals (Cleveland)—a large additive manufacturer. The combined lines include Rolfite 101, 303, 404 and 808 organometallic combustion catalysts, which are specifically geared to improve fuel efficiency.

Rolfite claims that many customers using the compounds report fuel savings of 5 to 10%. The chemicals have found roles in a number of chemical process industries, including pulp and paper, textiles, food, chemicals and petrochemicals.

Calgon makes a range of manga-

nese- and barium-based combustion improvers. These are cost-effective, even when efficiency is raised only slightly, avers the firm. No manufactured product can up fuel efficiency more than 3%, maintains a Calgon representative, and that's "pretty tough to get even under the best conditions, so a 1 or 2% improvement is very good."

Since the early 1960s, Ethyl Corp. has been making an organometallic combustion catalyst—methyl cyclopentadienyl manganese tricarbonyl—dubbed CI-2. The material can cut fuel bills by up to 5%, claims the firm. But some critics charge that the recommended dosage (1 gal/52,000 gal of oil) yields a manganese concentration of about 8 ppm, which is said to be too low to be effective. Cost of treatment is only about 60¢/1,000 gal, according to Ethyl.

DRY ADDITIVE—Apollo Chemical's Kukin claims that the firm has gone a step beyond conventional fuel additives with Coaltrol M, its new cold-end treatment. This consists of a powder, which, when injected directly into the flue gas (instead of being mixed with fuel oil), virtually eliminates sulfur trioxide. This allows the boiler to operate at lower exit-gas temperatures, and yet no corrosion takes place, says Kukin. For each drop of 10°F in stack-gas temperature, fuel requirements are cut about 0.25%, says the executive.

According to Apollo Chemical, a typical application, in which Coaltrol M was used in a 350-MW boiler fed with #6 fuel oil, indicates that substantial fuel savings can be made. By adding 5 lb of the material per 100 bbl of oil, a drop of 30°F in exit-gas temperature was obtained, cutting the fuel load by 14,000 bbl/yr.

Kukin claims that another Apollo Chemical product, the manganese-based MC-7, also cuts fuel bills. A recent application in an oil refinery, for example, is said to have resulted in fuel savings of about 3%, at Mn concentrations of 100 ppm.

Kukin reports the following net savings when the materials are used to treat fuel oil: 20 to 30¢/bbl with Coaltrol M, and 5 to 7¢/bbl with MC-7. A combined MC-7 and Coaltrol-M treatment would cost about 10¢/bbl, with potential savings running up to 50¢/bbl.

Larry J. Ricci

How to avoid problems of waste-heat

Peter Hinchley, Imperial Chemical Industries Ltd.*

☐ Substantial reductions in the running costs of many process plants can be obtained by using their waste heat to raise steam. This heat is removed in economizers/feedwater-heaters, in boilers and in superheaters. The range of size and duty is wide, but on single-stream plants for certain processes, the steam-raising capacity is comparable to that of a large, industrial, power-station boiler. For example, modern plants for making ammonia raise approximately 250,000 kg/h (550,000 lb/h) of steam at pressures of 100 to 120 bar (1,450 to 1,740 psi) and temperatures of 460 to 520°C (860 to 970°F).

In nitric and sulfuric acid plants, the steam is raised by cooling the process gas. This is also the case with ammonia, methanol, hydrogen and town-gas plants. Further some of the steam in these plants is raised by cooling the flue gases from furnaces within the plant. Frequently, these flue gases are supplemented by auxiliary firing, particularly during plant startup.

Recovering heat from flue gases

Typically, flue gas under a slight draft, at 900 to 1,200°C (1,600 to 2,200°F) is cooled down to between 150 and 200°C (300 to 400°F). In modern plants, the flue gases, almost without exception, are drawn across banks of tubes containing the fluids being heated. Depending on the flowsheet adopted, there may be banks of tubes for superheating steam, heating process fluids, generating steam, vaporizing feedstocks, heating boiler feedwater, and preheating combustion air for the furnace.

Steam is generated in watertube boilers that are connected to steam drums. The boilers are usually of the horizontal forced-circulation type (F/1) or the vertical natural-circulation type (F/2), the choice being dictated largely by plant layout. For natural circulation,

the drum should be some distance above the boiler to give sufficient driving force to the water and steam. The cost of the additional structure needed for elevating the steam drum to the required height may outweigh the savings in circulating pumps.

The actual disengagement of the steam from the boiling water takes place in the steam drum, which normally has a water level at or near the center line. The steam usually has to pass through a series of internal separators that remove entrained water before the steam reaches the outlet pipe. The complexity of the separators depends upon the required steam purity, which in turn depends on the type of plant. If the steam is passed through a superheater or into a turbine, very high purity is required to prevent the deposition of solids.

Recovering heat from process gases

Some process-gas boilers handle gases varying in pressure between 1 and 350 bar (15 and 5,000 psi) and in temperature between 400 and 1,200°C (750 and 2,200°F), and raise steam at pressures up to 120 bar (1,740 psi). They fall into two categories, firetube and watertube. The watertube type are generally for the higher outputs and pressures because of mechanical-design difficulties encountered with large, high-pressure, firetube boilers.

The principal features of a typical natural-circulation firetube boiler on a reformer are shown in F/3. Several companies design and make such boilers. However, there are significant differences between boilers with respect to maximum heat flux, maximum steam pressure, method of attaching the tubesheet to the shell and channel and of attaching tubes to the tubesheet, and method of protection of the tubesheet.

The manufacturers also differ in their ability to carry out a sufficiently rigorous stress analysis, bearing in mind the very high stresses that occur due to the high pressures involved and the differential expansion between tubes and shell.

---

* This article is adapted from a paper titled "Waste-Heat Boilers in the Chemical Industry," presented at a meeting on Energy Recovery in Process Plants, held by The Inst. of Mech. Engrs., 1 Birdcage Walk, Westminster, London SW1 H933. The paper will be published, along with others from the meeting, by IME.

Originally published September 1, 1975

boilers

Recovering heat from process gases, and from flue gases of associated furnaces, saves energy and money. Here is a review of the problems occurring in the various types of boilers, along with cases and solutions for actual plants.

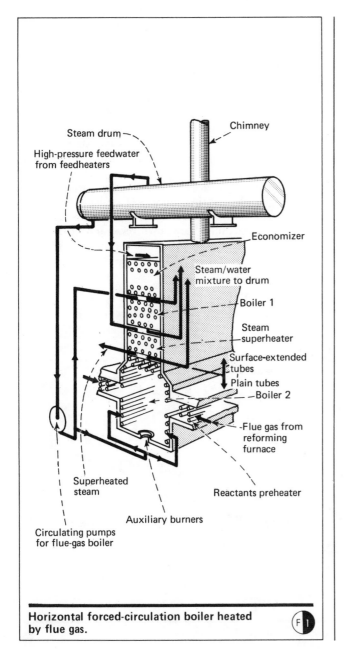

Horizontal forced-circulation boiler heated by flue gas. (F1)

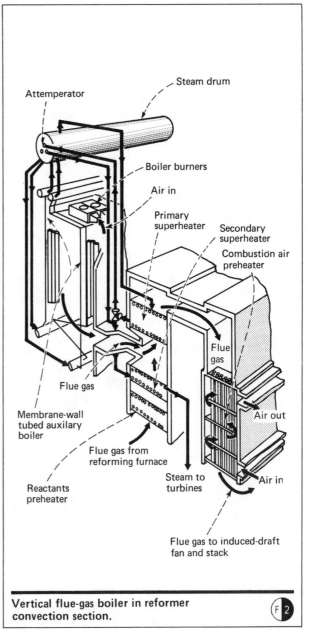

Vertical flue-gas boiler in reformer convection section. (F2)

The situation for watertube boilers heated by process gases is completely different, since a large number of companies have developed and patented their own designs. Some of these are dependent on pumped circulation, but others work on natural circulation. F/4 shows the principal features of one of the simpler designs, the vertical U-tube, which has been progressively developed for reformed-gas boiler duties by a U.S. company.

Whether the boilers are of the firetube or watertube type, they are generally connected to steam drums in the same way as flue-gas boilers, and commonly share the same drums. There are, however, occasional deviations from this arrangement. Some firetube boilers have their own integral steam spaces, and some watertube boilers are of the once-through type, where the water is converted directly into superheated steam in the tubes and there is no drum.

Failures in flue-gas boilers

In flue-gas-heated boilers, onstream corrosion has occurred due to excessively caustic water conditions together with a combination of heat flux, mass velocity and steam quality, which caused "dryout," i.e., inadequate wetting of the top of the tube [1]. The corrosion takes the form of deep gouging of the top of the tube. The problem has been solved by improving the control of water quality and by increasing the water circulation rate.

Several superheaters associated with flue-gas boilers have failed due to creep-rupture as a result of carrying over of boiler solids. The reasons include inadequate means of separating the steam from the water within the steam drum, and mechanical failure due to overload of good-quality primary and secondary separators. Another failure was caused by a badly made joint in a drum-type attemperator, which let boiler solids pass into the secondary superheater.

Some economizers associated with flue-gas boilers have failed as a result of oxygen pitting. Careful and thorough checking using stainless-steel sampling lines has shown that many proprietary deaerators fail to achieve their specified duties.

At least one steam economizer failed as a result of overheating caused by water starvation due to "excursive instability" or the so-called Ledinegg effect [2]. The problem was overcome by fitting distribution orifices at the inlets of the economizer tubes [4].

Failures in process-gas boilers

In firetube boilers, severe corrosion on the water side of the tubes has occurred just beyond the end of the heat-resisting ferrules in the tube inlets. This has happened in several plants, and has usually been caused by caustic or acid breakthrough into the boiler water. Naturally, boilers having the highest heat flux or poorest water distribution, or both, are most prone to failure should boiler-water quality deviate from the required optimum value.

Crevice cracking of the tubes within the tubesheet adjacent to the welds has been caused by failure to seal the crevices by post-weld expansion, and has occurred because the boiler-water quality deteriorated as a result of one of the dosing pumps being out of commission for one or two days.

Failure of the tubes and tubesheet as a result of overheating caused by the buildup of boiler solids behind the tubesheet has occurred on several boilers, as a result of either inadequate blowdown provision in the design

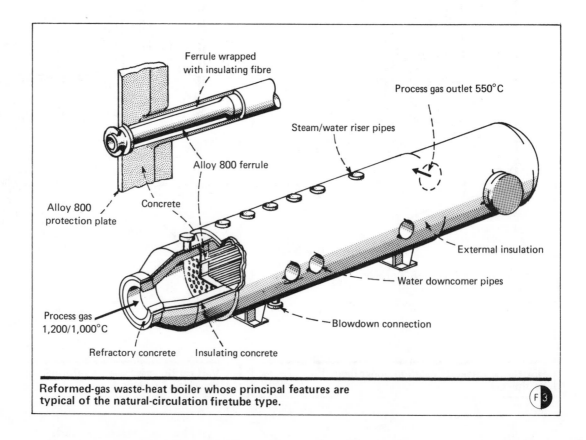

Reformed-gas waste-heat boiler whose principal features are typical of the natural-circulation firetube type.

F 3

or the failure to use the blowdown device provided.

Initial leakage of tubes on a sulfuric acid plant was a failure due to a combination of inadequate specification, poor workmanship and deficient quality control. The heat capacity of the sulfuric acid plant was such that even to repair one leaking tube a shutdown of 10 days was necessary. Furthermore, even a small leak could not be tolerated because of the immediate formation of corrosive acid.

The gas channels of several boilers have swollen as a result of gross overheating when external insulation on the shell has been extended on to a refractory-lined channel, or when there has been failure of the internal refractory lining.

Cracking and overheating of tubeplates may be due to their exposure to hot gases following the failure of their refractory protection.

Mechanical/thermal failure of tubes can occur when circulation stops following sudden depressurization of the steam system. In one case, this was caused by failure of the blowdown line.

Overheating of tubes within a tubesheet can result from using a thick tubesheet, inadequately protected by heat-resisting ferrules in the tube ends. The gaps between the ferrules and the tubes should be increased and filled with ceramic fiber paper and the inner ends of the ferrules should be swaged out to prevent gas tracking. (See detail on F/3)

Failure of the main tubesheet-to-shell welds occurred in one case during hydrostatic test, and in another after several months of operation. In both, there was a combination of unsatisfactory design, poor workmanship and inadequate quality control.

Failures in watertube boilers

It is not possible to go into the same amount of detail on watertube-boiler failures because the relevant details are for proprietary designs. Nevertheless, in order to present a balanced picture, an attempt will be made to indicate the reasons for the many failures that have occurred.

Horizontal-watertube boilers

One U-tube boiler handling gas at 980°C (1,800°F) had frequent failures at the top of the tubes as a result of dryout. The problem was eventually solved by increasing the water rate, changing to a higher grade of tubing, and fitting twisted tapes inside the tubes to ensure that the upper surfaces of the tubes were constantly wetted.

A similar U-tube boiler operating at much lower temperature had corrosion/erosion failures at the top of the tubes, caused by dryout—leading to a concentration of harmful chemicals in the boiler water. The problem was overcome by increasing the water-circulation rate.

A boiler for a nitric acid plant failed by overheating as a result of dryout, finally attributed to a low water velocity (which was made even lower than the design velocity as a result of blockage of some of the inlet distribution nozzles by excess hardness in the boiler feedwater). The hardness of the feedwater was improved and an additive was introduced to keep the solids in

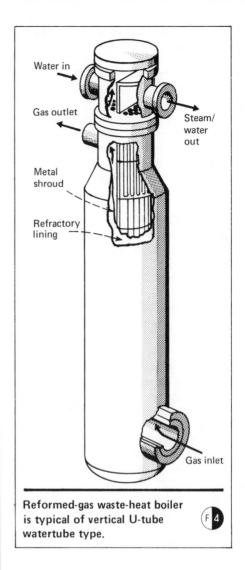

Reformed-gas waste-heat boiler is typical of vertical U-tube watertube type.

F 4

suspension. This appeared to have overcome the problem until two years later, when another spate of failures occurred. This time, it was shown that these failures were due to the additive, causing the boiler solids to be laid down in the tubes. Another additive is now in use and so far there have been no failures.

Another boiler for a nitric acid plant, with a circular inlet manifold, failed as a result of the buildup of solids at the remote ends of the manifold, thus blocking off the end-tubes in this region. The problem was solved by putting blowdown connections at the ends of the manifold and by improved control of the boiler-water quality.

Vertical-watertube boilers

Creep-rupture failures can be caused by the blocking of individual tubes by construction/fabrication debris, or by magnetite scale dislodged during an upset in water treatment.

On-load corrosion failures can be caused by the concentration of harmful chemicals (in the boiler water) on any deposits existing in the regions of the tubes subject to the highest heat flux.

There can be creep-rupture failures of the gas pressure shell because of the failure of the refractory lining—the latter usually being due to a combination of poor design, workmanship and quality control.

Water quality for all types of boilers

Many of the above failures have occurred as a result of breakthrough of acid or alkali from the water treatment, or by contamination of return condensate. Now it is becoming common practice to monitor continuously the conductivity of water entering deaerators and to dump automatically any water that is out of specification. The supply is then taken from a standby tank of treated water.

Choice of waste-heat boiler

It is not possible to give a general answer to the question, "Which is the best boiler?" However, the following points can be made for flue-gas boilers and process-gas boilers.

Flue-gas boilers—Vertical, natural-circulation boilers are intrinsically more reliable than horizontal, forced-circulation boilers. True, the vast majority of horizontal boilers are reliable, but there is always a potential risk of dryout that can cause overheating or corrosion of the tubes and other components if harmful chemicals are present in the boiler water [1].

Insufficient data exist on the conditions that cause dryout or departure from nucleate boiling (DNB), so that we cannot be certain at the design stage that this will not occur in horizontal boilers. With vertical waste-heat boilers, there is little risk of DNB because the critical heat flux is many times greater than that in horizontal boilers [1].

Process-gas boilers—If their mechanical design and fabrication are to high enough standards, horizontal, natural-circulation, firetube boilers are basically more reliable than forced-circulation, watertube boilers. The latter are prone to failure if the dirt or debris gets into the system, particularly with vertical boilers that have their low point at the point of highest temperature and heat flux. However, faulty water treatment can give rise to failures of either type of boiler; and if the tubes fail, then a spare watertube bundle can be fitted in a short time, whereas a firetube boiler requires a very long time to retube or replace. For some processes such as nitric acid manufacture, there are well-proven designs of horizontal-watertube boilers.

Cost and consequences of boiler failures

There is an obvious reluctance in industry to report the cost of specific failures. Nevertheless, it is vital to indicate some idea of the order of costs, so as to illustrate the magnitude of losses that have occurred and do occur.

In the past, plant outputs were much smaller, the plants frequently had multiple units of equipment, and any steam that was raised was at low pressure. Thus, failures were less frequent, and the financial losses were not great.

It is now common to have single-stream plants, having outputs ten times those of a decade ago, which contain boilers raising steam at 120 bar (1,740 psi) and 500°C (930°F). A recent survey of 27 U.S. ammonia plants analyzed the causes of plant shutdowns [3]. Boil-

ers were the second most frequent cause of shutdowns. One failure could be expected every three years, and on average this failure has caused the loss of five days' output. Thus, the average annual loss of profit due to boiler failures on an ammonia plant is $180,000.

The above statistics deal with normal plants, but there have been a number of plants where the financial losses were much greater. One pair of nitric acid plants had such frequent boiler troubles that in a three-year period there was a loss of over $2.4 million. Several ammonia plants have had single failures that have shut them down for up to 50 days, and one methanol plant was shut down for 106 days. These failures have caused millions of dollars in production losses because the plants were shut down while repairs were being made.

Conclusion

It will be apparent from this review of problems on waste-heat boilers that they are potentially a major cause of unreliability. As plants have become larger, the financial consequences of failures on single-stream plants have become more serious. One failure of a waste-heat boiler can cause a complete plant shutdown of between 5 and 100 days, and loss of millions of dollars.

It is not often appreciated by the boiler vendor that a single failure of his boiler can cause a loss of profit at least equal to the cost of that boiler, and frequently very much more.

From this, it will be clear that waste-heat boilers warrant very careful attention at all stages, from selection, design, fabrication, erection, through to commissioning, and also during normal operation.

Acknowledgements

The author acknowledges the helpful comments offered by his colleagues, and in particular the assistance given by one of his design engineers, J. W. Watson. Thanks are also due to the following who permitted the inclusion of illustrations based on equipment they have designed: Foster Wheeler; Humphreys & Glasgow; and Struthers Wells.

References

1. Collier, J. G., "Convective Boiling and Condensation," McGraw-Hill, New York, 1973.
2. Ledinegg, M., Instability of Flow During Natural and Forced Circulation, *Die Wärme*, Vol. 61.8 (AEC-TR-1861), 1954.
3. Sawyer, J. G. and Williams, G. P., Papers presented at AIChE Symposium on Safety in Air and Ammonia Plants, Atlantic City, N.J., 1971, and Vancouver, B.C., 1973.
4. Margetts, R. J., Excursive in Feedwater Coil, presented at Thirteenth National Heat Transfer Conference, AIChE/ASME, Denver, Colo., 1972.

The author

Peter Hinchley is furnace and boiler section manager of the agricultural div. of Imperial Chemical Industries Ltd., Billingham, England. He has worked for ICI since 1956 on equipment development, plant maintenance, engineering design and project engineering. He is a graduate of Sheffield University with a first class honors degree in chemical engineering, and is a member of the Institution of Mechanical Engineers.

Boiler heat recovery

To minimize fuel costs, heat can be recovered from both flue gases and boiler blowdown.

William G. Moran and **Guillermo H. Hoyos,** *
Engineering Experiment Station,
Georgia Institute of Technology

☐ Boiler flue gases are rejected to the stack at temperatures at least 100 to 150°F higher than the temperature of the generated steam. Obviously, recovering a portion of this heat will result in higher boiler efficiencies and reduced fuel consumption.

Heat recovery can be accomplished by using either an economizer to heat the water feedstream or an air preheater for the combustion air. Normally, adding an economizer is preferable to installing an air preheater on an existing boiler, although air preheaters should be given careful consideration in new installations.

Economizers are available that can be economically retrofitted to boilers as small as 100 hp (3,450 lb/h steam produced).

Fig. 1 can be used to estimate the amount of heat that can be recovered from flue gases. Two main assumptions have been made in developing this graph:

1. The boiler operates close to optimum excess-air levels. (It does not make sense to use an expensive heat-recovery system to correct for inefficiencies caused by improper boiler tuneup.)

2. The lowest temperature to which the flue gases can be cooled depends on the type of fuel used: 250°F for natural gas, 300°F for coal and low-sulfur-content fuel oils, and 350°F for high-sulfur-content fuel oils.

These limits are set by the flue-gas dewpoint, or by cold-end corrosion, or heat driving-force considerations.

Example

A boiler generates 45,000 lb/h of 150-psig steam by burning a No. 2 fuel oil that has a 1% sulfur content. Some of the condensate is returned to the boiler and mixed with fresh water to yield a 117°F boiler feed. The stack temperature is measured at 550°F.

Determine the annual savings (assuming 8,400 h/yr boiler operation) that will be achieved by installing an economizer in the stack.

Assume that fuel energy costs $3/million Btu.

*This material has been prepared by staff members of Georgia Tech's Industrial Energy Extension Service, which is a continuing energy-conservation program funded by the State of Georgia Office of Energy Resources.

Originally published December 3, 1979

Babcock & Wilcox Ltd.

From steam tables, the following heat values are available:

For 150°F saturated steam	1,195.50 Btu/lb
For 117°F feedwater	84.97 Btu/lb

The boiler heat output is calculated as follows:

$$\text{Output} = 45,000(1,195.50 - 84.97)$$
$$= 50 \text{ million Btu/lb.}$$

Using the curve for low-sulfur-content oils, the heat recovered that corresponds to a stack temperature of

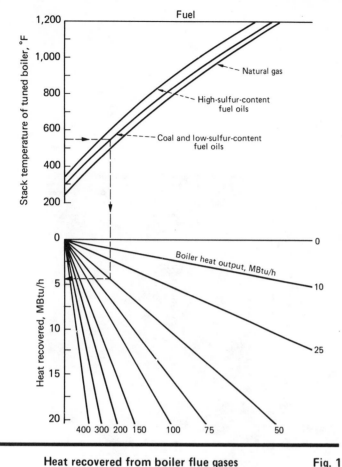

Heat recovered from boiler flue gases   **Fig. 1**

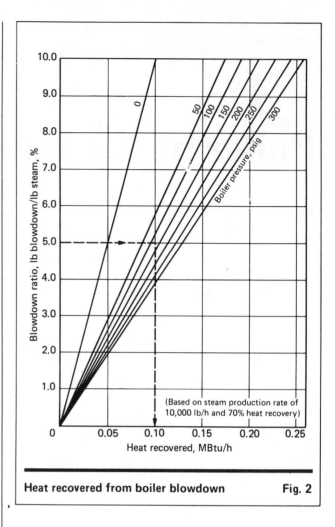

Heat recovered from boiler blowdown   **Fig. 2**

550°F and a boiler duty of 50 million Btu/h can be read from the graph as 4.3 million Btu/h. The annual savings are:

4.3 MBtu/h × $3/MBtu × 8,400 h/yr = $108,000/yr

Recovering heat from boiler blowdown

Heat can also be recovered from boiler blowdown, by preheating the boiler makeup water through use of a heat exchanger. This is done most conveniently in continuous-blowdown systems.

Example

In a plant where the cost of steam is $2.50/MBtu, 1,250 lb/h are continuously purged in order to avoid buildup of solids in the boiler tubes. Determine the yearly savings (assume 8,760 h of operation per year) if the blowdown is passed through a heat exchanger to raise the temperature of the makeup water. The boiler generates 125-psig steam at a rate of 25,000 lb/h.

First calculate the blowdown ratio:

$$\text{Blowdown ratio} = \frac{1,250 \text{ lb/h blowdown}}{25,000 \text{ lb/h steam}} = 5\%$$

From Fig. 2, the heat recovered, corresponding to 5% blowdown ratio and 125-psig boiler pressure, is 0.100 MBtu.

Since Fig. 2 is based on a steam production rate of

10,000 lb/h, the actual yearly savings for this plant are:

Savings:

$$= 0.100\frac{\text{MBtu}}{\text{h}} \times \frac{25,000 \text{ lb/h}}{10,000 \text{ lb/h}} \times 8,760\frac{\text{h}}{\text{yr}} \times \frac{\$2.5}{\text{MBtu}}$$

$$= \$5,500$$

The values obtained from Fig. 2 are based on a makeup water temperature of 70°F and a heat recovery of 70%.

Suggested actions

Determine stack temperature after boiler has been carefully tuned up. Then determine minimum temperature to which stack gases can be cooled and study economics of installing an economizer or air preheater. Also study savings from installing a heat exchanger on blowdown.

The authors

William G. Moran is a research engineer at the Engineering Experiment Station, Georgia Institute of Technology, Atlanta, GA 30332. His present fields of interest include industrial and commercial energy conservation, alternative energy sources, and research and development. He holds a B.S. from the University of Massachusetts and an M.S. from Rensselaer Polytechnic Institute, both in mechanical engineering, and is a registered professional engineer in Georgia.
Guillermo H. Hoyos was a research engineer with Georgia Tech's Engineering Experiment Station when these materials were prepared. He holds B.S. and M.S. degrees in chemical engineering from Georgia Institute of Technology. He now resides in Bogota, Colombia.

New heat-exchange units rely on enhanced transfer

Special fin and flute configurations, porous-surface tubes and ceramic-based units are becoming part of the CPI's all-out effort to improve the performance of heat exchangers.

☐ Although techniques for enhancing heat-transfer coefficients in order to boost exchanger efficiency have been around for a long time, the chemical process industries (CPI) have been pursuing new ones with greater urgency in the past few years, as the cost of materials and energy continues to go through the roof.

CPI development work is advancing along several fronts: enhanced tube surfaces (rough-textured) for use mainly within the boiling and condensation ranges; ceramic heat-transfer surfaces for high-temperature heat exchange and recovery, intended mostly for coal-gasification and magneto-hydrodynamics facilities; and new heat-exchanger designs—e.g., the rod-baffle configuration.

Tube fabricators have found ways to make enhanced-surface tubes from such diverse materials as carbon steel, stainless steel, titanium and tantalum. And CPI firms are looking closely at potential applications not necessarily requiring the use of "clean" fluids that would not clog the tubes' rough surfaces.

AUGMENTED SURFACES—UOP Inc.'s Wolverine Div. (Decatur, Ala.), which lays claim to pioneer work in finned tubing (for years the workhorse of augmented-surface heat exchange), is developing a porous-surface tube geared for large commercial refrigeration evaporators and condensers. Called Koro-Tex II, the tubing is of corrugated copper, with an outside bond of porous copper "foam."

According to Wolverine's engineers, the combination of turbulence caused by the corrugations inside the tube, and the increased nucleate boiling outside (generated by the porous layer), results in "very substantial efficiency gains." The firm adds that several clients have placed orders for large prototype bundles (300 to 400 tubes, 8 to 20 ft long and ¼ to ⅝ in. dia.) for testing.

ENTER SAW-TEETH—In 1976, after four years of development, Hitachi Cable, Ltd. (Tokyo) came out with its line of Thermoexcel tubes, to compete with the finned and porous-surface types.

The Japanese tubes are machined to provide sawtooth-shaped fins over the outside surface (see figure). Hitachi says these teeth are more efficient in dropping condensate than are smooth tubes or low-fin tubes. As a result, says the firm, the liquid film that remains on the tube stays thin, and thermal resistance stays low, resulting in a high heat-transfer efficiency. Hitachi has named this condensing-service tube the Thermoexcel-C.

For enhanced efficiency in boiling, the firm bends over the tips of the teeth on the Thermoexcel-C, forming tunnels running beneath the surface of the tube, with regularly spaced pores connecting the tunnels with the surrounding environment. This configuration is called Thermoexcel-E.

Though primarily used by Hitachi Cable's parent firm (Hitachi Ltd.) in refrigeration equipment, the tubes have recently attracted the attention of CPI firms, which are considering process heat-exchanger applications. While the Thermoexcel tubes cost about 30% more than smooth tubes, says Hitachi Cable, they save about 30% in energy costs.

LOW-FIN VARIATION—Hoping to challenge the Hitachi tubes is an entry by Wieland-Werke AG (Ulm, West Germany). For use in flooded evaporators, the so-called Gewa-T high-performance finned tube has spirally wound fins with a tee-shaped cross-section. This arrangement creates a spiral chamber with a narrow slit along the outer surface of the tube, enabling the surrounding fluid to become trapped inside the formed chamber. Upon evaporation, bubbles escape along the length of the slit, acting as nucleation sites to enhance boiling. However, the narrowness of the slit would not permit all the evaporated material to escape, thus maintaining constant, rather than intermittent, boiling.

A Wieland-Werke spokesman says that, overall, the heat-transfer coefficient of the tubes is around 30% higher than that of conventional low-fin ones, while the fabrication cost is only slightly higher.

NEW TRICKS—Even years-old designs of finned, fluted and corrugated tubing are not escaping the scrutiny of designers, and revisions often result in increased efficiency.

For example, double-fluted tubes (in which the inner and outer surfaces are longitudinally fluted) are finding increasing use in chemical and petro-chemical process heat exchangers, says the Tube Div. of Vereinigte Deutsche Metallwerke (VDM) of West Germany)—Western Europe's largest manufacturer of specialized heat-exchanger tubes.

Although double-fluted tubes made their debut in desalting plants in the mid-1960s, their widespread commercialization was delayed. Early on, says VDM, incorrect fluting contours caused the falling film of condensed steam on the outside of vertical desalting tubes to break off, decreasing efficiency. Considerable study and development work, notes VDM, has now established that a softer fluting contour will correct condensate-film breakoff, and that the optimal number of flutings is determined by the diameter of the tubing.

Bechtel, Inc. (San Francisco) has recently completed a design study for the Saudi Arabian Seawater Conversion Corp. that indicates that "rope" tubes (wherein the tubes are shallowly grooved internally) would not only enhance heat transfer by two to three times versus smooth tubes, but would save 25% in the capital cost of desalting equipment because of the smaller heat-transfer area needed.

J. S. Yampolsky, a senior technical advisor with General Atomic Co. (San Diego, Calif.), has invented a tube with spiral flutes both inside and outside. Yampolsky says the heat-transfer coefficient of one of his tubes is three times that of a smooth tube.

Originally published February 25, 1980

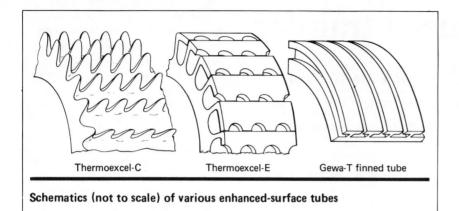

| Thermoexcel-C | Thermoexcel-E | Gewa-T finned tube |

Schematics (not to scale) of various enhanced-surface tubes

Also, he says, "it can be made from almost any material, and it looks like the cost will be reasonably close to that of a straight, welded tube of the same material."

General Atomics has applied for a patent on the tube, and is preparing for small-scale production to get a better idea of costs and possible manufacturing problems. Yampolsky says commercial use is probably one or two years away; first applications will probably be in condensers for chemical-processing or power plants.

HIGH-TEMP TRANSFER — "Intriguing" is the word used by Robert A. Lengemann, science and technology vice-president at UOP Inc. (Des Plaines, Ill.), to describe the development of silicon carbide ceramic heat exchangers for coal-gasification systems that operate around 2,500°F. Metal heat exchangers fail somewhere above 1,800°F, so the coal gas must be quenched, lowering efficiency and causing slag and ash to form on the heat exchangers. Units withstanding high temperatures could boost system efficiency.

The U.S. Dept. of Energy (DOE) is sponsoring a number of demonstration projects involving ceramic heat exchangers, mostly as recuperators. DOE's philosophy is that the demonstrations will show that ceramic heat exchangers have such distinct advantages over conventional types for certain applications (those that use and lose a lot of heat, such as aluminum smelting) that they will offer payback times of six to eight months.

In one program, General Telephone & Electronics Corp.'s Sylvania Div. (Stamford, Conn.) is developing a ceramic recuperator that can work up to 2,500°F and recover 63% of heat formerly lost in such operations as steelmill metal-soaking pits, glass-melting furnaces and molybdenum-reheating furnaces.

Clyde Engineering Service (West Knoxville, Tenn.), in conjunction with researchers at Alfred University Research Foundation (Alfred, N.Y.), has developed a technique for combining silicon carbide sponge with plates of the same material, to make a heat exchanger that surpasses the efficiency of one made only from the plates (*Chem. Eng.*, Mar. 26, 1979, pp. 105-106).

In a proposed unit, silicon carbide sponge would combine high porosity and surface area with the good thermal conductivity of silicon carbide. The sponge would be alternated with silicon carbide plates. Hot and cold fluids would flow countercurrently through alternate sections of sponge. The idea is that the sponge would induce turbulence, requiring the fluid flowing through it to follow a tortuous route to the plate surface. This eliminates laminar flow along the surface, enhancing heat transfer.

ROUNDABOUT ENHANCEMENT — Perhaps more energy-savers than true heat-transfer enhancers are the recently commercialized baffleless heat exchangers, such as the Rodbaffle of Phillips Petroleum Co., Bartlesville, Okla. (*Chem. Eng.*, July 17, 1978, pp. 91-92) and the Nests unit of Ecolaire, Inc., Malvern, Pa. (*Chem. Eng.*, Aug. 27, 1979, pp. 41-42). In a broad sense, both enhance performance by using rods or corrugated metal strips instead of baffle plates to support the tube bundle and prevent tube vibration. The units provide heat transfer at a much lower pressure drop than conventional heat exchangers. Because there are no low-flow areas on the shell side, where fouling is most likely to occur (as with baffle-plate units), the exchangers provide improved heat transfer over a long period of time, saving maintenance and downtime costs.

Philip M. Kohn

Continuous tube cleaning improves performance of condensers and heat exchangers

Continuous mechanical tube cleaning of condensers and heat exchangers through the recirculation of sponge-rubber balls results in improved performance and greater efficiency

William I. Kern, Amertap Corp.

☐ Because of higher fuel costs and shortages, pressure from environmentalists, and shrinking availability of sites and cooling media, it is very important to obtain greater reliability and efficiency from surface condensers and heat exchangers. Condensers for fossil-fueled turbine units continue to increase in surface area, and nuclear units now online or planned for the future are a quantum jump in size. Moreover, many more special functions have been imposed to a condenser, such as reheating, regenerative preheating, multipressure units, etc. Thus, designers and operators can no longer rely on standards of performance prevalent a decade ago.

Turbine performance

The most effective way to improve heat transfer is to improve turbine performance through better maintenance of design back-pressure. Back pressure, however, is a function of condenser efficiency. When the circulating cooling water produces an absolute pressure at the turbine flange that induces choking flow, the optimum heat rate is obtained. At a pressure above or below that point, there is a loss in heat transfer because, at choking flow, the steam passing through the turbine exit annulus has reached sonic velocity. When this occurs, no additional energy can be imparted to the turbine by lowering the absolute pressure [1].

Improved heat-transfer rates can result in both increased generation capacity and reduced fuel requirements. For example, consider a 600-MW generating unit. As little as 0.2 in. Hg absolute-pressure improvement in condenser back-pressure can equal a 0.5% reduction in turbine heat rate. This is equal to about three additional megawatts of generating capacity. In addition, significant fuel savings result. At a fuel cost of 40¢ per million Btu, and a 60% loading factor for this unit, savings of over $60,000/yr can be realized.

A condenser or heat exchanger that is kept free from corrosion, leakage and blockage is a more reliable unit. In addition to the cost of unnecessary fuel consumption, plant outages, chemical treatment, and inefficient use of cooling media, there are pressures applied to industry to meet environmental regulations.

Effect of water velocity

Inlet water temperature affects the back pressure of a condenser. If once-through cooling water is used, inlet temperature is not a factor that can be controlled, but mechanical-draft cooling towers can regulate the temperature. Where once-through water is used, seasonal variations magnify the effects of increased back pressure. One study [2] points out the seasonal effects of an increase as small as 0.1 in. Hg absolute pressure, occasioned by waterside deposits in the condenser of a 265-MW generating unit. Increased operating costs that resulted from this fouling were as high as $1,000/month during periods of high inlet temperature.

Waterside resistance accounts for 72% of the total resistance to heat transfer of a tube. There are two components to waterside resistance: the laminar film of stagnant water next to the tube wall, and the building sedimentary deposits called "fouling."

The laminar film has insulating qualities that can account for 39% of the total heat-transfer resistance of a condenser. In addition, it isolates the turbulent water within the tube from the fouling buildup on the tube surface, which contributes to a fast rate of deposit. Tests [4] have shown that there is a rapid decrease in heat transfer after a thorough manual cleaning of condenser tubes—as much as 15–20%—after only 10 h of operation on a 90,000-ft²-surface condenser. If the

Originally published October 13, 1975

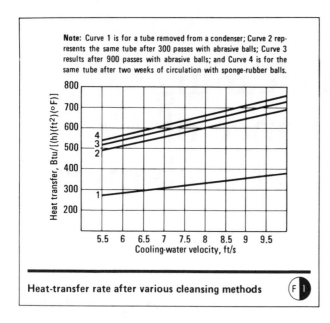

Note: Curve 1 is for a tube removed from a condenser; Curve 2 represents the same tube after 300 passes with abrasive balls; Curve 3 results after 900 passes with abrasive balls; and Curve 4 is for the same tube after two weeks of circulation with sponge-rubber balls.

Heat-transfer rate after various cleansing methods F 1

laminar film is disrupted by continuous mechanical cleaning, the heat transfer through the tube can be increased (F/1), and the rate of waterside fouling significantly reduced.

Tube-material considerations

Among the many factors that affect the type of deposit are dissolved solids, pH, chlorination, bacteria and water velocity. Deposits contribute to corrosion, particularly of stainless steel tubing [5,6,7]. The most troublesome deposits [6] have been oxygen-excluding patches of calcium carbonate containing chlorides, and the soft, porous precipitate of a hydrous oxide of manganese found in seaboard locations.

Another fouling source is organic in nature. It varies from bacterial action that produces a heavy slime, to the growth of marine organisms that thrive in the cooling media. Non-copper-bearing metals are especially susceptible to slime formation. The presence of bacteria can also increase the manganese content of cooling water. It has been noted [6] that, though a river may contain no detectable concentration of manganese, once the water is impounded, its content of manganese may increase as a result of biological activity.

Power plants and industrial facilities must rely more than ever on cooling water that is increasingly subjected to pollution. A large portion is either sea water or brackish water from estuaries that contain marine life. The result is either outright plugging of the tube or localized erosion leaks. As cooling water flows past an obstruction, it must necessarily increase in velocity, which results in rapid erosion in the vicinity. Copper-bearing alloys are highly vulnerable to this; stainless steel is affected to some extent. Only titanium seems resistant to localized velocity increases [8].

Cleaning methods

A number of methods have been used to increase condenser and heat-exchanger efficiency. They range from cleaning techniques, to new tube materials and cooling towers.

Various chemicals (acids, chlorine) have been used to reduce fouling and restore tube cleanliness. Acids may either be strong (which damages the equipment) or weak (citric, formic, sulfamic); these are less effective and probably more expensive. Acid cleaning is limited to once a year or less often; between cleanings, fouling can severely affect condenser performance. The use of chlorine is being cut back or eliminated in many regions by government regulations. Thus, mechanical cleaning of condenser tubing—manual or automatic—seems to be assuming more of the burden of maintaining condenser back pressure.

Manual methods include periodic cleanings with rubber plugs, nylon brushes, metal scrapers, or turbining tools. Where clean water exists and the deposits are light, manual cleaning can be effective in reducing fouling deposits, which can account for 33% of the heat transfer resistance. However, manual cleaning is becoming more expensive as condensers and heat exchangers grow in size, with higher hourly rates and downtime playing important roles. Moreover, it is difficult for manual cleaning to keep up with fouling and blockage where the cooling water has high deposition due to its content of chlorides and solid debris. The greatest drawback to manual cleaning is that it is intermittent; between cleanings, fouling builds up rapidly.

Automatic cleaning by means of sponge-rubber balls is economical in areas where deposition, pollutants, chlorides and other corrodents exist. These balls distribute themselves at random through the condenser, passing through a tube at an average of one every five minutes. Slightly larger in diameter than the tube, they wipe the surface clean of fouling and deposits.

In an effort to reduce solid debris ingestion, particularly in seawater locations, most facilities have used "traveling" or "sliding" screens. These have not been very effective because the screens sometimes act as handy conveyors for flotsam and jetsam right into the water mains.

There are also several filters available, but until recently none could be flushed out online. A debris filter now is available that can be located adjacent to the waterbox and can be flushed without interrupting the flow of screened water to the condenser.

As good sites and once-through cooling water become less available, industry is turning more and more to cooling towers, even though they do not solve all condenser problems. Among the drawbacks are a buildup of dissolved solids and chloride concentrations. This is accentuated as plants strive for zero discharge.

Rubber-ball cleaning

The basic principle of cleaning with sponge-rubber balls is to frequently wipe clean the inside of the tube while the unit is in operation. Since the balls are slightly larger in diameter than the tube, they are compressed as they travel the length of the tube. This constant rubbing action keeps the walls clean and virtually free from deposits of all types. Thus, suspended solids are kept moving, and not allowed to settle, while bacterial fouling is wiped quickly away. Pits do not form because deposits are prevented.

The balls are selected in accordance with the installation, their specific gravity being nearly equal to that of the cooling media. Therefore, they distribute themselves in a homogeneous fashion. They travel the length of the tube forced by the pressure differential between the inlet and the outlet. The ball's surface allows a certain amount of water to flow through the area of contact with the wall, flushing away accumulated deposits ahead of the ball. They are available in various degrees of resiliency, depending on the requirements.

An abrasive-coated ball is also available for situations where the cooling water tubes have already been heavily fouled. Here, the effect is a gentle scouring that removes the scale slowly but steadily, until the tube is ready to be maintained by the normal sponge-rubber ball. Heat-transfer efficiency climbs steadily throughout this treatment.

The balls are circulated in a closed loop, including the condenser or heat exchanger (F/2). At the discharge end, they are caught in a screen installed directly in the line. They are then rerouted through a collector back to the condenser ball-injection nozzles to ensure that the balls are uniformly distributed.

At the collector unit, the balls can be counted or checked for size. The number required for a particular service is a function of the number of cooling tubes. Naturally, some wear occurs so that the balls must eventually be replaced. The labor needed to count and check a charge is usually about 1 h/wk.

This cleaning system can be retrofitted into most existing condensers or heat exchangers, although some modification of piping or unit design may be required. The slight increase in pumping resistance due to the

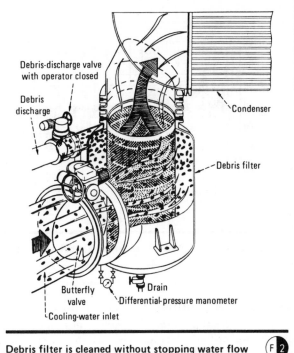

Debris filter is cleaned without stopping water flow F 2

pressure drop across the screening device is more than offset by the reduction in fouling resistance in the condenser or heat exchanger tubes. The most effective way to take advantage of this system is to provide for its installation at the design stage. A filter prevents solid debris from entering the waterbox of the condenser or heat exchanger. Located in the cooling-water inlet, it is flushed as needed without shutting down or bypassing the filter (F/3).

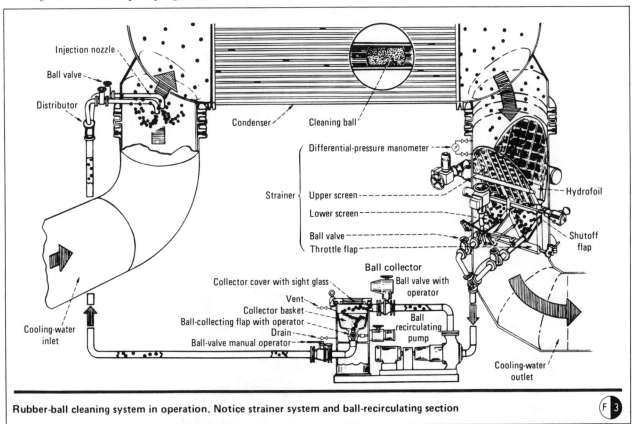

Rubber-ball cleaning system in operation. Notice strainer system and ball-recirculating section F 3

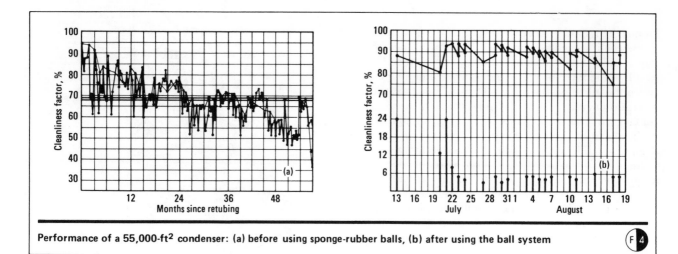

Performance of a 55,000-ft² condenser: (a) before using sponge-rubber balls, (b) after using the ball system

F 4

Examples of continuous tube cleaning

There are approximately 2,000 installations of sponge-rubber-ball cleaning systems in Europe, and about 200 in the U.S. All have achieved an outstanding record of success in maintaining condenser efficiency and reliability. A typical case is shown in the "before" and "after" graphs in F/4. Here, a 95-MW unit was retubed with stainless steel 304. Data kept for five years showed an average 27% deterioration in cleanliness from the original 94%, despite 69 manual cleanings. After a sponge-rubber system was installed in the same condenser, the cleanliness approached 95% within one week of operation.

One of the most extensive studies of continuous tube cleaning was conducted by United Illuminating at its English Station [11], which uses brackish, polluted water. It was found that the rubber-ball system can reduce the fouling rate of new brass tubes by as much as 50%, when compared with tubes operated without continuous cleaning.

Another instance involved stainless steel tubing, where the rubber-ball system maintained a cleanliness

factor of 98% and a back pressure of 1.49 in. Hg. After 1,800 h of operation, the tube-cleaning system was taken out of service for testing purposes. During a month of operation without cleaning, the condenser back pressure climbed to 1.65 in. Hg, and the cleanliness factor dropped from 98 to 81%. When the cleaning was restarted, the original back pressure and cleanliness were recovered in 10 days.

Extensive tests at the Tennessee Valley Authority Widows Creek Plant [12] for over a year showed that the continuous system was highly economical and produced superior performance over manual cleaning. Results from two nearly identical condensers were compared; continuous cleaning by ball recirculation showed a 17% better performance than manual cleaning. This projected to an annual net savings of fuel (in 1969) of $20,000.

Summarizing, continuous cleaning and filtering systems maintain a high level of condenser and heat-exchanger efficiency. The ball-cleaning scheme results in fuel savings, fewer outages, and the reduction or elimination of cleansing chemicals.

References

1. Stoker, R. J., and Seavy, E. F., The Selection of Large Steam Surface Condensers, *Combustion*, Sept. 1967.
2. Heidrich, A., Jr., Roosen, J. J., and Kunkle, R. J., "Calorimetric Evaluation of Heat Transfer by Admiralty Condenser Tubes," ASME Publication 65-WA/CT-2, presented in Chicago, Nov. 7–11, 1965.
3. Lustenader, E. L., and Staub, F. W., "Development Contribution to Compact Condenser Design," INCO Power Conference, Wrightsville Beach, N.C., May 5, 1964.
4. Detwiler, D. S., "Improving Condenser Performance With Continuous In-Service Cleaning of Tubes," American Soc. for Testing Materials, Publication STP 538.
5. Maurer, J. R., The Use of Stainless Steel Tubing in Condenser and Related Power Plant Equipment—A Progress Report, *Combustion*, July 1967.
6. Long, N. A., "Recent Operating Experiences With Stainless Steel Tubes," paper presented at the 1966 American Power Conference.
7. "Experiences With Stainless Steel Tubes in Utility Condensers," Nickel Topics, Vol. 24, No. 5, 1971, International Nickel Co.
8. Feige, N. C., "Titanium Tubing for Surface Condenser Heat Exchanger Service," Bull. SC-1, Timet, Div. of Titanium Metals Corp. of America, West Caldwell, N.J.
9. Papamarcos, J., Condenser Tube Design Directions, *Power Eng.*, Vol. 77, No. 7, July 1963.
10. O'Keefe, W., Better But Costlier Tube Metals Tackle Present and Future Condenser Problems, *Power*, Vol. 117, No. 8, Aug. 1973.
11. Kuester, C. K., and Lynch, C. E., "Amertap at English Station," ASME

Publication 66-WA/CT-1, contributed to the ASME by the Research Committee on Condenser Tubes, New York, Nov. 27 to Dec. 1, 1966.
12. Condenser Cleaning Improves Economics, *Electrical World*, Dec. 15, 1969.
13. Kuester, C. K., Here's How to Eliminate Debris from Heat Exchangers, *Electric Light and Power*, Energy/Generation ed., Aug. 1973.

The author

William I. Kern is General Manager of Amertap Corp. (Div. of Fa. Taprogge, West Germany), P.O. Box 151, Mineola, NY 11501, with responsibility for operations in the U.S. and Canada. Before, he was regional sales manager. He also worked as development and sales engineer for the Hamilton Standard Div., U.S. Aircraft Corp. He holds a B.S. degree in metallurgical engineering and a B.A., both from New York University, has presented papers on electron-beam welding of aerospace materials, and has conducted welding-development programs for the U.S. Air Force. He belongs to the Air Pollution Control Assn., National Assn. of Corrosion Engineers, and the American Soc. for Testing and Materials.

Heat more efficiently— with electric immersion heaters

Electric immersion devices are recommended for heating a wide variety of liquids. Different types of units are described, and their assets and liabilities listed. Several factors that must be considered before selecting an immersion heater are reviewed and explained.

David R. Martignon, Watlow Electric Manufacturing Co., Rochester, Mich.

☐ When heating water, oil, chemicals, molten metals, molasses, or any of a wide variety of liquids, the most efficient approach is the use of an electric immersion heater. That is because, compared with other sources, electric heat is usually more efficient. One kilowatthour can be converted to heat with an efficiency of 100%, whereas when such fuels as gas, oil and coal are converted into heat, the efficiency may range from 5% to 80% because of losses in burning and applying the heat.

This article describes a variety of electric immersion heaters for all types of application parameters. Advice is given on how to choose the right type of heater, how to select the proper sheath material, and how to calculate the watt density of the heating elements.

Key advantages

Electric immersion units can be controlled by automatic devices—there are several types of temperature controllers that are used with electric heating. Among these, the solid-state controls provide accurate performance and long life. With these, there are no mechanical parts that wear out, or valves or gates, as in gas and oil systems.

The heater's elements are simple, of rugged construction, and easily replaceable. They cost less and have a longer life than elements for fossil-fuel heaters. They also need less floor space, since they are placed directly in the processing medium.

Oil- and gas-fired heaters require large burners and exhaust systems. Often, large exhaust fans are needed to clear the fumes from the working area. In some cases, storage tanks for fuel must also be purchased. In addition, the device's air filter must be cleaned and replaced on a regular basis.

With combustion systems, insulation must be kept away from the flames, to prevent charring. Electric systems, on the other hand, are well insulated; they lose less heat, so the surrounding areas will be cooler. Be- cause there are no products of combustion, no ashes or dirt can soil the working area. And because there are also no fumes, fire and explosion hazards are greatly reduced.

Immersion heaters can be manufactured in accordance with ASME boiler codes, military guidelines, and other specifications. All of these illustrate how electric heating can save money and energy.

Available units

Immersion heaters are available in a variety of forms to fit all types of applications. The screw-plug immersion heater, for example, provides a permanent installation, which can nevertheless be removed without too much difficulty for cleaning or replacement. This unit simply screws into a threaded hole in the tank wall; if the wall is thick enough it can be drilled and tapped to accept the heater. If it is too thin, a pipe coupling can be welded or brazed in. Screw-plug units are normally available in 1–2.5-in. NPT sizes, and a variety of voltage and wattage values. Screw-plug elements are normally used when 20 kW or less are required to do the job.

Flanged immersion heaters are installed through the wall of the tank by bolting them to a matching pipe flange welded to the tank wall. The connection can be specially gasketed to withstand high pressures or contain penetrating liquids such as salts and ethylene glycol. Flanged immersion units are normally used where large kW packages are required. Flanged units are bulkier and somewhat more difficult to install than screw-plug heaters.

Over-the-side immersion units are ideal for applications in which easy installation and quick removal are desired. The lightweight portable immersion unit can be used for heating a wide variety of liquids in containers of different shapes and sizes. Hairpin-shaped tubular heaters are brazed or welded into a liquid-tight junction box. A riser takes the electrical connection up

Originally published May 23, 1977

Allowable watt densities for various liquids

Material being heated	Approximate temperature, °F	Allowable watt density, W/in^2
Water and solutions containing at least 80% water, including alkaline and acid cleaning solutions	210	40 to 300
Metals in liquid state, such as solder, lead, sodium and potassium	500	600
	700	500
	900	400
	1,100	300
	1,300	200
Cooking oil	400	30
Dowtherm A	500	20
Machine oil SAE-30	250	18
Therminol FR-2	500	12
Bunker C fuel oil	160	10
Asphalt	300	8
Molasses	100	5
Glue	100	No direct heating

(Cooking oil through Glue: Based on velocity of 1 ft/s)

to a liquid-tight terminal housing. The elements can be supplied straight or curved to fit a circular vessel.

Circulation heaters are available for oil, water and gas heating. They are used when indirect heating is required, or when processing will not permit placement of the immersion heater directly in the tank.

Indirect heating uses a heat-transfer medium that carries heat from the heater to the process material. Indirect heating units utilize a jacketed container for heating heavy, viscous materials that require extremely low watt-densities in order not to alter or ruin the process material. The immersion heaters are placed in the heat-transfer media, and can operate at watt densities that allow a more economical approach to heating those heavy, viscous materials.

Corrosion and watt density

These are factors that should be considered before buying an electric immersion heater.

The corrosion characteristics of the liquid must be weighed before selecting the proper sheath material for the heating elements. Copper is normally used in water applications in which high temperatures and pressures are not involved. Steel elements are employed for most oil applications. Incoloy is used for chemicals, some acid solutions, and also in steam generators and other applications in which high temperature of pressurized-water vessels is required. Other sheath materials are Monel, stainless steels, aluminum and titanium. These last materials are specially ordered, which may require long lead times. Copper, steel and Incoloy sheaths are standard, and can be obtained from most heater manufacturers in a reasonable time. Another factor that affects the life and performance of an immersion element is the watt density—the heat flux or total amount of heat being generated for each square inch of heated surface area of the element. The table gives some typical liquids and the recommended watt density for heating elements. Water allows for relatively high watt-density elements, whereas in oil and asphalts, the element must operate at a much lower watt density. If an immersion element is too high in watt density, the liquid will carbonize and insulate the sheath of the element, burning it out.

In the case of water heating, a normal expansion and contraction of the immersion element, when cycled on

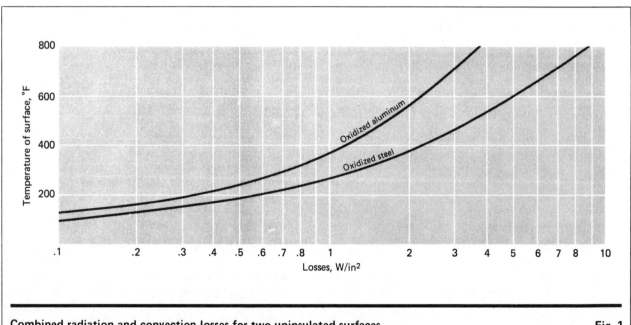

Combined radiation and convection losses for two uninsulated surfaces   **Fig. 1**

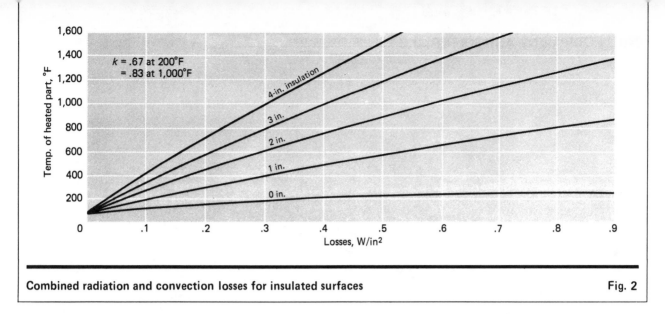

Combined radiation and convection losses for insulated surfaces   Fig. 2

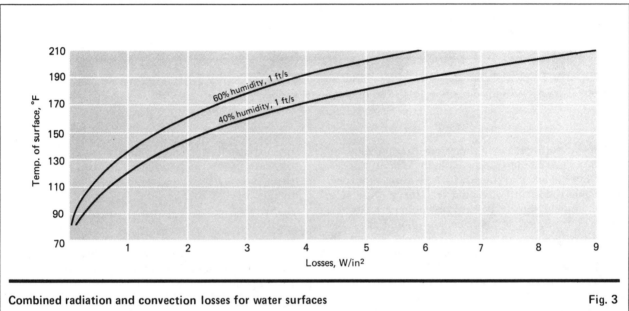

Combined radiation and convection losses for water surfaces   Fig. 3

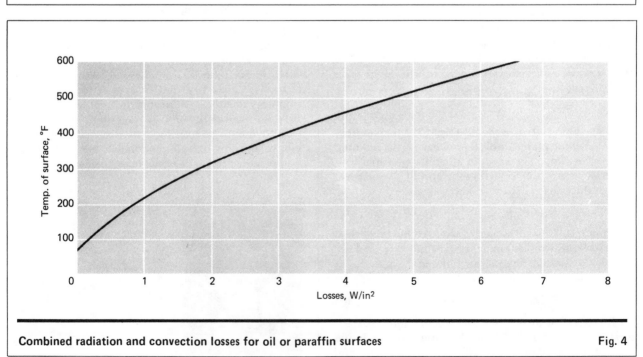

Combined radiation and convection losses for oil or paraffin surfaces   Fig. 4

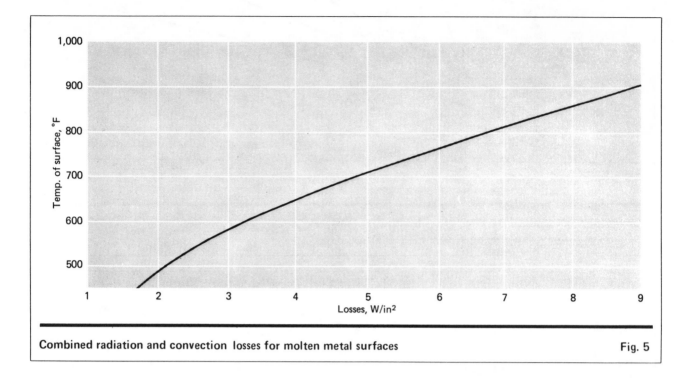

Combined radiation and convection losses for molten metal surfaces

Fig. 5

and off by controllers, tends to remove any lime deposits, so no maintenance is required. Where extreme hard-water conditions prevail, it is advisable to inspect the element after three months' use. If there is a lime buildup on the sheath, it should be cleaned off, and a periodic inspection of the elements should be made thereafter.

Check with the heater manufacturer and be sure to describe the material being heated. If it is a mixture of various chemicals, let the manufacturer know. Most suppliers have information allowing them to pick the correct sheath material and watt density, so that long life can be expected from the immersion element.

It is easy to determine the required wattage for an immersion-heater application. The power needed for any given use is either the watts required for startup, or the watts necessary for operation, whichever of these two is greater.

$$\text{Startup watts} = A + C + 3/4 \, D \qquad (1)$$

In Eq. 1, $A =$ watts absorbed in raising the temperature of the machine platen, tank, liquid, etc., in the required time; $C =$ watts absorbed in melting or vaporizing material in startup or operation; and $D =$ watts lost from surfaces by radiation, convection and conduction.

$$A = \text{lb/h} \times (c)(\Delta T)/3.412 \qquad (2)$$

In Eq. 2, lb/h = total weight divided by total time allowed for startup, or pounds of material processed per hour; $c =$ specific heat of the material; and $\Delta T =$ temperature rise of the material, °F.

$$C = \text{lb/h} \times r/3.412 \qquad (3)$$

In Eq. 3, $r =$ heat of fusion or vaporization.

D is the wattage lost from the surface of the liquid and/or the container by radiation, convection, and conduction. Several heater manufacturers have engi-

neering application guides containing curves that give the watts lost through radiation and convection for such surfaces as water, oil and steels (Fig. 1-5) at an ambient temperature of 70°F. In the case of watts lost by conduction:

$$D = (k)(a)(\Delta t)/491d \qquad (4)$$

In Eq. 4, $k =$ thermal conductivity, (Btu)(in.)/(h)(ft²)(°F); $a =$ area of conduction, in²; $\Delta t =$ difference in temperature across the insulator (usually an estimate), °F; and $d =$ thickness of insulator, in.

$$\text{Operating watts} = B + C + D \qquad (5)$$

In Eq. 5, $B =$ watts absorbed in raising the temperature of parts or material during the working cycle. Eq. 2 is used to estimate B.

If additional assistance is needed, most heater manufacturers maintain a staff of application and product engineers to help with problems.

As the availability of fossil fuels diminishes, the use of electric immersion heating as an easily controlled, reliable, clean, and overall economical means of heating liquids and gases will be increasing steadily.

The author

David R. Martignon is sales manager for Watlow Electric Manufacturing Co., at which he has also held the positions of chief applications engineer and sales representative. He has had previous experience in research and development, and in mold design. He holds an associate degree in Electrical Technology from Washington University (St. Louis).

Energy-efficient motors spark an old controversy

News that the U.S. Dept. of Energy plans to mandate standards for this type of unit has revived a traditional debate—whether the benefits provided by the motors outweigh the premium price buyers must pay.

☐ U.S. equipment makers and their clients are once more taking sides over the pros and cons of energy-efficient (EE) electric motors*. Stirring the controversy this time is an expected (at presstime) U.S. Dept. of Energy report—the revision of a study issued last February—that probably will mandate specific energy-efficient ratings for such motors within the 5.1-125-hp range†.

DOE is optimistic that this and other measures proposed by its study will save as much as 52 billion kWh on an annual basis by the year 2000. But the National Electrical Manufacturers Assn. (Washington, D.C.) is not so sure. Says NEMA president Bernard H. Falk, "We feel that the economic analysis of various options regarding electric motors is incomplete and in some cases inaccurate, and that the conclusions reached by DOE are therefore not justified."

The NEMA/DOE debate is not likely to win converts for EE motors among chemical process industries (CPI) buyers, some of whom believe that the units are not worth the 10-25% price premium they command, that it is time-consuming to compare them with standard motors, and that delivery times (reportedly up to 4 mo) are too long. In addition, say these critics, there is confusion about what distinguishes EE motors from others (the standard motor line of one maker may have the same efficiency as another's EE line).

* These achieve higher efficiencies through use of more copper in the windings to cut resistance losses, better silicon steel in the stator and rotor cores to improve magnetizing properties, and improved quality control to reduce stray load losses. The motors weigh about 10% more than standard models, and run quieter and cooler.

Another means of saving energy—the use of electronic motor controllers—will be covered in an upcoming issue.

† A bill that just cleared the Senate has a similar goal. Proposed by Sen. Howard M. Metzenbaum (D., Ohio), the measure sets mandatory efficiencies and market penetrations for electric motors.

General Electric uses a dynamometer to test its Energy Saver motors

Originally published November 17, 1980

New motor standard seeks to clear up efficiency debate

The new motor standard, NEMA MG 1-12.53a,b, is an attempt to standardize motor efficiency ratings with one definable, reproducible test. Previously, efficiencies were described by such terms as apparent efficiency, full load (4/4), nominal, calculated or average.

The test involves the use of a dynamometer (a load-bearing device that carefully measures the output of a motor) under partial and full motor loads. Stray-load losses—caused by such factors as stator and rotor geometry, air gaps and the quality of manufacturing—are calculated separately from resistance losses (generally termed I^2R losses). The stray-load losses are calculated at six motor load points between 25 and 150% of the motor's rated load, then averaged using a regression analysis.

Once the efficiency of a particular motor design is found, that rating becomes the "Nominal Efficiency" value. It is presumed that this value is the median point of all the motors of that design, with half above and half below it. Some manufacturers are now guaranteeing a "Minimum Efficiency," based on a bell-shaped distribution curve with the Nominal Efficiency at the apex of the curve. Thus, a NEMA Design B motor is now specified as having a Nominal Efficiency of 95.0%, and a guaranteed Minimum Efficiency of 94.1%.

The segregated calculation of stray-load losses is intended to make the motor rating more accurate; this calculation is the main difference between the new NEMA standard and standards written by such bodies as the International Electrotechnical Commission or the Japanese Electrotechnical Committee (JEC). In General Electric sales literature, mention is made of a test in which a 7.5-hp and a 20-hp motor were rated according to these various standards. The results showed that the NEMA test was the most stringent:

Standard	Full-load efficiency (%)	
	7.5 hp	20 hp
International (IEC 34-2)	82.3	89.4
British (BS-269)	82.3	89.4
Japanese (JEC-37)	85.0	90.4
NEMA MG 1-12.53a	80.3	86.9

For obvious reasons, U.S. manufacturers are urging their customers to specify the NEMA test when comparing motor efficiencies. An official of the Japan Electrical Manufacturers' Assn., which works closely with the JEC, commented, "We are aware of the foreign [i.e., U.S.] move toward a refined testing standard and some manufacturers' fears of unfair competition in the case of JEC's failure to follow suit. We make it our principle to use the uniform method worldwide."

These views, however, are not shared by many engineering firms and CPI companies that claim a substantial reduction in operating costs via use of EE equipment.

HUBBUB—The debate seems to have started in 1975 when the then Federal Energy Administration commissioned Arthur D. Little, Inc. (Cambridge, Mass.) to study the U.S. electric motor market. The ADL report showed a significant potential for energy conservation and took a swipe at how motor manufacturers rated the efficiencies of their products. "Reliable and consist-ent data on motor efficiency is [sic] not now available to motor appliers [users]. Data published by manufacturers appear to range from very conservative to cavalier," it stated.

NEMA confirmed these conclusions in a series of round-robin tests, in which the motors of one maker were tested in the facilities of another. The results showed that efficiencies varied by as much as ±1.9 percentage points from one test site to another. These variations can amount to thousands of dollars' difference in operating expense over the motors' life.

To correct the discrepancy, NEMA studied how best to segregate and quantify motor energy losses. The procedure settled on was adapted from the Institute of Electric and Electronic Engineers 112, Test Method B standard, and is now designated as NEMA MG 1-12.53a,b (see box). NEMA suggests that all its members adhere to this standard in describing their products.

However, in late 1978 DOE announced what it described as 11 "exceptional near-term opportunities" for conserving energy (*Chem. Eng.*, July 16, 1979, p. 49), and included EE motors as one of the 11. That precipitated another round of studies, which culminated in the February 1980 release of a preliminary report entitled "Classification and Evaluation of Electric Motors and Pumps." The document recommends that DOE carry out three steps to accelerate the distribution of EE motors in the marketplace:

■ The preparation and dissemination of a motor application manual describing the properties of motor designs and selection procedures;

■ A labeling program that would require efficiencies to be stamped and certified on all a.c., polyphase NEMA Design B motors in the 5.1-to-125 hp range (this range has been identified as having the greatest energy conservation potential);

■ A standards program requiring all such motors to have efficiencies of 90% in the 5.1-to-20 hp range. 92% in the 21-to-50 hp range, ana 93% in the 51-to-125 hp range by 1987.

Costs for the first two steps are estimated at over $8 million in administration expenses over the next decade, and $5.5 billion for compliance up to the year 2010. For the entire program, administration costs are estimated at more than $130 million over the next ten years, plus $9 billion for compliance up to the year 2010.

A CASE OF OVERKILL?—NEMA argues that there is no need to spend all that money to increase EE-motor market penetration, because equipment manufacturers already are doing a pretty good job. "Since the beginning of last year," says Falk, "more models of energy-efficient motors have been introduced than during 1970-78. At present, there are over 2,000 models of energy-efficient motors available from manufacturers."

In fact, NEMA claims that DOE underestimates the market penetration of such units, thereby overestimating the energy savings of mandatory standards. Even before DOE prepared its report, which assumes a flat 5% market share for EE motors in the near future, market penetration already had surpassed the 11% mark, says the Association.

The increasing availability of new EE models seems to support NEMA's contention. For example, General Electric Co. (Schenectady, N.Y.) introduced its "Energy Saver" line in 1977, and recently announced a second-generation line with efficiencies as high as 95.8%. Reliance Electric Co. (Cleveland) came out in January with its "Duty Master XE" series, and Westinghouse Electric Corp. (Pittsburgh) unveiled its "MAC II" line in June 1979.

But greater availability, says DOE, has not made it easier for buyers to choose an energy-efficient unit. For instance, some manufacturers—e.g., Baldor Electric Co. (Fort Smith, Ark.)—offer standard motors with nominal efficiencies of up to 94% that do not carry an EE label. Moreover, according to DOE (see figure), no one manufacturer offers the lowest-cost, highest-efficiency motor in more than one horsepower rating (the manufacturers are identified only with the letters A through E in the report). According to the agency's report, this means that "a motivated purchaser will require a time-consuming program of contacting manufacturers . . . and waiting weeks before sufficient information is received on which to make a valid choice."

SALES RESISTANCE—This problem has made some engineers wary of the manufacturers' claims about the cost-effectiveness of EE motors. "Small electric motors are so arbitrarily sized, and the time required to compare them with standard motors is such that there is no economic advantage in buying them," says Henry Kanervo, chief electrical engineer for Jacobs Engineering Group, Inc. (Pasadena, Calif.).

Adds Kanervo: "We have clients that are beginning to specify energy-efficient motors . . . but from an in-house standpoint it's not worth looking at economically. If you spend a couple of hundred dollars of a client's money to try to decide if he should save that two to three percent [the improvement in efficiency of many EE motors], you

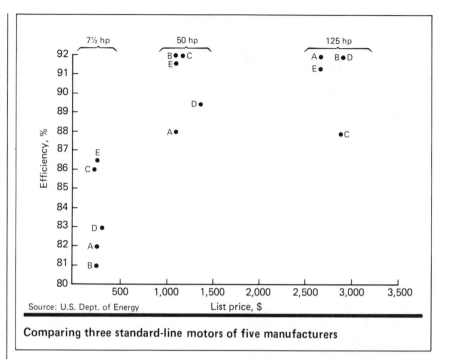

Source: U.S. Dept. of Energy   List price, $

Comparing three standard-line motors of five manufacturers

aren't doing him a favor."

"In the 40-hp-and-under range, one is more likely to consider a high-efficiency motor, but one has to realize that basically we are simply going back to the motors we bought in the 50s; they're merely putting more copper into them," is the opinion of George C. Briley, vice-president of Refrigeration Engineering Corp. (San Antonio, Texas).

POSITIVE OPINIONS—Other customers, particularly the larger CPI companies and engineering firms, disagree with these perspectives. "If you look at the operating costs of EE versus standard motors, you'll find that you couldn't afford standard motors," says Craig Druskins, an engineering manager for Dow Chemical Co. (Midland, Mich.).

Having compared several manufacturers' EE version of a $2,000, 75-hp motor with an operating life of 10 years, he has found that the difference in operating costs can be as high as $982. "And the numbers you see when you compare these motors with standard units are even more dramatic," adds Druskins.

Dow has taken an aggressive approach to evaluating its electric motors. It has a "best-buy" list of preferred motor makers, and has constructed a mobile laboratory at one of its Texas plants that includes dynamometers to test both motors and the loads they must sustain in operation.

And when Dow puts out bids for the construction of process equipment that includes motors, they ask the supplier to analyze his motor choice. "We're telling them that they have a choice of standard or EE motors; to let us know which they specify; and to justify economically why they're using it," says Druskins.

The attention paid to evaluating the economics of an EE motor is seconded by Grover Brown, a motors specialist for Ralph M. Parsons Co. (Pasadena, Calif.). "All the client has to do is come up with the predicted cost of power, and all we have to do is come up with alternative bids," he asserts. He said that the needed numbers are available and the calculations are not time-consuming. "Two to 5 years from now, it [consideration of EE motors] will be a must," concludes Brown.

The motor makers, of course, agree with this assessment and are trying to make their claims as concrete as possible. General Electric, for instance, has started a computer time-sharing program in which a sales representative connects with a computer over the telephone and performs a cost-benefits analysis in the customer's office. "The price premium of an EE motor can often be recovered in the first six months of operation," says David C. Montgomery, a manager of GE's Small Motors Dept.

Nicholas Basta

Energy-efficient motors gain wider interest

After some years of relatively unenthusiastic reception, energy-efficient motors are finally beginning to catch on, especially in the chemical process industries. The impetus: rising energy costs.

☐ Late last year, the U.S. Dept. of Energy (DOE) selected what it termed 11 "exceptional near-term opportunities" for conserving energy, with the aim of moving the technologies into the marketplace as quickly as possible. One technology chosen was energy-efficient electric motors.

Though by no means new, such motors had not been widely accepted because although more efficient than standard motors, by anywhere from around 9 to about 24%, they cost about 25% more.

Rising energy prices have done their share in changing attitudes toward these more costly motors, especially in the chemical process industries (CPI) and related industrial sectors, where such units get long and generally continuous use. In such cases, savings in operating costs due to the higher efficiency can often make up for the higher initial cost within months of installation. In other industries, where use is more intermittent and of shorter duration, the initial-cost payback is much longer, making substitution less appealing.

DOE's self-set goal for penetration by energy-efficient motors into the $580-million/yr motor market is 33% by 1985, and 74% by 1990. To implement this, the agency will begin, in the fourth quarter of this year, a program to make motor users and manufacturers more aware of the benefits offered by energy-efficient units. Both buyers and sellers now seem to consider only initial cost. Other efforts to stimulate demand will also be implemented.

For the time being, DOE's thrust will remain strictly educational. However, a study on motors and pumps now being done for the agency could recommend putting some legislative or regulatory teeth into the energy-efficient-motor program, such as issuing requirements for labeling, or establishing standards for various classes of motors.

In the meantime, however, motor makers are continuing to offer ever-wider lines of energy-efficient motors, and users are buying them in increasing numbers.

WHAT IS ENERGY-EFFICIENT?— The term "energy-efficient" is generally used to describe motors that are more efficient in converting electrical energy into mechanical energy than comparable standard models. There are any number of ways to accomplish this increase in efficiency, many of which have been known for years (see box). But incorporating the techniques or changes adds to the inherent cost of the motor. In addition, not all makers use

---

Other routes to better motor performance

Over and above replacing standard motors with energy-efficient ones, motor users and makers are looking at other ways of cutting costs by boosting efficiency. These efforts fall primarily in two areas: speed controllers, and power-factor controllers on lightly loaded units.

The big news on the speed-controller scene is Exxon Enterprises' (New York City) ACS (Alternating Current Simulator) device. Though Exxon will not reveal details on how the device works, it is known that ACS uses a microprocessor chip and several transistors to match motor speed to changing load by varying the voltage and frequency to the motor. This prevents the motor from running at a constant speed, which in most cases is matched to maximum expected load.

The concept is not new. Indeed, transistorized systems for varying motor speeds have been commercial for at least 15 years. But, as with energy-efficient motors, their acceptance has been slow because the energy savings they offered were not generally considered great enough to balance the initial cost. Exxon's breakthrough seems to lie in its claim that it will market ACS for between $7.50 and $37.50 per horsepower by the mid-1980s, a range that is only a fifth to a tenth as costly as that for current speed-control devices. Competing manufacturers are skeptical that Exxon can come through on this claim.

The power-factor controller is a device developed by the National Aeronautics and Space Administration (NASA). Basically a fixed-speed unit, it constantly senses the load on a motor by measuring its torque. If it senses no loading, the unit reduces the power to the motor, thereby idling the machine. DOE is working with NASA on the further development of the device and seems optimistic about its usefulness on certain types of lightly loaded motors.

Originally published July 16, 1979

No black magic in energy-efficient motors

A primary objective in energy-efficient electric-motor design is to reduce known losses that occur in standard a.c. electric motors. These include:

■ *Power losses.* Typically comprising 55 to 60% of total losses, these are heating losses that result from current passing through the stator and rotor.

■ *Core losses.* These losses—as a rule 20 to 25% of the total—are due to hysteresis effects and eddy currents in the stator and rotor, caused by 60-Hz magnetization of the core material.

■ *Stray load losses.* The most difficult to control, these losses, which are typically 11 to 14% of the total, are due to magnetic and electrical characteristics of the motor and its materials. These are often controllable only through careful processing procedures.

■ *Friction and windage losses.* Up to 5% of total losses, these result from bearing friction, windage, and circulation of air within the motor.

To reduce these losses, the following design changes are among those made, in any number and combination:

■ Lengthening the stator and rotor will reduce magnetic-flux density, cutting hysteresis effects and eddy currents.

■ Using silicon steel, rather than carbon steel, in the stator and rotor will also reduce hysteresis and eddy current effects.

■ Using thinner metal laminations in the rotor and stator will help reduce core losses by minimizing eddy currents.

■ Changing the winding configuration, using more copper and increasing the winding volume will improve conductivity, lower resistance and reduce power losses.

■ Optimizing the air-gap design, so that it is small enough to diminish current, yet large enough to reduce stray load losses, will also improve overall motor efficiency.

■ Using smaller, more-efficient fans will reduce windage losses.

■ Many of the above changes, by increasing the amount of core and winding material used, decrease internal flux density and leakage reactance, thereby boosting power factor as well as overall efficiency.

(For a more thorough look at motors and how they work, see *Chem. Eng.,* Mar. 12, 1979, pp. 85-91.)

the same techniques for upping motor efficiency, thus frustrating efforts to define a standard for energy efficiency in terms of physical criteria (number and volume of windings, length and mass of stator and rotor, type of steel used, etc.). Likewise, efforts to define efficiency by performance suffer from the lack of a standardized method of measurement.

The National Electrical Manufacturers Assn. (NEMA), Washington, D.C., is working on this problem. It has established a preferred method—developed by the Institute of Electrical and Electronics Engineers—to quantify efficiency and is now attempting to find a way to minimize the statistical variations in test results that have heretofore kept the method from being accepted. NEMA is also looking to establish guidelines for labeling motors with efficiency and power factor* data.

DOE EFFORTS—DOE is wholeheartedly behind NEMA, since one point in the former's educational program is to encourage manufacturers and associations to establish nomenclature for describing energy-efficient motors and what applications are most appropriate for such units. Among other actions planned:

■ Guidelines for purchasing energy-efficient motors will be published and made available to government agencies and the private sector.

■ Procedures will be established for these agencies to procure energy-effi-

*Power factor, the ratio of work-producing current consumed by a motor to total current consumed, also plays a role in energy efficiency. The higher the power factor, the more efficient the motor.

cient motors. This is expected to stimulate demand, as well as set an example for the private sector.

■ Trade associations will be urged to communicate to their members the advantage of using energy-efficient motors.

■ A request will be made to the U.S. Bureau of the Census to establish a reporting category for the motors.

■ DOE will publish and distribute literature describing the potential savings and payback periods for energy-efficient motors.

Albert J. Hayes, resource manager for DOE's electric motor program, says that the agency's thrust will be aimed at motors in the 1 to 25-hp range. He adds that although motors ranging up to 125 hp are the "workhorses" of industry, those larger than 25 hp are generally "custom" in that the purchaser has usually carefully calculated loading and sizing, and often will call for an energy-efficient model. Besides, he notes, large motors are generally quite energy-efficient to start with, and the potential for saving energy is smaller. (For example, a 1-hp motor might be 75% efficient, while a similar 200-hp motor might be 92% efficient.)

The chance for saving energy is much greater in the smaller unit, ranging as high as 24% in some cases, though the saving depends on such factors as horsepower, loading, and line-voltage variation. The price premium for energy-efficient motors in the smaller sizes is also proportionately greater, so there is often greater resistance to purchasing the more costly motor.

A particularly troublesome area is the so-called "hidden motor," according to Hayes. These are machines bought from motor manufacturers by OEMs (original-equipment manufacturers) who install the motors in other pieces of equipment. The ultimate purchaser of the equipment usually doesn't analyze the component parts, and often makes his purchase decision based on the original cost. Since the OEM business is highly competitive, points out Hayes, the OEMs often opt for the cheaper, standard motors rather than energy-efficient ones. DOE will try to change this by educating OEMs and their customers to look at the savings gained over the life of the equipment, rather than just looking at the purchase price.

WIDE-OPEN MARKET—About 76.3% of U.S. industry's electric consumption of 600 billion kWh/yr is accounted for by electric motors. Clear-cut success for energy-efficient units so far seems mostly limited to the CPI and other industrial sectors. Still, these sectors represent about 30% of the total U.S. motor market.

Motor makers see energy-efficient designs taking off at a "satisfying" pace, considering the very small base that they started from a few years ago. Thomas Costello, vice-president, motor divisions, Westinghouse Electric Corp. (Pittsburgh, Pa.), says that although energy-efficient models comprise only about 3 to 4% of his firm's motor shipments, their number has doubled each year since the line was introduced in 1977.

It is still too early to say who is the leader in energy-efficient motor technology. All manufacturers are shooting for the same market (1 to 200-hp motors), and no one firm seems to be specializing yet in any one type or application. It is generally conceded that about nine firms lead the group, with General Electric Corp. (Schenectady, N.Y.) being the largest in terms of motor sales. Next comes Westinghouse, with Reliance Electric Co. (Cleveland, Ohio) and U.S. Electric Motor Div. of Emerson Electric Co. (Milford, Conn.) close behind. After these, the firms are clumped, and include: Allis-Chalmers Corp. (Milwaukee, Wis.), Baldor Electric Co. (Ft. Smith, Ark.), Century Electric Div. of Gould Inc. (St. Louis, Mo.), Marathon Electric Co. (Wausau, Wis.), and Japan's Toshiba.

WHAT ABOUT THE USERS?—By far the largest numbers of energy-efficient motors in the chemical processing industries are used for pumps, compressors, fans and blowers. Du Pont, which has been using the units for power service and chemical-processing applications, says it is satisfied so far with their performance. Standard Oil Co. (Ind.) feels the same way.

Amoco Chemicals, which has completed an extensive study of energy-efficient motors at its Texas City facility, concluded that such units make real savings possible. The firm says it will evaluate the test results further, and make a decision on whether to replace standard motors with energy-efficient ones on a companywide basis.

Philip M. Kohn

Variable-speed drives can cut pumping costs

Operated at less than design flowrates, conventional single-speed-pump and throttling-valve systems waste energy. Pressure-drop losses can be avoided by driving pumps at variable speeds.

James D. Johnson, General Electric Co.

☐ Pumping systems exist that were installed with 10 to 15% extra capacity over design, and that include control valves rated at 35 to 50% of system frictional losses. These, admittedly, represent worst cases that do not often occur.

Pumps powered by a.c. variable-speed drives can handle such maximum conditions without the energy penalty incurred by the conventional arrangement of a single-speed centrifugal pump and throttling valve. Additionally, when flowrates range between 50 and 100% of design, and at least 50% of the pumping head consists of friction loss, variable-speed drives can substantially reduce operating energy costs, and even im-

For a defense of control valves in pumping systems, see Hans D. Baumann's Control valve vs. variable-speed pump, *Chem. Eng.*, June 29, 1981, p. 81.

prove system reliability, which in turn hikes production.

Variable-speed drives are normally suited for pump ratings of from 20 to 500 hp, and even larger. They make possible energy savings of as much as 57% in 100% frictional systems operated at 75% of design flowrate.

Factors that must be included in a comparative evaluation of the economics of the conventional control-valve and variable-speed-drive systems are:

- Magnitude of turndown (i.e., range of duty cycle compared to design flowrate).
- Duration of turndown.
- Comparative cost of equipment.
- Cost of electrical energy.
- Individual flow-system characteristics.

Pumping system hydraulics

Hydraulic systems are characterized by the ratio of frictional head to total head:

$$F = H_f/(H_s + H_f)$$

Here, H_f = design frictional head, and H_s = design static head.

Friction, F, is independent of the control-valve head loss; for example, if the frictional head = 75 ft, static

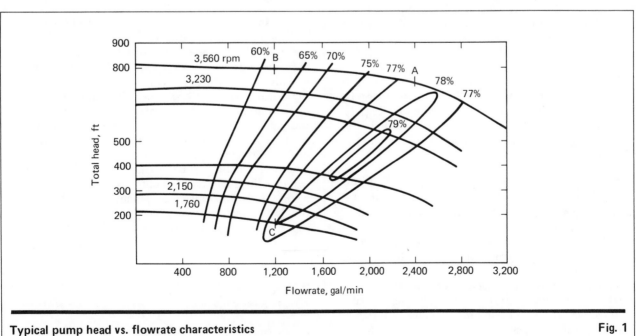

Typical pump head vs. flowrate characteristics

Fig. 1

Originally published August 10, 1981

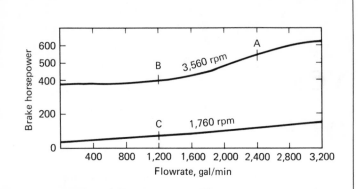

Typical pump brake-horsepower vs. flowrate characteristics   **Fig. 2**

	Energy savings at turndown conditions, %			
	System type			
Rated flow, %	**100% F**	**75% F**	**50% F**	**25% F**
100	21	16	9	3
87.5	38	31	20	12
75	57	47	35	20
50	81	69	55	41

Source: *Hydrocarbon Proc.*, September 1979.

head = 25 ft, and control valve loss = 30 ft, $F = 75/(75 + 25) = 0.75$. This system would be described as a 75% frictional system.

Unit horsepower at the design point is calculated via:

$$hp_u = \frac{H_s + H_f + H_v}{H_s + H_f}$$

Here, H_v = control valve head.

For the foregoing example, the unit horsepower is:

$$hp_u = \frac{75 + 25 + 30}{75 + 25} = 1.3$$

This shows that 30% additional horsepower is required for controllability at the system's design point.

Horsepower savings illustrated

How a.c. variable-speed drives can save energy can be shown by means of Fig. 1 and 2.

At a design operating point of 2,400 gal/min at 3,560 rpm (Point **A**, Fig. 1), the pump efficiency is 77.5% and the brake horsepower requirement is 550 hp (Point **A**, Fig. 2). Changing the flowrate to 1,200 gal/min without varying the pump speed will result in a horsepower requirement of 400 (Point **B**, Fig. 2) and an efficiency of 63% (Point **B**, Fig. 1).

If, however, the pump's speed is altered to accomplish the flow reduction, only 70 hp is required (Point **C**, Fig. 2), and the efficiency is 78% (Point **C**, Fig. 1). Therefore, the variable-speed drive provides a horsepower saving of 330 (400 − 70) and no loss in pumping efficiency.

Energy savings from adjusting pump speed

Percentages of energy savings at various flowrates for different types of frictional systems are summarized in the table. It should be noted that hydraulic systems are typically purchased at 15% above design conditions, and that the average turndown is close to 87.5%.

Energy savings are easily calculated by means of the table. The profitability of the higher investment for a variable-speed system can be evaluated by comparing the incremental cost for equipment against the reduction in operating expense.

For example, assume an installation of a centrifugal pump with a single-speed 200-hp motor in a 75% fric-

tional system, a 185-hp loading at the design flowrate, an electrical rate of $0.035/kWh, and a duty cycle of 15% operating time at 100% design flowrate, 65% at 87.5%, and 15% at 50%.

Designating hp_d as the design hp, op as the operating hours, c as the cost of energy, and E as the pump efficiency, the annual operating cost, $, of this system amounts to:

$$\$ = \left(\frac{0.746 \times hp_d}{E}\right) \times op \times c$$
$$= \left(\frac{0.746 \times 185}{0.92}\right) \times 8760(0.95) \times 0.035$$
$$= 43,693$$

The operating saving to be gained from installing a variable-speed drive to turn the pump is obtained by multiplying the foregoing annual operating cost of $43,693 by the percentage energy savings (taken from the table for the 75% frictional system and the appropriate rated-flow percentages) and by the operating time percentages stated in the duty cycle. Thus:

$$\$ = 43,693[(0.15 \times 0.16) + (0.65 \times 0.31) + (0.15 \times 0.69)$$
$$= 14,375$$

This results in an annual energy saving of 33%.

To gauge the profitability of the incremental investment for the variable-speed drive, one need consider only the difference between the costs for the inverter and the motor starter, because the cost of the motor, stepdown-isolation transformer (if required), breaker and installation will be approximately the same for either system. In new installations, savings will also be realized through the elimination of the conventional motor starter and the throttling valve.

Other benefits to be gained from variable-speed drives include a lower level of mechanical noise at operating speeds, and longer equipment service life, because the "soft" startup of variable-speed drives reduces the shock of peak loading from single-speed startups.

The author

James D. Johnson, as Manager — AC Drives Project of General Electric Co.'s Speed Variator Products Operation (Erie, PA 16531), is responsible for market and product development for a.c. adjustable-speed inverter drives. Previously, he was a sales engineer and an applications engineer in the company's industrial sales division. Holder of a B.S. in mechanical engineering from the University of Cincinnati, he is a registered engineer in the State of Ohio.

Motor controllers spell savings for the CPI

Devices that change a motor's speed or alter the power factor are attracting attention as energy savers. Some are said to cut power consumption by as much as 30-50%.

Variable-frequency unit is linked with motor during performance test

Originally published December 29, 1980

□ Besides looking at energy-efficient motors (*Chem. Eng.*, Nov. 20, p. 93), the chemical process industries (CPI) are trying out new motor controllers—the other half of the drive package—in their push for better energy utilization. And equipment manufacturers are filling this need mostly with a generous helping of low-cost, electronically updated versions of an old device (the variable-frequency controller), sprinkled with a dash of space-age technology (power-factor controllers).

The new units should come in handy, because there seems to be plenty of room for saving energy in the CPI via more-efficient motor controllers. According to a February 1980 report prepared for the U.S. Dept. of Energy by Arthur D. Little, Inc. (Cambridge, Mass.), the cost of energy to operate motors is, as a percentage of industry sales, at its highest in the CPI—2.2%, vs. an average of 1.0% for all manufacturing industries. Moreover, after an informal industry survey, ADL has found that about 20% of the shaft energy delivered by motors to pumps is wasted; of this percentage, as much as $12\frac{1}{2}\%$ can be lost via conventional throttling-control methods.

REPLACING THE TRADITIONAL—Until now, the CPI have been handling their flow-control needs with throttling valves or slip devices on pumps, and dampers on fans and blowers. But rising electricity costs are making these traditional methods less attractive (see graph).

This is one major reason why equipment makers are tempting CPI users to switch to variable-frequency hardware, which has been around for at least 20 years. Although the new lines are still expensive—the cost of variable-frequency controllers is estimated at $400/hp for small motors and about $150/hp for large (150 hp) motors—horsepower range has been extended to around 150 hp (other types of controllers are rated to 800 hp), and optional features that allow most of the motor's operation to be centralized in one control box are included.

Among the options are tachometers, overcurrent protection, reversing, self-diagnostics, etc. With most of these, the necessary circuit board is simply plugged into place—a feature said to make servicing easier. And some of the available circuit boards have been condensed onto large-scale integrated silicon chips similar to those used in computers and other sophisticated electronic equipment.

Among the manufacturers offering new lines of variable-frequency devices are Reliance Electric Corp. (Cleveland, Ohio), with its V*S drives; General Electric (Erie, Pa.), with its Aftrol I drive; Allen-Bradley (Milwaukee, Wis.), with Bulletin 1330, 1379 and 1381 drives; U.S. Electric Motors (Milford, Conn.), with the Ultra Torq II drive; and Parametrics

(Orange, Conn.), with the ParaJust and ParaMizer drives.

Two promising devices, said to lower the cost of a.c. controllers by 50% or more while providing 30-50% energy savings, have not yet entered the market because of unexplained delays. Exxon Enterprises (New York City), which announced its "Alternating Current Synthesizer" (ACS) last year (*Chem. Eng.*, July 16, 1979, p. 49), reports through a company spokesman that "the whole concept is still under development." And the Morse Chain Div. (Ithaca, N.Y.) of Borg-Warner Corp., which announced the "CF-1100" in 1979 (*Chem. Eng.*, Oct. 22, 1979, p. 82) and had expected to start distributing it early this year, has yet to begin production.

Both companies, however, remain sure of success. "There is no question that we will be able to offer the CPI significant cost savings," says Monte May, a marketing manager at Borg-Warner who envisages "valveless" plants in which modulating flow-con-

trol is done via variable-speed motors instead of throttling valves. "There will be a day when those valves will be unnecessary," he adds. "There will be shutoff valves for on-off operation . . . but I would not want to be in the throttling-valve business today."

The announcements of the Exxon and Borg-Warner controllers featured low costs. The ACS, said Exxon, would be offered at $7.50-$37.50/hp. As for the CF-1100, the word was that it would halve the costs of competitive models. Now, according to May, "Borg-Warner will meet its original target predictions for cost effectiveness in the 7.5-to-20 hp range."

The company also is testing controllers up to the 50-hp size. But beyond that, says May, "it's pure speculation whether the power transistors can be made to work."

POWER-FACTOR CONTROL—Another aspect of energy usage getting a closer look today is power-factor* control. A low power factor means that a motor runs with slightly less efficiency, and utilities that supply power often charge higher rates for plants where low-power-factor equipment predominates.

CPI operators are turning to capacitors as one means of correcting low power factor. These units reduce energy consumption by bringing the waveforms of voltage and current into closer alignment.

"We're using capacitors at many installations in our plants," says Larry Adamcik, an electrical specialist with Dow Chemical Co. (Midland, Mich.). "For some of our plants, the utility puts a surcharge on our power if the power factor goes below 0.9; usually, we're operating above that now, and we've seen a payback of 2-3 years on the equipment."

Most a.c. motors run with power factors anywhere between 0.5 and 0.9. Low power factor can be a more noticeable problem in cases where the motor has been oversized, so that it is never running at full load. The capacitors available now are sold in modules that the user can wire together to provide increments of power-factor correction to match the reactive current† his equipment generates.

An entirely new, much publicized method—the power factor controller (PFC)—has apparently not aroused

---

What you should know about variable-frequency units

As the name implies, variable-frequency controllers alter motor speed by changing the normal 50/60-Hz line frequency. A motor's speed is proportional to it, and most can operate reliably with changing frequency. However, the motor does need to maintain a steady elecromagnetic-flux density in the air gap between rotor and stator—a condition that requires a constant frequency/voltage ratio. Therefore, a variable-frequency controller is also a variable-voltage controller.

There are units available for both a.c. and d.c. motors, but a.c. units are of more interest to the CPI, since the industries use more a.c. than d.c. motors. This is because these require less maintenance and are more easily built with explosion- or drip-proof enclosures—a common requirement in hostile plant environments. At any rate, the combined cost and efficiency of a.c. drives (motor plus controller) is pretty close to that of d.c. drives.

Many available a.c. controllers are called inverters, but this is really a misnomer, since the inverter is not the system's sole component. An a.c. controller has three parts: a rectifier that transforms line current into d.c.; some means of regulating the d.c. current; and an inverter that switches the d.c. back to a.c. at the proper voltage and frequency. The components operate in tandem to take incoming current, "chop" it into parts, and synthesize these into the voltage and frequency required to maintain the desired motor speed.

None of the available units is completely successful in synthesizing perfect sine waves of voltage and current, so the discrepancies among the synthesized waveforms cause small power losses (a few percent). However, according to John Robachek, a product manager for Reliance Electric Co., "You're down to splitting hairs among the different control techniques, especially when you compare their differences to the tremendous difference between applying one of them to a fan or pump, and doing nothing at all."

---

*The ratio of power used, to power actually delivered to the motor.

How changing speeds can save money

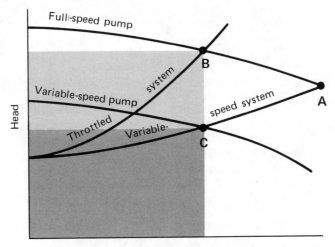

Flow produced by a pump can be lowered by (1) continuing to run it at full speed but throttling the output, or (2) leaving output unthrottled but lowering the speed. In the first case, the pump head/flow performance continues to be shown by the same pump curve ("Full-speed pump" above); but in the second case, the curve shifts downward (to "Variable-speed pump").

Let Point A represent full flow. In the throttling alternative, as the throttle valve closes, the flow approaches (say) Point B. But with a variable-speed pump, the same flow is obtained at Point C, with significantly less head. Since power used by a pump is proportional to (head)x(flow), the power saved is denoted by the shaded rectangle.

much CPI enthusiasm. This device is based on technology developed at the Marshall Space Center of the National Aeronautics and Space Administration (NASA), and is now licensed to over 75 manufacturers. The first products on the market this past summer were

†A.c. motors require a certain amount of current to magnetize the motor core. This "reactive" current does no useful work, in the sense of moving any mass. It shows up as a phase lag between the waveforms of the current and voltage; the power factor is inversely proportional to this phase lag. When a motor is lightly loaded—for instance, the motor propelling a conveyor belt that has unloaded its charge and is now running empty—power factor is lower.

single-phase controllers; various makers are now announcing plans for controllers for the more common three-phase motor, and at ratings up to 30 hp.

PFCs do not vary a motor's speed; they save energy by raising or lowering a motor's torque to match the load the motor is carrying.† The PFC uses a triac (in essence, two back-to-back silicon-controlled rectifiers, which are electronic "gate valves" that allow current to pass in one direction only) and a current-sensing element that moni-

tors the motor's load. Each time a.c. current passes through zero, the sensor checks motor load and, if it is less than full-load, interrupts the current momentarily.

According to Edward Yrisarri, president of Iveco, Inc. (Huntington Beach, Calif.), other advantages of the device are lowered total power demand (which could cut peak-demand surcharges of utilities) and longer motor life and lower air-conditioning needs because the motors run cooler. Iveco is currently offering PFCs at ratings up to 30 hp, and is gearing up production to offer 120-hp units by the end of the first quarter of 1981. Among other companies manufacturing PFCs are Electronic Relays, Inc., (Downer's Grove, Ill.), Energy Technology Corp. (Huntsville, Ala.), and the Energy Devices Div. of Scott & Fetzer Co. (Cleveland, Ohio).

Iveco's units range in price from $200 to $600 for three-phase models up to 30 hp. Yrisarri notes that CPI reaction to the new devices has been mixed so far. "People have been testing them," he says, "but in the CPI there is some hesitation to try something new if you have a critical process. However, I believe it's only a matter of getting some experience."

These misgivings aside, Iveco is expecting a banner year in 1981 for its PFCs. The company believes it will sell more than 50,000 single- and three-phase units, primarily to the air-conditioning equipment market. This is over eight times the 6,000-plus units sold this year.

Nicholas Basta

Saving energy and costs in pumping systems

For lowest cost in pumping systems, looking at first costs
is not enough. You must evaluate the overall system, including
flowrates and variable-capacity and material requirements.

John A. Reynolds, Union Carbide Corp.

☐ There are many ways to unknowingly squander
energy in pumping systems. Some of the more prominent ones will be discussed in this article. But first, we
should look at what energy is costing us, and find out
how we can save operating horsepower by investing in
a more efficient pump.

Energy costs

Let us compare two pumps of different manufacture
for water-booster service. Brand A requires 10 hp at the
specified performance conditions; Brand B requires 9
hp. The electric utility rate is 2¢/kWh. The direct operating cost for a ten-year-life project would be
$(2¢/kWh)(8,760 \text{ h/yr})(0.746 \text{ kW/hp}) (10 \text{ yr}) \div 0.85$
motor efficiency = $1,538/hp.

However, a dollar spent in the tenth year of the project is worth less than a dollar now, due to the time
value of money. (Using a discounted-cash-flow method

of analysis, one can enter the various elements of consideration, such as investment, direct costs, interest
rates, project life, etc., and arrive at a precise number.)
For simplicity in later comparisons, assume that $1,000
is the amount we can afford to spend in first cost to
save one operating horsepower, which we will call the
investment equivalent of the lifetime operating costs. Thus,
if Brand A pump costs $1,500 and Brand B costs
$1,700, Brand B would have the best overall econom-
ics, as shown by the comparison presented for Brand A
and Brand B in Table I.

Once we have established an investment equivalent
of utility usage, it is advisable to pass this value on to
the supplier (even to having an entry on the equipment
data sheet). This will help him to offer the pump with
the best overall economics; it will encourage him to
avoid limiting his bid to one that is lower in first cost
only.

Originally published January 5, 1976

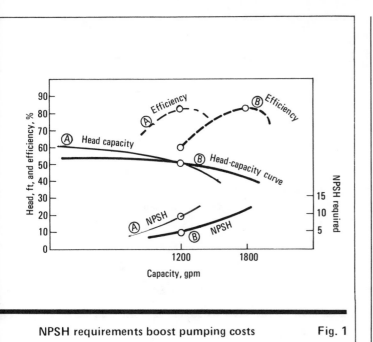

NPSH requirements boost pumping costs   Fig. 1

Calandria circulating pumps

Quite often, the design engineer decides upon the elevation above grade of his distillation column before he selects the calandria circulating pump. In the interest of economy, he makes the height as small as possible, not realizing that he will be forever paying an energy penalty for a smaller first-cost saving.

For example, he may need a 1,200-gpm, 50-ft-head pump. But with a small elevation, perhaps only 5 ft of NPSH (net positive suction head) will be available. Pump A, shown in Fig. 1, has the best efficiency for the

Total evaluated costs			Table I
Pump	First cost	Investment equivalent	Total cost
Brand A	$1,500	$1,000	$2,500
Brand B	1,700	0	1,700
		Difference in favor of Brand B	$ 800

Larger pump vs. added skirt height			Table II
Pump	First cost	Other costs	Total cost
A (18.3-hp rating)	$3,500	$1,200*	$4,700
B (25.3-hp rating)	4,500	7,000†	11,500
		Net savings of pump A is	$6,800

*for extra column skirt and piping
†cost of 7 hp at $1,000 hp (investment equivalent)

application, but requires 9 ft of NPSH at the rated capacity. Consequently, the vendor must offer an oversized and more expensive pump, operating at a lower efficiency to meet the 5 ft of NPSH available. (There must always be more NPSH available than is required by the pump, or it will cavitate.)

If an expensive alloy is required for the application, the first cost of the larger pump is even greater. The higher first cost of this pump may of itself outweigh the cost of additional height for a steel column. Table II presents a comparison of typical economies of pumps A and B. We can realize a $6,800 savings by adding height and buying the more efficient pump. It is important to check the NPSH requirements for a pumping application before establishing elevations above grade.

Variable capacity requirements

Some processes have capacity requirements that vary over a wide range. A common way of handling this is to buy a pump with a fixed-speed electric motor drive and install a control valve to reduce the flow during lower demand periods. In systems where the pressure drop is a significant part of the total head requirement, a variable-speed drive merits consideration. Take the example shown in Fig. 2: the normal capacity is 500 gpm, but at certain times up to 1,000 gpm is required.

Obviously, a 1,000-gpm pump with an electric motor and a control valve can handle the 500-gpm demand. But notice that at that rating, such a pump's efficiency is only 65%, as compared with 83% at the 1,000-gpm level. Also note that the actual head requirement of the system is only 30 ft, not the 100 ft that the pump is developing at the high rpm.

The head developed by a centrifugal pump varies as the square of the rpm ratio. Its capacity varies directly with the rpm ratio. Thus, it can "track" the system pressure-drop curve very nicely, while keeping at or near its best efficiency point, if the rpm can be varied. Looking at Table III, it is apparent that a variable-speed drive is the best choice.

Before leaving this example, let us look at the system's pressure-drop curve, keeping in mind the affinity laws concerning pump head and capacity versus rpm. If demand is equally divided between 500 and 1,000 gpm, with no in-between requirements, then a two-speed motor having a high rpm twice the lower rpm and costing about $1,000 would be the most economical choice. If there were a few in-between demands, a control valve could be used to handle them.

One last word about variable-speed drives for capacity control of centrifugal pumps: They can be used to vary the flow *exclusive* of a control valve, and can reduce horsepower requirements as much as 50% or more.

Proper pump selection

Often, the design engineer will specify a standard centrifugal pump for very small flowrates. Consider an example where there is a 3-gpm requirement for deionized water at a 324-ft differential head, and Type 316 stainless steel materials of construction are necessary. Table IV illustrates that a non-standard, small, high-speed centrifugal pump has the most attractive overall economics (when compared with a standard

Fixed-speed drive vs. variable-speed drive   Table III

Drive	Investment equivalent	Motor	Total cost
Fixed rpm* (19.4 hp)	$19,400	$ 500†	$19,900
Variable rpm* (4.6 hp)	4,600	4,500‡	9,100
Net savings for variable-rpm driver			$10,800

* At 500 gpm.
† Cost of standard 30-hp motor.
‡ Cost of 30-hp variable speed motor and controls.

Types of pumps—very small flowrates   Table IV

Pump*	First cost	Investment equivalent	Total cost
Standard centrifugal (10.9 hp at design)	$1,200	$10,900	$12,100
Metering (0.58 hp at design)	3,500	580	4,080
Small high-rpm centrifugal (1.5 hp at design)	1,300	1,500	2,800

* All pumps handle 3 gpm at 324 ft (140 psi) head.

Types of pumps — small flowrates   Table V

Pump*	First cost	Investment equivalent	Total cost
Standard centrifugal (17.2 hp at design)	$1,500	$17,200	$18,700
Small high-rpm centrifugal (4.7 hp at design)	1,500	4,700	6,200
Triplex plunger (1.7 hp at design)	5,500	1,700	7,200

* All pumps handle 12 gpm of deionized water at 400 ft (173 psi)

Pumping high-viscosity fluids   Table VI

Pump*	First cost	Investment equivalent	Total cost
Standard centrifugal (34 hp at design)	$1,600	$34,000	$35,600
Rotary-gear (4.5 hp at design)	1,600	4,500	6,100
Difference in favor of gear pump			$29,500

* All pumps handle 50 gpm, at 230 ft (100 psi) head of fluids with 500-cp viscosity.

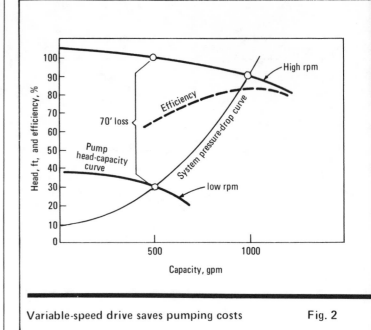

Variable-speed drive saves pumping costs   Fig. 2

centrifugal pump or a metering pump), although it is not low in first cost, nor is it the most efficient. (A gear pump would also have good economics, but has not been considered because of the poor lubricity of the water, which could cause galling of the 316 SS gears.)

If exact capacity regulation is required for the flow of the deionized water, then the costs of both centrifugal pumps must include that of a measuring and regulating system. This system could be omitted from the metering-pump cost since that pump has an inherent accuracy of flow control, and since a pneumatic stroke-control is included in the cost shown. Thus, in this example, the metering pump may turn out to have the best economics.

If we take another example, where we want to pump 12 gpm of deionized water at 400 ft head, the capacity needed is greater than that of a metering pump, and we might consider a triplex plunger pump, a standard centrifugal, or a small high-speed centrifugal. Table V shows how the small high-speed centrifugal pump has the best economics.

However, if carbon-steel were an acceptable material, the triplex pump would cost much less, e.g. $1,800, which would give it the lowest overall cost. On the other hand, if capacity control were again a requirement, the added cost of a variable-speed drive or of bypass provisions for the triplex might cause the small centrifugal pump to be the best choice.

Our final example will illustrate the economics of selecting pumps for high-viscosity fluids. Let us say we want to pump 50 gpm of a clean product having a 500-cp viscosity at 230 ft of differential head. In Table VI, the first costs of a centrifugal pump and a gear pump put them at a standoff. But the required horsepower for the centrifugal at that rating makes a clear case for the rotary gear pump, even if a variable-speed drive or a bypass to suction is needed for flow variations.

These various examples show that the entire system

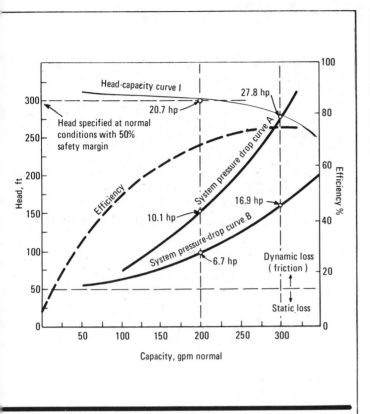

Safety margin makes valve a power consumer   Fig. 3

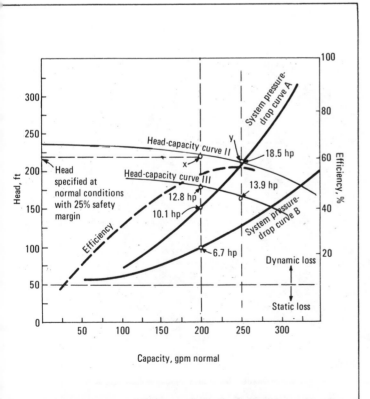

25% overdesign reduces power requirements   Fig. 4

must be considered, in order to make a sound decision—looking at the type of pump alone is not enough.

Rated vs. normal flows—valves and lines

The flow for which a centrifugal pump is rated is the calculated normal flow, plus an added margin of safety. This margin of safety can cause a waste of power in two ways, if it is overly generous. First, the selection is usually based on a pump that would be near its highest efficiency at the rated flow; and subsequent operation at normal flow would be away from this highest efficiency. Second, the large margin of safety in flow results in a much higher calculated head, since the pressure drop of the system increases exponentially with increase in flow.

In the system of Fig. 3, for example, curve B is made up of 50 ft of static loss (difference in elevation between suction and discharge vessels) plus a variable dynamic loss (friction in lines, heat exchangers, etc.) which varies with flowrate. Curve A consists of curve B plus the pressure drop through the control valve in the pump discharge. One accepted method for sizing such valves is to make the valve-pressure drop equal to the sum of all the other dynamic pressure drops in the system. Thus, a 50% margin of safety and very conservative control-valve sizing criteria would lead one to specify a pump having a 30-hp motor and using 20.7 hp at the normal flowrate. The dynamic loss over the control valve becomes more than five times all other dynamic losses at that flowrate.

Curve II of Fig. 4, by contrast, shows that if a 25% safety margin is used, a 20-hp pump with a lower head will do the job, even when the same conservative control-valve sizing criteria are used. Furthermore, by making the pressure drop through the control valve only half all other dynamic losses, we can use a pump with a head-capacity curve like curve III. This pump requires a 15-hp motor and uses 12.8 hp at normal flow.

In many pumping systems, the pressure drop of the piping is the major dynamic loss. We have seen that the control-valve pressure drop is predicted on such dynamic losses. Therefore, one of the first analyses to be made in pump selection is the balance between a low first cost of small piping versus a higher operating cost caused by high dynamic losses. After that, a choice of reasonable margins of safety and control-valve sizing criteria should bring some significant power savings.

The author

John A. Reynolds is a staff engineer in the Chemicals and Plastics Engineering Dept., Union Carbide Corp., South Charleston, WV 25303. He is responsible for specification, procurement and consultation to plants on pumps, seals and related equipment. Mr. Reynolds holds a B.S. degree in mechanical engineering from West Virginia University and has been active in mechanical evaluation of machinery since 1951. He has specialized in pumping equipment since 1962.

Select pumps to cut energy cost

Specifications that are out-of-date and too restrictive can prevent engineers from selecting energy-efficient pumps. A guide based on pump specific speed indicates the type of pump to choose.

John H. Doolin, Worthington Pump Inc.

☐ Considerable energy savings can be made in pump systems. Of course, the first place to look for these savings is in system design [*1*]. However, even after system requirements have been reduced to the minimum and hydraulic conditions have been fixed, attention should be given to selecting the most efficient pumps for the system.

Most engineers fully regard the efficiencies quoted by vendors. This, however, may not be enough, because company specifications—such as on operating speed, number of stages and impeller configuration—may preclude the supplier from offering the most efficient pump.

Although pump technology has advanced considerably during the past 25 years, many specifications are still based on experience prior to then. Such specifications can lead to the selection of inefficient pumps in terms of power consumption at a time when a savings of 1 hp can justify an investment of $1,000 [*2*].

A guide for selecting efficient pump type

The most efficient type of pump for a particular application could be single-stage, multistage, high-speed, or even reciprocating. Too many specifications, especially those based on out-of-date experience, limit the possibility of energy-efficient pumps being selected

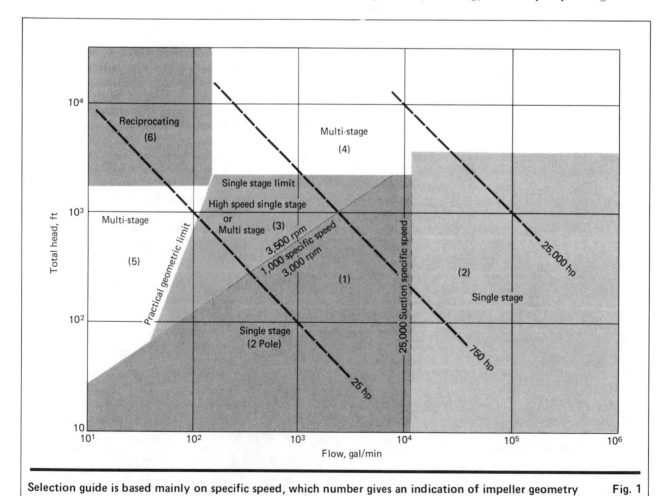

Selection guide is based mainly on specific speed, which number gives an indication of impeller geometry   Fig. 1

Originally published January 17, 1977.

Inducer fits into suction opening of impeller   Fig. 2

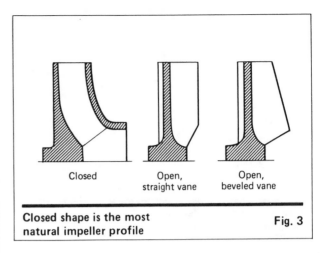

Closed   Open, straight vane   Open, beveled vane

Closed shape is the most
natural impeller profile   Fig. 3

by restrictive specifying, such as "single-stage centrifugal," "two-stage centrifugal," or "multistage."

Fig. 1 offers a guide to efficient pumps ranging in capacity to 100,000 gal/min and in total head to 10,000 ft. The selection guide is based primarily on the characteristic of pump specific speed: $N_s = NQ^{1/2}/H^{3/4}$. (In this equation, N = rotating speed, rpm; Q = capacity, gal/min; and H = total head, ft.)

Fig. 1 is mapped into six areas, each of which indicates the type of pump that should be selected for the highest energy efficiency, as follows:

Area 1: single-stage, 3,500 rpm.

Area 2: single-stage, 1,750 rpm or lower.

Area 3: single-stage, above 3,500 rpm, or multistage, 3,500 rpm.

Area 4: multistage.

Area 5: multistage.

Area 6: reciprocating.

When the value of N_s for any condition is less than 1,000, the operating efficiency of single-stage centrifugal pumps falls off drastically, in which case either multistage or higher-speed pumps offer the best efficiency.

Area 1 is the densest, crowded both with pumps operating at 1,750 rpm and 3,500 rpm, because years ago 3,500-rpm pumps were not thought to be as durable as 1,750-rpm ones. Since the adoption of the AVS standard in 1960 (superseded by ANSI B73.1), pumps with stiffer shafts have been proven reliable.

Also responsible for many 1,750 rpm pumps in Area 1 has been the impression that the higher (3,500 rpm) speed caused pumps to wear out faster. However, because impeller tip speed is the same at both 3,500 and 1,750 rpm (as, for example, a 6-in. impeller at 3,500 rpm and a 12-in. one at 1,750 rpm), so is the fluid velocity, and so should be the erosion of metal surface. Another reason for not limiting operating speed is that improved impeller inlet design allows operation at 3,500 rpm to capacities of 5,000 gal/min, and higher.

Evaluate limits on suction performance

Choice of operating speed may also be indirectly limited by specifications pertaining to suction performance, such as that fixing the top suction specific speed directly, or indirectly by choice of Sigma constant or by reliance on Hydraulic Institute charts.

Suction specific speed is defined as: $S = NQ^{1/2}/H_s^{3/4}$. (In this equation, N = rotating speed, rpm; Q = capacity, gal/min; and H_s = net positive suction head, ft.)

Values of S below 8,000 to 10,000 have long been accepted for avoiding cavitation. However, since the development of the inducer (Fig. 2), S values in the range of 20,000 to 25,000 have become commonplace, and values as high as 50,000 have become practical.

The Sigma constant, which relates NPSH to total head, is little used today, and Hydraulic Institute charts (which are being revised) are conservative.

In light of today's designs and materials, past restrictions due to suction performance limitations should be re-evaluated or eliminated entirely.

Consider off-peak operation

Even if the most efficient pump has been selected, there are a number of circumstances in which it may not be operated at peak efficiency. Today's cost of energy has made these considerations more important.

A centrifugal pump, being a hydrodynamic machine, is designed for a single peak operating-point of capacity and total head. Operation at other than this best efficiency point (bep) reduces efficiency. Specifications now should account such factors as:

1. A need for a larger number of smaller pumps. When a process operates over a wide range of capacities, as many do, pumps will often work at less than full capacity, hence at lower efficiency. This can be avoided by installing two or three pumps in parallel, in place of a single large one, so that when operations are at a low rate one of the smaller pumps can handle the flow.

2. Allowance for present capacity. Pump systems are frequently designed for full flow at some time in the future. Before this time arrives, the pumps will operate far from their best efficiency points. Even if this interim period lasts only two or three years, it may be more economical to install a smaller pump initially and to replace it later with a full-capacity one.

3. Inefficient impeller size. Some specifications call for pump impeller diameter to be no larger than 90 or 95% of the size that a pump could take, so as to provide reserve head. If this reserve is used only 5% of the time,

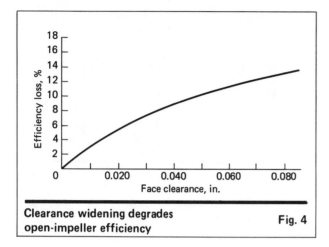

Clearance widening degrades open-impeller efficiency Fig. 4

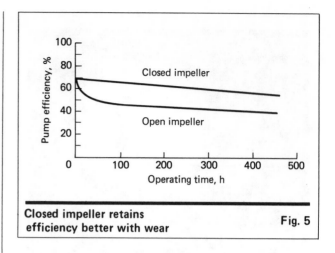

Closed impeller retains efficiency better with wear Fig. 5

all such pumps will be operating at less than full efficiency most of the time.

4. Advantages of allowing operation to the right of the best efficiency point. Some specifications, the result of such thinking as that which provides reserve head, prohibit the selection of pumps that would operate to the right of the best efficiency point. This eliminates half of the pumps that might be selected, and results in oversized pumps operating at lower efficiency.

If an open impeller is carefully machined, it can be as efficient as a closed one. Because of the manufacturing problems of getting smooth hydraulic profiles (even though numerically controlled machine tools now make contoured open impellers), closed-impeller pumps tend to be more efficient.

The closed shape is the most natural profile for impellers (Fig. 3), being the easiest to achieve with cast surfaces. Open impellers are typically straight-sided or beveled, because these shapes are simple to machine.

Closed impellers are also more efficient because the efficiency of open impellers is very dependent on the amount of face clearance between the impeller and casing wall. Although an open-impeller pump can be built with only a 0.015-in. clearance, it is not uncommon for this clearance to enlarge to 0.050 in. after a short time in service. Many studies have revealed this, including an NASA report, which shows efficiency drops 10% at a 0.050-in. clearance (Fig. 4) [3].

Of course, closed-impeller clearances also widen with service. However, the loss in efficiency is less. An accelerated wear test of open and closed impellers of otherwise identical geometry showed that when the clearances of both impellers opened to 0.050 in., the efficiency of the open impeller dropped 28%, whereas that of the closed impeller fell only 14% (Fig. 5) [4].

If possible, avoid special pumps

Sometimes, the unusual requirements of a system dictate the selection of special pumps, which are often inherently lower in efficiency. Such requirements should be carefully evaluated to determine if a special pump is really necessary and worth the loss of efficiency. The following are examples of such special pumps:

■ Self-priming pumps are designed with suction and discharge chambers that generate significant friction loss. Recirculation also reduces efficiency. A vertical

wet-pit pump could be considered as a substitute.

■ Canned motor pumps, which are often installed when zero leakage is mandatory, are less efficient because the magnetic gap must be wider to allow for the chamber that seals the motor.

■ Hydrodynamically-sealed pumps, another form of zero-leakage pump, prevents leakage by the back-pumping of a second impeller. The pumping of the second impeller penalizes power efficiency.

■ Pumps for handling solids are usually specified to be oversized or of low-efficiency design so as to handle fairly large solids without plugging. Rather than select such a pump, it may be more economical to unplug a more efficient pump that clogs occasionally.

■ Pumps of proprietary mechanical design are sometimes chosen because the special design may best fit the unusual requirements of a system. Nevertheless, the likely loss of efficiency should be carefully evaluated before such a pump is specified.

In general, avoid restrictive specifications that exclude more-efficient pumps. Allow manufacturers latitude in offering pumps by following such a guide as Fig. 1. In fact, take an affirmative stance and preface specifications with a statement that efficiency is given primary consideration and that inefficiency is penalized at $1,000 per horsepower.

References

1. Karassik, I. J., Design and Operate Your Fluid System for Improved Efficiency, *Pump World,* Summer, 1975, Worthington Pump Inc.
2. Reynolds, J., Saving energy and costs in pumping systems, *Chem. Eng.,* Jan. 5, 1976.
3. NASA Report No. CR-120815.
4. Pumping Abrasive Fluids, *Plant Eng.,* Nov., 1972.

The author

John H. Doolin is Director of Product Development for Worthington's Standard Pump Div. (14 Fourth Ave., East Orange, NJ 07017). Holder of several patents on pump design and author of eight articles on centrifugal pumps, he is Worthington's representative to the Hydraulic Institute, and a representative on ANSI B-73 (pump standards), Chairman, Subcommittee 1 on horizontal end suction pumps. He received Bachelor and Master of Science degrees from Newark College of Engineering, and is a licensed engineer in the State of New Jersey.

Practical process design of particulate scrubbers

Energy consumed by a scrubber reveals much about its performance—regardless of its geometry or internals. Here the author tells how to predict collection efficiency, using the "contacting power" technique.

Konrad T. Semrau, SRI International

☐ Only 30 years ago, it was widely held that scrubbers were ineffective for capturing fine particles of less than 2 to 3 micrometers. The assertion was not true even then, for units having such capabilities were known at least as far back as the 1890s, although most could not meet such standards.

Today, many scrubbers suitable for fine-particle collection are available. However, progress has resulted not from the introduction of new devices but from the use of a new design method that relates power dissipation to scrubbing efficiency. The method provides a systematic procedure for setting performance.

Lapple and Kamack [1] first demonstrated that, for a given particulate, there is a functional relationship between the collection efficiency of a scrubber and the energy expended in the gas-liquid contacting process. Their work was confirmed and extended in a series of three papers [2–4] by the present author and coworkers. It was concluded not only that efficiency is determined by power dissipation, but also that it is relatively independent of scrubber geometry, and of the way that the power is applied to the gas-liquid contacting. The validity of this conclusion is still not fully explored, but it is at least a good approximation.

With the importance of specific design configurations discounted, manufacturers and users of scrubbers have in recent years turned mostly to a few relatively simple designs. Venturi and orifice scrubbers having no essential design differences are now produced by most of the major manufacturers. One feature common to many of these units is a variable-area throat with provisions for automatically controlling the gas-pressure drop—and hence, the collection efficiency—at a constant level.

Grouping scrubbers by power input

The number and variety of scrubber designs that are in use or that have appeared in the past 100 years are so great as to defy any scheme of classification that is at once self-consistent and practical. However, power consumption is a salient characteristic of scrubbers. Hence,

Informal grouping of scrubbers according to mechanism of power input — Table I

Mechanism	Scrubber[1]
Gas-phase contacting power[2]	Plate (sieve, valve, bubblecap, etc.)
	Gas-atomized spray[3]
	Centrifugal
	Baffle
	Impingement-and-entrainment (self-induced spray)
	Moving bed[7]
Liquid-phase contacting power[4]	Preformed spray[5]
	Ejector venturi[6]
Mechanical contacting power	Mechanically aided devices (eg., disintegrator scrubber)[8]
Wet-film collection (no spray formation)	Massive and fibrous packing[9]

Notes:

1. See Aug. 29, 1977 issue, p. 54, for further description of the devices listed here.
2. This category claims the bulk of particulate scrubbers in use. Many designs get appreciable contributions of liquid-phase contacting power from spray nozzles.
3. Includes venturi and orifice scrubbers, and rod-bank devices.
4. All of these units have some gas-phase friction losses, and hence incorporate some gas-phase contacting power. Most are low or medium-energy. To obtain high-energy scrubbing, one must specify high liquid-gas ratios and/or high liquid-feed pressures.
5. Includes hydraulic (or pressure) spray nozzles, rotating nozzles, spinning-disk atomizers, as well as the simple spray chamber. Cyclonic spray devices (centrifugal) are also included if liquid-phase contacting power dominates.
6. Major subgroup of preformed spray scrubbers (also known as water-jet scrubbers). Two-fluid nozzles are sometimes used.
7. Other names: fluidized packing, mobile packing, floating or "ping-pong" type devices, and turbulent contacting absorber (TCA).
8. Includes all devices that use a motor-driven rotor to assist gas-liquid contacting.
9. Operated below floodpoint. Above flooding, spray formation contributes to contacting power mechanisms.

Originally published September 26, 1977

Formulas for predicting effective friction loss and contacting power   Table II

	Symbol	English units	Metric units	Dimensional formula* English units	Dimensional formula* Metric units
Effective friction loss	F_E	in. H_2O	cm H_2O	Δp†	Δp†
Gas-phase contacting power	P_G	hp/1,000 ft³/min**	kWh/1,000 m³**	$0.1575\,F_E$	$0.02724\,F_E$
Liquid-phase contacting power	P_L	hp/1,000 ft³/min	kWh/1,000 m³	$0.583\,p_f\,(L/G)$††	$0.02815\,p_f\,(L/G)$††
Mechanical contacting power	P_M	hp/1,000 ft³/min	kWh/1,000 m³	$1,000\,(W_s/G)$††	$16.67\,(W_s/G)$††
Total contacting power	P_T	hp/1,000 ft³/min	kWh/1,000 m³	$P_G + P_L + P_M$	$P_G + P_L + P_M$

*See nomenclature.
† Effective friction loss is usually approximately equal to the scrubber pressure drop, Δp.
** 1.0 kWh/1,000 m³ = 2.278 hp/1,000 ft³/min.
†† This quantity is actually power input and represents an estimate of contacting power.

---

it is convenient to characterize (if not to classify) scrubbers by the way, or ways, that power is introduced.

That scrubbers of widely different designs have about the same relationship of collection efficiency to contacting power implies that the underlying mechanisms are essentially the same. "High energy" scrubbers used to collect fine particles are not fundamentally different from other scrubbers, but merely incorporate mechanical arrangements that aid power input.

Energy for gas-liquid contacting can be supplied in three ways (Table I):
- From the energy of the gas stream.
- From the energy of the liquid stream.
- From a mechanically driven rotor.

There are a number of variations on these. Liquid stream energy may be supplied from the pressure head, as in hydraulic nozzles; from a second, compressible fluid (air or steam) as in two-fluid nozzles; or from a mechanically driven atomizer, such as a spinning disk. In at least one scrubber, the high-pressure water is also superheated, so that the water-jet emerging from the nozzle is given kinetic energy by the flashing of part of the liquid into steam.

In ejector units, the liquid supplies energy directly, and also supplies, via the draft, energy to the gas stream that is partly used in scrubbing.

If a scrubber is supplied with power in a single way, one can correlate collection efficiency as a function of gross power input. But for better understanding, one should identify the component of power consumption that relates to collection efficiency. This is a practical necessity if power is being supplied via two or more paths, and the separate components are to be combined on a consistent basis to give the total power. The gross power input includes losses in motors, drive shafts, fans and pumps that obviously should be unrelated to scrubber performance.

The basic quantity that does correlate with scrubber efficiency is "contacting power," power per unit of volumetric gas flowrate that is dissipated in contacting and is ultimately converted into heat [4]. The correlation clearly implies that collection mechanisms are associated with fluid turbulence.

Calculation of contacting power

Although contacting power is easily defined in abstract terms, it is less easily specified, and still less easily measured. In the simplest case, when all the energy is obtained from the gas stream as pressure drop, contacting power is equivalent to friction loss across the wetted equipment (rather than to pressure drop), an observation which may reflect kinetic energy changes rather than energy dissipation.

Although friction loss across a scrubber is often nearly equal to the pressure drop, the distinction between the two is important. Pressure drops due solely to kinetic energy changes in the gas stream will not correlate with performance. Furthermore, any friction losses taking place across equipment that is operating dry obviously do not contribute to gas-liquid contacting and likewise do not correlate with performance [7].

The author has adopted "effective friction loss" to denote friction loss across the wetted system; the loss is measured in units of gas pressure-drop, in. or cm of H_2O. Conversion to contacting power requires only a change of units to power per unit of volumetric gas flowrate, hp/1,000 ft³/min, or kWh/1,000 m³.

Nomenclature

F_E Effective friction loss, in. or cm water
G Gas flowrate, ft³/min or m³/min
L Liquid flowrate, gal/min or L/min
N_t Number of transfer units, dimensionless
p_f Liquid feed pressure, lb/in² or atm, gage
Δp Gas-pressure drop, in. or cm water
P_G Gas-phase contacting power, hp/1,000 ft³/min, or kWh/1,000 m³
P_L Liquid-phase contacting power, hp/1,000 ft³/min, or kWh/1,000 m³
P_M Mechanical contacting power, hp/1,000 ft³/min, or kWh/1,000 m³
P_T Total contacting power, hp/1,000 ft³/min, or kWh/1,000 m³
W_s Net mechanical power input, hp or kW
α Coefficient of P_T in Eq. (6), [hp/1,000 ft³/min]$^{-\gamma}$ or [kWh/1,000 m³]$^{-\gamma}$
γ Exponent of P_T in Eq. (6), dimensionless
η Fractional collection efficiency, dimensionless
$1 - \eta$ Fractional penetration, dimensionless

Comparing gas- and liquid-phase inputs

This "gas-phase contacting power" is easily determined, since pressure drops across dry equipment can be measured and subtracted from the total, and those resulting from kinetic energy changes can be estimated or measured. However, "liquid-phase contacting power" supplied from the stream of scrubbing liquid, and the "mechanical contacting power" supplied from a mechanically driven rotor, are not readily measured. One is generally obliged to compute a theoretical contacting power supplied to the gas-liquid contacting process rather than measure actual contacting power consumed.

For example, if hydraulic spray nozzles are used, theoretical liquid-phase contacting power can be taken as the theoretical kinetic energy of the liquid jet emerging from the nozzle, which is equal to the product of the volumetric liquid flowrate and the gage pressure of the liquid upstream of the nozzle. However, some of this energy may be lost by friction in the nozzle itself, and after the liquid jet emerges from the nozzle, some of its remaining kinetic energy may be lost in ways that do not contribute to contacting. If a rotor is used, theoretical mechanical-contacting power delivered to the gas stream is equal to the input to the motor, minus losses in the motor, couplings and drive shaft.

If a two-fluid nozzle is used, analogy suggests that the corresponding theoretical contacting power be taken as the sum of the separate kinetic energies of the liquid stream and the compressible-fluid stream as they emerge from the nozzle. An approximate value for the kinetic energy of the compressible fluid can be calculated by assuming an isentropic expansion of the fluid through the nozzle. Unfortunately, almost no data are available to show the validity of this procedure.

In ejector-type scrubbers, and usually in scrubbers with mechanically driven rotors, part of the energy supplied is not dissipated but produces a rise in the pressure of the gas stream, and therefore is by definition not contacting power. To predict contacting power, power equivalent to the gas-pressure rise must be subtracted from theoretical power input to the scrubber.

A summary of formulas for computing actual or estimated contacting power in English and metric units is presented in Table II.

Correlation of efficiency data

Collection efficiency of a scrubber is an exponential function of contacting power:

$$\eta = 1 - \exp[-f(P_T)] \qquad (1)$$

where η is fractional collection efficiency, and $f(P_T)$ is a function of total contacting power, P_T. Collection efficiency is thus an insensitive function for purposes of data correlation, particularly in its high range. The penetration

$$1 - \eta = \exp[-f(P_T)] \qquad (2)$$

is better, but the number of transfer units, defined in this case by

$$N_t = \ln [1/(1 - \eta)] \qquad (3)$$

or

$$\eta = 1 - \exp(-N_t) \qquad (4)$$

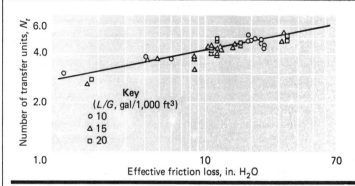

Performance curve for a venturi scrubber collecting fly ash **Fig. 1**

is still more convenient for correlating performance data.

From these equations, it is seen that for a scrubber:

$$N_t = f(P_T) \qquad (5)$$

Data from many different sources [2–4,8,9] have shown that

$$N_t = \alpha \, P_T{}^\gamma \qquad (6)$$

where α and γ are empirical constants that depend on the aerosol (dust or mist) collected, and are little affected by the scrubber size or geometry, or the manner of applying the contacting power. This is the "contacting power rule."

Recent data

In a log-log plot of N_t vs P_T, γ is the slope of the curve, and α is the intercept at $P_T = 1$. The relationship is illustrated for an actual case in Fig. 1, which presents data for a pilot-plant venturi scrubber operating on fly ash from a power boiler fired with pulverized coal. The curve is fairly typical, although the slope is flatter than with most particulates. The data also show that the liquid-to-gas ratio in itself had essentially no influence on performance.

The author has adopted the term "scrubber performance curve" for plots such as Fig. 1. One must appreciate that scatter in the data does not result solely from experimental errors, but also from actual variations in the particle-size characteristics of the dust itself. In fact, it is difficult to maintain consistency even in test aerosols generated under controlled laboratory conditions. The curve for an industrial particulate necessarily represents an average material. Thus, for best definition, it should cover as wide a range of contacting power as is practical. Obviously, such data can be obtained only with pilot-plant equipment that has the flexibility to operate over a wide range of conditions.

Aerosol efficiencies

The efficiency of a test scrubber at a fixed contacting power—or, better, the performance curve—is the most sensitive indicator of variations in an aerosol, although it does not provide quantitative measurement of changes. For practical design, the performance curve is more useful than particle-size distribution. Especially for finding *in situ* aerosol particle size in waste gases, measurements of particle size are tedious and still subject to serious errors, some of which are not even fully defined yet.

Performance curves for an industrial particulate are straightforward, provided a small pilot-plant scrubber is used. A common error is to carry out tests with a large and inflexible unit. If a scrubber is to be used for the first time in a large, critical application, operational experience with a semiworks unit of reasonable size is appropriate. However, such a unit is *not* appropriate for gathering basic efficiency data.

Past experience [2–4] shows that scrubber size has little or no effect on the performance curve, at least down to 75 to 100 ft³/min capacity. The relationship may break down at some lower scrubber size, but if so, the critical size has not been established. At least three companies have employed microscale scrubbers for such test work [11–13].

Experimental caveats

Often, workers will measure only outlet dust concentration from a scrubber and correlate this as a function of scrubber pressure drop, without measuring inlet concentration. This is equivalent to correlating penetration through the scrubber on the basis of a constant inlet concentration [4].

This method is justifiable only with certain dusts. Those from iron blast furnaces, for example, may contain two dissimilar fractions: a fine-fume fraction with moderate concentration variation, and a second, larger fraction of coarse material that varies widely in concentration. Outlet concentration will be determined by inlet concentration of the fine fume, whereas the coarse fraction will be collected with virtually 100% efficiency at all scrubber operating conditions.

In such a circumstance, overall collection efficiency based on total input dust-concentration will be virtually meaningless for assessing performance.

Unresolved are the conditions at which gas flowrate should be evaluated in contacting-power correlations. The problem arises in calculating liquid-phase and mechanical contacting power, where gas flowrate appears explicitly. Where there are large changes in gas pressure, temperature, and water-vapor content across the scrubber, there can be large changes in volume (and considerable changes in mass) during passage of the gas.

In the case where all contacting power is gas-phase and is measured by gas pressure drop or effective friction loss, gas volume is implicit only, but represents an average value across the scrubber. It also happens to be the case for which the contacting-power correlation is best verified.

Hence, by inference, the average of inlet and outlet volumetric flowrates seems appropriate when the gas rate appears explicitly—or is at least consistent with the above computation of gas-phase contacting power.

What these data mean

The basis of the contacting power rule is almost entirely empirical, although there are fundamental reasons for expecting relationships between power consumption and collection efficiency for various scrubbing

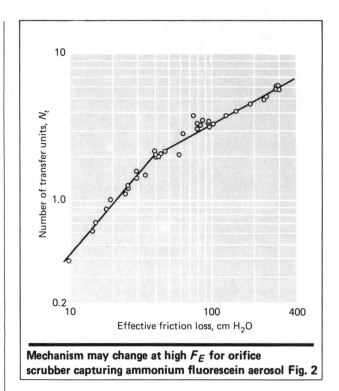

Mechanism may change at high F_E for orifice scrubber capturing ammonium fluorescein aerosol Fig. 2

devices; for example, particle collection by both inertial and diffusional mechanisms should increase with turbulence. However, there is no obvious reason to expect the relationship to be so nearly the same for radically different devices—as has been found by experiment. An interesting parallel occurs in the case of cocurrent gas absorbers, where investigators have correlated mass-transfer coefficients with energy dissipation [14,15].

The dominant collection mechanism in particulate scrubbers is inertial. On theoretical grounds, diffusion would be expected to become operative only at less than 0.5 μm, and to become a major factor only for particles under perhaps 0.1 μm, for which diffusivities are relatively high. There is little evidence that industrial aerosols contain appreciable mass fractions of particles under 0.1 μm; such aerosols coagulate rapidly except at very low concentrations. There has been no evidence that particulate scrubber performance has been appreciably affected by gas residence time, as would be expected if diffusion were important.

Recent investigations [8,9] have provided new insights into the efficiency/contacting power relationship, but have also produced new questions. Interpretation of the slope of performance curves has always presented difficulties [4]. A value of $\gamma < 1$ indicates that the aerosol becomes more difficult to collect at higher efficiencies. In keeping with this fact, scrubbers preferentially collect larger particles in a polydisperse aerosol (one containing particles of varying sizes), and higher contacting power is needed to collect finer particles. If diffusion were the dominant mechanism and also increased with contacting power, finer particles would collect preferentially, and collection would become harder with increasing efficiency and contacting power. This would again lead to a value of $\gamma < 1$.

For a monodisperse aerosol (all particles the same

size), one would logically expect $\gamma = 1$. It is difficult to imagine a physical process that would give a value of $\gamma > 1$, which indicates that collection becomes easier as collection efficiency increases. Yet correlations of data from scrubbing of some aerosols have produced such performance curves [3,4].

In recent investigations with small pilot-plant orifice-scrubbers, performance curves with two branches (Fig. 2) were obtained for fine polydisperse synthetic aerosols of ammonium fluorescein [8,9]. The lower branch had a value of $\gamma > 1$, and the upper branch a value of $\gamma < 1$. If investigations previously reported [3,4] had been extended to high-enough contacting-power levels, the performance curves with slopes greater than 1 might have been followed by additional branches with slopes less than 1.

Among aerosols that yield double-branched performance curves, or curves with slopes greater than 1, the only common feature is that all have large weight fractions composed of submicrometer particles. The steep lower branch of the curve suggests a region in which both inertial and diffusional collection increase with contacting power; the upper branch suggests a region in which inertial mechanisms dominate.

Gas- vs. liquid-phase contacting power

The precise degree to which contacting power supplied by each of the three basic methods is equivalent has still not been fixed. Previous correlations [2–4] indicate that liquid-phase contacting power supplied by hydraulic spray nozzles was generally equivalent to gas-phase contacting power, but the data were relatively few and were not generally obtained under closely controlled conditions. Performances of a number of widely different pressure-spray nozzles [9] were compared with that of the standard orifice contactor, using fine synthetic, ammonium-fluorescein aerosols under lab conditions. Spray-nozzle contactors always consumed some gas-phase contacting power, and the gas-phase and liquid-phase contacting powers varied widely, both absolutely and relative to each other. Some deviations, mostly moderate, appeared among spray nozzles, but all performed more or less inferior to the orifice contactor.

Deviation of spray-nozzle performance from that of the orifice scrubber was found to be correlated by a simple parameter, *the ratio of liquid-phase contacting power to total contacting power for the scrubber with spray contactor.* In addition, relative deviation of spray-contactor performance from that of the orifice contactor rose with a drop in the particle size of the test aerosol. It is still unclear how much of the apparent difference between gas-phase and liquid-phase contacting power is real and how much stems from the difficulty of estimating true liquid-phase contacting power.

Gas-phase power vs. other mechanisms

No similar comparison has yet been made between gas-phase and mechanical contacting power. Such data as have been available for mechanical scrubbers [3,4] show that these perform about as well as units using gas-phase contacting power. The mechanically driven rotor in such a device is commonly designed to promote draft as well as gas scrubbing, but in its latter role it is a turbulence promoter comparable to a venturi or orifice contactor.

To date, it appears that a venturi or orifice scrubber will give as good performance as can be expected from "normal" scrubbing, and that most other units using gas-phase contacting power will give essentially the same performance. Occasionally, a device has performed poorly, but none has done better [7]. Where superior performance has occurred (sometimes in conventional devices), it has followed some additional collection mechanism that is not ordinarily operative in scrubbing. A gas-liquid contactor composed of a packed bed of metal fibers gave superior performance, apparently because filtration was coupled to scrubbing [7]. Condensation of water vapor boosts efficiency in conventional venturi and orifice scrubbers [2,4,8].

It must be recognized, however, that condensation scrubbing requires transfer of heat energy, with its attendant costs, and may in some cases also demand heat energy that could be used in another way [4,8]. Hence, economics of condensation scrubbing will generally be decided by the needs of the associated process system. Condensation scrubbing may be practical when gases must be dehumidified as well as cooled and cleaned (e.g., blast-furnace gas, feed gas for contact sulfuric-acid plants), or when there may be a use for low-level heat, as in some kraft pulp mills.

References

1. Lapple, C. E., and Kamack, H. J., *Chem. Eng. Progr.*, Vol. 51, No. 3, 1955, p. 110.
2. Semrau, K. T., Marynowski, C. W., Lunde, K. E., and Lapple, C. E., *Ind. Eng. Chem.*, Vol. 50, 1958, p. 1,915.
3. Semrau, K. T., *J. Air Poll. Control Assn.*, Vol. 10, 1960, p. 200.
4. Semrau, K. T., *J. Air Poll. Control Assn.*, Vol. 13, 1963, p. 587.
5. Walker, A. B., *J. Air Poll. Control Assn.*, Vol. 13, 1963, p. 622.
6. Walker, A. B., and Hall, R. M., *J. Air Poll. Control Assn.*, Vol. 18, 1968, p. 319.
7. Semrau, K. T., Unpublished data, SRI International, Menlo Park, Calif., 1952–53.
8. Semrau, K. T., and Witham, C. L., "Condensation and Evaporation Effects in Particulate Scrubbing," Paper 75-30.1, Air Pollution Control Assn., Boston, Mass., June 1975.
9. Semrau, K. T., Witham, C. L., and Kerlin, W. W., "Relationships of Collection Efficiency and Energy Dissipation in Particulate Scrubbers," Second Fine Particle Scrubber Symposium, New Orleans, La., May 1977.
10. Raben, I. A., "Use of Scrubbers for Control of Emissions from Power Boilers—U.S.," Paper 13, U.S.-U.S.S.R. Symposium on Control of Fine-Particulate Emissions from Industrial Sources, San Francisco, Calif., Jan. 1974.
11. Hardison, L. C., *Amer. Ceram. Soc. Bull.*, Vol. 49, 1970, p. 978.
12. Riley Environeering Inc., Bull. 112, Skokie, Ill., Nov. 1976.
13. McIlvaine Co., "The McIlvaine Mini-Scrubber," Northbrook, Ill. undated.
14. Jepson, J. C., *AIChE Journal*, Vol. 16, 1970, p. 705.
15. Reiss, L. P., *I & EC Proc. Des. Develop.*, Vol. 6, 1967 p. 486.

The author

Konrad T. Semrau is a Senior Chemical Engineer at SRI International, 333 Ravenswood Ave., Menlo Park, CA 94025, where he is chiefly concerned with control of air pollution from stationary sources. His long interest in particulate scrubbers encompasses a major pilot-plant program. A graduate of the University of California at Berkeley, he holds an M.S. in chemical engineering. He is active in the Air Pollution Control Assn., ACS and AIChE.

Low-cost evaporation method

*A unique multiple-stage
process liquor or waste
chemical or to incinerate*

William G. Farin,

☐ The direct-contact multiple-effect system combines the best features of a direct-contact evaporator and a multiple-effect evaporator.

By itself, the direct-contact evaporator (DCE) is a high·consumer of energy because the liquor to be concentrated is directly contacted by hot combustion gases. The resulting latent heat in the evaporated water dissipates to the atmosphere. However, the DCE is a low capital-cost unit capable of handling difficult liquors of high solids content. In directly firing the fuel into an evaporator, the operating cost per million Btu can be as low as half that of a million Btu of steam from a boiler.

The multiple-effect evaporator, on the other hand, is a high-capital-cost combination, because a boiler to produce steam, as well as considerable heat-transfer surfaces in several stages (effects), are required. The steam from boiler passes into the first stage to evaporate the liquor; vapor from the first stage then passes to a second stage to evaporate liquor at a lower pressure and temperature; and so on. With this system, energy savings are appreciable.

The direct-contact multiple-effect systems now being used [1-4], an example of which is shown in Fig. 1, include a:

"First Effect" of direct-contact evaporation, in which the combustion gases from burning a fuel come into contact with the liquor directly in a venturi scrubber to evaporate water, which saturates the flue gases in the conventional manner.

"Second Effect" provides heat recovery by vacuum evaporation. Condensate heated by countercurrent scrubbing contacts the saturated flue gas to remove its latent heat. The heated condensate passes through a heat exchanger to heat liquor that is circulated and flash-evaporated in a vacuum system. This procedure uses the latent heat for one or more effects of vacuum evaporation.

"Third Effect" of air evaporation heats liquor in the vacuum-evaporator's surface condenser. Additional heat is obtained by further cooling the flue gas with condensate and passing the heated condensate through a heat exchanger to heat the liquor. The heated liquor is circulated to an air evaporator, where it is contacted with air—thereby heating and saturating the air for added liquor evaporation and cooling.

The techniques can be applied to evaporation systems to use:

- Any fuel economically.
- Heat generated in burning waste liquors.
- Flue heat from recovery or waste-heat boilers.
- Waste heat contained in a plume.

Originally published March 1, 1976

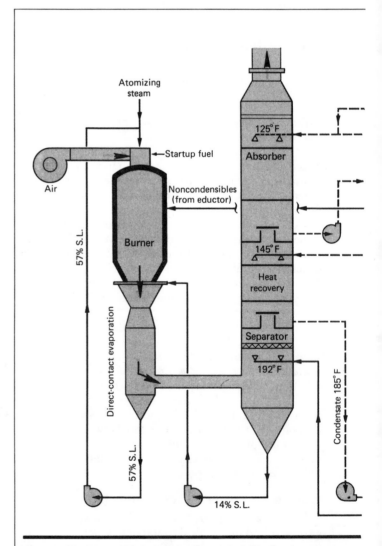

Direct-contact triple-effect system concentrates and incinerates

saves energy by reusing heat

*evaporation system concentrates any
stream to effect the recovery of a
and dispose of a waste product.*

Marathon Engineering Inc.

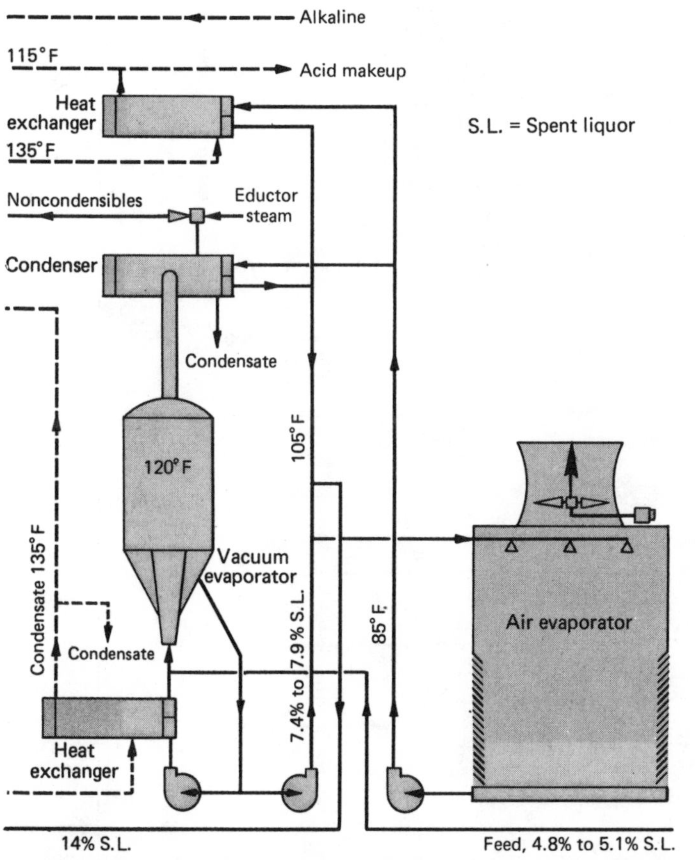

■ Heat from a condensing system for vacuum or air evaporation.

Furthermore, the system eliminates the steam normally needed for heating, and the water for condensing, by providing basic methods for heat-recovery evaporation that can be effective for energy conservation and environmental control in the chemical process industries.

Direct-contact evaporation

The direct-contact evaporator is particularly useful in handling liquors high in soluble or precipitated solids. It avoids the problems of low heat transfer due to high viscosity, scaling of heat-exchanger surfaces, and high boiling-point rise (where liquor temperatures exceed their vapor temperatures and result in a reduction of the available temperature differential and capacity in multiple-stage steam evaporators).

The design of the DCE is quite critical in order to provide complete saturation at minimum pressure drop, to avoid wet and dry buildups, and to control scaling, foaming and entrained carryover to the heat-recovery stages.

When incinerating a product, separation of the ash may require cooling the flue gas below 1,600° F to solidify the ash, facilitate separation and prevent carbonization of the product being direct-contact evaporated. This is achieved in one mill [7] by quenching the 5% liquor being concentrated in the combustion chamber with the 5% solids also being incinerated. At another mill, flue gas is recycled after scrubbing, to reduce its temperature to 900° F before it goes to the DCE [8]. Ash from the system is a molten smelt that discharges to a flaker for chemical recovery. Flue-gas temperatures have also been controlled with excess air. However, this reduces the potential for heat recovery.

Heat-recovery vacuum evaporation

Heat-recovery rates for vacuum evaporation often exceed direct-contact evaporation rates, due to the utilization of latent heat. Vacuum evaporation is particularly useful following a recovery boiler, fluidized-bed reactor, or incinerator burning wastes high in water content where it can be two to three times as high as that for direct-contact evaporation.

The amount of latent heat can be determined from Fig. 2, where the water-vapor content along with the heat content is shown per pound of dry flue gas at various saturation temperatures. The molecular weight of flue gas (about 31) compared to that for air (about 29) proportionally reduces the water-vapor content at a

(text continues on p. 318)

Alkaline
115° F
Acid makeup
Heat exchanger
135° F
S.L. = Spent liquor
Noncondensibles
Eductor steam
Condenser
Condensate
Condensate 135° F
120° F
105° F
Vacuum evaporator
85° F.
Air evaporator
7.4% to 7.9% S.L.
Condensate
Heat exchanger
14% S.L.
Feed, 4.8% to 5.1% S.L.

combustible solids from feedstream　　　　　**Fig. 1**

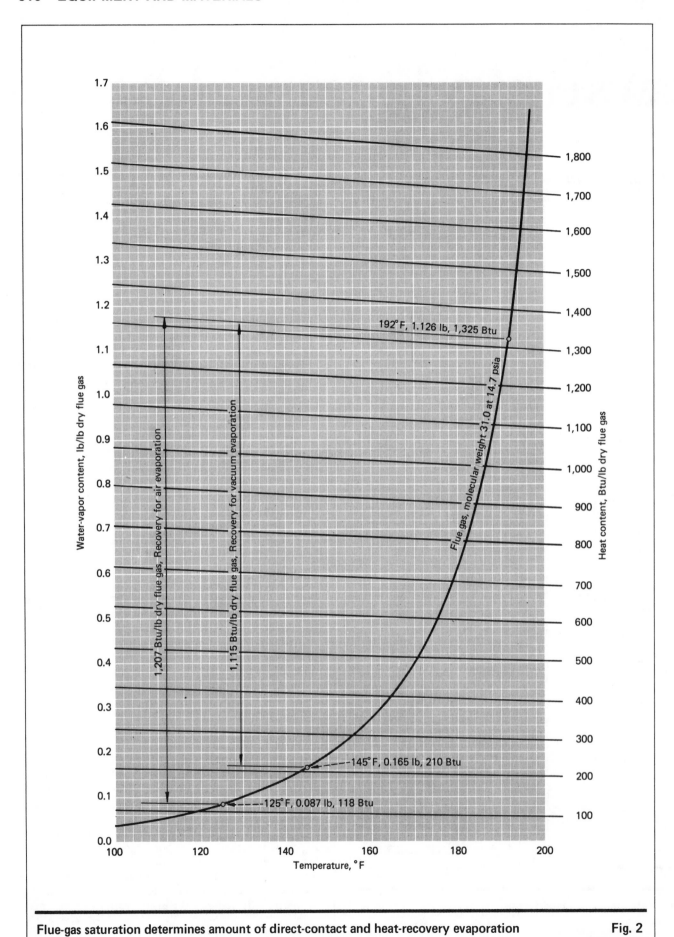

Flue-gas saturation determines amount of direct-contact and heat-recovery evaporation   Fig. 2

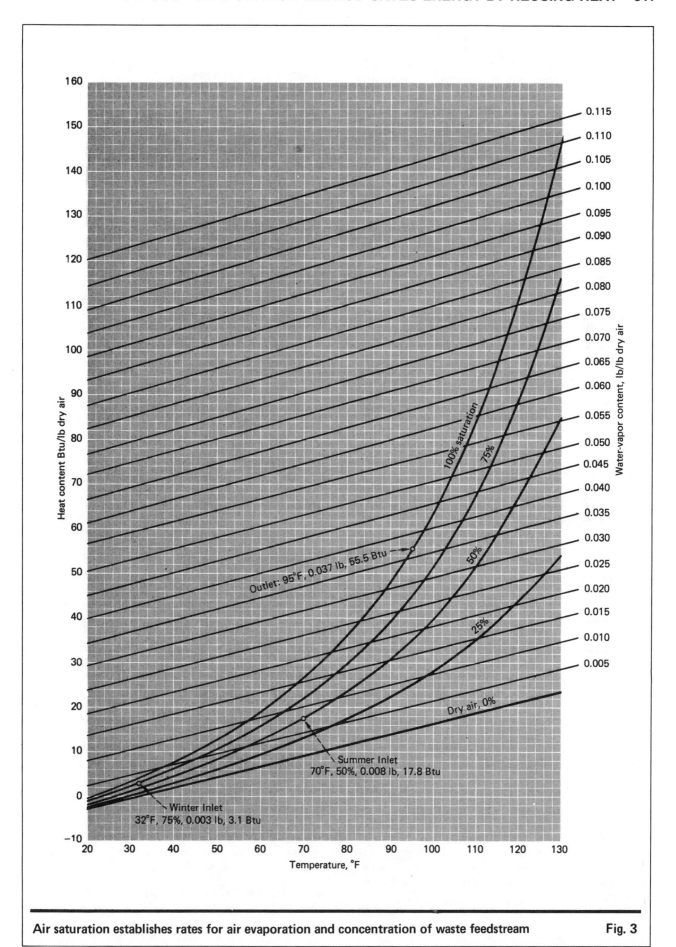

Air saturation establishes rates for air evaporation and concentration of waste feedstream   Fig. 3

given temperature, and thereby increases the saturation temperature. Increased pressure can also reduce water-vapor content and thereby increase saturation temperature. Both factors aid in providing an increased temperature differential for heat recovery.

If a sufficient temperature differential exists, two stages of vacuum evaporation can be considered. A multiple-effect flash-type evaporator may also be used for greater economy [1].

The type of evaporator can be varied to suit the product. When the final concentration is done in the DCE, the vacuum evaporator can operate at low concentrations and temperatures. This provides good heat-transfer rates with reduced scaling and corrosion. This unit can also be a final concentration stage or a vacuum evaporator-crystallizer for chemical recovery.

Heat-recovery air evaporation

The air evaporator is an effective tool for using low-quality heat for evaporation and for eliminating the need of cooling water for condensation. It can provide an added effect on any multistage evaporator by using heat from the surface condesnser. It can also utilize heat removed in cooling the flue gas to lower temperatures, and can exceed the evaporation possible with the DCE and sometimes that obtained by vacuum evaporation. Air-evaporation rates can readily be determined from Fig. 3.

The air evaporator cannot be used for concentrating liquors containing volatile chemicals that will cause air pollution. It has been successfully applied to pulp-mill sulfite-waste liquor. Provisions for vacuum stripping and preneutralization (originally provided as a safeguard) have proven unnecessary in this application. Concentrations up to 20% have been successfully maintained. The air evaporator is a good stripper [5], and will oxidize liquors such as Kraft-pulp waste [6]. This oxidation controls the release of odorous sulfur compounds during direct-contact evaporation and burning.

For the air evaporator, we can use standard cooling-tower equipment built of corrosion-resistant materials, exercising special care when providing additional mist-collection facilities. The design is also modified to prevent foam and gasification of circulated liquor, and to handle any solids encountered. In certain applications, a scrubber may also be necessary.

Scrubbing requirements

All scrubbing functions after the DCE are ordinarily carried out in a single multipurpose tower. In addition to recovery of heat in this tower, chemicals may be added to the condensate for scrubbing and chemical recovery if condensate dilution is not a problem. More often, scrubbing will follow cooling in an added absorption stage.

The heat-recovery scrubber should have low pressure drop, and should prevent condensate gasification, avoid loss of condensate back to the DCE, and avoid entrainment.

Particulates can normally be removed in the DCE and heat-recovery scrubber. However, when submicron particulates are present, a high-energy venturi can be added. Wet-wall precipitators or wet fiber-glass filters can also be used.

Waste-heat recovery

Whether the full direct-contact, triple-effect evaporator or just one or two of its components can be used for a particular heat-recovery application will depend on the characteristics of the flue gas and the liquor to be concentrated. The most economical method will depend on total heat content of the flue gas, and on water-vapor content and heat content per pound of dry flue gas.

Heat recovery with a waste-heat or recovery boiler, producing steam from incineration of wastes high in water-vapor, will ordinarily result in the discharge of 25 to 50% of the heat to the stack. Most of this heat is in the latent form from product moisture, steam atomization, combustion moisture, soot blowing, etc.

For example, in one application, in which discharge is at 450° F, the stack gas contains 365 Btu/lb of dry flue gas due to a water-vapor content of 0.21 lb/lb of dry flue gas. Or, for every million Btu ordinarily vented, this low-quality heat can still evaporate the following amounts of water:

By direct-contact evaporation	233 lb
By vacuum evaporation	616 lb
By air evaporation	580 lb
Total evaporation	1,429 lb

In addition, the flue gas is cooled and prepared for chemical scrubbing and air-pollution control.

Direct-contact evaporation depends on a high-temperature flue gas (at least 350° F) because latent heat is created but not used in this step. For the same heat content, dry flue gas provides the greatest potential for DCE since more of its sensible heat can be transformed to latent heat by evaporation.

On the other hand, flue gas high in moisture content has the greatest potential for heat-recovery evaporation even though the temperature may be well below 200° F. In these cases, the DCE would be eliminated and the type of evaporation determined by the saturation temperature (see Fig. 2). For vacuum evaporation, a saturation temperature above 150° F is required; a second stage of air evaporation provides an added potential. For air evaporation, a saturation temperature above 135° F is necessary, the amount depending on the quantity of flue gas available and evaporation requirements.

Heat-recovery evaporation provides an energy-saving, low-cost method for any process liquor or waste liquor being concentrated for chemical recovery, incineration or disposal. The oxidizing and stripping potential of the air evaporator as well as its ability for evaporation without producing condensate provides us with a variety of additional methods for evaporation, treatment and disposal that may be effectively used to meet present and future environmental requirements.

Example illustrates techniques

A direct-contact triple-effect evaporator will concentrate a feed liquor containing about 5% solids, having a dry-solids heat value of 7,500 Btu/lb, to a final

Typical heat and material balances in a direct-contact triple-effect evaporator system   **Table I**

Item	Balance	Quantity*
1.00	Solids burned	
.01	Feed to burner, 57% solids	1.754 lb/lb
.02	Atomizing steam	0.3 lb/lb
.03	Air required	5.697 lb/lb
.04	Dry flue gas	6.165 lb/lb
2.00	Heat generated in burning	
.01	Combustion heat	7,500 Btu/lb
.02	Heat in liquor feed	208 Btu/lb
.03	Heat in atomizing steam	358 Btu/lb
.04	Heat in air	176 Btu/lb
.05	Radiation loss	−346 Btu/lb
.06	Heat in flue to direct-contact evaporator	7,896 Btu/lb
3.00	Water vapor from burning	
.01	Water from steam	0.300 lb/lb
.02	Water in air	0.088 lb/lb
.03	Water in liquor	0.754 lb/lb
.04	Combustion moisture	0.432 lb/lb
.05	Total moisture content	1.574 lb/lb
4.00	Direct-contact evaporation	
.01	Heat in feed	483 Btu/lb
.02	Heat from flue	7,896 Btu/lb
.03	Heat loss in product	−208 Btu/lb
.04	Heat for direct-contact evaporation	8,171 Btu/lb
.05	Heat in dry flue gas	1,325 Btu/lb dry flue gas
.06	Saturation temperature	192°F
.07	Water-vapor content of dry flue gas (Fig. 2)	1.126 lb/lb dry flue gas
.08	Water vapor	6.942 lb/lb
.09	Original moisture content	1.574 lb/lb
.10	Direct-contact evaporation	5.368 lb/lb
.11	Feed at 105°F, 14% solids	7.122 lb/lb
5.00	Heat recovery for vacuum evaporation	
.01	Flue heat, temperature out	145°F
.02	Heat content, (Fig. 2)	210 Btu/lb dry flue gas
.03	Heat content	1,294.7 Btu/lb
.04	Water-vapor content (Fig. 2)	0.165 lb/lb dry flue gas
.05	Water-vapor content	1.017 lb/lb
.06	Original water content	6.942 lb/lb
.07	Condensation	5.925 lb/lb
.08	Heat lost in condensate discharged	−610 Btu/lb
.09	Heat from direct-contact evaporator	8,171 Btu/lb
.10	Flue heat discharged	−1,294.7 Btu/lb
.11	Heat to heat exchanger	6,266.3 Btu/lb
6.00	Vacuum evaporation	
.01	Heat in feed at 170°F	2,704.8 Btu/lb
.02	Heat in product at 120°F	−1,109.7 Btu/lb
.03	Heat from heat exchanger	6,266.3 Btu/lb
.04	Total heat for vacuum evaporator	7,867.5 Btu/lb
.05	Heat for evaporation	1,113.3 Btu/lb water
.06	Vacuum evaporation	7.060 lb/lb
7.00	Heat recovery for air evaporation	
.01	Flue heat, temperature out	125°F
.02	Heat content (Fig. 2)	118 Btu/lb dry flue gas
.03	Heat content	727.5 Btu/lb
.04	Water-vapor content (Fig. 2)	0.087 lb/lb dry flue gas
.05	Water vapor	0.536 lb/lb
.06	Original water content	1.017 lb/lb
.07	Condensation	0.481 lb/lb
.08	Heat lost in condensate discharged	−39.9 Btu/lb
.09	Heat of inlet flue gas	1,294.7 Btu/lb
.10	Heat in discharged flue gas	−727.5 Btu/lb
.11	Heat from surface condenser	7,244.4 Btu/lb
.12	Recovered heat for air evaporation	7,771.7 Btu/lb
8.00	Heat-recovery air evaporation	
.01	Heat in feed	1,103.7 Btu/lb
.02	Heat in product	−483.0 Btu/lb
.03	Recovered heat for air evaporation	7,771.7 Btu/lb
.04	Total heat for air evaporation	8,392.4 Btu/lb
.05	Heat needed for summer evaporation (Fig. 3)	1,300 Btu/lb water
.06	Summer evaporation	6.45 lb/lb
.07	Summer feed to air evaporator, 7.4% solids	13.572 lb/lb
.08	Summer feed to vacuum evaporator, 4.8% solids	20.632 lb/lb
.09	Heat needed for winter evaporation (Fig. 3)	1,541 Btu/lb water
.10	Winter evaporation	5.44 lb/lb
.11	Winter feed to air evaporator, 7.9% solids	12.562 lb/lb
.12	Winter feed to vacuum evaporator, 5.1% solids	19.622 lb/lb

*lb/lb indicates lb/lb of solids burned.

content of 57% solids. The system whose flowsheet is shown in Fig. 1 provides for air evaporation, vacuum evaporation, scrubbing, combustion and direct-contact evaporation, along with the necessary heat recovery.

We will begin our step-by-step analysis of this system by starting with the incineration of the 57%-solids liquor in the burner at the conditions shown in Table I under Item 1.00. Heat is generated and water vapor

formed (per Items 2.00 and 3.00). Flue gas discharges to the venturi, directly contacts the circulating liquor, and concentrates it to the 57% burning consistency from the 14% feed concentration (Item 4.00).

Flue gas saturated at 192° F is cooled to 145° F (Item 5.00) by countercurrent scrubbing with 135° F condensate, which is then heated to 185° F by the flue gas. The 185° condensate is sent through a heat exchanger to heat liquor that is circulated and flash-evaporated in a vacuum evaporator. This provides a second stage by vacuum evaporation at 120° F (Item 6.00 in Table I). The feed liquor is concentrated from the 5% to 7½% range.

The condensate, now cooled to 135° F, is recycled to the scrubber. Excess condensate (from condensed water vapor) is discharged. Noncondensibles from the vacuum evaporator are injected into the burner for combustion in order to enable recovery of chemicals and utilize the latent heat.

The flue gas is cooled further to 125° F in the absorption section by countercurrent scrubbing (Item 7.00) with 115°-F condensate. The condensate, in turn, is heated to 135° F by the flue gas, and also absorbs sulfur dioxide. The heated condensate is circulated through a heat exchanger, where cool (85° F) liquor is heated to 105° F. This liquor is now circulated through the air evaporator and cooled to 85° F, and it also becomes the coolant for the surface condenser of the vacuum evaporator. Heat from both exchangers serves to concentrate the liquor from the 7½% to 14% range by air evaporation (Item 8.00).

Calculation procedures

To find evaporation rates for the several stages of this system, we will use the data plotted as charts for flue gases and for air. From Fig. 2, we can quickly determine the rates for direct-contact evaporation and for heat-recovery evaporation; and from Fig. 3, the rates for air evaporation.

Direct-contact evaporation is calculated by totaling the heat input per pound of dry flue gas. The heat generated in burning is listed in Table I as Item 2.00. Direct-contact evaporation rates are shown in Item 4.00. Each pound of solids burned provides 8,171 Btu, or 1,325 Btu/lb of dry flue gas.

The flue gas following direct-contact evaporation will have a temperature of 192° F at 14.7 psia, and a saturated-water content of 1.126 lb/lb dry flue gas (Fig. 2), or 6.942 lb/lb of dry solids. Subtracting the 1.574 lb of water originally contained in the flue gas (Items 3.00 and 4.00), a direct-contact evaporation rate of 5.368 lb/lb dry solids is achieved. Feed to the direct-contact evaporator will have an added 1.754 lb./lb of the product sent to the burner, or 7.122 lb/lb of solids at 14% concentration.

For heat recovery during vacuum evaporation, flue gas is cooled to 145° F. As shown in Fig. 2, this reduces the water content to 0.165 lb/lb of dry flue gas, or 1.017 lb/lb of solids. The heat content is reduced to 210 Btu/lb of dry flue gas, or 1,294.7 Btu/lb of solids. This recovers 6,876.3 Btu/lb of solids, with the condensate heated from 135° F to 185° F. Of this total, 610 Btu/lb of solids is lost in the condensate (5.925 lb/lb of dry solids) discharged at 135°F. The balance (6,266.3 Btu/lb of solids) is available for heat transfer to the vacuum evaporator (Item 5.00).

The vacuum evaporator receives additional heat from the feed liquor (see Item 6.00 in Table I). This provides 7,867.5 Btu/lb of solids for the vacuum evaporation of 7.06 lb of water/lb of solids. Feed liquor is thus concentrated from 5% to 7½% solids.

For heat recovery during air evaporation, the flue gas is further cooled to 125° F (Item 7.00), thereby reducing the water-vapor content to 0.087 lb/lb of dry flue gas, or 0.536 lb/lb of solids. The heat content is reduced to 118 Btu/lb of dry flue gas, or 727.5 Btu/lb of solids. This recovers 567.2 Btu/lb of solids and condenses 0.481 lb of water/lb of solids. A loss of 39.9 Btu/lb solids occurs in the condensate discharged to acid makeup at 115° F. Added to 7,244.4 Btu/lb solids recovered in the surface condenser, a total of 7,771.7 Btu/lb solids is available for air evaporation.

Air-evaporation rates are obtained by determining the relative changes in heat content and water-vapor content of the air going in and out of the air evaporator.

Referring to Fig. 3, we find that, under summer conditions, air at 70° F and 50% saturated has a heat content of 17.8 Btu/lb of air and a water content of 0.008 lb/lb of air. If this air is heated to 95° F and saturated, its heat content rises to 55.5 Btu/lb and water content to 0.037 lb/lb air. The 0.029 lb of water evaporated into the heated air requires 37.7 Btu, or 1,300 Btu/lb water evaporated.

Under winter conditions at 32° F and 75% saturated, the initial heat content of air is 3.1 Btu/lb air, with 0.003 lb water vapor. The 0.034 lb of water evaporated requires 52.4 Btu, or 1,541 Btu/lb water evaporated.

The heat in the product raises the heat for air evaporation to 8,392.4 Btu/lb solids (Item 8.00). This provides a summer evaporation rate of 6.45 lb/lb of solids, and a winter rate of 5.44 lb/lb of solids. Thus, the concentration of the liquor stream increases from 7.4 to 14% in the summer, and from 7.9 to 14% in the winter.

Direct-contact double- and triple-effect evaporation systems are already operating for processing pulpmill waste liquors for evaporation and disposal [7, 8]. These systems provide for double and triple heat utilization and have very low capital and operating costs.

References

1. Farin, W. G., U.S. Patent 3,425,477 (Feb. 4, 1969).
2. Farin, W. G., U.S. Patent 3,638,708 (Feb. 1, 1972).
3. Farin, W. G., Canadian Patent 906,394 (Aug. 1, 1972).
4. Farin, W. G., *Tappi*, Sept. 1973, p. 69.
5. Estridge, B. G., Turner, B. G., Smathers, R. L. and Thibodeaux, L. J., *Tappi*, Jan. 1971, p. 53.
6. Sarkanen, K. V., Hrutfiord, B. J., Johanson, L. N and Gardner, H. S., *Tappi*, May 1970, p. 766.
7. MacLeod, M., *Pulp & Paper*, Oct. 1974, p. 58.
8. Evans, J. C., *Pulp & Paper*, Oct. 1975, p. 63.

The author

William G. Farin is President of Marathon Engineering Inc., a consulting and engineering firm, P. O. Box 335, Menasha, WI 54952. He obtained his experience in evaporation with E. I. du Pont de Nemours & Co., as chief engineer of the Krystal Div. of Struthers Wells, and with Marathon Div. of American Can Co. His experience on evaporation and burning includes systems for many paper companies. He has written extensively on the processing of spent liquors and other chemicals for the pulp and paper industry.

INDEX